Geomorphology in the Digital Era

Geomorphology in the Digital Era

Guest Editors

Anselme Muzirafuti
Giovanni Randazzo
Stefania Lanza

Basel • Beijing • Wuhan • Barcelona • Belgrade • Novi Sad • Cluj • Manchester

Guest Editors

Anselme Muzirafuti
Department of Mathematics,
Computer Sciences, Physics
and Earth Sciences
University of Messina
Messina
Italy

Giovanni Randazzo
Department of Mathematics,
Computer Sciences, Physics
and Earth Sciences
University of Messina
Messina
Italy

Stefania Lanza
Department of Mathematics,
Computer Sciences, Physics
and Earth Sciences
University of Messina
Messina
Italy

Editorial Office
MDPI AG
Grosspeteranlage 5
4052 Basel, Switzerland

This is a reprint of the Special Issue, published open access by the journal *Applied Sciences* (ISSN 2076-3417), freely accessible at: www.mdpi.com/journal/applsci/special_issues/Geomorphology.

For citation purposes, cite each article independently as indicated on the article page online and using the guide below:

Lastname, A.A.; Lastname, B.B. Article Title. *Journal Name* **Year**, *Volume Number*, Page Range.

ISBN 978-3-7258-3276-7 (Hbk)
ISBN 978-3-7258-3275-0 (PDF)
https://doi.org/10.3390/books978-3-7258-3275-0

Cover image courtesy of Anselme Muzirafuti

Contents

Said El Makrini, Mustapha Boualoul, Younes Mamouch, Hassane El Makrini, Abdelhamid Allaoui and Giovanni Randazzo et al.
Vertical Electrical Sounding (VES) Technique to Map Potential Aquifers of the Guigou Plain (Middle Atlas, Morocco): Hydrogeological Implications
Reprinted from: *Appl. Sci.* **2022**, *12*, 12829, https://doi.org/10.3390/app122412829 . . . . **1**

Dicky Harishidayat, Abdullatif Al-Shuhail, Giovanni Randazzo, Stefania Lanza and Anselme Muzirafuti
Reconstruction of Land and Marine Features by Seismic and Surface Geomorphology Techniques
Reprinted from: *Appl. Sci.* **2022**, *12*, 9611, https://doi.org/10.3390/app12199611 . . . . **17**

Tommaso Orusa, Duke Cammareri and Enrico Borgogno Mondino
A Scalable Earth Observation Service to Map Land Cover in Geomorphological Complex Areas beyond the Dynamic World: An Application in Aosta Valley (NW Italy)
Reprinted from: *Appl. Sci.* **2022**, *13*, 390, https://doi.org/10.3390/app13010390 . . . . **39**

Emmanouil Manoutsoglou, Ilias Lazos, Emmanouil Steiakakis and Antonios Vafeidis
The Geomorphological and Geological Structure of the Samaria Gorge, Crete, Greece—Geological Models Comprehensive Review and the Link with the Geomorphological Evolution
Reprinted from: *Appl. Sci.* **2022**, *12*, 10670, https://doi.org/10.3390/app122010670 . . . . **64**

Shaker Ahmed, Adel El-Shazly, Fanar Abed and Wael Ahmed
The Influence of Flight Direction and Camera Orientation on the Quality Products of UAV-Based SfM-Photogrammetry
Reprinted from: *Appl. Sci.* **2022**, *12*, 10492, https://doi.org/10.3390/app122010492 . . . . **77**

Adrian Jarzyna, Maciej Babel, Damian Ługowski and Firouz Vladi
Morphology of Dome- and Tepee-like Landforms Generated by Expansive Hydration of Weathering Anhydrite: A Case Study at Dingwall, Nova Scotia, Canada
Reprinted from: *Appl. Sci.* **2022**, *12*, 7374, https://doi.org/10.3390/app12157374 . . . . **92**

Marian Rybansky and Josef Rada
The Influence of the Quality of Digital Elevation Data on the Modelling of Terrain Vehicle Movement
Reprinted from: *Appl. Sci.* **2022**, *12*, 6178, https://doi.org/10.3390/app12126178 . . . . **126**

Zhonghua Hong, Yahui Yang, Jun Liu, Shenlu Jiang, Haiyan Pan and Ruyan Zhou et al.
Enhancing 3D Reconstruction Model by Deep Learning and Its Application in Building Damage Assessment after Earthquake
Reprinted from: *Appl. Sci.* **2022**, *12*, 9790, https://doi.org/10.3390/app12199790 . . . . **150**

Jasper Knight and Mohamed A. M. Abd Elbasit
Characterisation of Coastal Sediment Properties from Spectral Reflectance Data
Reprinted from: *Appl. Sci.* **2022**, *12*, 6826, https://doi.org/10.3390/app12136826 . . . . **164**

Priscilla Indira Osa, Anne-Laure Beck, Louis Kleverman and Antoine Mangin
Multi-Classifier Pipeline for Olive Groves Detection
Reprinted from: *Appl. Sci.* **2022**, *13*, 420, https://doi.org/10.3390/app13010420 . . . . **189**

Chengzhe Lv, Yuefeng Lu, Miao Lu, Xinyi Feng, Huadan Fan and Changqing Xu et al.
A Classification Feature Optimization Method for Remote Sensing Imagery Based on Fisher
Score and mRMR
Reprinted from: *Appl. Sci.* **2022**, *12*, 8845, https://doi.org/10.3390/app12178845 . . . . **200**

Marian Rybansky
Determination of Forest Structure from Remote Sensing Data for Modeling the Navigation of
Rescue Vehicles
Reprinted from: *Appl. Sci.* **2022**, *12*, 3939, https://doi.org/10.3390/app12083939 . . . . **219**

**Ayoub Soulaimani, Saïd Chakiri, Saâd Soulaimani, Zohra Bejjaji, Abdelhalim Miftah and
Ahmed Manar**
Gradiometry Processing Techniques for Large-Scale of Aeromagnetic Data for Structural and
Mining Implications: The Case Study of Bou Azzer Inlier, Central Anti-Atlas, Morocco
Reprinted from: *Appl. Sci.* **2023**, *13*, 9962, https://doi.org/10.3390/app13179962 . . . . **233**

Bilawal Abbasi, Zhihao Qin, Wenhui Du, Jinlong Fan, Shifeng Li and Chunliang Zhao
Spatiotemporal Variation of Land Surface Temperature Retrieved from FY-3D MERSI-II Data in
Pakistan
Reprinted from: *Appl. Sci.* **2022**, *12*, 10458, https://doi.org/10.3390/app122010458 . . . . **254**

Myriam Benkirane, Nour-Eddine Laftouhi, Saïd Khabba and África de la Hera-Portillo
Hydro Statistical Assessment of TRMM and GPM Precipitation Products against Ground
Precipitation over a Mediterranean Mountainous Watershed (in the Moroccan High Atlas)
Reprinted from: *Appl. Sci.* **2022**, *12*, 8309, https://doi.org/10.3390/app12168309 . . . . **275**

Xiaomin Lu and Haowen Yan
An Algorithm to Generate a Weighted Network Voronoi Diagram Based on Improved PCNN
Reprinted from: *Appl. Sci.* **2022**, *12*, 6011, https://doi.org/10.3390/app12126011 . . . . **292**

Article

Vertical Electrical Sounding (VES) Technique to Map Potential Aquifers of the Guigou Plain (Middle Atlas, Morocco): Hydrogeological Implications

Said El Makrini [1,*], Mustapha Boualoul [1], Younes Mamouch [2,*], Hassane El Makrini [1], Abdelhamid Allaoui [1], Giovanni Randazzo [3], Allal Roubil [4], Mohammed El Hafyani [4,5], Stefania Lanza [6] and Anselme Muzirafuti [3,*]

[1] CartoTec Team, Faculty of Sciences, Moulay Ismaïl University, BP 11201 Zitoune, Meknes 50000, Morocco
[2] Laboratory Physico-Chemistry of Processes and Materials, Research Team Geology of the Mining and Energetics Resources, Faculty of Sciences and Technology, Hassan First University of Settat, Settat 26002, Morocco
[3] Dipartimento Scienze Matematiche e Informatiche, Scienze Fisiche e Scienze della Terra, Università Degli Studi di Messina, Via F. Stagno d'Alcontres, 31-98166 Messina, Italy
[4] Laboratory of Geoengineering and Environment, Research Group "Water Sciences and Environment Engineering", Department of Geology, Faculty of Sciences, Moulay Ismail University, BP11201, Meknes 50050, Morocco
[5] ULiège (Gembloux Agro-Bio Tech), Terra Research Center, Water-Soil-Plant Exchange, 5030 Gembloux, Belgium
[6] GeoloGIS s.r.l., Dipartimento di Scienze Matematiche e Informatiche, Scienze Fisiche e Scienze Della Terra, Università Degli Studi di Messina, Via F. Stagno d'Alcontres, 31-98166 Messina, Italy
* Correspondence: said.elmakrini@edu.umi.ac.ma (S.E.M.); y.mamouch@uhp.ac.ma (Y.M.); anselme.muzirafuti@unime.it (A.M.)

Citation: El Makrini, S.; Boualoul, M.; Mamouch, Y.; El Makrini, H.; Allaoui, A.; Randazzo, G.; Roubil, A.; El Hafyani, M.; Lanza, S.; Muzirafuti, A. Vertical Electrical Sounding (VES) Technique to Map Potential Aquifers of the Guigou Plain (Middle Atlas, Morocco): Hydrogeological Implications. *Appl. Sci.* 2022, 12, 12829. https://doi.org/10.3390/app122412829

Academic Editor: Filippos Vallianatos

Received: 11 November 2022
Accepted: 12 December 2022
Published: 14 December 2022

Publisher's Note: MDPI stays neutral with regard to jurisdictional claims in published maps and institutional affiliations.

Abstract: Vertical electrical sounding (VES) as a geoelectrical method has proven its effectiveness throughout the history of groundwater geophysical investigation. In this sense, VES was carried out 47 in the study area with the aim of determining the geometry and limits of Quaternary basaltic aquifer formations and, above all, the location of electrical discontinuities in the area located in the north of Morocco, between the center of Almis Guigou and the city of Timahdite. This area is experiencing an overexploitation of the groundwater due to excessive pumping and the development of intensive agriculture activities, resulting in a continuous decrease in piezometric levels. The processing of the diagrams by WINSEV software showed the presence of an electrically resistant surface level, attributed to basaltic formations, of the Quaternary age, whose thicknesses reach at least 150 m to the SW of the area. This level is superimposed on a moderately conductive horizon which, according to local geology, corresponds to Pliocene marl and limestone alternations. The correlation of VES interpretation models allowed us to elaborate thematic maps and geoelectrical sections which illustrate the vertical and lateral extension of the basaltic reservoir as well as its thickness, which decreases in general from the south-west to the north-east; however, the main electrical discontinuities also correspond to faults and fractures, and they show a NE–SW direction sub-parallel to the major accidents of the Middle Atlas. A prospectivity map of the local aquifer was generated, coinciding with regional fault lines and confirmed by the alignment of very good flowing water boreholes. This geophysical study by electrical sounding shed light on the geometry and extension of the aquifer and opened avenues to draw further conclusions on its physical and hydrodynamic characteristics, as well as to optimize the future siting of groundwater exploitation boreholes through the elaboration of the local aquifer prospectivity map.

Keywords: vertical electrical sounding; groundwater; hydrogeophysics; Tabular Middle Atlas; applied geophysics; agriculture; Morocco

---

1. Introduction

The plain of Guigou, located in the center of Morocco between the town of Timahdite and the village of Almis Guigou, has seen an important development in agricultural activities in recent decades, especially in the production of potatoes and onions, which are very well known and in demand in the Moroccan market. Indeed, this crop, which is very demanding in terms of irrigation water, is at the origin of the overexploitation of the water table in the said region. This need is manifested by the decrease in pumped flows and, consequently, the phenomenal drop in the piezometric level [1].

In this respect, detailed knowledge of the aquifer system is of great interest for the optimal management of groundwater resources in such a region [2–6]. The aquifer exploited in the region is formed by Plio-Quaternary basalts with Miocene and Cretaceous marl and limestone formations as a semi-impermeable bedrock. The fracturing affecting these basalts plays a key role in the circulation of groundwater and impacts the productivity of hydrogeological boreholes. In addition, there are few hydrogeological studies of the Plio-Quaternary basaltic aquifer in the region [1,7], and there are none based on geophysical surveys, which means that this aquifer is still poorly known and needs to be explored in detail.

The geoelectrical method, with its three most widely used techniques (tomography, profiling and vertical drilling), has long been considered the most widely used geophysical method for the characterization of aquifer systems in the world [8–21].

Due to its simplicity of implementation and cost-effectiveness compared to other methods, the vertical electrical sounding (VES) technique has proven to be very useful in mapping aquifer systems, the geological layers forming their impermeable bedrock and the detection of structural anomalies corresponding to faults and fractures [4,22–25].

In this study, we used the vertical electrical sounding (VES) technique to map the aquifer system of the Guigou plain. The main objectives of this study included the following: (i) to identify the electrical horizons of the subsoil; (ii) to determine the geological layers forming the aquifer; (iii) to determine the geometry of the aquifer system; (iv) to estimate the depth of the impermeable bedrock; and (v) to locate the geoelectrical discontinuities and determine their role in groundwater drainage.

2. Geological Background

The Moroccan Middle Atlas, shown in Figure 1, where our study area is located, is a chain structured during the Alpine orogeny. It is subdivided into two units: to the west is the Tabular Middle Atlas (TMA) unit, and to the east is the Folded Middle Atlas (FMA) unit [26,27]. The passage between these two units is underlined by a network of faults called the North Middle Atlas accident (NMAA). The Guigou plain is located at the southeastern edge of the TMA and along the NE–SW NMAA [27–29]. It corresponds to a collapse ditch about 5 km wide and 40 km long, essentially filled with volcanic lava and traversed by Oued Guigou, which constitutes its alluvial plain [1,7,30]. The chronological succession of the geological formations of the Middle Atlas begins with the sandstone-pelitic series constituting a base attributed to the Paleozoic era, which is structured by the Hercynian orogeny [26,31–35]. In angular discordance, a thick series from the Triassic age composed of lower and upper argillites and framing a basaltic complex of doleritic-type rocks rests on this base [28,35–37]. This series is underlain by the carbonate formations of the Lias, which are interspersed with marl and limestone of the Dogger age to the Mio–Pliocene age [32]. Locally at TMA, the Quaternary period is represented by volcanic lavas of a basaltic nature. These lavas come from craters located to the north of the study area, the most important of which are Jbel Habri, Chedifat and Bou-Ahsine [28,38]. At their exit, following the example of our study area in the Guigou plain, these lava flows used the slopes to occupy the depressions [28,38]. In this plain, in addition to these basalt flows which predominantly outcrop, there are also some marl and limestone formations from the Mio-Pliocene period which outcrop mainly in the NE part in the form of small plateaus at the foot of the folded Middle Atlas [7,34].

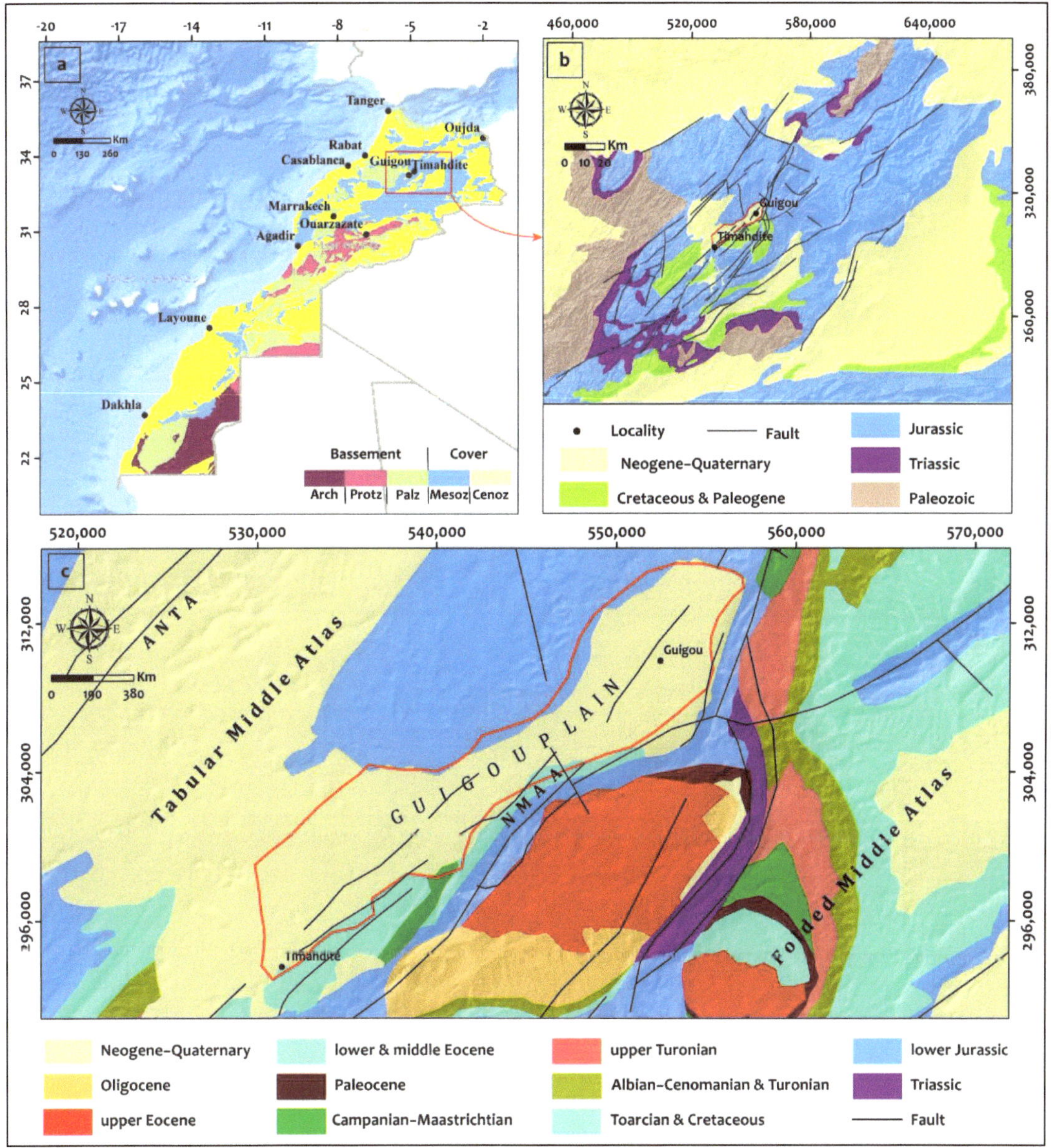

Figure 1. (**a**) General geological map of Morocco showing the location of the Middle Atlas Mountain; (**b**) Simplified geological map of the Middle Atlas extracted from the geological map of Morocco at 1/200,000, showing the location of the study area; (**c**) Sketch of the geological map of the Middle Atlas where the study area marked by the red polygon.

Tectonically, the TMA is characterized by brittle rather than folded deformation [7,28,33,39]. Indeed, two major fault networks can be distinguished, namely the NMAA fault network and the Tizi-n-Tretten (TNTA) fault network with a NE–SW direction. These two major accidents are part of the same fault system inherited from the Hercynian orogeny, which was replayed several times during the Alpine orogeny [26,40,41]. These faulted structures that shape the TMA are easily mapped on the ground, except in areas covered by basaltic flows [28,38].

3. Materials and Methods

3.1. Vertical Electrical Sounding (VES) Basic Principle and Data Acquisition

The geoelectrical method, using the vertical electrical sounding (VES) technique, consists of measuring the variations in apparent resistivity (ρa) as a function of depth. In principle, the measurement protocol consists of injecting an electric current of intensity (I) through two current electrodes (A and B) and measuring the potential differences (ΔV) created between the two receiving electrodes (M and N), called potential electrodes (Figure 2).

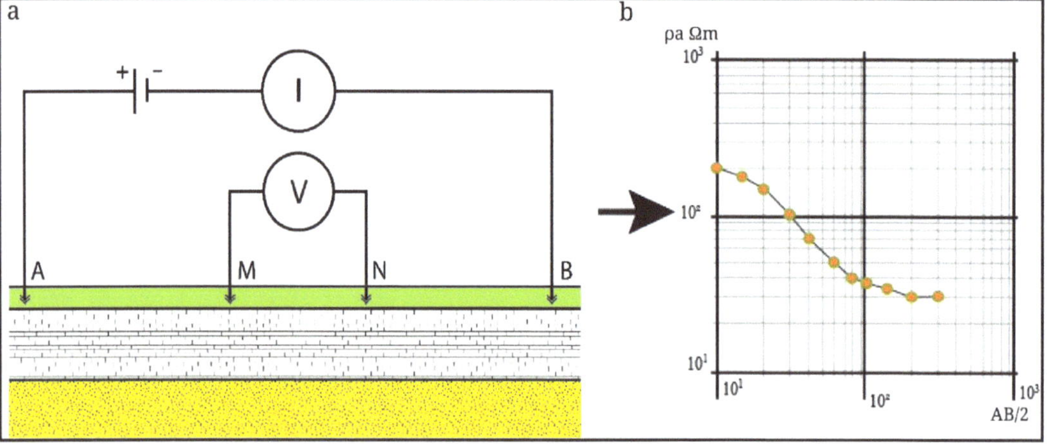

Figure 2. (a) General scheme of a soil-resistivity measurement using the Schlumberger configuration with a four−electrode device (ABMN), **(b)** bi−logarithmic diagram for the representation of VES measurements.

According to Ohm's law, the apparent resistivity is a function of ΔV, I and the geometric coefficient (K). It is calculated by the following formula:

$$\rho_a = \frac{\Delta V}{I} * K \text{ And } K = 2\pi \left(\frac{1}{\frac{1}{AM} + \frac{1}{AN} + \frac{1}{BM} - \frac{1}{BN}} \right)$$

The curve $\rho_a = f\left(\frac{AB}{2}\right)$ is obtained by plotting the apparent resistivity values ρ_a against AB/2 (half spacing of the current electrodes, which can reach up to 10^1 km) in a bi-logarithmic scale.

In the present study, the measurement of VES data at forty-seven stations was carried out and arranged according to seven profiles that were generally oriented NW–SE, perpendicular to the general direction of the flow of the volcanic lava. The coordinates of the measurement stations were taken by a Garmin MAP-64 s GPS. A Syscal Pro resistivity meter was used to acquire geoelectrical data. This automated instrument is powerful in DC electrical readings with the transmitter and receiver integrated in the same instrument. The measurements were made using the Schlumberger configuration. The distance between the current injection electrodes (AB) varied logarithmically from 6 to 1000 m for each measuring station. The position of the measuring stations was organized according to the profiles with a spacing of 3 to 5 km between them (Figure 3). The direction of the spread of the power cables was NE–SW, in the same direction as the regional fault system, to avoid polarity reversals created by the presence of faults or anomalous geological contacts. At the same time, measurements of the piezometric level of the water table using a 200 m piezometric probe were acquired.

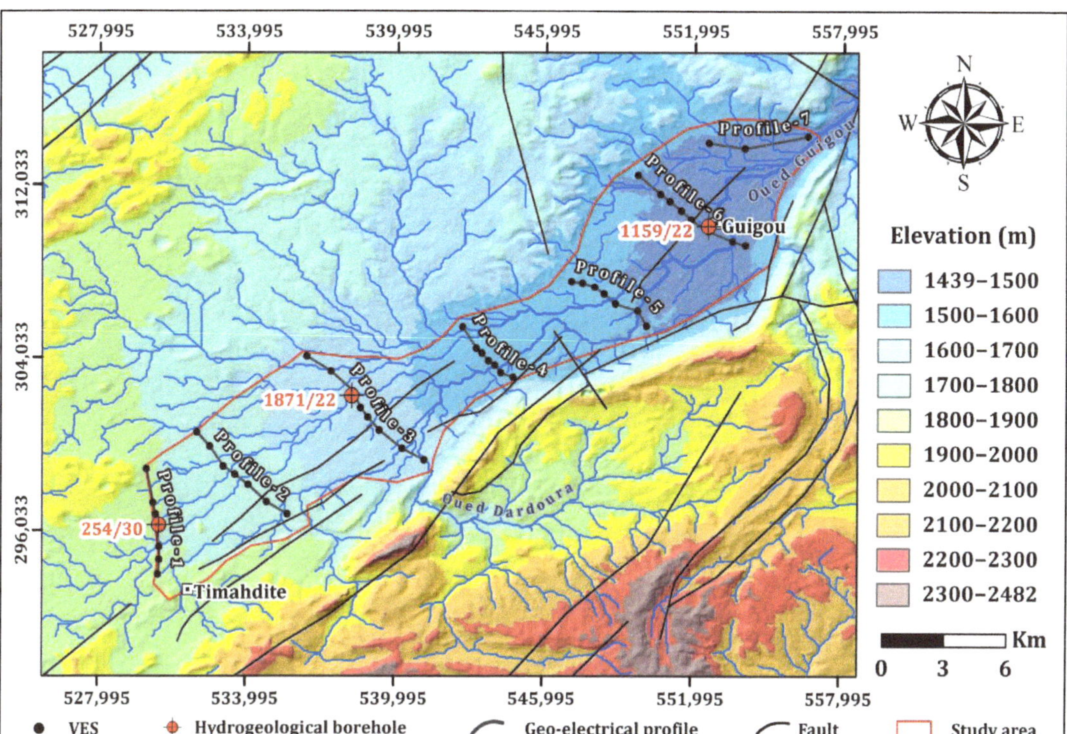

Figure 3. Position plan of VES profiles on digital terrain model.

3.2. Data Processing and Resistivity Interpretation

The VES data acquired were subjected to a series of processing steps to facilitate their interpretation, summarized in four steps. In the first step, the VES diagrams were first smoothed to eliminate all outliers. They were then inverted using Geosoft's Winscv software, which allows each diagram to be broken down into well-defined electrical levels in terms of thickness and resistivity. The models of the VES diagrams were calibrated by lithological logs of the three existing hydrogeological boreholes (Figure 3). In the second step, the geoelectric levels of the VES models of each profile (Figure 3) were correlated horizontally. Consequently, four geoelectrical sections were drawn up to follow the evolution of the resistivity and the thickness of the formations crossed in both the vertical and lateral directions. Then, the third step was to interpolate the VES data using the inverse distance weighting (IDW) method. Four thematic maps were produced, including two iso-resistivity maps, a bedrock-depth map and a thickness map of the main reservoir. Finally, based on the main geoelectrical characteristics extracted from the geoelectrical sections and maps, coupled with the available geological information, the fourth step concerned the elaboration of the groundwater prospectivity map. Figure 4 shows the methodological flowchart applied in this work, described in the four steps above.

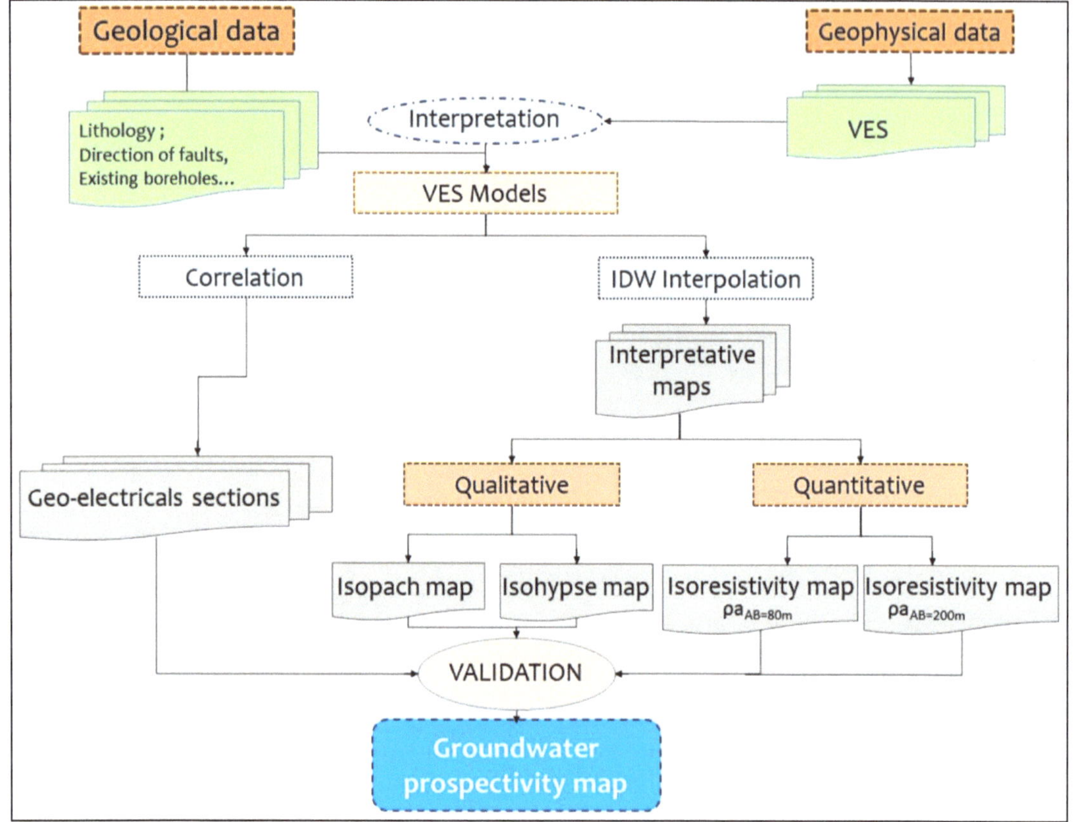

Figure 4. The methodological flowchart used in this study.

4. Results and Discussion

4.1. Geological Significance of Resistivities and VES Categories

The processing and analysis of the electrical boreholes made it possible to identify three categories of VES (C1, C2 and C3), characterizing the whole study area, based on the shape of the curve and the succession of electrical levels (resistant, conductive, two-layer intermediates, etc.). The table below gives a summary of the inversion results and the interpretation of typical VES in each category (Figure 5).

The first category, C1, was the most dominant, containing a total of 30 VES measurements. A typical VES curve is represented by the 4P1 diagram (Figure 5). The interpretation of the latter is based on the lithological data from borehole 254/30. The diagram of VES4P1 shows, from bottom to top, the presence of the following geoelectric levels (Figure 5): (i) a relatively conductive bedrock with a resistivity of 170 Ωm, located at a depth of 152 m and attributed to fairly compact Pliocene marlstone; (ii) a moderately resistant level of 450 Ωm resistivity and a thickness of 75 m that can be made to correspond to water-bearing basalts; (iii) a very resistant complex formed by two levels, a first lower level of with a resistivity of 2043 Ωm and a thickness if 55 m, which could correspond to dry and fairly compact basalts, surmounted by an upper level with a resistivity of 1347 Ωm and a thickness of 21 m, most probably attributed to weathered basalts on the surface; and (iv) a superficial level that is 2 m thick, representing the vegetal soil.

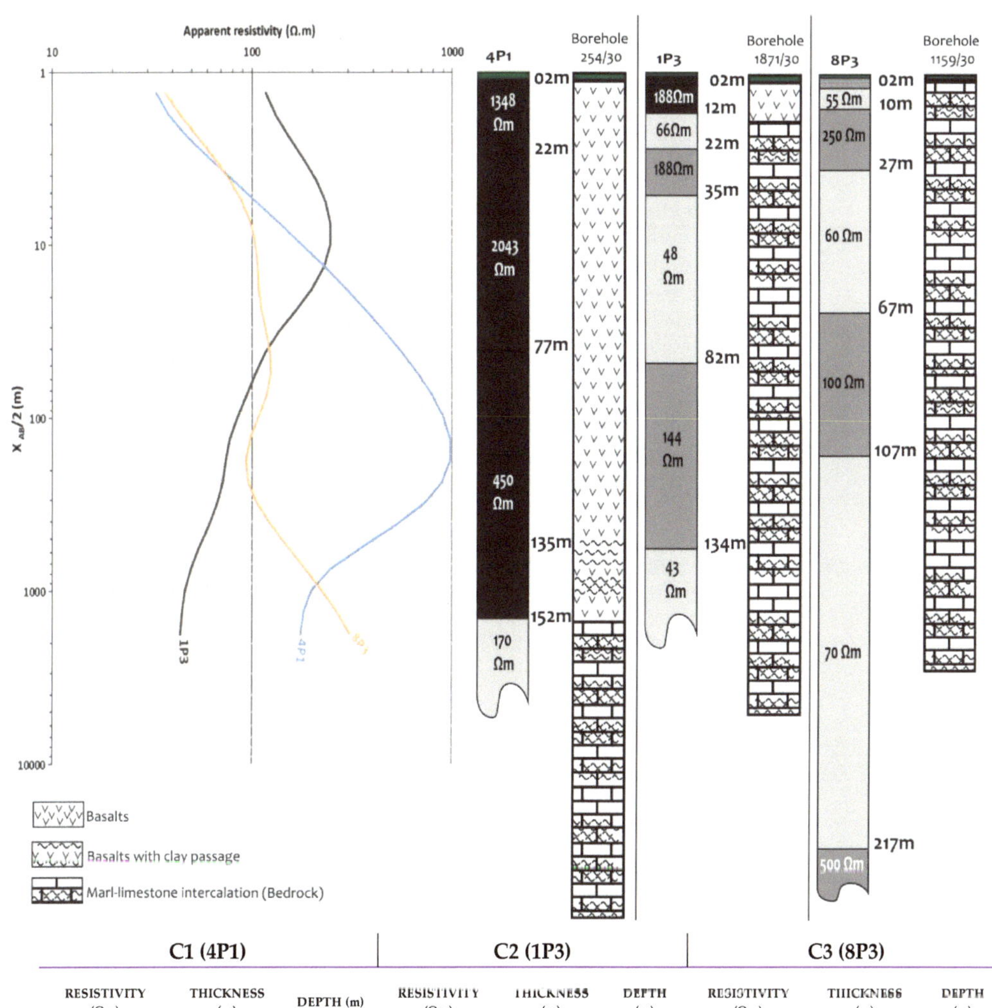

	C1 (4P1)			C2 (1P3)			C3 (8P3)		
	RESISTIVITY (Ωm)	THICKNESS (m)	DEPTH (m)	RESISTIVITY (Ωm)	THICKNESS (m)	DEPTH (m)	RESISTIVITY (Ωm)	THICKNESS (m)	DEPTH (m)
$\varrho1$	28	1.4		100	1.2		30	1	
$\varrho2$	1347	21	1.4	500	2.5	1.2	200	3	1
$\varrho3$	2043	55	22	150	8	3.7	55	6	4
$\varrho4$	450	75	64	52	10	12	250	17	10
$\varrho5$	170		152	177	13	22	60	40	27
$\varrho6$	–	–	–	42	47	35	100	40	67
$\varrho7$	–	–	–	144	52	82	70	100	107
$\varrho8$	–	–	–	46		134	500		207

Figure 5. Diagrams and interpretation models of typical SEV of each category and the borehole data with its corresponding lithological logs.

The second category (C2) contained a total of 9 VES measurements. This category is represented by the diagram of 1P3, the typical shape of the curves of which is given in Figure 5. An examination of the 1P3 diagram shows, from bottom to top, the presence of the following geoelectrical levels: (i) a succession of moderately conductive to resistant levels, with resistivity varying between 40 Ωm and 177 Ωm, and thickness being 122 m, which corresponds to the marl–limestone intercalation of the Pliocene; and (ii) a 12 m thick resistant ensemble representing the Quaternary basalts, which were altered at the beginning, then dry before becoming slightly damp afterwards.

The third category (C3) contained a total of 08 VES measurements. The curve of the electrical borehole 8P3 (Figure 5), which is the representative of this category, shows, from bottom to top, the presence of the following geoelectrical levels: (i) a very resistant substratum with a resistivity of 500 Ωm. It is also very deep, located at a depth of 207 m, and could be the equivalent of Pliocene limestone and sandstone or any underlying formation (e.g., Jurassic limestones); (ii) a succession of moderately conductive to resistant levels, from resistivities of 50 Ωm to 250 Ωm, and 206 m thick, which would correspond to Pliocene marl and limestone; and (ii) a superficial conductive level, 1 m thick, which can be made to correspond to the vegetal soil.

4.2. Geoelectrical Section Analysis

Geoelectrical sections (GSs) yield the visualization of lateral and vertical variations of resistive and conductive horizons, as well as possible electrical discontinuities. Based on the correlation between the interpreted and inverted VES patterns, seven GSs were performed along the N–S, NW–SE and E–W directions.

It should be noted that due to the sometimes-large distance between electrical soundings in the same profile, correlations may be influenced by the presence of structural discontinuities, such as faults and fractures, making it difficult to correlate, which requires calibration with existing borehole data.

Only the four most representative cuts of these directions are presented and discussed in this work, namely GS1 (Profile 1), GS2 (Profile 3), GS3 (Profile 5) and GS4 (Profile 7). Figures 6–9 show these geoelectrical cross-sections and for each one, a simplified cross-section has been drawn with the same horizontal and vertical scale.

Section GS1, grouping the VES of profile 1 (Figure 6), shows the gradual plunge of the marl–limestone bedrock towards the north and, consequently, an increase in the thickness of the basaltic flow from the south to the north. The presence of the North Middle Atlas accident (NMAA) and the corresponding satellite faults to the south of the profile is a physical argument that justifies the electrical discontinuities between holes 1P1 and 2P1 and between holes 2P1 and 3P1. It is highly likely that because of the faulting in this corridor, the marl and limestone bedrock revealed by boreholes 1P1 and 2P1 may have risen. The rising bedrock prevented the basaltic flow from flowing southwards. Section GS1 has been correlated with data from borehole NIRE 254/30.

Section GS2 (Figure 7) groups the electrical soundings carried out in profile 3. It shows a gradual sinking of the marl and limestone bedrock towards the south, unlike the previous section, which generated a synclinal depression into which the basalts flowed with a thickness that increased from north to south. The existence of electrical discontinuities between boreholes 1P3 and 2P3 and between 2P3 and 3P3 corresponds to the continuity of NMAA satellite faults, leading to the uplift of the marl–limestone bedrock at the level of borehole 2P3, according to a horst structure, which prevented the flow of basaltic flows towards the south.

Section GS3 (Figure 8) at the level of profile 5, shows that the top of the marl–limestone bedrock forms a bowl structure where boreholes 4P5 and 5P5 show the lowest points. Towards the two north-western and south-eastern extremities, a rise in the roof is observed, which implies a decrease in the thickness of the overlying basaltic flows. A geophysical discontinuity was recorded between electrical boreholes 5P5 and 6P5, which may represent a fault or a simple flexure of the bedrock.

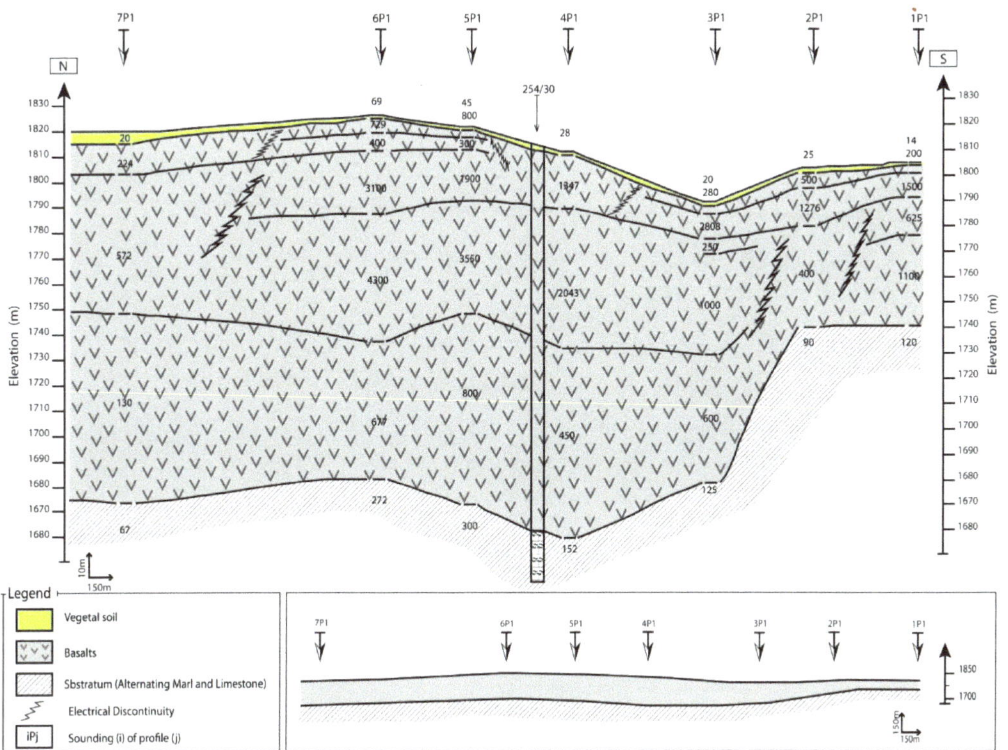

Figure 6. GS1 geoelectrical section of profile 1.

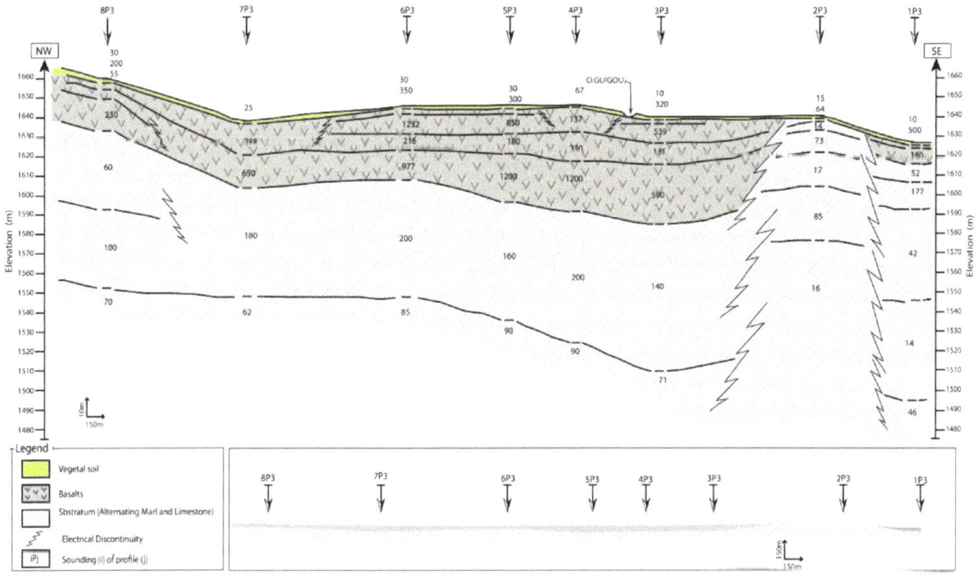

Figure 7. GS2 geoelectrical section of profile 3.

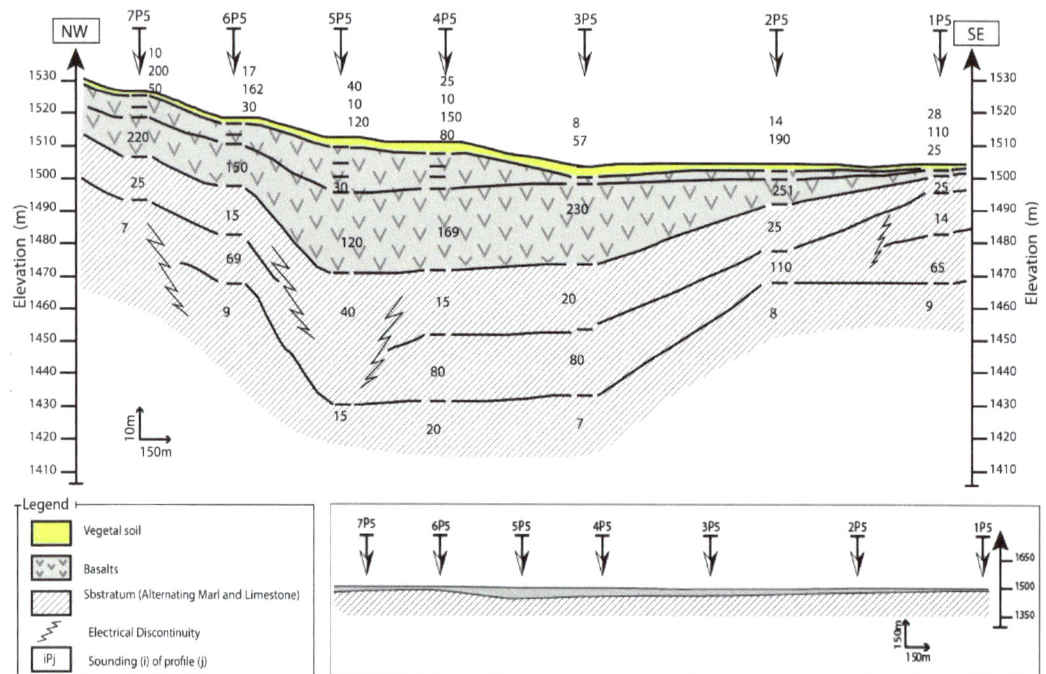

Figure 8. GS3 geoelectrical section of profile 5.

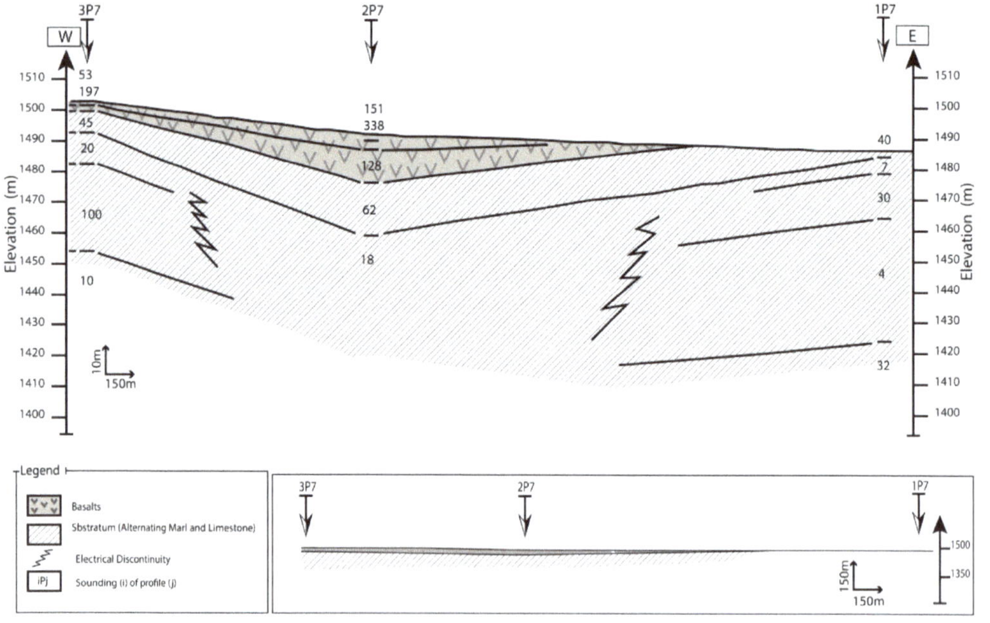

Figure 9. GS4 geoelectrical section of profile 7.

Section GS4 (Figure 9) is oriented E–W towards the north-eastern edge of the study area and follows profile 7. In this section, the thickness and extent of the basalts are clearly reduced. The maximum thickness reached by the basalts is 20 m, which was recorded in the middle of the section. Furthermore, a decrease in the thickness of these basalts is highlighted towards the west side where it reaches less than 5 m on the one hand. On the other hand, going towards the east end, this thickness progressively decreases until the bedrock outcrops.

4.3. Interpretative Map Analysis

The apparent iso-resistivity maps show the dispersion and lateral variation of the apparent resistivity (ρa) in the study area for different slices of the ground. Two iso-resistivity maps have been developed ($\rho a_{AB=80\,m}$ and $\rho a_{AB=200\,m}$). The analysis of the apparent resistivity map $\rho a_{AB=80\,m}$ presented in Figure 10a for a line length of AB = 80 m, corresponding to an investigation depth of about 15 m, allowed us to distinguish the following areas: (i) highly resistive areas ($\rho a > 200$ Ωm), corresponding to zones where the basalts are quite thick, notably in the SW of the study area; (ii) quite conductive areas ($\rho a \leq 100$ Ωm), located in the east and north-east of the study area. This decrease in apparent resistivity (ρa) is most probably linked to the absence of basalts on the surface; (iii) the rest of the map is occupied by intermediate resistivity ($100 \leq \rho a \leq 200$ Ωm), occupying large areas and corresponding to zones where basalts are present with small thicknesses, or they are more altered. For a deeper slice, the map $\rho a_{AB=200\,m}$ (Figure 10b) corresponds to a depth of investigation of about 30 m, and this map roughly follows the pattern of the previous map $\rho a_{AB=80\,m}$ and shows a very resistant zone ($\rho a \geq 500$ Ωm), located SE of the study area. This increase in ρa reflects the presence of very resistant terrain corresponding to basalts. A relatively conductive zone ($\rho a \leq 100$ Ωm) is located to the north-east of the surveyed area. This decrease in electrical resistivity is due to the presence of Pliocene marl and marl–limestone soils from the surface. The zones of intermediate resistivity ($100 \leq \rho a \leq 500$ Ωm) correspond to areas where the basalts are of low thickness above the Pliocene marl–limestone bedrock.

The isohypse map of the marl–limestone bedrock (Figure 10c) shows that the roof altitude varies between 1800 m, recorded at the upstream end of the plain, and 1460 m, the minimum value recorded downstream. It shows, in a general way, the behavior of the roof of the marly–limestone substratum, which gradually plunges from the south-west to the north-east in accordance with the flow of Oued Guigou and the water table. The isopach map represents the thickness distribution of the basaltic aquifer. The examination of this map (Figure 10d) shows that the maximum values are located to the south-west of the study area, mainly at the level of the electrical boreholes of profile 1 where the thickness values are close to 150 m. Overall, the thickness of the basaltic formations decreases progressively from the south-west to the north-east, with a slight increase in profiles 5 and 6.

4.4. Tectonic and Hydrogeological Implications

This work, based on geophysical reconnaissance by VES, shows the existence at depths of a more or less conductive level corresponding to calcareous marl alternations, surmounted by a very resistant level attributed to Quaternary basalts with a thickness that reaches at least 150 m in the south-western part of the study area. These thicknesses are consistent with those found by A. Bentayeb and C. Leclerc [38]. These basaltic formations have good hydrodynamic characteristics, which can yield significant flows depending on their degree of fracturing, and their rate of recharge [1].

The production of qualitative geoelectrical maps in terms of apparent resistivity made it possible to identify conductive areas and resistant areas that were interpreted differently according to the different lengths of the injection line. The conductive patches were attributed to the sub-cropping of the marl and limestone bedrock formations, while the resistant patches reflected the presence of fairly thick basalts. The quantitative isohypse map of the marl–limestone bedrock roof shows, as do the geoelectrical sections, that this roof gradually dips from the south-west to the north-east. The maximum elevation of a roof at this level would be to the order of 1900 m towards the south-west of the surveyed area, and it records the coast of 1480 m towards the north-east with a difference in altitude of 420 m for a distance of 32,000 m and a gradient of 1.3%.

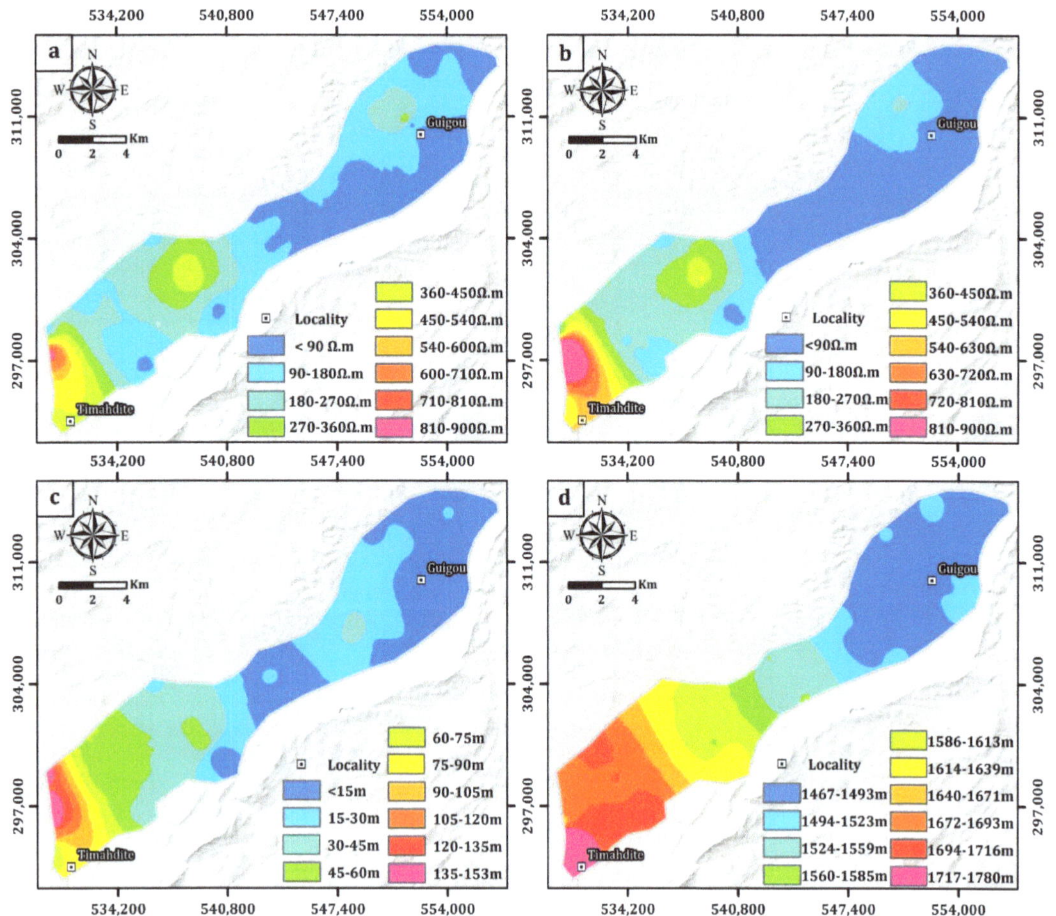

Figure 10. Thematic maps: (**a**): Apparent resistivity map for AB = 80 m (**b**): Apparent resistivity map for AB = 200 m (**c**): Isohypse map of the bedrock of marl and limestone (**d**): Map of Quaternary basalt isopach.

The main geoelectrical characteristics extracted from the geoelectrical sections and maps, coupled with the geological information available on this area, allowed us to elaborate on the aquifer prospectivity map presented in Figure 11. The alignment of electrical discontinuities allowed the continuity of lineaments between the geoelectrical sections to be estimated and the orientation and interpretation of these lineaments in terms of faults or fractures to be deduced. These fractures show a NE–SW direction sub-parallel to the major accidents of the Middle Atlas, highlighted by numerous old works [1,7]. The work of Amrani and Hinaj (2016) demonstrated that groundwater flows in the Plio-Quaternary aquifer system follow the NE–SW trend, with multi-gap faults affecting the collapsed zone of the Guigou Plain. In this sense, these physical discontinuities play a major hydrogeological role in the preferential circulation of groundwater. The local aquifer prospectivity map shown in Figure 11 outlines the potential alignment of regional faults deduced from the electrical discontinuities (EDs). The zones of passage of these faults between the electrical sections are confirmed by the alignment of the boreholes with particularly good water flows.

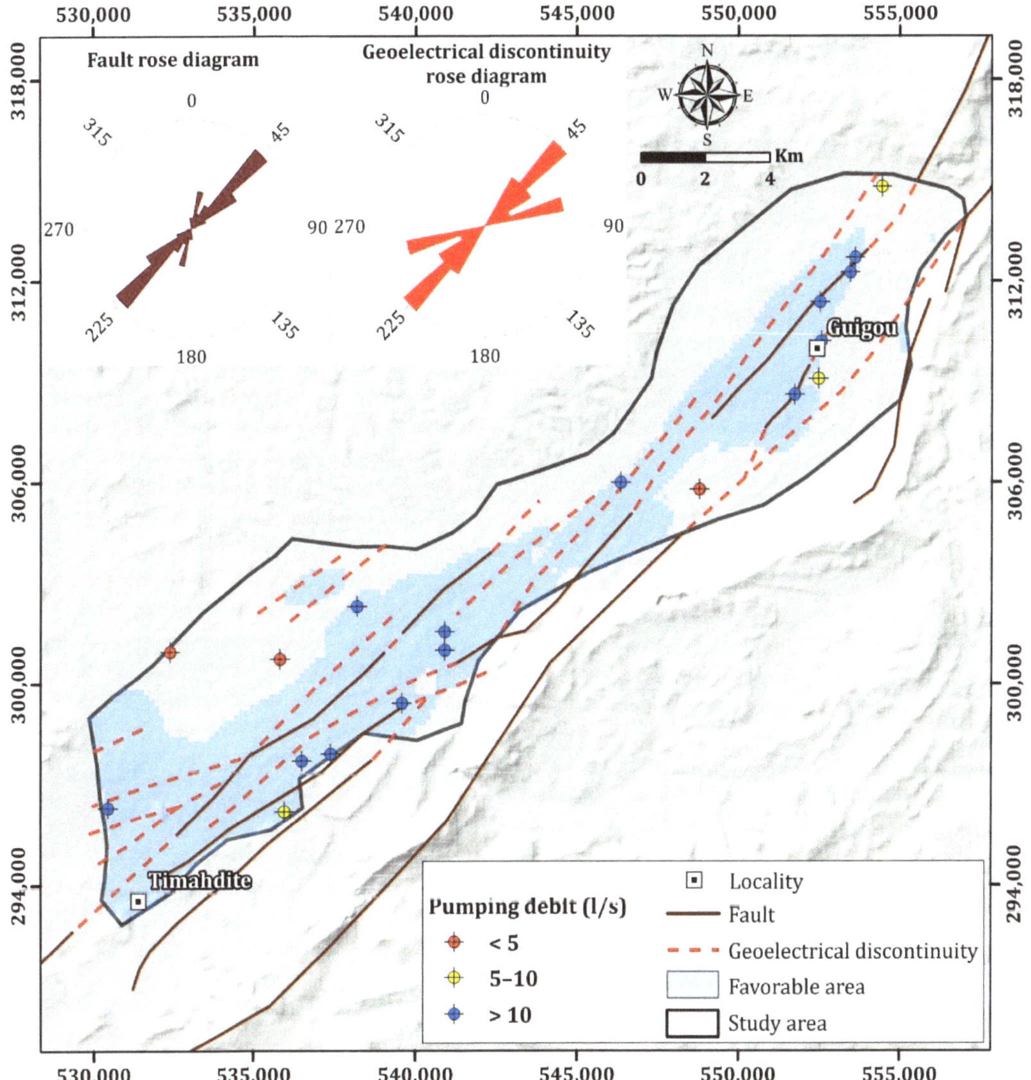

Figure 11. Aquifer prospectivity map showing the most favorable areas for drilling.

The lineaments in red correspond to the zones of passage of the electrical discontinuities. On the other hand, those in brown reflect the faulted structures as they are mapped on the ground from the structural and geological maps of the study area. Subsequently, the superposition of the results obtained allowed us to identify the most favorable areas for the installation of groundwater exploitation wells. To justify this choice, a hydrogeological survey was carried out, which showed that the position of the boreholes with a good flow rate (Q > 10 L/s) coincide with the favorable zones (see Figure 11). On the other hand, low-flow boreholes (Q < 5 L/s) are located far from favorable areas and medium-flow boreholes (5 L/s < Q < 10 L/s) are located near favorable areas. The results of this study will serve as a guide to optimize the location of new wells and/or boreholes.

5. Conclusions

The present work demonstrates the importance of using VES geoelectrical data in the characterization of aquifers. This technique was applied to target areas of high groundwater-flow potential in the Guigou plain between the town of Timahdite and the village of Almis Guigou. The results obtained led to detailed map of the electrical discontinuities corresponding to fracture zones affecting the basalts forming the main aquifer. Iso-resistivity, the marl–limestone bedrock and basalt thickness maps were generated to characterize this aquifer.

The analysis and Interpretation of all the VES measurements show the presence of a very resistant upper level with a thickness varying between 0 and 150 m, attributed to the basaltic formations of the Quaternary age, followed by a moderately conductive horizon with a resistivity of about 90 Ωm. According to local geological data, this conductive level corresponds to Pliocene marl and limestone alternations. The correlation between the geophysical models obtained from the VES interpretations and their confrontation with local geological data and the lithological sections of the mechanical drillings made it possible to draw a certain number of geoelectrical sections, which reflect the evolution of the thickness and resistivity of the basalts above the marl–limestone bedrock. Indeed, the marl and limestone formations generally plunge from the south-west to the north-east with the presence of geophysical discontinuities, which locally interrupt this plunge. These detected electrical discontinuities could be interpreted in terms of a manifestation of the scarping of the satellite faults of the North Middle Atlas accident, which blocked the flow of basalts southwards and formed this basaltic aquifer, preventing the flow of basalts.

The integration of geophysical and geological field knowledge has led to a more informed regional tectonic interpretation and assessment of the hydrogeological prospects of the region. To this end, the results of this geophysical survey have made it possible to better characterize the geometry of the Quaternary basaltic aquifer and the electrical discontinuities and its Pliocene calcareous marl substratum. Thus, this study constitutes a basic document to help decision makers better manage the siting of water boreholes in order to ensure the good integrated management of the region's groundwater resources.

Author Contributions: Conceptualization, S.E.M., A.R., A.M. and M.B.; Methodology, S.E.M., Y.M., A.M. and A.R.; Software, S.E.M., Y.M., A.R. and A.M.; Validation, S.E.M., Y.M., G.R., M.B. and A.M.; Formal analysis, S.E.M., Y.M., H.E.M., A.A., A.M. and A.R.; Investigation, S.E.M., M.E.H., S.L. and A.M.; Resources, A.M.; Data curation, S.E.M., Y.M., H.E.M., A.A. and A.M.; Writing—original draft, S.E.M., Y.M., A.M. and H.E.M.; Writing—review & editing, S.E.M. and A.M.; Visualization, S.E.M., G.R., M.E.H., A.M. and M.B.; Supervision, M.B.; Project administration, M.B.; Funding acquisition, A.M. All authors have read and agreed to the published version of the manuscript.

Funding: This research received no external funding.

Institutional Review Board Statement: Not applicable.

Informed Consent Statement: Not applicable.

Data Availability Statement: Not applicable.

Acknowledgments: The authors would like to thank the Editors of the Special Issue as well as the anonymous reviewers for their valuable comments on this article, which allowed us to improve the scientific quality of this research.

Conflicts of Interest: The authors declare no conflict of interest.

Appl. Sci. **2022**, *12*, 12829

References

1. Amrani Hydrodynamisme. Hydrogéochimie et Vulnérabilité de La Nappe d'Eau Superficielle et Leur Relation Avec La Tectonique Cassante Dans La Zone Effondrée Timahdite—Almis Guigou (Moyen Atlas, Maroc). Available online: http://www.secheresse.info/spip.php?article845642016 (accessed on 1 December 2022).
2. Todd, D.K.; Mays, L.W. *Groundwater Hydrology*; John Wiley & Sons: Hoboken, NJ, USA, 2004; ISBN 0471059374.
3. Fetter, C.W. *Applied Hydrogeology*; Waveland Press: Long Grove, IL, USA, 2018; ISBN 1478637447.
4. Jha, M.K.; Kumar, S.; Chowdhury, A. Vertical Electrical Sounding Survey and Resistivity Inversion Using Genetic Algorithm Optimization Technique. *J. Hydrol.* **2008**, *359*, 71–87. [CrossRef]
5. Raju, N.J.; Reddy, T.V.K. Fracture Pattern and Electrical Resistivity Studies for Groundwater Exploration. *Environ. Geol.* **1998**, *34*, 175–182. [CrossRef]
6. Mohamed, A.; Al Deep, M.; Othman, A.; Taha Al Alshehri, F.; Abdelrady, A. Integrated Geophysical Assessment of Groundwater Potential in Southwestern Saudi Arabia. *Front. Earth Sci* **2022**, *10*, 937402. [CrossRef]
7. Hinaje, S. Tectonique Cassante et Paléochamps de Contraintes Dans Le Moyen Atlas et le Haut Atlas Central (Midelt-Errachidia) Depuis Le Trias Jusqu'à l'actuel. Unpublished Thesis, Mohamed V University, Rabat, Morocco, 2004.
8. Christensen, N.B.; Christiansen, A. V Using Geophysical Survey Results in the Inference of Aquifer Vulnerability Measures. *Near Surf. Geophys.* **2021**, *19*, 505–521. [CrossRef]
9. Bangalore Nagaraj, P.; Mandalagiri Subbarayappa, M.K.; Jean-Michel, V.; Hoareau, J. Estimation of Anisotropic Hydraulic Conductivity Using Geophysical Data in a Coastal Aquifer of Karnataka, India. *Hydrol. Process.* **2021**, *35*, e14395. [CrossRef]
10. Nugraha, G.U.; Nur, A.A.; Pranantya, P.A.; Lubis, R.F.; Bakti, H. Analysis of Groundwater Potential Zones Using Dar-Zarrouk Parameters in Pangkalpinang City, Indonesia. *Environ. Dev. Sustain.* **2022**, 1–23. [CrossRef]
11. Puttiwongrak, A.; Men, R.; Vann, S.; Hashimoto, K.; Suteerasak, T. Application of Geoelectrical Survey and Time-Lapse Resistivity with Groundwater Data in Delineating a Groundwater Potential Map: A Case Study from Phuket Island, Thailand. *Sustainability* **2022**, *14*, 397. [CrossRef]
12. Rhubango, H.-A.; Mwangaza, N. Geo-Electrical Investigation for Groundwater Resources in a Part of Butembo Area (North Kivu Province; Democratic Republic of Congo). *Am. Acad. Sci. Res. J. Eng. Technol. Sci.* **2022**, *85*, 157–169.
13. Essahlaoui, A.; Sahbi, H.; El Yamine, N. Application de La Géophysique (Méthode Géoélectrique) à La Reconnaissance Du Plateau de Meknès (Bassin de Saïss), Maroc. *Geol. Belg.* **2000**. [CrossRef]
14. Tijani, M.N.; Obini, N.; Inim, I.J. Estimation of Aquifer Hydraulic Parameters and Protective Capacity in Basement Aquifer of South-Western Nigeria Using Geophysical Techniques. *Environ. Earth Sci.* **2021**, *80*, 466. [CrossRef]
15. Ben Fraj, A.; Gabtni, H. Quaternary Alluvial Aquifer Study Using Integrated Geophysical Approach in Zaghouan Plain (Northeastern Tunisia). *Nat. Resour. Res.* **2021**, *30*, 307–319. [CrossRef]
16. Mohamed, A.; Gonçalvès, J. Hydro-Geophysical Monitoring of the North Western Sahara Aquifer System's Groundwater Resources Using Gravity Data. *J. Afr. Earth Sci.* **2021**, *178*, 104188. [CrossRef]
17. Obiora, D.N.; Ibuot, J.C. Geophysical Assessment of Aquifer Vulnerability and Management: A Case Study of University of Nigeria, Nsukka, Enugu State. *Appl. Water Sci.* **2020**, *10*, 29. [CrossRef]
18. Doetsch, J.; Linde, N.; Coscia, I.; Greenhalgh, S.A.; Green, A.G. Zonation for 3D Aquifer Characterization Based on Joint Inversions of Multimethod Crosshole Geophysical Data. *Geophysics* **2010**, *75*, G53–G64. [CrossRef]
19. Bayowa, O.G.; Afolabi, O.A.; Akinluyi, F.O.; Oshonaiye, A.O.; Adelere, I.O.; Mudashir, A.W. Integrated Geoelectrics and Hydrogeochemistry Investigation for Potential Groundwater Contamination around a Reclaimed Dumpsite in Taraa, Ogbomoso, Southwestern Nigeria. *Int. J. Energy Water Resour.* **2022**, 1–22. [CrossRef]
20. Azffri, S.L.; Azaman, A.; Sukri, R.S.; Jaafar, S.M.; Ibrahim, M.F.; Schirmer, M.; Gödeke, S.H. Soil and Groundwater Investigation for Sustainable Agricultural Development: A Case Study from Brunei Darussalam. *Sustainability* **2022**, *14*, 1388. [CrossRef]
21. Ebraheem, A.A.; Sherif, M.; Mulla, M.A.; Alghafli, K.; Sefelnasr, A. Assessment of Groundwater Resources in Water Spring Areas Using Geophysical Methods, Northern UAE. In *Wadi Flash Floods*; Springer: Singapore, 2022; pp. 493–508.
22. Shishaye, H.A.; Tait, D.R.; Befus, K.M.; Maher, D.T. An Integrated Approach for Aquifer Characterization and Groundwater Productivity Evaluation in the Lake Haramaya Watershed, Ethiopia. *Hydrogeol. J.* **2019**, *27*, 2121–2136. [CrossRef]
23. Hamzah, U.; Samsudin, A.R.; Malim, E.P. Groundwater Investigation in Kuala Selangor Using Vertical Electrical Sounding (VES) Surveys. *Environ. Geol.* **2007**, *51*, 1349–1359. [CrossRef]
24. Troisi, S.; Fallico, C.; Straface, S.; Migliari, E. Application of Kriging with External Drift to Estimate Hydraulic Conductivity from Electrical-Resistivity Data in Unconsolidated Deposits near Montalto Uffugo, Italy. *Hydrogeol. J.* **2000**, *8*, 356–367. [CrossRef]
25. Mamouch, Y.; Attou, A.; Miftah, A.; Ouchchen, M.; Dadi, B.; Achkouch, L.; Et-tayea, Y.; Allaoui, A.; Boualoul, M.; Randazzo, G.; et al. Mapping of Hydrothermal Alteration Zones in the Kelâat M'Gouna Region Using Airborne Gamma-Ray Spectrometry and Remote Sensing Data: Mining Implications (Eastern Anti-Atlas, Morocco). *Appl. Sci.* **2022**, *12*, 957. [CrossRef]
26. Bernini, M.; Boccaletti, M.; Gelati, R.; Moratti, G.; Papani, G.; Mokhtari, J. El Tectonics and Sedimentation in the Taza-Guercif Basin, Northern Morocco: Implications for the Neogene Evolution of the Rif-Middle Atlas Orogenic System. *J. Pet. Geol.* **1999**, *22*, 115–128. [CrossRef]
27. Termier, H. Etudes Géologiques Sur Le Maroc Central et Le Moyen-Atlas Septentrional. 1936. Available online: https://www.persee.fr/doc/rga_0035-1121_1937_num_25_3_3982_t1_0525_0000_1 (accessed on 1 December 2022).

28. Colo, G. Contribution a L'étude Du Jurassique Du Moyen Atlas Septentrional: Atlas de Planches Hors Texte; Éd. de la Division de la géologie, Direction, Ministère, Royaume du Maroc. 1961. Available online: https://bibliotheques.mnhn.fr/medias/detailstatic.aspx?INSTANCE=exploitation&RSC_BASE=HORIZON&RSC_DOCID=379840 (accessed on 1 December 2022).

29. Gentil, L. Notice Sur Les Titres et Travaux Scientifiques, E. Larose. 1918. Available online: https://gallica.bnf.fr/ark:/12148/bpt6k90008k.texteImage (accessed on 1 December 2022).

30. Martin, J. Le Moyen Atlas Central, Étude Géomorphologique. 1981. Available online: http://pascal-francis.inist.fr/vibad/index.php?action=getRecordDetail&idt=PASCALGEODEBRGM8220223851 (accessed on 1 December 2022).

31. Dresnay, R. du Recent Data on the Geology of the Middle-Atlas (Morocco). In *The Atlas System of Morocco*; Springer: Berlin/Heidelberg, Germany, 1988; pp. 293–320.

32. Sachse, V.F.; Leythaeuser, D.; Grobe, A.; Rachidi, M.; Littke, R. Organic Geochemistry and Petrology of a Lower Jurassic (Pliensbachian) Petroleum Source Rock from Aït Moussa, Middle Atlas, Morocco. *J. Pet. Geol.* **2012**, *35*, 5–23. [CrossRef]

33. Termier, H.; Dubar, G. *Carte Géologique Provisoire Du Moyen-Atlas Septentrional Au 1/200.000 e: Notice Explicative*; Imprimerie Officielle: Radès, Tunisia, 1940.

34. Charroud, M. Evolution Géodynamique de La Partie Sud-Ouest Du Moyen Atlas Durant Le Passage Jurassique-Crétacé, Le Crétacé et Le Paléogène: Un Exemple d'évolution Intraplaque. These 3eme Cycle. Université Mohammed V, Rabat, Morocco, 1990.

35. Ouarhache, D.; Charriere, A.; Chalot-Prat, F.; Wartiti, M.E.L. Triassic to Early Liassic Continental Rifting Chronology and Process at the Southwest Margin of the Alpine Tethys (Middle Atlas and High Moulouya, Morocco); Correlations with the Atlantic Rifting, Synchronous and Diachronous. *Bull. Société Géologique Fr.* **2012**, *183*, 233–249. [CrossRef]

36. Hamidi, E.M.; Boulangé, B.; Colin, F. Altération d'un Basalte Triasique de La Région d'Elhajeb, Moyen Atlas, Maroc. *J. Afr. Earth Sci.* **1997**, *24*, 141–151. [CrossRef]

37. Muzirafuti, A.; Boualoul, M.; Barreca, G.; Allaoui, A.; Bouikbane, H.; Lanza, S.; Crupi, A.; Randazzo, G. Fusion of Remote Sensing and Applied Geophysics for Sinkholes Identification in Tabular Middle Atlas of Morocco (the Causse of El Hajeb): Impact on the Protection of Water Resource. *Resources* **2020**, *9*, 51. [CrossRef]

38. Bentayeb, A.; Leclerc, C. Le Causse Moyen Atlasique. In *Ressources en Eaux du Maroc, Tome3, Domaines Atlasiques Sud-Atlasiques*; 1977; pp. 37–66. Available online: https://www.scribd.com/document/494028246/Ressources-en-Eau-Du-MAROC-Tome-III (accessed on 1 December 2022).

39. Ellouz, N.; Patriat, M.; Gaulier, J.-M.; Bouatmani, R.; Sabounji, S. From Rifting to Alpine Inversion: Mesozoic and Cenozoic Subsidence History of Some Moroccan Basins. *Sediment. Geol.* **2003**, *156*, 185–212. [CrossRef]

40. Arboleya, M.L.; Teixell, A.; Charroud, M.; Julivert, M. A Structural Transect through the High and Middle Atlas of Morocco. *J. Afr. Earth Sci.* **2004**, *39*, 319–327. [CrossRef]

41. Gomez, F.; Barazangi, M.; Bensaid, M. Active Tectonism in the Intracontinental Middle Atlas Mountains of Morocco: Synchronous Crustal Shortening and Extension. *J. Geol. Soc. Lond.* **1996**, *153*, 389–402. [CrossRef]

applied
sciences

Review

Reconstruction of Land and Marine Features by Seismic and Surface Geomorphology Techniques

Dicky Harishidayat [1,*], Abdullatif Al-Shuhail [1], Giovanni Randazzo [2], Stefania Lanza [3] and Anselme Muzirafuti [2,*]

1   Department of Geosciences, College of Petroleum Engineering and Geosciences, King Fahd University of Petroleum and Minerals, Dhahran 31261, Saudi Arabia
2   Department of Mathematics, Computer Sciences, Physics and Earth Sciences, University of Messina, 31-98166 Messina, Italy
3   GeoloGIS s.r.l., Università degli Studi di Messina, Via F. Stagno d'Alcontres, 31-98166 Messina, Italy
*   Correspondence: dicky.hidayat@kfupm.edu.sa or dickyharishidayat@gmail.com (D.H.); anselme.muzirafuti@unime.it (A.M.)

Abstract: Seismic reflection utilizes sound waves transmitted into the subsurface, reflected at rock boundaries, and recorded at the surface. Interpretation of their travel times and amplitudes are the key for reconstructing various geomorphological features across geological time (e.g., reefs, dunes, and channels). Furthermore, the integration of surface geomorphology technique mapping, such as digital elevation models, with seismic geomorphology can increase land and marine feature modelling and reduce data uncertainty, as well. This paper presents an overview of seismic and surface geomorphology techniques and proposes an integrated workflow for better geological mapping, 3D surface imaging, and reconstruction. We intend to identify which techniques are more often used and which approaches are more appropriate for better output results. We noticed that an integration of surface and subsurface geomorphology techniques could be beneficial for society in landscape mapping, reservoir characterization, and city/regional planning.

Keywords: quantitative geomorphology; seismic geomorphology; seismic reflection; 3D imaging; earth surface reconstruction; remote sensing; aerial photogrammetry; geological mapping; integrated geomorphology

Citation: Harishidayat, D.; Al-Shuhail, A.; Randazzo, G.; Lanza, S.; Muzirafuti, A. Reconstruction of Land and Marine Features by Seismic and Surface Geomorphology Techniques. *Appl. Sci.* **2022**, 12, 9611. https://doi.org/10.3390/app12199611

Academic Editors: Ricardo Castedo and Fabrizio Dalsamo

Received: 8 August 2022
Accepted: 22 September 2022
Published: 24 September 2022

Publisher's Note: MDPI stays neutral with regard to jurisdictional claims in published maps and institutional affiliations.

1. Introduction

Seismic reflection interpretation has existed for decades, beginning with a two-dimensional (2D) seismic reflection method and developing into a more intensive three-dimensional (3D) method. The primary components of seismic interpretation are seismic reflection data coupled with geologic depositional and tectonic models. These provide a framework for integrating borehole, microseismic, outcrop, and modern landscape analogue data that result in a realistic earth surface (geomorphology), subsurface reconstruction, and reservoir model [1–10]. Meanwhile, a proper understanding of seismic reflection data is a necessary precursor to successful interpretation [1,11–23]. The seismic reflection data utilize the transmission of a sound wave (triggered by an air gun, vibroseis, etc.) that propagates into the subsurface and is reflected to the surface when encountering an interface between two different rock properties, such as density and velocity. The reflected sound wave is recorded at the surface by receivers that measure its amplitude and arrival time (two-way travel time). Later, the recorded data are processed utilizing mathematical and signal processing techniques to produce an image of the subsurface. The seismic processing workflow could be different from one dataset to another depending on the geological conditions, target, processor, etc. The processing of seismic reflection data could be utilized in commercial software, e.g., Vista™ (Schlumberger), SeisSpace ProMax™ (Halliburton), Echos™ (Paradigm), Geovation™ (CGG), and many more. In addition, open source and free license software are also available, e.g., Madagascar, Seismic UNIX (Colorado School of Mines), SEPlib (University of Stanford), any many more.

The advancements in seismic acquisition and processing have provided significant improvements to the quality and resolution of seismic data; thus, the extraction of detailed geological information (and ultimately the development of a realistic 3D geological model) is becoming a reality. Plan-view images of the depositional elements and depositional systems (on a large scale) were provided by 3D seismic data, in which the morphology of these elements could be extracted from a seismic cube. Analysis of such images could significantly enhance predictions of the spatial and temporal distributions of subsurface lithology (reservoirs, sources, and seals), fluids, compartmentalization, and stratigraphic trapping capabilities. Furthermore, it can contribute to an enhanced understanding of process sedimentology, sequence stratigraphy, and tectonics [15,20,24–42].

Seismic geomorphology itself is the study of the subsurface using plan-view images from three-dimensional seismic data and aims to extract geomorphological, depositional, and other geologically significant features [12,15,20,27,29,43]. Seismic geomorphology, which depends on the interpretation of plan-view seismic images, is rapidly developing on several fronts, including [29]:

- Understanding the development of seascapes and landscapes in clastic and carbonate settings;
- Advances in workflows directed toward lithological prediction through the integration of seismic stratigraphy and seismic geomorphology;
- Revising and improving sequence stratigraphic models;
- Development of new and increasingly sophisticated analytical techniques.

Furthermore, the quantification of geomorphology from seismic reflection data features morphometric analysis of (for instance) sediment conduits that play an important role in the quantitative interpretation of sedimentary processes and paleoenvironments [39,44–53]. Morphometry includes:

- Height, defined as the vertical distance within a sediment conduit from its base to spill point;
- Top width, defined as the horizontal distance between two spill points;
- Base width, defined as the horizontal distance between two points in its floor;
- Cross-sectional area (CSA), defined as the area of a sediment conduit perpendicular to its axis;
- Aspect ratio, defined as the ratio between width and height of the sediment conduit's CSA;
- Sinuosity, defined as the ratio between a reference point and the sediment conduit's axis;
- Gradient, calculated from depth changes along the sediment conduit.

Moreover, surface geomorphology is the study of earth's physical land–marine surface features: its forms, processes, origin, development, and evolution that finally form a land–marine feature [54–61]. Nevertheless, shallow subsurface processes, other than surface and atmospherics processes, also play an important role on shaping the earth's land–marine forms (the evolution of topography). This subsurface process is mainly related to the geological processes including tectonics, volcanic activity, earthquakes, sedimentation, groundwater activity, sea-level changes, etc.

Furthermore, surface geomorphological data acquisition techniques, such as:

1. Quantitative geomorphology approaches have shown a great potential for identifying the location of geomorphological boundaries [62–64], the distance, surface, and the volume of geomorphological processes evolution [62,65–67].

2. Remote sensing techniques from space using radar sensors for supporting geomorphological interpretation of slow-moving coastal geohazards [68,69] or for monitoring subsurface deformation for interferometric analysis on a regional [70,71] or continental scale [72].

3. In situ and proxy geomorphological mapping techniques using unmanned aerial vehicles (UAV) photogrammetry, optical survey (for fine-scale topographic data), LIDAR

(Light Detection and Ranging), and any geographic information systems whose mapping provides reliable surface geomorphology data [73,74].

Apart from techniques and approaches used for data acquisition, surface geomorphology data processing algorithms and software have recently been developed and improved. Most of the earth observation satellite images available in Google Earth Engine (GEE) can be analyzed based on algorithms and a software interface of GEE Python application programming interface [75]. van Natijne et al. [76] used SAR images available in GEE to assess the potential of Interferometric Synthetic Aperture Radar (InSAR)-based deformation tracking; the authors demonstrated that the deformation could be detected on at least 91% of the global landslide-prone slopes. Other open-source software such as QGIS and SNAP desktop have been developed for the analysis of Optic and Radar images. The development of aerial photogrammetry technology has been accompanied by the launch of powerful Geographic Information System (GIS) software [77,78] such as Pix4D Mapper [79], Agisoft PhotoScan [80], and ArcMap for LiDAR and orthophoto quantitative analysis and geobody extraction.

Recent advances in technology (especially remote sensing) have provided a wider and larger area coverage of the earth's surface with high spatial resolution and free access (with term and conditions). In addition, near-surface or deep-surface acquisition data tools like seismic reflection data could provide results of deeper and wider areas (compared to other subsurface data) when it comes to depth of penetration (km scale) and lateral area (km^2 scale), respectively. Therefore, the aim of this study is to review seismic and surface geomorphology techniques already existing in the energy industry and academia and relate them to 3D earth subsurface imaging, reconstruction, sedimentary architecture, and geomorphological feature extraction from seismic reflection and surface geomorphology data. These techniques that are part of seismic geomorphology are grouped into the following main themes: seismic attributes, seismic sedimentology (including slicing techniques), volume rendering and geo-body extraction, and machine learning. In addition, the integration of seismic geomorphology and surface geomorphology techniques is proposed through an integrated surface and subsurface workflow toward a reliable 3D model of the earth and its geomorphology.

It is important to note that this research was conducted in the framework of Recovery Assistance for Cohesion and the Territories of Europe (REACT-EU) for Italian National Operational Program" PON Research and Innovation 2014-2020" projects on innovation and green issues (DM 1062/2021) granted in December 2021. The project was financed for green research project aiming at studying quantitative geomorphology from images. This research review intends to answer some questions such as where and which seismic and surface geomorphology techniques are used more often in identifying marine and land surface features? which approaches and technologies give better results? information in-dicating the possible combinations of technologies to obtain the best quality results and the ratios of different technologies were also analyzed. In addition, more details about the use of software for data processing are provided. In the future, the results of this research review will be based on during marine deep sediments identification for a new strategy of coastal protection in Sicily and for Italian coastal protection in general.

2. Review Method and Protocol

This review follows the guidelines of general literature review papers by Mohamed Shaffril et al. [81], Munn et al. [82], Snyder [83], and Xiao and Watson [84], where there are two main stages of the literature review method that have been implemented:

1. Planning the review: identifying why this review paper is needed and identification of research questions.
2. Conducting the review: selection of primary research, data extraction, and result reporting.

There are many studies about the reconstruction of past land and marine geomorphological features using seismic reflection data. However, there lacks an in-depth compilation of these techniques when they are combined with surface geomorphological techniques for present reconstruction (surface and shallow subsurface). Meanwhile, the utilization of other geophysical techniques (e.g., ground penetrating radar, gravity, magnetic, resistivity, electrical resistivity tomography, passive seismic, and seismic refraction) has indicated a significance increase during the last decade for present surface and shallow subsurface geomorphological reconstruction [85–87]. Furthermore, this compilation is important for research, as well as the industrial community to give a holistic view on seismic reflection and surface geomorphological techniques available for reconstructing the land and marine geomorphological features. Thus, the acquired data would be utilized in maximal and effective ways to be able to have better results and decrease the cost of production. Therefore, some research questions are:

- R.Q.1 What is the meaning of seismic geomorphology?
- R.Q.2 What are the available seismic geomorphology techniques?
- R.Q.3 How is the integration of seismic and surface geomorphology techniques accomplished?
- R.Q.4 Which technique is most often used and which approach gives better results?

This review utilizes research papers from the period of 2000 to 2022 in worldwide databases (Table 1). The paper-searching strategy included both manual and automatic search strategies to retrieve seismic geomorphology and surface geomorphology techniques from online databases including Scopus, Web of Science, GeoScienceWorld, and Multidisciplinary Digital Publishing Institute (MDPI). The manual search was based on the authors experience and expertise having worked with these techniques for many years, while the automatic search utilized specific keywords related to the research question. The keywords were "seismic geomorphology", "surface geomorphology technique", and "integrated seismic and surface geomorphology" with no filters on affiliation, country, or funding sponsor. However, the following filters were applied on the automatic searching strategy: subject area (earth science), document type (article and book chapter), and language (English).

Table 1. Summary of seismic and surface geomorphology research papers used in this review with the source of database.

Database	Keywords	Records	Total
Scopus	Seismic geomorphology	1197	
	Surface geomorphology technique	590	1841
	Integrated seismic and surface geomorphology	54	
Web of Sciences	Seismic geomorphology	1229	
	Surface geomorphology technique	613	1926
	Integrated seismic and surface geomorphology	58	
Geoscience World	Seismic geomorphology	3990	
	Surface geomorphology technique	3498	10,031
	Integrated seismic and surface geomorphology	2543	
Google Scholar	Seismic geomorphology	3480	
	Surface geomorphology technique	12,700	19,040
	Integrated seismic and surface geomorphology	2860	
MDPI	Seismic geomorphology	45	
	Surface geomorphology technique	43	90
	Integrated seismic and surface geomorphology	2	

The primary paper selection criterion utilized screening the titles and recognizable authors in seismic geomorphology and surface geomorphology techniques. This created restrictions to only select the original articles published in high-quality journals. In addition, duplicate results on the different keywords and irrelevance with keywords (e.g., only considering single keywords) were also removed. The secondary paper selection criterion was based on the following eligibility (exclusion and inclusion) criteria:

Inclusion Criteria:

- Research paper is published in peer-reviewed and good journal, represented in major indices with high impact factor.
- Research paper is accessible.
- Research paper has relevant content to seismic geomorphology and surface geomorphology techniques.

Exclusion Criteria:

- Research paper is published in non-peer-reviewed journal.
- Research paper is inaccessible.
- Research paper has no relevant content to seismic geomorphology and surface geomorphology techniques.

3. Seismic Attributes

A seismic attribute is any measure (quantitatively) of seismic data that helps interpreters to visually enhance or quantify features and geomorphologies of interpretive interest [88–95] (Figure 1). The development of seismic attributes has been integrated with seismic reflection interpretation and has roots extending back to the 1930s when geophysicists started to interpret (pick) travel-times with coherent reflections on recorded seismic field data [95].

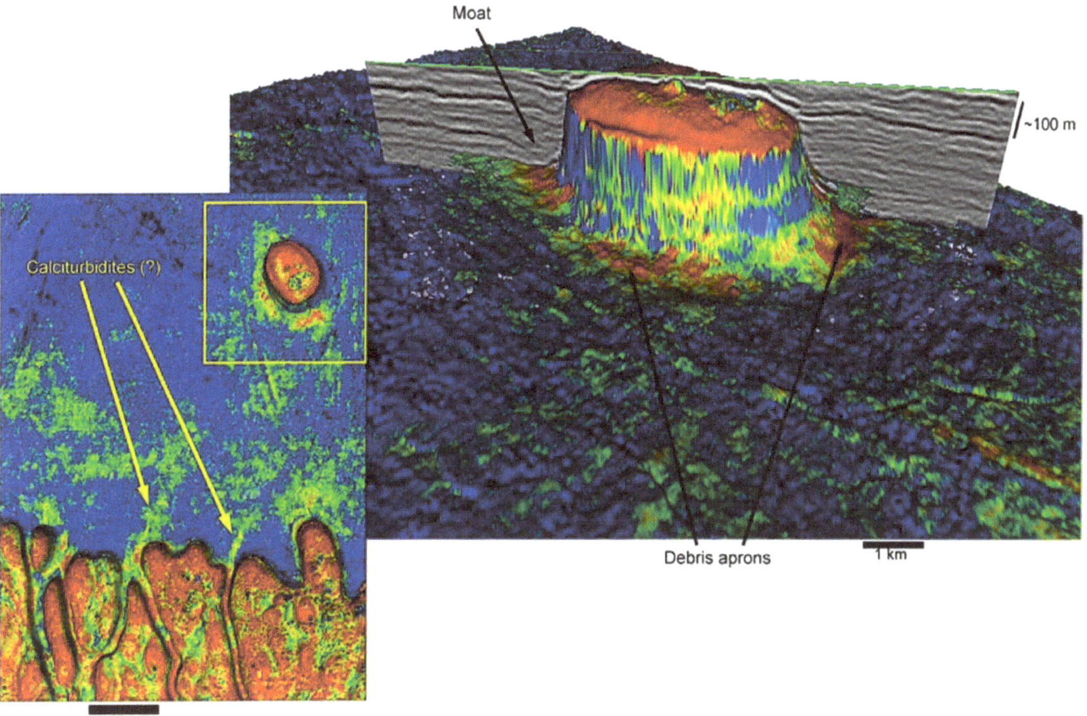

Figure 1. Seismic attributes (Root-Mean-Square Amplitude) showing geomorphology of the paleoseabed reef features. Source: courtesy of Virtual Seismic Atlas (VSA), 2015 (www.seismicatlas.org, accessed on 1 January 2015). (VSA author: Henry Posamentier).

The development of seismic technology has always preceded the development of seismic attributes that could be a powerful tool in delineating geomorphological features, identifying prospective hydrocarbon volumes, etc. Hundreds of seismic attributes are in existence today, and this number will increase over time. Several seismic attributes duplicate one another, while others are obscure, unstable, or unreliable; other seismic attributes are purely mathematical quantities or are not truly attributes [89,95–97].

Many authors have different classifications of seismic attributes; for example, Taner [93] divided the attributes into the following two general categories: geometrical attributes and physical attributes. The purpose of geometrical attributes is to enhance the visibility of the geometrical characteristics or reflection characteristics of the seismic data such as seismic reflection configuration (including seismic facies), the reflection intensity of seismic events, dip, azimuth, and continuity. These geometrical attributes have the main function of enhancing seismic interpretation of sequence stratigraphy, seismic stratigraphy, fault, and structural interpretation. On the other hand, the physical attributes pertain to the physical parameters of the subsurface, and, therefore, relate to lithology and pore fluids (reservoir characterization). These include instantaneous phase, correlation coefficient, instantaneous frequency, (attributes derived from analytical seismic traces), interval velocity, amplitude versus offset, and normal moveout (attributes derived from pre-stack data).

Brown [90] classified seismic attributes using a tree structure in which time, amplitude, frequency, and attenuation were the main branches that further branched out into post-stack and pre-stack categories with horizon and window mediums. Time attributes are very useful on providing faults and structural geological information; these include residual, dip azimuth and magnitude, curvature, edge, illumination, coherence, semblance, covariance, trace difference, etc. Amplitude attributes work very well on enhancing stratigraphic interpretation and reservoir information, including amplitude versus offset, reflection amplitude, relative impedance, reflection strength, amplitude ratio, root-mean-square (RMS) amplitude, average energy, variance of amplitude, maximum amplitude, etc. Frequency attributes are very useful for identifying and extracting reservoir information (e.g., sandstone reservoir with gas); these include instantaneous frequency, time derivative frequency, spectral decomposition, arc length, dominant frequency, etc. Lastly, attenuation attributes could help us in identifying and extracting other reservoir information (e.g., permeability information); these include instantaneous Q factor, slope instantaneous frequency, slope spectral frequency, etc.

Liner et al. [98] introduced a general seismic attribute that was developed following a singularity analysis of migrated seismic data and wavelet transform decomposition. This attribute provided a dense layer model of the subsurface that contained many structural and stratigraphic details (complement coherence and impedance seismic attributes) where these kinds of geological features were associated with singularities (discontinuities in seismic impedance). In addition, this seismic attribute does not require well controls for enhancing reservoir interpretation of migrated seismic data; nevertheless, well controls give an advance petroleum reservoir characterization.

Sidney and references therein [99] provided a classification of seismic attributes based on (1) wave kinematics or dynamic categories and (2) geologic reservoir feature categories. These include:

(1) Amplitude (reflection strength, RMS amplitude, etc.), waveshape (apparent polarity and maximum peak amplitude), frequency (instantaneous frequency and average zero crossing), attenuation (amplitude slope and attenuation of sensitive bandwidth), phase (instantaneous phase and response phase), correlation (length and average), energy (reflection strength and vibration energy), and ratios (ratio of adjacent peak amplitudes).

(2) Bright and dim spots (slope of reflection strength), unconformity traps (average correlation), oil and gas bearing anomalies (instantaneous real amplitude), thin layer reservoirs (finite frequency–bandwidth energy), stratigraphic discontinuity (apparent polarity), clastic–carbonate differentiation (ratio of adjacent peak amplitudes), structural discontinuity (maximum–minimum correlation), and lithology pinch-out (cosine of instantaneous phase).

Furthermore, seismic attributes are often considered a form of "inversion", a process widely used to transform seismic reflection data into valuable and meaningful geomorphological elements and reservoir properties that can later be integrated into detailed geomorphology, reservoir geology, and simulation modeling. This inversion method integrates seismic and wellbore data from which an attribute such as acoustic impedance is derived from sonic and density logs and is subsequently used to populate their properties into the seismic cube [26,100–102].

Al-Shuhail, Al-Dossary, and Mousa [89] introduced "digital image processing" as a complement for seismic attribute analysis where advances in digital image-processing algorithms and computing technology are able to identify and delineate geological and geomorphological features from seismic reflection data. In addition, the removal of random noises and artifacts from seismic reflection data such as velocity push-down or pull-ups, seabed multiples, acquisition and processing footprints, etc., can be solved utilizing digital image-processing techniques including edge/structure-preserving smoothing algorithms [89]. This kind of technique also provides automatic interpretation based on seismic attribute analysis. In particular, these techniques performed very well on fault and channel detection using several seismic attributes such as edge detection, coherence, dip, curvature, randomness, and spectral decomposition.

4. Seismic Sedimentology

Seismic sedimentology is the use of seismic data in the study of sedimentary rocks (lithology, thickness, and fluid properties) and the depositional processes by which they were formed by revealing their sedimentary and erosional geomorphology and their relationship with preserved landforms [43,103–105]. The main tools of seismic sedimentology are ninety-degree phasing of the seismic data, seismic lithology, seismic slicing, and seismic geomorphology [43]. In addition, a display of seismic attributes on geologic time surfaces is another important tool of seismic sedimentology. However, there are strict limitations and conditions under which reservoir geometries and geomorphology can be optimally delineated on time and/or horizon slices [43,106]. This seismic sedimentology approach, which attempts to resolve the resolution of seismic reflection data, is a supplement for the existing seismic stratigraphy where the seismic response to sedimentary layers and geomorphological surfaces may act differently at low and high resolutions [43].

The ninety-degree phasing of seismic data attempts to provide the best correlation between the seismic trace, wireline lithologic logs, and stratigraphic architecture (especially in a thin-bed depositional unit) by providing a symmetrical waveform to be tied directly with the acoustic impedance profile [43,104,107,108]. The unique and symmetrical ninety-degree phasing of seismic data will eliminate the dual polarity of the thin-bed response (less amplitude distortion) that created better seismic image of thin-bed reflection termination and configuration (seismic facies), lithology, impedance profiles, and stratigraphy [43,107,108].

Seismic lithology is a reservoir geophysics technique that converts seismic traces into acoustic impedance logs (seismic inversion) to produce an acoustic impedance volume (valid impedance model) for lithologic and stratigraphic imaging at high resolutions [43,104,106]. This seismic sedimentology technique is very useful for identification and characterization of the thin-bed reservoir from 3D seismic reflection data. The thin-bed parameters might be consisting of (but not limited to) sand/shale ratio, lithofacies, shale (sand) content (as a pseudo-log), thickness of sandstone, etc. In addition, Dvorkin, Gutierrez, and Grana [21] provide the relationships between lithology (sandstone and shale), fluid, porosity, clay content, and acoustic impedance that can validate seismic lithology results, which are:

- Shale with medium-porosity gas sand.
- Shale with low-porosity gas sand.
- Gas sand and wet sand.
- Wet sand and shale.

The seismic slicing technique usually consists of time or depth, horizon, and stratal slicing types in 3D seismic reflection data that (together with seismic attribute) provide the geomorphology of geological features (Figure 2). The time or depth and horizon seismic slicing are self-explanatory, whereas the stratal slicing technique utilizes the horizontal seismic resolution of 3D seismic reflection data and spatially correlates the geological interpretation (especially at reservoir scale) in a geological timeline (Chronostratigraphy). Nevertheless, Zeng [109,110] suggests that some of parameters need to be understood to make this technique valid, which are:

- Good quality geologic-time framework should be in place.
- Depositional system should be linear with lateral changes in thickness.
- No significant angular unconformity.

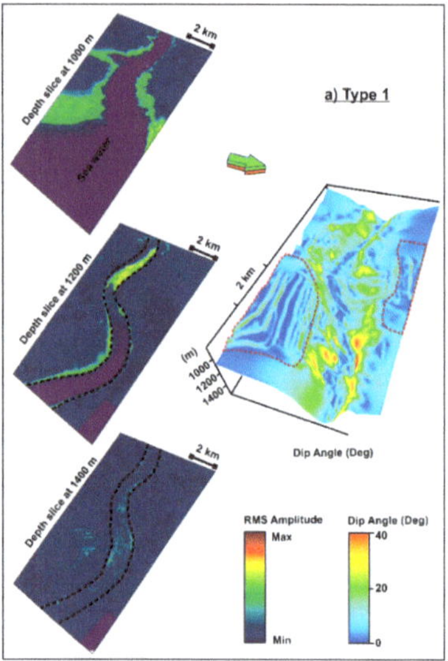
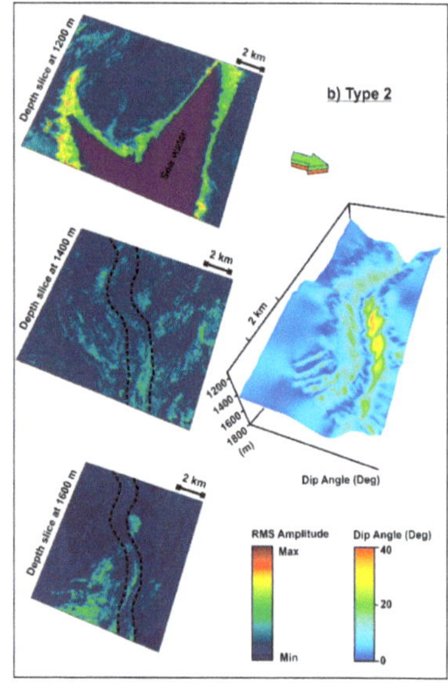
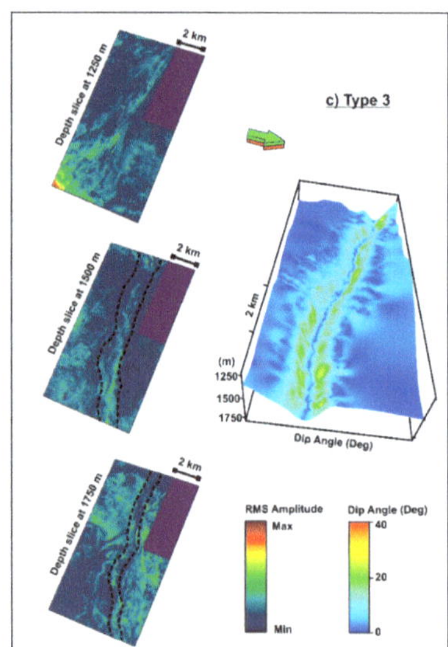
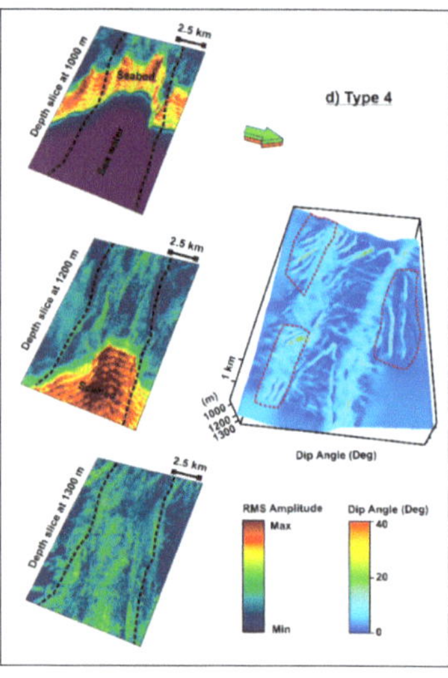

Figure 2. 3D view of dip angle (**right** images) and depth root-mean-square (RMS) amplitude slice from 3D seismic reflection data (**left** images) of (**a**) Type 1, (**b**) Type 2, (**c**) Type 3, and (**d**) Type 4 channel complexes in the Canterbury Basin, offshore of New Zealand [46]. Note that the black dashed line is the channel outline, while the red dashed line is the area of the contourite deposit.

Based on the analytical technique, seismic geomorphology involves the extraction and study of preserving subsurface landforms that utilizes plan-view images from 3D seismic reflection data by seismic slicing and seismic attribute techniques [27,29,43,111]. As part of the seismic sedimentology techniques, seismic geomorphology needs to be integrated (complementary) and evaluated with other seismic sedimentology techniques (i.e., the ninety-degree phasing of seismic data, seismic lithology, and seismic slicing) for maximum benefit on the extracting geology, geophysics, and reservoir information out of 3D seismic reflection data [24,26,43]. A study conducted by Deiana et al. [112] revealed that paleo-landscape geomorphological attributes can be extracted from seismic data acquired by MBES. By analyzing these data, the authors noticed that it was possible to identify the geometrical position of a beach rock located at 45 m of water depth (Figure 3).

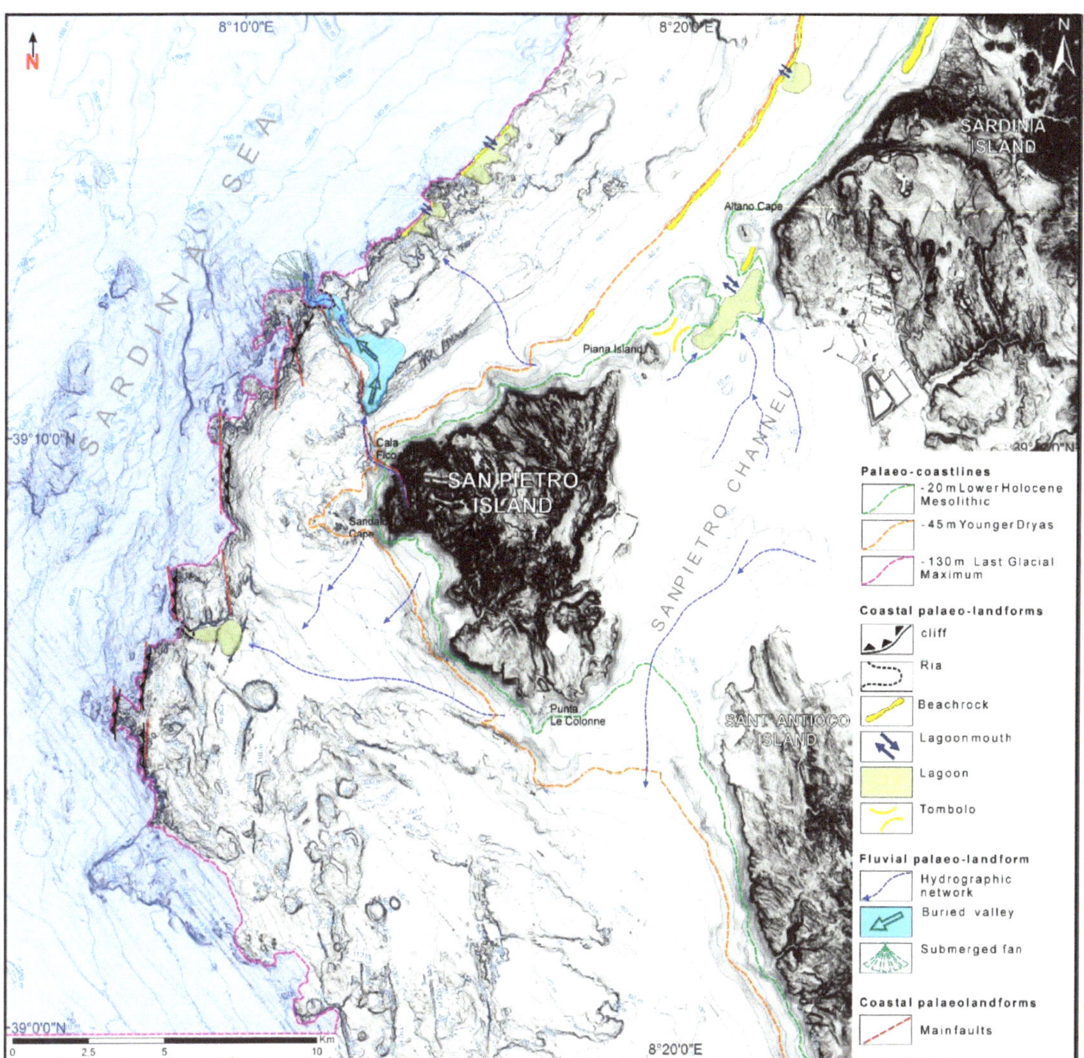

Figure 3. Geomorphological sketch of the San Pietro continental shelf. Submerged paleo-landscape from LGM (20 ka) to 9 ka [112].

5. Volume Rendering and Geobody Extractions

The earth has always been three dimensional (3D). Such dimensions can be acquired using UAV [113] and analyzed for paleo-seismic offset studies (Figure 4). Today, seismic technology is able to image a small portion of the earth using 3D seismic reflection data (primarily for energy purposes) to identify, isolate, and extract seismic anomalies (e.g., geomorphology, reservoirs, fluids, volcanic features, salt, etc.). Volume rendering and the red-green-blue (RGB) blending method as part of the seismic geomorphology technique allow users to interactively blend multiple seismic reflection volumes, identify and isolate areas of interest, and extract any relevant geologic and geomorphologic features from a 3D object called a geobody [25,26,114–121].

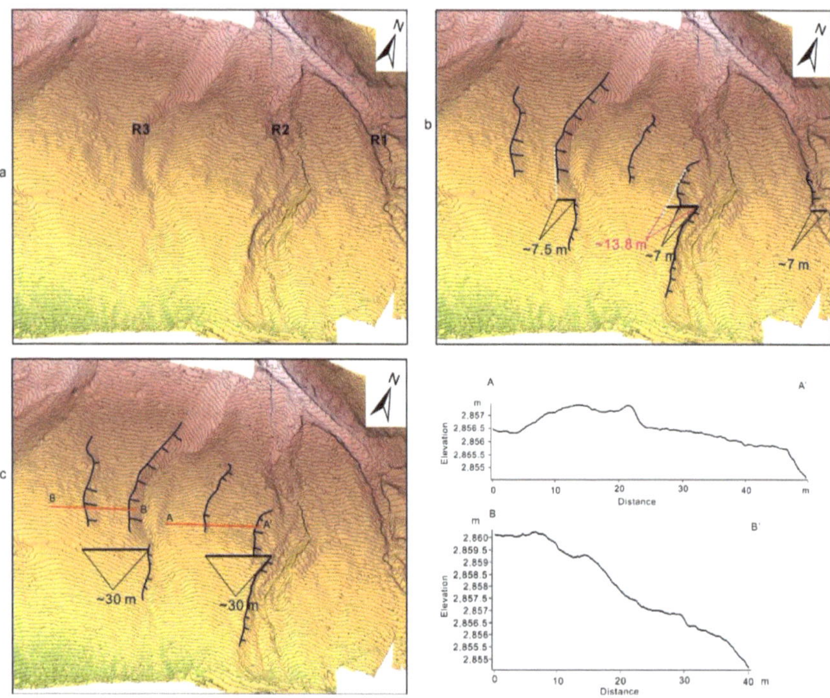

Figure 4. (**a**) Digital elevation model with contour lines (0.25 m interval) for R1, R2 and R3. (**b**) Interpretation of terrace risers and reconstruction of the ~7 m offset. (**c**) Reconstruction of the ~30 m offset [113].

In volume rendering, the volume is considered as a semi-transparent medium for each pixel of the screen, and the computation of a volume rendering integral accumulates the voxels' contribution along a virtual viewing ray [114–117,119–123]. On the other hand, the seismic volume rendering process is a display of all data within a seismic volume at the same time. By rendering a seismic volume and working with the opacity to make it partially opaque (high amplitude) and partially transparent (crossover amplitudes), it is possible to identify hidden structural, geomorphological, or depositional features.

However, at times, it is difficult to identify areas of interest with volume rendering. Therefore, the RGB method is the best option for this situation because it allows for the blending of different seismic attributes, and its opacity scale uses the primary colors (red, green, and blue), which facilitates better visualization of geological and geomorphological features [114].

Furthermore, Chopra and Marfurt [116] suggested that co-rendering seismic attributes (blending two or more seismic attributes) into a single 3D seismic volume could demonstrate the maximum value of volumetric interpretation of seismic reflection data. In addition, a false-color guidance of red-green-blue blending on co-rendering seismic attributes are as follows [116]:

- Red for lower values (less significant geological features).
- Green for intermediate values that represent geological features.
- Blue for higher values that represent more geological features.

After visualizing and isolating a 3D object through seismic volume rendering or RGB blending, the geobody can then be extracted (Figure 5). The volumetric of this geobody can then be calculated, or it can be directly sampled in a geological model as a discrete object to condition the petrophysical modelling. The resulting property can then be used in a similar way as a facies model to condition the petrophysical property models [114,121]. Therefore, the geobody provides anomalous subsurface geological features of interest to rapidly visualize the orientation, geometry, and extent in three-dimensional ways [119]. In addition, this geobody extraction also visualizes the geomorphological features with rock or physical properties from 3D seismic reflection data, and once these features are depth converted (from millisecond to meter), then such metric calculations could be possible with other supporting parameters.

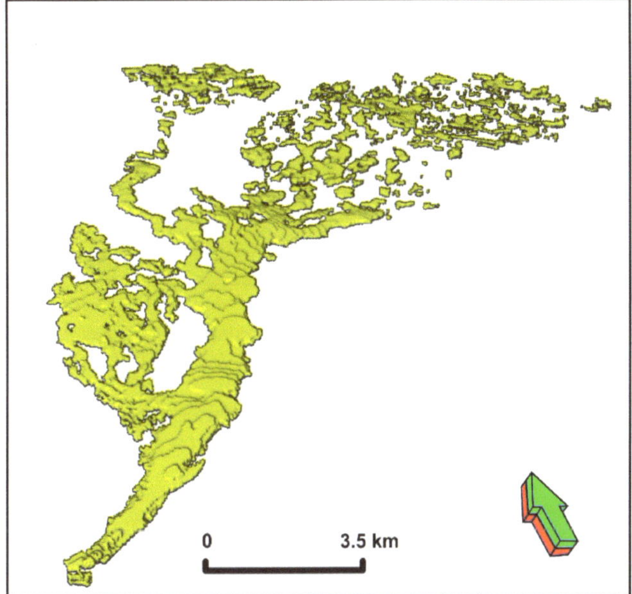

Figure 5. Geobody extraction of deep-water channel-lobe system from seismic geomorphology technique. Note that 3D red-green block is pointing toward the north direction.

Nowadays, computer technology, as well as the ability to process large-scale seismic reflection data, makes volume rendering and geobody extraction a common tool in the visualization of geology, geophysics, and reservoir features out of 3D seismic reflection data. In addition, volume rendering and geobody extraction provide more realistic subsurface features with some uncertainties based on data resolution and the workflow that was implemented in the study. Nevertheless, geological understanding of the studied area as well as integration with other subsurface datasets (e.g., wells, gravity, magnetic, etc.) are needed for obtaining more geologic interpretations (higher levels of confidence in interpretations) and low uncertainty on the subsurface model.

6. Machine Learning

Seismic reflection interpretation is usually a time-consuming process (especially with low-quality data) when the interpreter analyzes seismic data at a standard industrial workstation. This long process (potentially months) also influences the result, since the interpreter is a human that could have emotional biases when working on a project for long period of time, which might lead to less objectivity. Machine learning in seismic interpretation (seismic geomorphology, in particular) utilizes applied statistics that build computational models using various machine learning techniques such as random forests, decision trees, support-vector machines, convolutional neural networks, deep neural networks, and generative adversarial networks [124–128]. This requires input data that will be processed for training using applied statistics with computational algorithms to produce reliable outputs (Figure 6).

With today's trend toward digitalization and automation, this process will significantly reduce seismic interpretation to a very short period, possibly hours or minutes depending on the size of the seismic reflection data. Furthermore, the utilization of machine learning in seismic interpretation processes covers most of the subsurface areas, including surfaces, geomorphology and facies interpretation, e.g., [129–141], faults and fracture interpretation, e.g., [139,142–147], and geological volume-global interpretation, e.g., [128,148–150].

Machine learning in horizon seismic interpretation is an important aspect on identification geomorphology, faults, reservoir, etc. Lou, Zhang, Lin, and Cao [137] introduced that seismic horizons follow the reflector dip, thus having similar instantaneous phase values (the same horizons). Therefore, automatic horizon-interpretation algorithms need to implement the integration between the reflector dip and instantaneous phase attributes [137]. Another type of machine learning on horizon seismic interpretation is utilizing dislocated horizons (faulted, truncated, etc.) where dynamic time warping and unwrapped instantaneous phase-constraint are used to correlate horizon grids in a 3D window [133].

Geomorphological interpretation using 3D seismic reflection data is usually time-consuming, and machine learning with automation processes definitely helps to reduce the working hours of an interpreter. Infante-Paez and Marfurt [134] and La Marca and Bedle [140] presented unsupervised machine learning (self-organizing maps) on identification of deep-water and volcanic seismic facies, geomorphology, and architectural elements. Kumar and Sain [132] presented supervised learning based on an artificial neural network to automate the identification and delineation of mass-transport deposits' geomorphological surfaces and bodies out of 3D seismic reflection data (offshore).

Manual seismic facies interpretation makes the results very subjective, depending on the experience and knowledge of the interpreter. Zhang, Chen, Liu, Zhang, and Liu [131] presented automatic seismic facies interpretation utilizing deep learning, the convolutional neural network, and encoder–decoder architecture, whereas Singh, Tsvankin, and Naeini [141] utilized Bayesian inference on supervised and semi-supervised deep learning on a shallow marine prograding delta offshore of the Netherlands.

Faults and geological volume interpretations take most of the interpreter's time in manual 3D seismic interpretation. Wu, Liang, Shi, and Fomel [145] introduced "FaultSeg3D", an automatic and machine-learning tool that used an end-to-end convolutional neural network to produce an image-to-image fault segmentation. Di and AlRegib [147] presented a semi-automatic fault or fracture seismic interpretation not using the conventional-based fault interpretation on displacement, but utilizing seismic geometry analysis. Furthermore, 3D seismic interpretation is usually done using a 2D interpretation view with dependency on the human interpreter. de Groot [148] and de Groot, et al. [151] introduced "global seismic interpretation" as a seismic volume interpretation where the algorithm correlates amplitude and time lines in the pre-calculated seismic dip field.

Nevertheless, this automation and time reduction produces results that the interpreter needs to validate. We have to keep in mind that machine learning is only a tool that helps interpreters to work effectively and should not be relied on entirely.

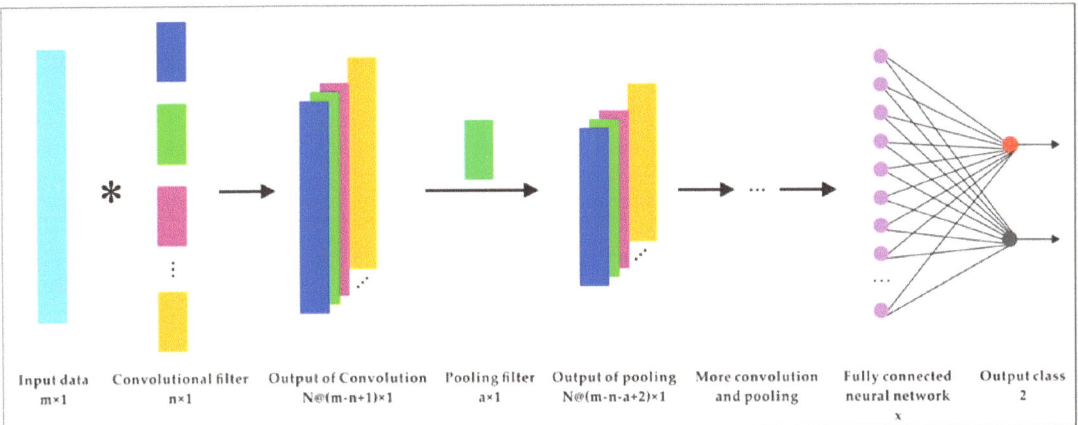

Figure 6. Convolutional Neural Network workflow [152].

7. Integrated Seismic Geomorphology with Surface Geomorphological Techniques

Seismic geomorphology, which consists of seismic attributes, seismic sedimentology, volume rendering, and geobody extraction, is a unique seismic interpretation method that produces reliable geomorphology (both surface and geobody) out of seismic reflection data. Nevertheless, the integration of surface and subsurface techniques such as digital elevation models (i.e., LIDAR, photogrammetry, and surface outcrops) and seismic geomorphology could lead to reduced data uncertainty and better geological mapping, 3D earth surface imaging (geomorphology), and reconstruction.

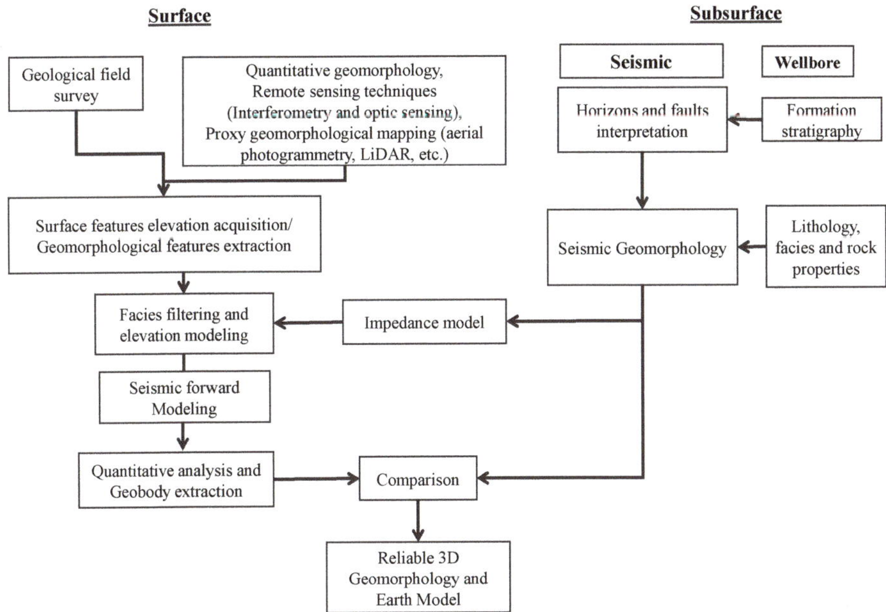

Figure 7. Proposed integrated seismic (with addition of wellbore data, if available) and surface geomorphology workflow for a reliable 3D model of the earth and its geomorphology.

This integration could be beneficial for society in various aspects, including mapping and mitigating landslides and earthquake hazards, engineering geology for building, surface-groundwater identification and utilization, city planning (land-use mapping and landscape planning), halt and reverse land degradation, characterization of sustainable energy on the land surface, better environmental management, etc. Seismic reflection data and geology itself are often difficult to correlate [4,153–160]. This is because of the different factors used to describe the two in terms of units, scales, time, and resolutions. In addition, surface data also have different factors compared to subsurface data. Therefore, we propose an integrated workflow that consists of geological fieldwork, remote sensing techniques, quantitative geomorphology, an aerial photogrammetry-based point-cloud (digital outcrop modelling), surface outcrops, LIDAR (surface data), and seismic reflection data (i.e., seismic geomorphology for subsurface data) to bridge the gap between seismic reflection data (geology) and surface outcrops to produce a reliable 3D model of the earth and its geomorphology (Figure 7).

Subsurface data mainly consists of seismic and wellbore data (and other data, e.g., gravity, magnetic, ground-penetrating radar, etc). The basic interpretations from this subsurface dataset are horizon (utilizing formation stratigraphy from wellbore) and fault interpretations that lead to seismic geomorphology, lithology, facies, and rock properties (Figure 7). The surface data is a combination of geological field work data (lithology, faults, geomorphology, etc.) and surface geomorphological techniques, e.g., drone photogrammetry (digital outcrop model), LIDAR, etc. The basic interpretation from this surface geomorphology technique is surface geomorphological feature identification and characterization. This feature usually extends and relates to the subsurface, and the seismic geomorphology technique is needed to image the subsurface and correlate this surface feature with its subsurface counterpart. Since there is a difference domain between two datasets (depth, amplitude values, etc.), an impedance model (consisting of mainly density and velocity data) with facies and a rock model consisting of rock properties (e.g., density, porosity, permeability, etc.) will be a bridge between seismic geomorphology and surface geomorphology (seismic forward modelling). The final surface geomorphology product is a geomorphology volume that could render and extract specific geomorphology feature (Geobody). Furthermore, the seismic geomorphology and geobody need to be compared to produce a reliable 3D model of the earth and its geomorphology. In addition, results from all the techniques will depend on the quality of the data that has limitations as functions of vertical and horizontal data resolutions. Therefore, the results and technological preference are very dependent on data resolution. Nevertheless, all techniques should be implemented in order to reduce the uncertainty coming from the data.

In addition, this kind of integration is also beneficial for reducing the uncertainty that is produced by surface and subsurface data, whereas more accuracy is always needed utilizing updated technology for surface and subsurface data. Finally, our integrated workflow that helps for better geomorphological mapping as well as 3D earth surface imaging and reconstruction could be applicable with available and similar datasets worldwide.

8. Discussions and Results Overview

Several seismic and surface geomorphology techniques exist for delineating, identifying, and extracting the geomorphology features from both seismic reflection and surface geomorphology data (Table 2).

Table 2. Seismic geomorphology technical analysis.

Techniques	Geomorphological Analysis and Frequency	Scale and Resolution	Results	Ratios (In Time) and References
Seismic attribute	Most cases, often used	Vertical (20–30 m) & horizontal (10–20 m)	Good (Depends on data quality)	1980s—now, see chapter 3
Seismic sedimentology	In special cases (e.g., thin bedded)	Vertical (2–10 m) & horizontal (10–20 m)	Good for thin bedded	2010s—now, see chapter 4
Volume rendering and geobody extraction	In special cases (e.g., (3D seismic reflection data)	Vertical (20–30 m) & horizontal (10–20 m)	Good (Depends on data quality)	2000s—now, see chapter 5
Machine learning	Becoming often (Last decade)	Vertical (20–30 m) & horizontal (10–20 m)	Good (Needs human validation)	2010s—now, see chapter 6
Integrated seismic and surface geomorphology	Not often, lack of reference	Vertical (2–10 m) & horizontal (10–20 m)	Better (Reducing result uncertainty)	Proposed (This study)

The seismic attribute technique (including geometrical attribute: seismic facies) is widely used as the first approach of quantitatively identifying geological and geomorphological features from seismic reflection data. Therefore, this kind of seismic geomorphology technique is often used in industry and academia since its introduction back in the 1980s. The results offered by this technique really depend on the quality of seismic reflection data together with human knowledge and experience (Table 2). The seismic sedimentology technique is a relatively new technique that deals with thin-bedded geology and geomorphology (Table 2), thus, it needs the human knowledge and experience to provide a good result. Thin-bedded features could be in a range of 2 to 10 m thick with horizontal resolution of the conventional seismic reflection data. This kind of technique is usually ignored when the seismic attributes technique could solve the problem of imaging geomorphological features from seismic reflection data. Furthermore, volume rendering and geobody extraction depend on the availability of 3D seismic reflection data (Table 2). Application of this technique will also be dependent on the objective of the project; it is mainly used for volume calculation of the geological and geomorphological features. The results offered would depend on the quality of the seismic reflection data (Table 2). The machine learning technique could also be applied on the seismic geomorphology technique to speed up the interpretation processes. Nowadays, this kind of technique is often used, but the results offered always need to be validated by human experience and knowledge (Table 2). The surface geomorphology techniques combining a large number of emerging space and proxy-remote sensing technologies are presented as the best techniques used for local, regional, and continental geomorphological studies (Table 3). Such techniques improve the ability to acquire the surface feature of elevation and to extract geomorphological features. In addition, the proliferation of a large number of GEE and GIS software allow for facies filtering and elevation modeling in quantitative analysis and geobody extraction studies, and eventually allowing the development of a reliable 3D geomorphological earth surface model.

Table 3. Surface geomorphology technical analysis.

Techniques	Geomorphological Analysis	Scale of Studied Area	Results	References
Remote sensing techniques	Surface Depressions Surface processes Surface deformation	Local, Reginal, Continental	Information on the location, distance, and volume	Melis et al. 2021 [59] Muzirafuti et al. 2020 [52] Borzi et al. 2021 [64] Bianco et al. 2021 [65] Randazzo et al. 2020 [67] Cigna et al. 2021 [68] Mantovani et al. 2016 [69] Jiang et al. 2021 [71] Crosetto et al. 2020 [72] van Natijne et al. 2022 [76]
Proxy geomorphological mapping (aerial photogrammetry, LiDAR.)	Surface and marine processes Surface and marine features	Regional local	Information on the location, distance, and volume, 3D models	Randazzo et al.2020 [67] Anders et al. 2021 [62] Bonasera et al. 2022 [63] Muzirafuti et al. 2021 [73] Deiana et al. 2021 [112] Gao et al. 2017 [113]
Geological field survey	Surface and marine processes	Local	Information on the location, distance, and volume	Bonasera et al. 2022 [63] Taufani et al. 2021 [155]
Quantitative geomorphology	Marine sedimentary features	Local	Information on the location, distance, and volume	Distefano et al. 2021 [51] Muzirafuti et al. 2021 [73]

Finally, the integration of the seismic and surface geomorphology workflow is proposed to give better results for reconstructing land and marine features (Figure 7 and Table 1). The data quality together with human knowledge and experience will influence the results, but this integrated workflow will reduce the uncertainty about the results since it combines several seismic and surface geomorphology techniques that can complement each other.

9. Conclusions

We presented a review of seismic and surface geomorphology techniques for imaging and reconstructing land and marine geomorphology features, and we have revealed that:

- Seismic geomorphology is a subsurface (including near surface) study that extracts geomorphology features out of 3D seismic reflection data.
- Active proxy of surface geomorphology techniques and remote sensing techniques have huge potential in vertical and horizontal deformation monitoring.
- The reconstruction of high-resolution images of land and marine surface features by surface and subsurface geomorphology techniques is reliable through several techniques, including seismic sedimentology, volume rendering, geobody extraction, quantitative geomorphology approaches, and mapping.
- The integration of surface and subsurface techniques provides more realistic and suitable 3D models of the earth and its geomorphology. In addition, it enhances the interpretation of sedimentary processes, geomorphology, the earth's surface, the paleoenvironment, economical prospective, natural hazards, etc. Therefore, we propose a workflow that integrates surface and subsurface techniques to provide realistic and acceptable earth models.

Author Contributions: Conceptualization, A.A.-S., A.M., and D.H.; methodology, D.H. and A.M.; software, D.H. and A.M.; validation, A.A.-S., D.H., A.M., G.R., and S.L.; formal analysis, D.H. and A.M.; investigation, A.A.-S., D.H., A.M., G.R., and S.L.; resources, A.A.-S., A.M.; data curation, D.H. and A.M.; writing—original draft preparation, A.A.-S., D.H., and A.M.; writing—review and editing, A.A.-S., D.H., A.M., G.R., and S.L.; visualization, D.H., A.M., G.R., and S.L.; supervision, A.A.-S., G.R. and A.M.; project ad-ministration, A.A.-S. and G.R.; funding acquisition, A.A.-S. and A.M. All authors have read and agreed to the published version of the manuscript.

Funding: This research was funded by King Fahd University of Petroleum & Minerals through College of Petroleum Engineering & Geosciences Startup grant number SF18060 and the APC was funded also by the same grant. The research was conducted in the framework of REACT EU-National opera-tional program "PON Research and Innovation" 2014-2020 on innovation and green issues (DM 1062/2021) granted in December 2021 for quantitative geomorphology from images project.

Institutional Review Board Statement: Not applicable.

Informed Consent Statement: Not applicable.

Data Availability Statement: Not applicable.

Acknowledgments: The authors thank Department of Geosciences, College of Petroleum Engineering and Geosciences, King Fahd University of Petroleum and Minerals (KFUPM), Kingdom of Saudi Arabia for supporting this work. This publication is based upon work supported by King Fahd University of Petroleum & Minerals and the author(s) at KFUPM acknowledge the support. This paper is part of Special Issue: "Geomorphology in the Digital Era" on the MDPI Applied Science section: "Earth Sciences and Geography".

Conflicts of Interest: The authors declare no conflict of interest.

References

1. Chopra, S.; Marfurt, K.J. Evolution of seismic interpretation during the last three decades. *Lead. Edge* **2012**, *31*, 654–676. [CrossRef]
2. Doust, H. The exploration play: What do we mean by it? *AAPG Bull.* **2010**, *94*, 1657–1672. [CrossRef]
3. Weimer, P.; Slatt, R.M.; Bouroullec, R.; Fillon, R.; Pettingill, H.; Pranter, M.; Tari, G. Reservoir Quality of Deepwater Deposits. In *Introduction to the Petroleum Geology of Deepwater Setting*; American Association of Petroleum Geologists: Tulsa, Ok, USA, 2006; Volume 57.
4. Harishidayat, D.; Taufani, L.; Fardiansyah, I.; Wibowo, A. Integrated Digital Outcrop Model (DOM) and Seismic Forward Modeling: Application to Enhance Subsurface Interpretation in Deltaic System of Kutai Basin, East Kalimantan. *Indones. Pet. Assoc.* **2017**. [CrossRef]
5. Grammer, G.M.; Harris, P.M.M.; Eberli, G.P. *Integration of Outcrop and Modern Analogs in Reservoir Modeling*; American Association of Petroleum Geologists: Tulsa, Ok, USA, 2004.
6. Sullivan, M.D.; Foreman, J.L.; Jennette, D.C.; Stern, D.; Jensen, G.N.; Goulding, F.J.; Grammer, G.M.; Harris, P.M.M.; Eberli, G.P. An Integrated Approach to Characterization and Modeling of Deep-water Reservoirs, Diana Field, Western Gulf of Mexico. In *Integration of Outcrop and Modern Analogs in Reservoir Modeling*; American Association of Petroleum Geologists: Tulsa, Ok, USA, 2004.
7. Gluyas, J.; Swarbrick, R. *Petroleum Geoscience*; John Wiley & Sons: Hoboken, NJ, USA, 2013.
8. Harishidayat, D. *Architecture, Morphometry and Development of Subaqueous Sediment Conduits*; Norwegian University of Science and Technology: Trondheim, Norway, 2021.
9. Kemper, M.; Abel, M.; De Ros, L.; Hansford, J. Integration of Lithological Data for Advanced Seismic Inversion. In *Third EAGE/SBGf Workshop 2016*; European Association of Geoscientists & Engineers: Amsterdam, The Netherlands, 2016.
10. Sams, M.; Millar, I.; Satriawan, W.; Saussus, D.; Bhattacharyya, S. Integration of geology and geophysics through geostatistical inversion: A case study. *First Break* **2011**, *29*, 47–56. [CrossRef]
11. Brown, A.R. *Interpretation of Three-Dimensional Seismic Data*, 7th ed.; Society of Exploration Geophysicists: Houston, TX, USA; The American Association of Petroleum Geologists: Tulsa, Ok, USA, 2011.
12. Cartwright, J.; Huuse, M. 3D seismic technology: The geological 'Hubble'. *Basin Res.* **2005**, *17*, 1–20. [CrossRef]
13. Davies, R.J.; Posamentier, H.W. Geologic processes in sedimentary basins inferred from three-dimensional seismic imaging. *GSA Today* **2005**, *15*, 4. [CrossRef]
14. Herron, D.A. *First Steps in Seismic Interpretation*; Society of Exploration Geophysicists: Houston, TX, USA, 2011.
15. Posamentier, H.W.; Paumard, V.; Lang, S.C. Principles of seismic stratigraphy and seismic geomorphology I: Extracting geologic insights from seismic data. *Earth-Sci. Rev.* **2022**, *228*, 103963. [CrossRef]
16. Avseth, P.; Mukerji, T.; Mavko, G. *Quantitative Seismic Interpretation: Applying Rock Physics Tools to Reduce Interpretation Risk*; Cambridge University Press: Cambridge, UK, 2010.
17. Cox, D.R.; Newton, A.M.; Huuse, M. An introduction to seismic reflection data: Acquisition, processing and interpretation. In *Regional Geology and Tectonics*; Elsevier: Amsterdam, The Netherlands, 2020; pp. 571–603.
18. Badley, M. *Advanced Practical Seismic Interpretation*; IHRDC Press: Boston, MA, USA, 1992.
19. Badley, M.E. *Practical Seismic Interpretation*; IHRDC Press: Boston, MA, USA, 1985.
20. Davies, R.J.; Posamentier, H.W.; Wood, L.J.; Cartwright, J.A. *Seismic Geomorphology: Applications to Hydrocarbon Exploration and Production*; GSL Special Publications: London, UK, 2006; p. 276.
21. Dvorkin, J.; Gutierrez, M.A.; Grana, D. *Seismic Reflections of Rock Properties*; Cambridge University Press: Cambridge, UK, 2014.
22. Schultz, P. *The Seismic Velocity Model as an Interpretation Asset*; Society of Exploration Geophysicists: Houston, TX, USA, 1998.
23. Sheriff, R.E.; Geldart, L.P. *Exploration Seismology*; Cambridge University Press: Cambridge, UK, 1995.

24. Omosanya, K.O.; Harishidayat, D. Seismic geomorphology of Cenozoic slope deposits and deltaic clinoforms in the Great South Basin (GSB) offshore New Zealand. *Geo-Mar. Lett.* **2019**, *39*, 77–99. [CrossRef]
25. Harishidayat, D.; Johansen, S.E.; Batchelor, C.; Omosanya, K.O.; Ottaviani, L. Pliocene–Pleistocene glacimarine shelf to slope processes in the south-western Barents Sea. *Basin Res.* **2021**, *33*, 1315–1336. [CrossRef]
26. Harishidayat, D.; Omosanya, K.O.; Johansen, S.E. 3D seismic interpretation of the depositional morphology of the Middle to Late Triassic fluvial system in Eastern Hammerfest Basin, Barents Sea. *Mar. Pet. Geol.* **2015**, *68*, 470–479. [CrossRef]
27. Posamentier, H.W. Seismic Geomorphology: Imaging Elements of Depositional Systems from Shelf to Deep Basin Using 3D Seismic Data: Implications for Exploration and Development. *Geol. Soc. Lond. Mem.* **2004**, *29*, 11–24. [CrossRef]
28. Posamentier, H.W. Application of 3D seismic visualization techniques for seismic stratigraphy, seismic geomorphology and depositional systems analysis: Examples from fluvial to deep-marine depositional environments. *Geol. Soc. Lond. Pet. Geol. Conf. Ser.* **2005**, *6*, 1565–1576. [CrossRef]
29. Posamentier, H.W.; Davies, R.J.; Cartwright, J.A.; Wood, L. Seismic geomorphology–An overview. *Geol. Soc. Lond. Spec. Publ.* **2007**, *277*, 1–14. [CrossRef]
30. Harishidayat, D.; Farouk, S.; Abioui, M.; Aziz, O.A. Subsurface Fluid Flow Feature as Hydrocarbon Indicator in the Alamein Basin, Onshore Egypt; Seismic Attribute Perspective. *Energies* **2022**, *15*, 3048. [CrossRef]
31. Harishidayat, D.; Emmel, B.U.; De Jager, G.; Johansen, S.E. Assessment of continental margin clinoform systems in the Sørvestsnaget Basin, western Barents Sea: From clinoform parameters towards paleo-water depth. *Mar. Geophys. Res.* **2022**, *43*, 22. [CrossRef]
32. Alfaro, E.; Holz, M. Seismic geomorphological analysis of deepwater gravity-driven deposits on a slope system of the southern Colombian Caribbean margin. *Mar. Pet. Geol.* **2014**, *57*, 294–311. [CrossRef]
33. Bourget, J.; Ainsworth, R.B.; Thompson, S. Seismic stratigraphy and geomorphology of a tide or wave dominated shelf-edge delta (NW Australia): Process-based classification from 3D seismic attributes and implications for the prediction of deep-water sands. *Mar. Pet. Geol.* **2014**, *57*, 359–384. [CrossRef]
34. Burton, D.; Wood, L.J. Seismic geomorphology and tectonostratigraphic fill of half grabens, West Natuna Basin, Indonesia. *AAPG Bull.* **2010**, *94*, 1695–1712. [CrossRef]
35. Dalla Valle, G.; Gamberi, F.; Rocchini, P.; Minisini, D.; Errera, A.; Baglioni, L.; Trincardi, F. 3D seismic geomorphology of mass transport complexes in a foredeep basin: Examples from the Pleistocene of the Central Adriatic Basin (Mediterranean Sea). *Sediment. Geol.* **2013**, *294*, 127–141. [CrossRef]
36. Gee, M.J.R.; Gawthorpe, R.L.; Bakke, K.; Friedmann, S.J. Seismic Geomorphology and Evolution of Submarine Channels from the Angolan Continental Margin. *J. Sediment. Res.* **2007**, *77*, 433–446. [CrossRef]
37. Hadler-Jacobsen, F.; Gardner, M.H.; Borer, J.M. Seismic stratigraphic and geomorphic analysis of deep-marine deposition along the West African continental margin. *Geol. Soc. Lond. Spec. Publ.* **2007**, *277*, 47–84. [CrossRef]
38. Infante-Paez, L.; Marfurt, K.J. Seismic expression and geomorphology of igneous bodies: A Taranaki Basin, New Zealand, case study. *Interpretation* **2017**, *5*, SK121–SK140. [CrossRef]
39. Qin, Y.; Alves, T.M.; Constantine, J.; Gamboa, D. Quantitative seismic geomorphology of a submarine channel system in SE Brazil (Espírito Santo Basin): Scale comparison with other submarine channel systems. *Mar. Pet. Geol.* **2016**, *78*, 455–473. [CrossRef]
40. Rankey, E.C. Seismic architecture and seismic geomorphology of heterozoan carbonates: Eocene-Oligocene, Browse Basin, Northwest Shelf, Australia. *Mar. Pet. Geol.* **2017**, *82*, 424–443. [CrossRef]
41. Scarselli, N.; McClay, K.; Elders, C. Seismic geomorphology of cretaceous megaslides offshore Namibia (Orange Basin): Insights into segmentation and degradation of gravity-driven linked systems. *Mar. Pet. Geol.* **2016**, *75*, 151–180. [CrossRef]
42. Bischoff, A.; Planke, S.; Holford, S.; Nicol, A. Seismic geomorphology, architecture and stratigraphy of volcanoes buried in sedimentary basins. In *Updates in Volcanology-Transdisciplinary Nature of Volcano Science*; IntechOpen: London, UK, 2021.
43. Zeng, H. What is seismic sedimentology? A tutorial. *Interpretation* **2018**, *6*, SD1–SD12. [CrossRef]
44. Omosanya, K.O.; Harishidayat, D.; Marheni, L.; Johansen, S.E.; Felix, M.; Abrahamson, P. Recurrent mass-wasting in the Sørvestsnaget Basin Southwestern Barents Sea: A test of multiple hypotheses. *Mar. Geol.* **2016**, *376*, 175–193. [CrossRef]
45. Harishidayat, D.; Omosanya, K.O.; Johansen, S.E.; Eruteya, O.E.; Niyazi, Y. Morphometric analysis of sediment conduits on a bathymetric high: Implications for palaeoenvironment and hydrocarbon prospectivity. *Basin Res.* **2018**, *30*, 1015–1041. [CrossRef]
46. Harishidayat, D.; Raja, W.R. Quantitative Seismic Geomorphology of Four Different Types of the Continental Slope Channel Complexes in the Canterbury Basin, New Zealand. *Appl. Sci.* **2022**, *12*, 4386. [CrossRef]
47. Shumaker, L.E.; Jobe, Z.R.; Johnstone, S.A.; Pettinga, L.A.; Cai, D.; Moody, J.D. Controls on submarine channel-modifying processes identified through morphometric scaling relationships. *Geosphere* **2018**, *14*, 2171–2187. [CrossRef]
48. Kristensen, T.B.; Huuse, M.; Piotrowski, J.A.; Clausen, O.R. A morphometric analysis of tunnel valleys in the eastern North Sea based on 3D seismic data. *J. Quat. Sci.* **2007**, *22*, 801–815. [CrossRef]
49. Niyazi, Y.; Eruteya, O.E.; Omosanya, K.O.; Harishidayat, D.; Johansen, S.E.; Waldmann, N. Seismic geomorphology of submarine channel-belt complexes in the Pliocene of the Levant Basin, offshore central Israel. *Mar. Geol.* **2018**, *403*, 123–138. [CrossRef]
50. Wood, L.J. Quantitative Seismic Geomorphology of Pliocene and Miocene Fluvial Systems in the Northern Gulf of Mexico, U.S.A. *J. Sediment. Res.* **2007**, *77*, 713–730. [CrossRef]
51. Distefano, S.; Gamberi, F.; Borzì, L.; Di Stefano, A. Quaternary coastal landscape evolution and sea-level rise: An example from south-east sicily. *Geosciences* **2021**, *11*, 506. [CrossRef]

52. Muzirafuti, A.; Barreca, G.; Crupi, A.; Faina, G.; Paltrinieri, D.; Lanza, S.; Randazzo, G. The contribution of multispectral satellite image to shallowwater bathymetry mapping on the Coast of Misano Adriatico, Italy. *J. Mar. Sci. Eng.* **2020**, *8*, 126. [CrossRef]
53. Muzirafuti, A.; Crupi, A.; Lanza, S.; Barreca, G.; Randazzo, G. Shallow water bathymetry by satellite image: A case study on the coast of San Vito Lo Capo Peninsula, Northwestern Sicily, Italy. In Proceedings of the 2019 IMEKO TC19 International Workshop on Metrology for the Sea: Learning to Measure Sea Health Parameters, MetroSea 2019, Genoa, Italy, 3–5 October 2019; pp. 129–134.
54. Harvey, A. *Introducing Geomorphology: A Guide to Landforms and Processes*; Dunedin Academic Press Ltd.: Edinburgh, UK, 2012.
55. Huggett, R.J. *Fundamentals of Geomorphology*; Routledge: London, UK, 2016.
56. Holden, J. *An Introduction to Physical Geography and the Environment*; Pearson Education: London, UK, 2017.
57. Bierman, P.R.; Montgomery, D.R. *Key Concepts in Geomorphology*; W. H. Freeman and Company Publisher: New York, NY, USA, 2020.
58. Pieri, D. Geomorphology. In *Encyclopedia of Remote Sensing*; Njoku, E.G., Ed.; Springer: New York, NY, USA, 2014; pp. 237–241.
59. Melis, M.T.; Pisani, L.; De Waele, J. On the Use of Tri-Stereo Pleiades Images for the Morphometric Measurement of Dolines in the Basaltic Plateau of Azrou (Middle Atlas, Morocco). *Remote Sens.* **2021**, *13*, 4087. [CrossRef]
60. Muzirafuti, A.; Boualoul, M.; Barreca, G.; Allaoui, A.; Bouikbane, H.; Lanza, S.; Crupi, A.; Randazzo, G. Fusion of Remote Sensing and Applied Geophysics for Sinkholes Identification in Tabular Middle Atlas of Morocco (the Causse of El Hajeb): Impact on the Protection of Water Resource. *Resources* **2020**, *9*, 51. [CrossRef]
61. Randazzo, G.; Italiano, F.; Micallef, A.; Tomasello, A.; Cassetti, F.P.; Zammit, A.; D'Amico, S.; Saliba, O.; Cascio, M.; Cavallaro, F.; et al. WebGIS Implementation for Dynamic Mapping and Visualization of Coastal Geospatial Data: A Case Study of BESS Project. *Appl. Sci.* **2021**, *11*, 8233. [CrossRef]
62. Anders, K.; Winiwarter, L.; Mara, H.; Lindenbergh, R.; Vos, S.E.; Höfle, B. Fully automatic spatiotemporal segmentation of 3D LiDAR time series for the extraction of natural surface changes. *ISPRS J. Photogramm. Remote Sens.* **2021**, *173*, 297–308. [CrossRef]
63. Bonasera, M.; Cerrone, C.; Caso, F.; Lanza, S.; Fubelli, G.; Randazzo, G. Geomorphological and Structural Assessment of the Coastal Area of Capo Faro Promontory, NE Salina (Aeolian Islands, Italy). *Land* **2022**, *11*, 1106. [CrossRef]
64. Borzì, L.; Anfuso, G.; Manno, G.; Distefano, S.; Urso, S.; Chiarella, D.; Di Stefano, A. Shoreline evolution and environmental changes at the nw area of the gulf of gela (Sicily, Italy). *Land* **2021**, *10*, 1034. [CrossRef]
65. Bianco, F.; Conti, P.; García-Ayllon, S.; Pranzini, E. An integrated approach to analyze sedimentary stock and coastal erosion in vulnerable areas: Resilience assessment of san vicenzo's coast (Italy). *Water* **2020**, *12*, 805. [CrossRef]
66. Cinelli, I.; Anfuso, G.; Bartoletti, E.; Rossi, L.; Pranzini, E. The making of a gravel beach (Cavo, Elba Island, Italy). *J. Mar. Sci. Eng.* **2021**, *9*, 1148. [CrossRef]
67. Randazzo, G.; Barreca, G.; Cascio, M.; Crupi, A.; Fontana, M.; Gregorio, F.; Lanza, S.; Muzirafuti, A. Analysis of very high spatial resolution images for automatic shoreline extraction and satellite-derived bathymetry mapping. *Geosciences* **2020**, *10*, 172. [CrossRef]
68. Cigna, F.; Tapete, D. Sentinel-1 Big Data Processing with P-SBAS InSAR in the Geohazards Exploitation Platform: An Experiment on Coastal Land Subsidence and Landslides in Italy. *Remote Sens.* **2021**, *13*, 885. [CrossRef]
69. Mantovani, M.; Devoto, S.; Piacentini, D.; Prampolini, M.; Soldati, M.; Pasuto, A. Advanced SAR interferometric analysis to support geomorphological interpretation of slow-moving coastal landslides (Malta, Mediterranean Sea). *Remote Sens.* **2016**, *8*, 443. [CrossRef]
70. Cigna, F.; Tapete, D. Urban growth and land subsidence: Multi-decadal investigation using human settlement data and satellite InSAR in Morelia, Mexico. *Sci. Total Environ.* **2022**, *811*, 152211. [CrossRef]
71. Jiang, H.; Balz, T.; Cigna, F.; Tapete, D. Land subsidence in wuhan revealed using a non-linear PSInSAR approach with long time series of COSMO-SkyMed SAR data. *Remote Sens.* **2021**, *13*, 1256. [CrossRef]
72. Crosetto, M.; Solari, L.; Mróz, M.; Balasis-Levinsen, J.; Casagli, N.; Frei, M.; Oyen, A.; Moldestad, D.A.; Bateson, L.; Guerrieri, L.; et al. The evolution of wide-area DInSAR: From regional and national services to the European ground motion service. *Remote Sens.* **2020**, *12*, 2043. [CrossRef]
73. Muzirafuti, A.; Randazzo, G.; Maria, C.; Lanza, S. *UAV Photogrammetry-Based Mapping of the Pocket Beaches of Isola Bella Bay, Taormina (Eastern Sicily)*; The Institute of Electrical and Electronics Engineers (IEEE): Calabria, Italy, 2021.
74. Makama, A.; Kuladinithi, K.; Timm-Giel, A. Wireless Geophone Networks for Land Seismic Data Acquisition: A Survey, Tutorial and Performance Evaluation. *Sensors* **2021**, *21*, 5171. [CrossRef] [PubMed]
75. Canty, M.J.; Nielsen, A.A.; Conradsen, K.; Skriver, H. Statistical Analysis of Changes in Sentinel-1 Time Series on the Google Earth Engine. *Remote Sens.* **2020**, *12*, 46. [CrossRef]
76. Van Natijne, A.L.; Bogaard, T.A.; van Leijen, F.J.; Hanssen, R.F.; Lindenbergh, R.C. World-wide InSAR sensitivity index for landslide deformation tracking. *Int. J. Appl. Earth Obs. Geoinf.* **2022**, *111*, 102829. [CrossRef]
77. Garcia, G.P.B.; Grohmann, C.H. DEM-based geomorphological mapping and landforms characterization of a tropical karst environment in southeastern Brazil. *J. South Am. Earth Sci.* **2019**, *93*, 14–22. [CrossRef]
78. Magliulo, P.; Valente, A. GIS-Based geomorphological map of the Calore River floodplain near Benevento (Southern Italy) overflooded by the 15th october 2015 event. *Water* **2020**, *12*, 148. [CrossRef]
79. Conforti, M.; Mercuri, M.; Borrelli, L. Morphological changes detection of a large earthflow using archived images, lidar-derived dtm, and uav-based remote sensing. *Remote Sens.* **2021**, *13*, 120. [CrossRef]

80. Brunier, G.; Fleury, J.; Anthony, E.J.; Pothin, V.; Vella, C.; Dussouillez, P.; Gardel, A.; Michaud, E. Structure-From-Motion photogrammetry for highresolution coastal and fluvial geomorphic surveys. *Geomorphol. Relief Processus Environ.* **2016**, *22*, 147–161. [CrossRef]

81. Mohamed Shaffril, H.A.; Samsuddin, S.F.; Abu Samah, A. The ABC of systematic literature review: The basic methodological guidance for beginners. *Qual. Quant.* **2021**, *55*, 1319–1346. [CrossRef]

82. Munn, Z.; Peters, M.D.J.; Stern, C.; Tufanaru, C.; McArthur, A.; Aromataris, E. Systematic review or scoping review? Guidance for authors when choosing between a systematic or scoping review approach. *BMC Med. Res. Methodol.* **2018**, *18*, 143. [CrossRef] [PubMed]

83. Snyder, H. Literature review as a research methodology: An overview and guidelines. *J. Bus. Res.* **2019**, *104*, 333–339. [CrossRef]

84. Xiao, Y.; Watson, M. Guidance on conducting a systematic literature review. *J. Plan. Educ. Res.* **2019**, *39*, 93–112. [CrossRef]

85. Cook, K.L.; Dietze, M. Seismic Advances in Process Geomorphology. *Annu. Rev. Earth Planet. Sci.* **2022**, *50*, 183–204. [CrossRef]

86. Schrott, L.; Sass, O. Application of field geophysics in geomorphology: Advances and limitations exemplified by case studies. *Geomorphology* **2008**, *93*, 55–73. [CrossRef]

87. Kruse, S. 3.5 Near-Surface Geophysics in Geomorphology. In *Treatise on Geomorphology*; Shroder, J.F., Ed.; Academic Press: San Diego, CA, USA, 2013; pp. 103–129.

88. Chopra, S.; Marfurt, K. Curvature attribute applications to 3D surface seismic data. *Lead. Edge* **2007**, *26*, 404–414. [CrossRef]

89. Al-Shuhail, A.A.; Al-Dossary, S.A.; Mousa, W.A. *Seismic Data Interpretation Using Digital Image Processing*; John Wiley & Sons: Hoboken, NJ, USA, 2017.

90. Brown, A.R. 8a. Horizon and Formation Attributes. In *Interpretation of Three-Dimensional Seismic Data*; Society of Exploration Geophysicists: Houston, TX, USA; American Association of Petroleum Geologists: Tulsa, Ok, USA, 2011; pp. 247–282.

91. Herron, D.A. 3. Seismic Attributes. In *First Steps in Seismic Interpretation*; Society of Exploration Geophysicists: Houston, TX, USA, 2011; pp. 21–34.

92. Marfurt, K.J. *Seismic Attributes as the Framework for Data Integration throughout the Oilfield Life Cycle*; Society of Exploration Geophysicists: Houston, TX, USA, 2018.

93. Taner, M.T. Seismic attributes. *CSEG Rec.* **2001**, *26*, 48–56.

94. Barnes, A.E. Seismic attributes in your facies. *CSEG Rec.* **2001**, *26*, 41–47.

95. Chopra, S.; Marfurt, K.J. Seismic attributes—A historical perspective. *Geophysics* **2005**, *70*, 3SO–28SO. [CrossRef]

96. Barnes, A.E. Redundant and useless seismic attributes. *Geophysics* **2007**, *72*, P33–P38. [CrossRef]

97. Li, M.; Zhao, Y. Chapter 5–Seismic Attribute Analysis. In *Geophysical Exploration Technology*; Li, M., Zhao, Y., Eds.; Elsevier: Oxford, UK, 2014; pp. 103–131.

98. Liner, C.; Li, C.F.; Gersztenkorn, A.; Smythe, J. SPICE: A new general seismic attribute. In *SEG Technical Program Expanded Abstracts*; Society of Exploration Geophysicists: Houston, TX, USA, 2004; pp. 433–436.

99. Sidney, S. The Back Page; Chen and Sidney (1997) seismic attribute history. *Lead. Edge* **2007**, *26*, 1488. [CrossRef]

100. Huuse, M.; Feary, D.A. Seismic inversion for acoustic impedance and porosity of Cenozoic cool-water carbonates on the upper continental slope of the Great Australian Bight. *Mar. Geol.* **2005**, *215*, 123–134. [CrossRef]

101. Veeken, P.C.; Priezzhev, I.I.; Shmaryan, L.E.; Shteyn, Y.I.; Barkov, A.Y.; Ampilov, Y.P. Nonlinear multitrace genetic inversion applied on seismic data across the Shtokman field, offshore northern Russia. *Geophysics* **2009**, *74*, WCD49–WCD59. [CrossRef]

102. Chopra, S.; Marfurt, K.J. Seismic attributes for prospect identification and reservoir characterization. In *SEG Geophysical Development Series*; Society of Exploration Geophysicists: Houston, TX, USA; European Association of Geoscientists and Engineers: Tulsa, OK, USA, 2007.

103. Lutome, M.S.; Lin, C.; Chunmei, D.; Zhang, X.; Harishidayat, D. Seismic sedimentology of lacustrine delta-fed turbidite systems: Implications for paleoenvironment reconstruction and reservoir prediction. *Mar. Pet. Geol.* **2020**, *113*, 104159. [CrossRef]

104. Zeng, H.; Yuan, B. Seismic sedimentology: Concepts and challenges. In Proceedings of the Beijing 2009 International Geophysical Conference and Exposition, Beijing, China, 24–27 April 2009; Society of Exploration Geophysicists: Beijing, China, 2009; p. 280.

105. Zeng, H.; Hentz, T.F. High-frequency sequence stratigraphy from seismic sedimentology: Applied to Miocene, Vermilion Block 50, Tiger Shoal area, offshore Louisiana. *AAPG Bull.* **2004**, *88*, 153–174. [CrossRef]

106. Zeng, H.; Ambrose, W.A.; Villalta, E. Seismic sedimentology and regional depositional systems in Mioceno Norte, Lake Maracaibo, Venezuela. *Lead. Edge* **2001**, *20*, 1260–1269. [CrossRef]

107. Zeng, H.; Backus, M.M. Interpretive advantages of 90°-phase wavelets: Part 1—Modeling. *Geophysics* **2005**, *70*, C7–C15. [CrossRef]

108. Zeng, H.; Backus, M.M. Interpretive advantages of 90°-phase wavelets: Part 2—Seismic applications. *Geophysics* **2005**, *70*, C17–C24. [CrossRef]

109. Zeng, H. Stratal slicing: Benefits and challenges. *Lead. Edge* **2010**, *29*, 1040–1047. [CrossRef]

110. Zeng, H. Stratal slice: The next generation. *Lead. Edge* **2013**, *32*, 140–144. [CrossRef]

111. Posamentier, H.W.; Kolla, V. Seismic Geomorphology and Stratigraphy of Depositional Elements in Deep-Water Settings. *J. Sediment. Res.* **2003**, *73*, 367–388. [CrossRef]

112. Deiana, G.; Lecca, L.; Melis, R.T.; Soldati, M.; Demurtas, V.; Orrù, P.E. Submarine Geomorphology of the Southwestern Sardinian Continental Shelf (Mediterranean Sea): Insights into the Last Glacial Maximum Sea-Level Changes and Related Environments. *Water* **2021**, *13*, 155. [CrossRef]

113. Gao, M.; Xu, X.; Klinger, Y.; van der Woerd, J.; Tapponnier, P. High-resolution mapping based on an Unmanned Aerial Vehicle (UAV) to capture paleoseismic offsets along the Altyn-Tagh fault, China. *Sci. Rep.* **2017**, *7*, 8281. [CrossRef]
114. Chaves, M.U.; Oliver, F.; Kawakami, G.; Di Marco, L. Visualization of Geological Features Using Seismic Volume Rendering, Rgb Blending and Geobody Extraction. In Proceedings of the 12th International Congress of the Brazilian Geophysical Society, Janeiro, Brazil, 15–18 August 2011.
115. Alves, T.M.; Omosanya, K.O.; Gowling, P. Volume rendering of enigmatic high-amplitude anomalies in southeast Brazil: A workflow to distinguish lithologic features from fluid accumulations. *Interpretation* **2015**, *3*, A1–A14. [CrossRef]
116. Chopra, S.; Marfurt, K. Blended data renders visual value. *AAPG Explor. Geophys. Corner* **2011**, *3*, 38–40.
117. Zeng, H. RGB blending of frequency panels: A new useful tool for high-resolution 3D stratigraphic imaging. In Proceedings of the 2017 SEG International Exposition and Annual Meeting, Houston, TX, USA, 24–29 September 2017.
118. Mohammedyasin, S.M.; Lippard, S.J.; Omosanya, K.O.; Johansen, S.E.; Harishidayat, D. Deep-seated faults and hydrocarbon leakage in the Snøhvit Gas Field, Hammerfest Basin, Southwestern Barents Sea. *Mar. Pet. Geol.* **2016**, *77*, 160–178. [CrossRef]
119. Marfurt, K.; Wen, R.; Zhang, R. 3-D Visualization and Geobody Picking of Amplitude Anomalies in Deepwater Seismic Data. *AAPG Explor.* **2021**, *09*, 10–11.
120. Castanie, L.; Levy, B.; Bosquet, F. VolumeExplorer: Roaming large volumes to couple visualization and data processing for oil and gas exploration. In Proceedings of the VIS 05. IEEE Visualization, 2005, Minneapolis, MN, USA, 23–28 October 2005; pp. 247–254.
121. Chopra, S.; Marfurt, K.J. Detecting stratigraphic features via crossplotting of seismic discontinuity attributes and their volume visualization. *Lead. Edge* **2009**, *28*, 1422–1426. [CrossRef]
122. Castanie, L.; Bosquet, F.; Levy, B. Advances in seismic interpretation using new volume visualization techniques. *First Break* **2005**, *23*, 1365–2397. [CrossRef]
123. Lomask, J.; Guitton, A. Volumetric flattening: An interpretation tool. *Lead. Edge* **2007**, *26*, 888–897. [CrossRef]
124. Dramsch, J.S. Chapter One–70 years of machine learning in geoscience in review. In *Advances in Geophysics*; Moseley, B., Krischer, L., Eds.; Elsevier: Amsterdam, The Netherlands, 2020; Volume 61, pp. 1–55.
125. Karpatne, A.; Ebert-Uphoff, I.; Ravela, S.; Babaie, H.A.; Kumar, V. Machine learning for the geosciences: Challenges and opportunities. *IEEE Trans. Knowl. Data Eng.* **2018**, *31*, 1544–1554. [CrossRef]
126. Bishop, C. *Pattern Recognition and Machine Learning*; Springer: Amsterdam, The Netherlands, 2006; Volume 4.
127. Niyazi, Y.; Warne, M.; Ierodiaconou, D. Machine learning delineation of buried igneous features from the offshore Otway Basin in southeast Australia. *Interpretation* **2022**, *10*, 1–70. [CrossRef]
128. Hoyes, J.; Cheret, T. A review of "global" interpretation methods for automated 3D horizon picking. *Lead. Edge* **2011**, *30*, 38–47. [CrossRef]
129. Liu, M.; Jervis, M.; Li, W.; Nivlet, P. Seismic facies classification using supervised convolutional neural networks and semisupervised generative adversarial networks. *Geophysics* **2020**, *85*, O47–O58. [CrossRef]
130. Wrona, T.; Pan, I.; Gawthorpe, R.L.; Fossen, H. Seismic facies analysis using machine learning. *Geophysics* **2018**, *83*, O83–O95. [CrossRef]
131. Zhang, H.; Chen, T.; Liu, Y.; Zhang, Y.; Liu, J. Automatic seismic facies interpretation using supervised deep learning. *Geophysics* **2021**, *86*, IM15–IM33. [CrossRef]
132. Kumar, P.C.; Sain, K. A machine learning tool for interpretation of Mass Transport Deposits from seismic data. *Sci. Rep.* **2020**, *10*, 14134. [CrossRef]
133. Bugge, A.J.; Lie, J.E.; Evensen, A.K.; Faleide, J.I.; Clark, S. Automatic extraction of dislocated horizons from 3D seismic data using nonlocal trace matching. *Geophysics* **2019**, *84*, IM77–IM86. [CrossRef]
134. Infante-Paez, L.; Marfurt, K.J. Using machine learning as an aid to seismic geomorphology, which attributes are the best input? *Interpretation* **2019**, *7*, SE1–SE18. [CrossRef]
135. Liu, Z.; Song, C.; Li, K.; She, B.; Yao, X.; Qian, F.; Hu, G. Horizon extraction using ordered clustering on a directed and colored graph. *Interpretation* **2019**, *8*, T1–T11. [CrossRef]
136. Peters, B.; Granek, J.; Haber, E. Multiresolution neural networks for tracking seismic horizons from few training images. *Interpretation* **2019**, *7*, SE201–SE213. [CrossRef]
137. Lou, Y.; Zhang, B.; Lin, T.; Cao, D. Seismic horizon picking by integrating reflector dip and instantaneous phase attributes. *Geophysics* **2020**, *85*, O37–O45. [CrossRef]
138. Yang, L.; Sun, S.Z. Seismic horizon tracking using a deep convolutional neural network. *J. Pet. Sci. Eng.* **2020**, *187*, 106709. [CrossRef]
139. Bi, Z.; Wu, X.; Geng, Z.; Li, H. Deep Relative Geologic Time: A Deep Learning Method for Simultaneously Interpreting 3-D Seismic Horizons and Faults. *J. Geophys. Res. Solid Earth* **2021**, *126*, e2021JB021882. [CrossRef]
140. La Marca, K.; Bedle, H. Deepwater seismic facies and architectural element interpretation aided with unsupervised machine learning techniques: Taranaki basin, New Zealand. *Mar. Pet. Geol.* **2022**, *136*, 105427. [CrossRef]
141. Singh, S.; Tsvankin, I.; Naeini, E.Z. Facies prediction with Bayesian inference: Application of supervised and semisupervised deep learning. *Interpretation* **2022**, *10*, T279–T290. [CrossRef]
142. AlRegib, G.; Deriche, M.; Long, Z.; Di, H.; Wang, Z.; Alaudah, Y.; Shafiq, M.A.; Alfarraj, M. Subsurface Structure Analysis Using Computational Interpretation and Learning: A Visual Signal Processing Perspective. *IEEE Signal Process. Mag.* **2018**, *35*, 82–98. [CrossRef]

143. An, Y.; Guo, J.; Ye, Q.; Childs, C.; Walsh, J.; Dong, R. Deep convolutional neural network for automatic fault recognition from 3D seismic datasets. *Comput. Geosci.* **2021**, *153*, 104776. [CrossRef]
144. Laudon, C.; Qi, J.; Rondon, A.; Rouis, L.; Kabazi, H. An enhanced fault detection workflow combining machine learning and seismic attributes yields an improved fault model for Caspian Sea asset. *First Break* **2021**, *39*, 53–60. [CrossRef]
145. Wu, X.; Liang, L.; Shi, Y.; Fomel, S. FaultSeg3D: Using synthetic data sets to train an end-to-end convolutional neural network for 3D seismic fault segmentation. *Geophysics* **2019**, *84*, IM35–IM45. [CrossRef]
146. Yan, Z.; Zhang, Z.; Liu, S. Improving Performance of Seismic Fault Detection by Fine-Tuning the Convolutional Neural Network Pre-Trained with Synthetic Samples. *Energies* **2021**, *14*, 3650. [CrossRef]
147. Di, H.; AlRegib, G. Semi-automatic fault/fracture interpretation based on seismic geometry analysis. *Geophys. Prospect.* **2019**, *67*, 1379–1391. [CrossRef]
148. De Groot, P. Global Seismic Interpretation Techniques Are Coming of Age. *ASEG Ext. Abstr.* **2013**, *2013*, 1–4. [CrossRef]
149. Gao, D. 3D seismic volume visualization and interpretation: An integrated workflow with case studies. *Geophysics* **2009**, *74*, W1–W12. [CrossRef]
150. Paumard, V.; Bourget, J.; Durot, B.; Lacaze, S.; Payenberg, T.; George, A.D.; Lang, S. Full-volume 3D seismic interpretation methods: A new step towards high-resolution seismic stratigraphy. *Interpretation* **2019**, *7*, B33–B47. [CrossRef]
151. De Groot, P.; Bruin, G.; McBeath, K. OpendTect SSIS–Sequence stratigraphic interpretation system. *Drill. Explor. World* **2006**, *15*, 31–34.
152. Sun, T.; Li, H.; Wu, K.; Chen, F.; Zhu, Z.; Hu, Z. Data-Driven Predictive Modelling of Mineral Prospectivity Using Machine Learning and Deep Learning Methods: A Case Study from Southern Jiangxi Province, China. *Minerals* **2020**, *10*, 102. [CrossRef]
153. Johansen, S.; Granberg, E.; Mellere, D.; Arntsen, B.; Olsen, T. Decoupling of seismic reflectors and stratigraphic timelines: A modeling study of Tertiary strata from Svalbard. *Geophysics* **2007**, *72*, SM273–SM280. [CrossRef]
154. Bakke, K.; Gjelberg, J.; Agerlin Petersen, S. Compound seismic modelling of the Ainsa II turbidite system, Spain: Application to deep-water channel systems offshore Angola. *Mar. Pet. Geol.* **2008**, *25*, 1058–1073. [CrossRef]
155. Taufani, L.; Harishidayat, D.; Rohmana, R.C.; Fardiansyah, I.; Purnama, Y.S.; Indriyanto, I.B. *Utilization of Digital Mapping and Outcrop Model to Assess Reservoir Characterization and Quality Index: Study Case from Ngrayong Formation in the Randugunting Block, East Java*; Indonesian Petroleum Association: South Jakarta, Indonesia, 5–8 September 2021.
156. Bakke, K.; Petersen, S.A.; Martinsen, O.J.; Johansen, T.; Lien, T.; Thurmond, J. Seismic modeling of outcrop analogues: Techniques and applications. *Outcrops Revital. Tools Tech. Appl. Tulsa Okla. SEPM Concepts Sedimentol. Paleontol.* **2011**, *10*, 69–86.
157. Lecomte, I.; Lavadera, P.L.; Anell, I.; Buckley, S.J.; Schmid, D.W.; Heeremans, M. Ray-based seismic modeling of geologic models: Understanding and analyzing seismic images efficiently. *Interpretation* **2015**, *3*, SAC71–SAC89. [CrossRef]
158. Grippa, A.; Hurst, A.; Palladino, G.; Iacopini, D.; Lecomte, I.; Huuse, M. Seismic imaging of complex geometry: Forward modeling of sandstone intrusions. *Earth Planet. Sci. Lett.* **2019**, *513*, 51–63. [CrossRef]
159. Harishidayat, D.; Johansen, S.E.; Puigdefabregas, C.; Omosanya, K.O. Compound Seismic Forward Modeling of the Atiart Submarine Canyon Outcrop, Spain: Application to the Submarine Canyon System on the Subsurface Loppa High, Barents Sea. In Proceedings of the American Association of Petroleum Geologists (AAPG)–Annual Conference and Exhibition (ACE), Salt Lake City, UT, USA, 11–13 May 2018.
160. Rohmana, R.C.; Fardiansyah, I.; Taufani, L.; Harishidayat, D. Depositional processes and facies architecture of Balikpapan sandstone formation, application of 3D Digital Outcrop Model (DOM) to identify reservoir geometry and distribution in deltaic system. *Sci. Contrib. Oil Gas* **2019**, *42*, 35–42. [CrossRef]

Article

A Scalable Earth Observation Service to Map Land Cover in Geomorphological Complex Areas beyond the Dynamic World: An Application in Aosta Valley (NW Italy)

Tommaso Orusa [1,2,*], **Duke Cammareri** [2] and **Enrico Borgogno Mondino** [1]

[1] Department of Agricultural, Forest and Food Sciences (DISAFA), GEO4Agri DISAFA Lab, Università Degli Studi di Torino, Largo Paolo Braccini 2, 10095 Grugliasco, Italy
[2] Earth Observation Valle d'Aosta—eoVdA, Località L'Île-Blonde, 5, 11020 Brissogne, Italy
* Correspondence: tommaso.orusa@unito.it

Abstract: Earth Observation services guarantee continuous land cover mapping and are becoming of great interest worldwide. The Google Earth Engine Dynamic World represents a planetary example. This work aims to develop a land cover mapping service in geomorphological complex areas in the Aosta Valley in NW Italy, according to the newest European EAGLE legend starting in the year 2020. Sentinel-2 data were processed in the Google Earth Engine, particularly the summer yearly median composite for each band and their standard deviation with multispectral indexes, which were used to perform a k-nearest neighbor classification. To better map some classes, a minimum distance classification involving NDVI and NDRE yearly filtered and regularized stacks were computed to map the agronomical classes. Furthermore, SAR Sentinel-1 SLC data were processed in the SNAP to map urban and water surfaces to improve optical classification. Additionally, deep learning and GIS updated datasets involving urban components were adopted beginning with an aerial orthophoto. GNSS ground truth data were used to define the training and the validation sets. In order to test the effectiveness of the implemented service and its methodology, the overall accuracy was compared to other approaches. A mixed hierarchical approach represented the best solution to effectively map geomorphological complex areas to overcome the remote sensing limitations. In conclusion, this service may help in the implementation of European and local policies concerning land cover surveys both at high spatial and temporal resolutions, empowering the technological transfer in alpine realities.

Keywords: land cover; Sentinel-1 SAR; Sentinel-2; deep learning; Google Earth Engine; SAGA GIS; ESRI ArcGIS Pro; ESA SNAP; mountains; EAGLE; geomorphological complex areas

check for **updates**

Citation: Orusa, T.; Cammareri, D.; Borgogno Mondino, E. A Scalable Earth Observation Service to Map Land Cover in Geomorphological Complex Areas beyond the Dynamic World: An Application in Aosta Valley (NW Italy). *Appl. Sci.* **2023**, *13*, 390. https://doi.org/10.3390/app13010390

Academic Editors: Giovanni Randazzo, Stefania Lanza and Anselme Muzirafuti

Received: 11 November 2022
Revised: 15 December 2022
Accepted: 23 December 2022
Published: 28 December 2022

1. Introduction

Earth Observation (EO) data services are becoming very popular because of the significant increase in satellite missions and geospatial cloud-based platforms such as the Google Earth Engine and Microsoft Planetary [1,2]. New investments in the space economy have boosted the technological transfer in different fields opening new opportunities in terms of applied science [3–5]. It is worth noting that both the public and private sectors in alpine and rural areas are still far behind. Therefore, it is more crucial to fill this gap by realizing and exporting useful EO tools to strengthen the monitoring and study of the biophysical components of different territories and use public funds more efficiently and effectively [6–8]. This would permit mountainous areas to keep up, stay competitive and bring innovation even in apparently distant contexts. In particular, the Copernicus program, as well as many other scientific EO programs around the world, provide vast amounts of geographical datasets that may aid in achieving European and international technological transfer goals of to face considerable issues such as climate change, sustainable development and social inclusion worldwide [9–11].

Recently, Google announced its realization of the Dynamic World project. Dynamic World is a near real-time 10 m spatial resolution global land use land cover (LULC) dataset produced using deep learning and is freely available and openly licensed. It is the result of a partnership between Google and the World Resources Institute to produce a dynamic dataset of the physical material on the surface of the Earth. Dynamic World is intended to be used as a data product for users to add custom rules and assign final class values, producing derivative land cover maps. The main key innovation of Dynamic World is represented by near-real time image enabling the mapping of LULC every 5 days depending on the location and adopting Sentinel-2 top-of-atmosphere per-pixel probabilities across nine land cover classes with a 10 m GSD.

This EO service based on the Google Earth Engine is very powerful but regarding the accuracy in mountainous areas such as the Alps, it presents considerable criticalities. First, there is a strong confusion between the concept of land coverage and use (the first can be mapped by satellite, the second can only generally be used for certain uses such as mowing). Second, the system uses a probabilistic approach, not a deterministic one. Therefore, at a planning level, some problems can be encountered (e.g., a misleading biophysical component defined as an incorrect class for a forest or built up areas defined as water after snowmelt in alpine areas). Third, the Dynamic World training set distribution is almost completely absent in mountainous and alpine areas. Typically, these areas are the most complex to map for remote sensing. Furthermore, the classes are designed to map global changes at high resolution using fewer classes which do not answer to the local needs for land covers that adopt the new European EAGLE guidelines and that have local robust accuracies that are only obtainable with a continuous ground truth data validation set.

The EIONET Action Group on land monitoring in Europe (known as EAGLE group) is an open assembly of technical experts from different European Economic Area (EEA) Member States, mostly in their roles as national reference center (NRC) on LC. Currently, the development of the EAGLE concept and methodology is being funded by the EEA within the framework of the Copernicus program. In Italy, LC monitoring is performed by the Istituto Superiore per la Protezione e la Ricerca Ambientale (ISPRA). For their activities and policies, many user communities such as decision-makers, non-governmental organizations, European communities, scientists and researchers require various sorts of LC information [12]. LC data, for example, are used to assess the progress toward the UN Sustainable Development Goals (SDGs) targets [13], such as target 15.3 which relates to achieving land degradation neutrality (LDN) by 2030 [14]. Despite the importance of LC data in environmental monitoring and planning, the number of accessible national products is limited and their qualities are not always appropriate. The EAGLE idea was built on a clear difference between land cover and land use, and it is described in a matrix consisting of three descriptors: land cover components (LCC), land use attributes (LUA) and additional characteristics (LCH). The descriptors can be merged to develop unique categorization systems for various needs or to detect correspondences with existing classes while maintaining the independence of the three descriptors [15]. There are numerous algorithms for examining land cover, starting with the classification of satellite images. The most appropriate approach is determined by factors such as the type of data, class distribution, research interests and classifier interpretability, as well as the balance between the objectives and the available resources. In general, the automatic classification systems are more time consuming as the data dimension and volume grow and data interpretation might become problematic at times [16]. Supervised classifications may be used to analyze large amounts of data as they are based on selecting a sufficient number of training samples with known values [17], which are then used to predict unknown values in the testing data [18]. As a result, it's critical to choose examples that fully depict the diversity of the studied territory's characteristics.

Nowadays, the CORINE Land Cover (CLC) ensures a high level of thematic detail and a lengthy historical series, but it is limited in terms of geographical detail and updating frequency (https://land.copernicus.eu/pan-european/corine-land-cover, viewed on 6

November 2022). In recent years, the high-resolution layers (HRLs) made it possible to describe the principal land cover classes in high spatial detail while maintaining a multi-year update frequency. However, the low accuracies in the classification of alpine areas still persist [19]. At the same time, Copernicus national-scale products are still available with high thematic and spatial depth, including Urban Atlas, Riparian Zones, Coastal Zones and Natura 2000, but these are only for specific locations. The most recent official product is the CORINE LAND COVER v.2018 (CLC). However, this product does not permit a detailed mapping of the territory especially in alpine areas [20]. Nevertheless, in recent years, some institutions, academic centers and private enterprises have tried to overcome the spatial resolution issue by creating prototypal products at the national level by adopting Copernicus EO Data [21,22]. New evidence is represented by the 10 m land cover produced by ESRI on a global scale with 10 classes even if there are strong limitations and errors in the alpine area due to the absence of a local confusion matrix [23]. Another example linked to the Italian context is the prototypal LC of the whole Italian territory performed by the ISPRA for the year 2018. This LC proposes a methodology with the joint use of the optical, multispectral and radar data of Sentinel 1 and Sentinel 2 [12]. However, following the choice of the adopted input data and the need to map an entire territory, it has strong limits in mountainous areas. Compared to the validation set, these areas are low and are not adequate for mapping mountainous territories in detail as suggested by [19].

As previously mentioned, a robust local EO service only based on satellite remote sensing data can only map land cover (hereinafter called LC) with a higher accuracy.

In particular, only the definition of LC is fundamental, because in many existing classifications and legends it is confused with land use. LC is defined as the observed (bio)physical cover on the Earth's surface. According to this definition, land covers include forests, agricultural areas, human settlements, glaciers, water and wetlands per the Directive 2007/02 of the European Commission [24,25]. When considering LC in a very pure and strict sense, it should be confined to describe vegetation and man-made features. Consequently, the areas where the surface consists of bare rock or bare soil are describing land itself rather than LC. Additionally, it is disputable whether water surfaces are considered land cover. However, in practice, the scientific community usually describes those aspects under the term LC as well as some agricultural types of cover such as orchards, vineyards and pastures.

The application of Sentinel-2, Sentinel-1 and PlanetScope in land cover mapping have rapidly spread since 2020 [26]. Earth Observation data applications from multi-platform sensors, in particular from Sentinel-2, Sentinel-1 and PlanetScope, are mainly focused on plain areas [27]. There is still a lack of local application mountainous area. Most of classification approaches are based on single one-shot classifications or combined approaches with two supervised classifications from optical and SAR data [28]. Others are focused on multimodal remote sensing data fusion from open-access and commercial EO data in cloud platform, such as the Google Earth Engine, and in deep learning classification (mainly focused on classifying a single geometrical land cover component such as urban areas) [29]. Nevertheless, a mixed hierarchical approach adopting different sensors and several classifications to improve the land cover quality on mountainous areas still continues to be minutely explored and exploited.

Under this scenario, the development of a local LC EO continuous service according to the new EAGLE guidelines is becoming of great interest to the public administration at the alpine level and beyond. In fact, at the national level in Italy, the Istituto Superiore Per la Protezione e Ricerca Ambientale (ISPRA) produced and updated the national land consumption map, as well as several national land use and LC maps [21,22], but they do not necessary fit with alpine needs. Conversely, other regional products are frequently produced with CLC and are not up to date [30,31] or, in the case of high resolution global updated products such as Dynamic World, they do not answer to the local needs in terms of the accuracies, methodologies followed and legends. Therefore, in order to fill the gap of a lack of a LC at high resolution and answering to the EAGLE requests, the Aosta Valley autonomous region charged the Regional Cartographic Office to develop a new map with a 10 m GSD. This body commissioned the regional public company INVA spa, in particular the GIS unit, to carry out this work. The intent was to create a static product and service capable of dynamically mapping the Aosta Valley territory according to the required needs.

Therefore, the principal aim of this work is to present the new Aosta Valley LC and create a scalable and economically sustainable local EO service capable of mapping LC according to the EAGLE guidelines. The EO service developed adopted SAR Sentinel-1; Sentinel-2; PlanetScope and updated GIS local datasets to overcome the common troubles that remote sensing (RS) has in mountainous areas due to the topography, weather condition and shadows, as well as uniform LC class distributions.

2. Study Area

The EO Geospatial service was developed in the Aosta Valley autonomous region in NW Italy. It is the smallest Italian region in terms of surface extent, located in the mid-west of the Alps. It is surrounded by the four highest mountain massifs in Italy: Mont Blanc, which is also the highest peak in Europe, the Cervino-Matterhorn (4478 m), Monte Rosa (4634 m) and Gran Paradiso (4061 m). The conformation of the entire regional territory is the result of the work of many glaciations [32,33]. Therefore, considering the mountainous topography of the whole Aosta Valley territory, a specific EO geospatial land cover procedure was developed based on the EAGLE guidelines (see Figure 1).

Figure 1. EO geospatial service area of interest. Aosta Valley autonomous region in NW Italy (ED50 UTM 32N) EPSG:23032.

3. Materials

The development of the present EO geospatial local service is scalable to other geomorphological complex areas, such as other mountain territories, and is based on the following datasets:

- Copernicus Sentinel-2A surface reflectance data to map the land cover components
- Copernicus Sentinel-1A and B SLC and GRD data to map urban and water components, respectively
- PlanetScope four band data to define a part of the training and validation set
- GIS updated datasets such as GNSS ground truth data to define both the training set and the validation set.

3.1. Multispectral Optical Datasets

The **Sentinel-2** (hereinafter called S2) mission is part of the European Copernicus programme. The satellite acquires multispectral optical data with a spatial resolution between 10–20 m as a function of the considered band. The temporal resolution is 5 days due to two twin satellites, S2A and S2B. The multispectral optical data were obtained and processed in the Google Earth Engine (GEE) referring to the COPERNICUS/S2_SR collection. Sentinel-2 is a high-resolution, broad-spectrum, multispectral optical mission that supports the Copernicus Land Monitoring Service, including monitoring vegetation, soil and water cover and observing inland waterways and coastal areas. Sentinel-2 L2 data were downloaded from the Copernicus SciHub (the official distribution portal of the Earth Observation data in question). The images were pre-processed in Sen2cor (the official tool released by the European Space Agency—ESA). The EO data S2 that was pre-processed in Sen2cor contained 12 spectral bands of UINT16 (see Table 1). The images were ortho-projected in WGS84 and were in ground reflectance rescaled in dimensionless values from 0 to 10,000 starting from the existing DN to calculate the ground reflectance by removing the atmospheric contribution. There are also three QA bands for each scene, one of which (QA60) is a bitmask band with cloud mask information. In GEE, clouds can be removed as an alternative to using pixels in QA quality using COPERNICUS/S2_CLOUD_PROBABILITY. In this case, the QA bands were used.

Table 1. A simple overview of the Sentinel-2 surface reflectance product collection composition in the Google Earth Engine.

Bands	Description	Spatial Resolution (m)
B1	Aerosols	60
B2	Blue	10
B3	Green	10
B4	Red	10
B5	Red Edge 1	20
B6	Red Edge 2	20
B7	Red Edge 3	20
B8	NIR	10
B8A	Red Edge 4	20
B9	Water vapor	60
B11	SWIR 1	20
B12	SWIR 2	20
SCL	Mask	10
MSK_CLD_PRB	Cloud probability mask	20
QA10-60	Cloud mask	10–60

A yearly median composite imagery ranging from 1 May 2020 to 30 September 2020 without clouds and shadows was realized. The S2 data were used to create yearly harmonized and filtered NDVI and NDRE stacks with a 10 days step to map woody crops.

PlanetScope, as part of the private space program Planet acquired by Google with its ultra-high spatial resolution microsatellites, is increasingly becoming a reference reality in remote sensing activities due to the possibility of accessing the data free of charge for education and research purposes (https://www.planet.com/markets/education-and-research, last access on 6 November 2022). Starting with the daily data acquired by PlanetScope, a stack was created, including all the acquisitions in the reference period of study. With a self-developed GEE algorithm, a composite imagery was generated covering the same period as S2. This image was adopted as an extra product in the validation phase and in the definition of the training sets during a photo-interpretation phase. It is worth noting that the PlanetScope micro-satellites acquire multispectral optical data on a daily basis in four bands with a ground sample distance (GSD) of around 3 m with various levels of processing. In this case, geo-referenced and atmospheric calibrated products in surface reflectance were adopted. Considering that these data are not open-access and have a fee for use except in scientific purposes, they can be considered optional in the development of a fully free EO geospatial local service. However, they represent a useful tool during the suggested workflow.

3.2. Sentinel-1 SAR Dataset

The **Sentinel-1** mission is part of the European Copernicus programme. The satellite acquires radar data with a spatial resolution between 5–40 m depending on the acquisition mode. The temporal resolution is 5 days due to two twin satellites, S1A and S1B.

The radar data were retrieved from the NASA Alaska Satellite Facility (ASF, https://asf.alaska.edu/, last accessed 7 November 2022) and processed in the SNAP v.8.0.0 [34] and the Google Earth Engine (GEE) [1,35].

The Sentinel-1 (hereinafter called S1) mission provides data from a dual-polarized C-band SAR (Synthetic Aperture Radar) instrument. The Google Earth Engine provided only the Sentinel-1 ground range detected (GRD) collection. Each scene was preprocessed with the Sentinel-1 Toolbox in the SNAP using the following steps: (1) thermal and other noise removal, (2) Speckle–Lee filter application, (3) radiometric calibration, (4) ground correction using DTM 10m VDA (normally SRTM 30m worldwide) and (5), the final corrected values for the ground were converted into decibels via log scaling ($10 \times \log10(x)$).

The level-1 data were processed into either single look complex (SLC) and/or ground range detected (GRD) products. The SLC products preserved the phase information and were processed at the natural pixel spacing whereas the GRD products contained the detected amplitude and were multi-looked to reduce the impact of speckle. In particular. the level-1 SLC (IW) interferometric wide products (IW) were adopted [36].

The IW swath mode was the main acquisition mode over land and satisfied the majority of the service requirements (Richards 2009 [37]). As mentioned before, the SLC IW data were adopted by creating two separate datasets with the same orbit, frame and path of the scene in the study area. The two time series stacks, including all scenes ranging from 1 January 2020 to 31 December 2020 in ascending and descending mode, were considered. Those characteristics are reported in Table 2. As reported by [38], the main distortion in SAR data was the elevation displacement. In a radar image, the displacement was toward the sensor and became quite large when the sensor was nearly overhead. The displacement increased with a decreasing incidence angle. The characteristics resulting from the geometric relationship between the sensor and the terrain that were unique to radar imagery were foreshortening, layover and shadowing. The topographic features such as mountains and artificial targets such as tall buildings were displaced from their desired orthographic position. The effect was removed from an image through independent knowledge of the terrain profile.

Table 2. SAR stacks parameters criteria.

Absolute Orbit Number	Polarization	Frame	Path	Flight Direction
24,789	VV+VH	146	88	ASCENDING
24,417	VV+VH	441	66	DESCENDING

The ascending and descending values were both processed in the SNAP v.8.0.0 and then imported into the GEE to create a mosaicked-median composite to reduce the geometric distortions in the slopes where, normally, a given acquisition mode occurs.

3.3. GIS Products and Ground Data

In this EO service, other datasets were also considered.

First, the **digital terrain model (DTM)** from the Aosta Valley autonomous region with a 2 m GSD was resampled in SAGA GIS with a nearest neighbor algorithm in order to perfectly overlay the Sentinel imagery. It is worth noting that the DTM was acquired with flight lidar sensors in 2008.

Second, the *training set* was defined as different ESRI shapefile polygons per each class in order to train the classifier. This dataset was defined from one side by object-based segmentation (OBIA) using red, green, blue, near-infrared, red edge and shortwave bands and considering the spectral signatures and the photo-interpretation analysis by adopting the ground truth data polygon (GTDP).

Third, the *validation set* was defined through the ESRI shapefile polygons to validate the classification. This dataset was obtained both through photo-interpretation and in-the-field GTDP. The validation was carried out in two phases: the first by calculating the confusion matrix by adopting the dataset obtained from S1 and S2 processing bands and finally by assessing the classification accuracy after merging each part.

It is worth noting that a Garmin 64S and Lemon GPS smartphone application developed by the Italian GeneGIS company were also used to define the GTDP.

As mentioned before, the collection of such data allowed us to populate both the training set and the validation set. In particular, a random GTDP selection was performed in SAGA GIS vers. 8.2.0. with an allocation of 70% of the GTDP to the training set and 30% of the GTDP to the validation set.

Finally, the Italian AGEA (Agency for Disbursements in Agriculture) yearly *air flights imagery* coupled with the Aosta Valley 2018 Orthophoto were used to perform deep learning on built-up areas and refine the final product with a minimum mapping unit of 100 m^2 in order to keep the product coeval with the Sentinel datasets.

Generally, the tools adopted were the GEE [1], the SNAP vers. 8.0.0 to obtain and calibrate the data during the pre-processing phase, Orfeo Toolbox vers 8.0.0 [39,40], SAGA GIS vers.8.0.0 [41] to perform the classification during the processing phase and QGIS with GRASS and R v.3.0.1 [42–44] during the post-processing phase to prepare the final product.

4. Methods

4.1. Sentinel-2

The S2 data were obtained from the GEE. In particular, the collection COPERNI-CUS/S2_SR was used. A self-developed algorithm performed in the GEE was adopted to create the median composites. The S2 composite stack included bands, spectral indices and standard deviations. These input parameters are reported in Table 3. The S2 stack, including the DTM aspect and slope, was adopted as the input data during the classification while the S1 output layers served to better refine the urban and water classes. Each composite image was generated starting with the EO data available every 10 days for the period from 1 May 2020 to 30 September 2020 (t), i.e., the summer weather season, in order to correctly map the glacial surface of the territory falling within the ablation period and observe the vegetation during the phenological active season. It is worth noting that the generated composite images consisted of the median value for each pixel in the reference period t. For S2, we considered all the images that satisfied the condition in which each pixel had cloud cover equal to zero (the clouds and shadows were suitably masked and the pixel, if cloudy, was considered in the definition of the median value of the reflectance of each band). The S2 input data were reported in Table 3 as the input dataset for the k-nearest neighbor supervised classification considering all the classes. The input dataset was normalized.

Table 3. S2 input datasets.

ID	Bands/Index	Description
1	"B2"	Blue
2	"B3"	Green
3	"B4"	Red
4	"B5"	Vegetation Red Edge 1
5	"B6"	Vegetation Red Edge 2
6	"B7"	Vegetation Red Edge 3
7	"B8"	NIR

Table 3. *Cont.*

ID	Bands/Index	Description
8	"B8A"	Vegetation Red Edge 4
9	"B11"	SWIR 1
10	"B12"	SWIR 2
11	"B2_STD"	Standard deviation Blue
12	"B3_STD"	Standard deviation Green
13	"B4_STD"	Standard deviation Red
14	"B5_STD"	Standard deviation Red Edge 1
15	"B6_STD"	Standard deviation Red Edge 2
16	"B7_STD"	Standard deviation Red Edge 3
17	"B8_STD"	Standard deviation NIR
18	"B8A_STD"	Standard deviation Red Edge 4
19	"B11_STD"	Standard deviation SWIR 1
20	"B12_STD"	Standard deviation SWIR 2
21	"NDVI"	Normalized Difference Vegetation Index
22	"NDVI_STD"	Standard deviation Normalized Difference Vegetation Index
23	"BSI"	Bare Soil Index
24	"BSI_STD"	Standard deviation Bare Soil Index
25	"NDWI"	Normalized Difference Water Index
26	"NDWI_STD"	Standard deviation Normalized Difference Water Index
27	"NDSI"	Normalized Difference Snow Index
28	"NDSI_STD"	Standard deviation Normalized Difference Snow Index
29	"TCB"	Tasseled Cap Brightness
30	"TCB_STD"	Standard deviation Tasseled Cap Brightness
31	"TCG"	Tasseled Cap Greenness
32	"TCG_STD"	Standard deviation Tasseled Cap Greenness
33	"TCW"	Tasseled Cap Wetness
34	"TCW_STD"	Standard deviation Tasseled Cap Wetness
43	DTM	Digital Terrain Model 10 m
44	Slope	Terrain Slope
45	Aspect	Terrain aspect

The spectral indexes reported in Table 3 were calculated as follows using the S2 coefficient reported in (https://www.indexdatabase.de, last access 7 November 2022):

NDVI Normalized Difference Vegetation Index [45–49]

$$\text{NDVI} = \frac{\text{NIR} - \text{RED}}{\text{NIR} + \text{RED}}$$

BSI Bare Soil Index [50]

$$\text{BSI} = \frac{(\text{SWIR 1} + \text{RED}) - (\text{NIR} + \text{BLUE})}{(\text{SWIR 1} + \text{RED}) + (\text{NIR} + \text{BLUE})}$$

NDWI Normalized Difference Water Index [51,52]

$$\text{NDWI} = \frac{\text{NIR} - \text{SWIR 1}}{\text{NIR} + \text{SWIR 1}}$$

NDSI Normalized Difference Snow Index [53–56]

$$\text{NDSI} = \frac{\text{NIR} - \text{SWIR 1}}{\text{NIR} + \text{SWIR 1}}$$

TCB (Tasseled Cap Brightness) [57–60]

$$(BLUE * 0.3037) + (GREEN * 0.2793 + (RED * 0.4743) + (NIR * 0.5585) + (SWIR1 * 0.5082) + (SWIR2 * 0.1863)$$

$$TCG \text{ (Tasseled Cap Greenness) [57–60]}$$

$$(BLUE * -0.2848) + (GREEN * -0.243) + (RED * -0.5436) + (NIR * 0.7243) + (SWIR1 * -0.0840) + (SWIR2 * -0.1800)$$

$$TCW \text{ (Tasseled Cap Wetness) [57–60]}$$

$$((BLUE * 0.1509) + (GREEN * 0.1973) + (RED * 0.3279) + (NIR * 0.3406) + (SWIR1 * -0.7112) + (SWIR2 * -0.4572))$$

4.2. Sentinel-1

S1 SAR images were used only to map urban and water components in addition to the optics. The other classes were mapped only with optical remote sensing due to the fact that SAR distortions in mountainous areas do not permit higher accuracy land cover mapping. Therefore, the data offered by optical remote sensing are the only data in alpine environments that are truly capable of offering consistent and reliable mapping, despite being bound to atmospheric conditions. However, due to the composite in land cover, it is possible for it to be overcome.

In order to create a mask for urban areas, as first step, pairs of S1 SLC images were downloaded from the NASA ASF. In particular, to achieve interferometry with an exact repeated coverage, only images derived from the same satellite sensor in the exact acquisition mode were used (ascending or descending see Table 2). Due to the low rate of urbanization in recent years in the Aosta Valley (in terms of an increase in built-up structures), the changes in the urban footprint observed within the last couple of years can be neglected if considering the spatial resolution of the Sentinel-1 SAR sensors (deep learning was performed to refine this). Therefore, we can consider the urban footprint as a constant value for all the Sentinel-1 images acquired within a single-year time frame. The use and interpretation of SAR imagery require a series of complex pre-processing procedures, which we ran on ESA's SNAP v.8.0.0 software. Such procedures refer to the standard preprocessing commonly applied to Sentinel-1 products to derive interferometric coherence [61,62]. The interferometry was conducted only on those images pairs which had a perpendicular baseline possibly more of 130 m within the year (in the e.g., 2020) and a temporal baseline lower than 10 days. We reported the available adopted pairs from the ASF in Table 4.

Table 4. SAR S1 images pairs with the distance baseline and days.

S1 Pairs Ascending Orbit (Product n°, Baseline, Temproal Distance in Days between the Two Acqusitions)			
S1A_IW_SLC__1SDV_20200430T172327_20200430T172354_032360_03BEE8_2356	S1B_IW_SLC__1SDV_20200506T172238_20200506T172305_021464_028C15_773E	136 m	5
S1B_IW_SLC__1SDV_20200530T172240_20200530T172307_021814_029680_5539	S1A_IW_SLC__1SDV_20200605T172329_20200605T172356_032885_03CF21_34AB	152 m	7
S1A_IW_SLC__1SDV_20200804T172333_20200804T172400_033760_03E9BC_E6AD	S1B_IW_SLC__1SDV_20200810T172255_20200810T172322_022864_02B66E_1179	152 m	6
S1A_IW_SLC__1SDV_20200828T172334_20200828T172401_034110_03F5FE_8B79	S1B_IW_SLC__1SDV_20200903T172253_20200903T172320_023214_02C15A_3F08	162 m	6

Table 4. *Cont.*

S1 Pairs Ascending Orbit (Product n°, Baseline, Temproal Distance in Days between the Two Acqusitions)			
S1B_IW_SLC__1SDV_20200903T172 253_20200903T172320_023214_02C15 A_3F08	S1A_IW_SLC__1SDV_2020 0909T172335_20200909T172 402_034285_03FC20_A288	159 m	6
S1B_IW_SLC__1SDV_20201009T172 254_20201009T172321_023739_02D1 C8_57D8	S1A_IW_SLC__1SDV_2020 1015T172336_20201015T172 402_034810_040E9B_A403	134 m	6
S1B_IW_SLC__1SDV_20201114T172 240_20201114T172307_024264_02E22 F_E4D7	S1A_IW_SLC__1SDV_2020 1120T172335_20201120T172 402_035335_0420C3_E828	144 m	7
S1 Pairs Ascending orbit			
S1A_IW_SLC__1SDV_20200112T053 523_20200112T053550_030763_03871 C_D73E	S1B_IW_SLC__1SDV_20200 118T053455_20200118T0535 22_019867_02592E_ADC0	165 m	5
S1B_IW_SLC__1SDV_20200211T053 455_20200211T053522_020217_02647 9_497E	S1A_IW_SLC__1SDV_20200 217T053522_20200217T0535 48_031288_03996E_2722	155 m	7
S1A_IW_SLC__1SDV_20200324T053 522_20200324T053549_031813_03AB B5_4955	S1B_IW_SLC__1SDV_20200 330T053455_20200330T0535 22_020917_027ABA_DC4C	129 m	5
S1B_IW_SLC__1SDV_20200505T053 456_20200505T053523_021442_028B5 C_A52F	S1A_IW_SLC__1SDV_20200 511T053523_20200511T0535 50_032513_03C3F4_2251	138 m	7
S1B_IW_SLC__1SDV_20200118T053 455_20200118T053522_019867_02592 E_ADC0	S1A_IW_SLC__1SDV_2020 0124T053522_20200124T053 549_030938_038D40_8123	147 m	7

In particular, we adopted the approach described by the ESA guidelines available in [63–65] by introducing a variation in the type of classification. In this case, the maximum likelihood was not chosen. Instead, random forest and batch processing were created to involve all the selected pairs. It is worth noting that co-registration and terrain-shadow correction were performed in the ESA SNAP v.8.0.0.0 toolbox. See more detail in Figure 2.

We used both that polarizations (VH and VV) on all the SAR input data, hence the output coherence image consisted of two separate raster files related to the different polarizations.

In terms of the processing procedure, we selected only the bursts that covered our study area (the Aosta Valley autonomous region) from the original product. In addition, we computed the coherence estimation using a range window size of 10 pixels. Finally, we employed the Range–Doppler terrain correction method, which used the 10 m Aosta Valley DTM implemented in the SNAP repository, selecting ED50-UTM 32 N (EPSG: 23032) as the projected reference system, and selecting an average output resolution of 10 m. The output coherence image consisted of two different bands, reporting interferometric coherence values (from 0 to 1) for the two polarizations (VH and VV). It is worth noting that the coherence between the two SAR images expressed the similarity of the radar reflection between them. Any changes in the complex reflectivity function of the scene were manifested as a decorrelation in the phase of the appropriate pixels between the two images.

Within this type of raster, it was possible to extract the urban footprint by applying supervised (and unsupervised) classification algorithms. In this case, a supervised classification was performed starting with the training set. Since we were interested in distinguishing only two different classes, i.e., urban and non-urban areas, we aggregated all non-urban land cover types into the same class (such as glaciers, lawn pastures, needle forests. etc.). A random forest classifier was performed in the SNAP. We specified the maximum number of decision trees in the RF classifier at 500 as the optimal value to achieve good noise removal and a homogeneous response [63–65]. Following the instructions in the ESA online material [66], we applied the interferometric coherence processing methodology outlined in the previous section to a set of S1 data obtained from January to December 2020.

The classification images produced from S1 imagery consisted of a discrete raster, with all the pixels classified into either "urban" or "non-urban" values (with values of 1 and 0, respectively) and water or "non-water".

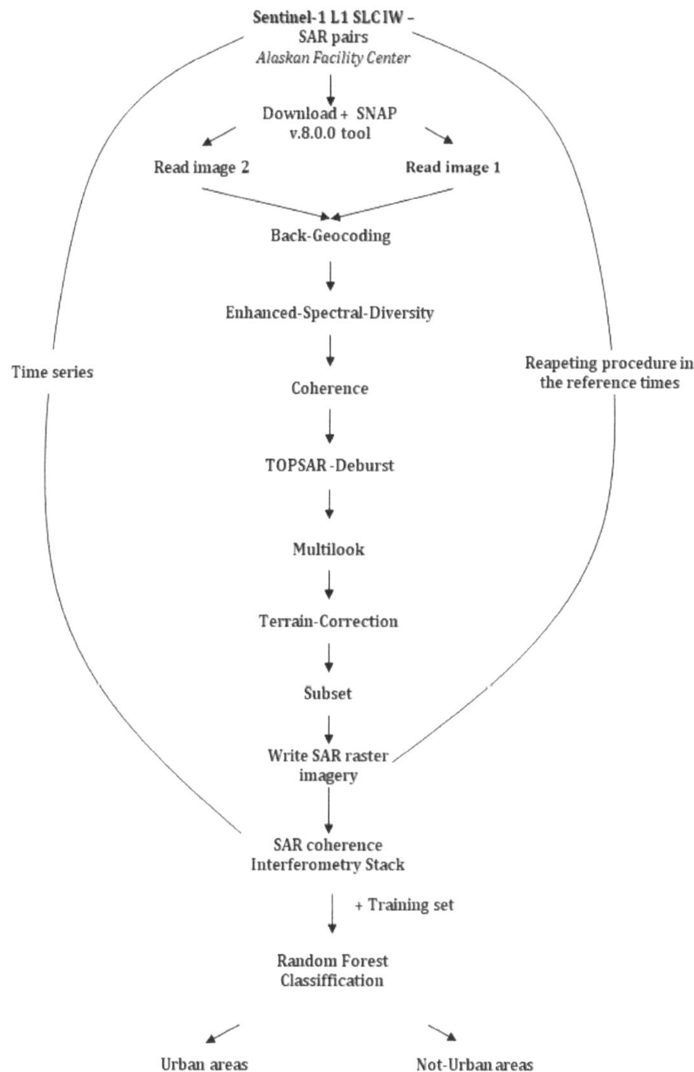

Figure 2. Interferometry and classification in the ESA SNAP tool.

4.2.1. Water Mask

The calibration process was done following the [67] approach. Additionally, the normalized difference polarization index (NDPI) and the cross ratio (CR) were calculated to examine water and humid areas.

Four S1 stacks were created defining the GRD Sigma0 dB product, considering the ascending and descending modes for the VV and VH bands used in the NDPI and CR computations (see Table 1 in the materials section). In places where SAR geometrical distortion often impacts a portion of the imagery taken in the ascending or descending mode, these stacks of bands were finally trimmed using an aspect layer recovered by the 10 m DTM VDA. The angle of view and the aspect layer were taken into consideration beginning with the ancillary and metadata files during the clipping to exclude areas affected by significant distortions in both the ascending and descending mode as described by [38].

In order to fill the gaps left in each stack by the removal of the sections that were severely impacted by the distortions, the stacks were finally mosaicked. In the case of both the distortions, we evaluated those portions with a higher incidence angle in accordance with [38,62]. SAGA GIS was used for this task. The finished stack was then uploaded into the GEE to produce an annual SAR synthetic composite in order to compute the NDPI and CR. As indicated earlier, the SAR composite was employed to map the water component more accurately.

To assess the water area components, the following SAR bands and indexes (Table 5) were adopted after a pre-processing phase and the creation of a composite to reduce the SAR distortions.

Table 5. SAR Sentinel-1 GRD bands in water mapping.

	S1 GRD	
ID	**Bands/Index**	**Description**
1	"VV	Single co-polarization, vertical transmit/vertical receive
2	"VH"	Dual-band cross-polarization, vertical transmit/horizontal receive
3	"VV_STD"	Standard deviation single co-polarization, vertical transmit/vertical receive
4	"VH_STD"	Standard deviation dual-band cross-polarization, vertical transmit/horizontal receive
5	"NDPI"	Normalized Difference Polarization Index
6	"NDPI_STD"	Standard deviation Normalized Difference Polarization Index
7	"CR"	Cross ratio
8	"CR_STD"	Standard deviation cross ratio

NDPI and CR has been calculated as follows:
NDPI Normalized Difference Polarization Index [68]

$$\text{NDPI} = \frac{\text{VH} - \text{VV}}{\text{VH} + \text{VV}}$$

CR Cross ratio [68]

$$\text{CR} = \frac{\text{VH}}{\text{VV}}$$

As demonstrated by [69] in a complex morphological context, the SAT approach was more effective than the Otsu thresholding method. Therefore, these bands were included to map surface water areas through a robust stepwise automatic thresholding (SAT) approach [69]. The SAT approach consisted of the following steps. (1) SAR data was pre-processed to create a backscattering coefficient that was georeferenced with high resolution LiDAR-derived DEM (in this case the Aosta Valley DEM with 2 m step resampled at 10 m). (2) SAT for relief displacement and de-speckle filtering was used to reduce noise in the data. (3) The conversion to dB was performed in the SNAP vers.8.0.0. In fact, the strength of the radar signal reflected from a unit area on the corresponding point in the scene determined the pixel value of the SAR image. The backscatter coefficient $\beta0$ was calibrated and employed to convert the values from the digital number to the reflectivity of the surface objects. The target's radar cross-section per surface unit with regard to the local incidence angle was parameter $\beta0$. After that, all SAR data were transformed from raw data to power units (decibels-dB). The de-speckle filter was used to eliminate the salt and pepper noise while keeping the edges and the textural structures prior to the data analysis due to the speckle effect created by the coherent radiation used by radar systems. A Speckle–Lee filter with a 5-pixel by 5-pixel window was adopted, resulting in a unique valley-hill pattern in the histogram that represented a better distinction between water and non-water surfaces. Additionally, a normalization between the incident angles were performed. In order to identify a proper threshold, a set of third-order polynomials was employed to fit the histogram in a manner of moving steps. The reason for this was that the third-order polynomial had a shape that best described the histogram of the backscattering coefficient after de-speckling and was easier to identify the turning points compared to the higher order polynomials.

Each pixel in the SAR image was identified as either land or water after the threshold was established, depending on whether its value was smaller or greater than the threshold. Through an iterative method that maximized between-class variations while simultaneously minimizing within-class variance, the threshold value was established. Finally, using the primary input pre-processed S1 GRD dataset and splitting the training set into water and non-water areas, a supervised classification (random forest) was carried out in the SNAP v.8.0.0 to improve the mapping of water areas.

4.2.2. Land Cover Legend Definition

The reference legend of the new Aosta Valley land cover and relative EO geospatial continuous service was agreed with the ISPRA and in particular the Land Remote Sensing Unit. The legend proposed perfectly reflected the new European guidelines defined by the EAGLE group. The EAGLE legend foresees more detailed levels at high resolution than those proposed in Dynamic World with a deterministic and probabilistic approach, allowing for detailed mapping of the various biomes at least at a European level. In particular, the new EAGLE legend moves away from the old Corine Land Cover which is tied to a mixture of cover and use similar to Dynamic World. In particular, given the characteristics of the mountainous areas, an expansion and more detailed definition for certain classes deemed of interest by local stakeholders was proposed. In Appendix A, the EAGLE–ISPRA legend was reported along with the agreements from the ISPRA for geomorphological complex areas such as the Aosta Valley region.

4.3. Training Set and Validation Set Definition

In order to better understand the spatial extent distribution of each class and determine the ideal number of training areas for each class in the training set, a K-means unsupervised classification with 15 classes was conducted after constructing the initial input dataset. Regarding the last criteria, there needed to be enough training pixels for each spectral class to enable accurate estimations of the components of the covariance matrix and the class conditional mean vector. The covariance matrix for an N-dimensional multispectral space was symmetric and has a size of N*N. Therefore, it required an estimation from the training data for $1/2N$ (N + 1) unique elements. It took at least N (N + 1) independent samples to keep the matrix from being singular. The good news was that each N-dimensional pixel vector contained N samples (one for each waveband). As a result, only (N + 1) independent training pixels were necessary. Since it was challenging to guarantee the independence of the pixels, more than the minimum amount was chosen. [37,70] advocate for using as many as 100 N training pixels per class, with 10 N being the lowest practical number. Therefore, a minimum of 250 polygons (containing a minimum of 5 pixels) were computed for this categorization, taking only the spectral bands and relative indices without the standard deviation (see Figure 3).

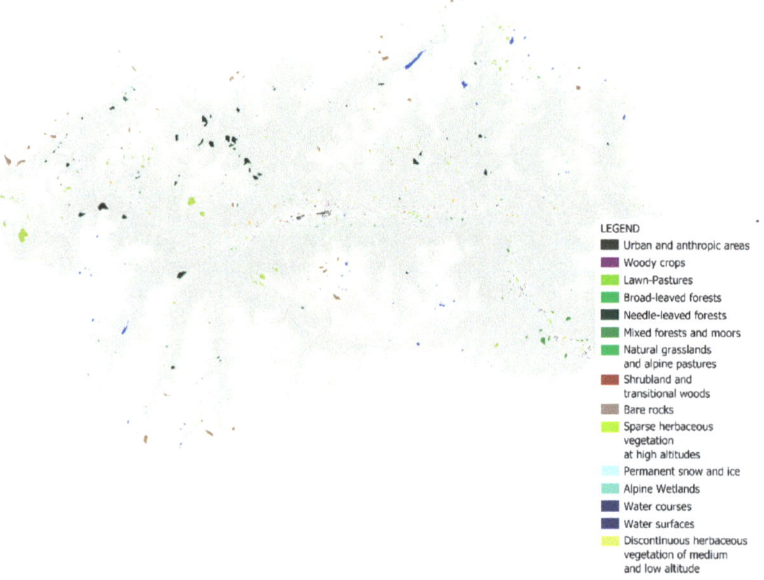

Figure 3. ROIs considering both the training and validation sets.

The object-based segmentation (OBS) approach was performed using the mean shift algorithm available on the Orfeo Toolbox software v.8.0.0 [71,72]. The OBS algorithms aimed at minimizing the spectral heterogeneity of the polygons by comparing the relative spectral properties of the neighboring pixels. The resulting segmentation vector layer (SVL) was generated according to a previously defined minimum mapping unit of 300 m^2. In particular, the segmentation was performed with reference to the S2 bands performing a bilinear resampling on those without a native 10 GSD. Then, the images were segmented based on an internally homogeneous spectral response. The segments were then vectorized to generate the corresponding vector layer. During the segmentation, the required parameters were set to the values shown in Table 6. The SEG was then used to explore the internal features other than the spectral signatures, such as the recurrent radiometric patterns (texture) and the shape. Some of these polygons were then randomly extracted and others were created by analyzing the signatures of the entire stack to define the training areas, including GTDP.

Table 6. Segmentation settings in SAGA GIS.

Segmentation Parameter	Settings
Spatial radius	3 pixels
Range radius	100 DN
Mode convergence threshold	0.1
Maximum numerous of iterations	200
Minimum region size	3 pixels

As previously mentioned, the regions of interests (ROI) per each class were defined mostly on the field and partially by applying both a segmentation and a spectral signature-photo interpretation phase. Figure 3 depicts the distribution of the ROIs in the study area. Each ROI per class had a number of polygons up to 250. An overall of 4300 ROIs were defined and 70% of them were adopted as the training set 30% as the validation set. In the developed EO local service, the ROI detection and the relative changes through time were performed by coupling a self-developed semi-automatic technique. The fact that the technique required a manual check in case of an anomaly was linked to the reason that a simple probabilistic approach such as those developed in Dynamic World by Google did not permit the mapping of the real changes that happen with a high degree of accuracy, especially in alpine areas.

In the first phase a pixel-based analysis into each of the ROIs were performed in the GEE by analyzing the variance of each of the $S2$ bands. If the median value per each band received a variance value at the time $t + 1$ up to 1.5 of its previous variances at the time t (considering the same seasons, in the example the summer meteorological season) a second phase followed. In this phase, a photo-interpretation with different EO images such as PlanetScope was performed as well as the ground detection and a change in the ROI in the necessary training polygon class. This empirical formula was developed to analyze the specific case of the Aosta Valley autonomous region. It is worth noting that the S1 data showed an intrinsic time-phase decorrelation in the case of the SLC product and geomorphological effects due to the territory in SLC and GRD products. Therefore, the radar backscatter is not recommended to be considered in this procedure regarding the entire ROIs LC components.

$$\sum_{n=\lambda_{S2}}^{1} \sigma^2_{\lambda(t0)}(\lambda_{S2}) \geq \sum_{n=\lambda_{S2}}^{1} 1.5\, \sigma^2_{\lambda(t0+1)}(\lambda_{S2})$$

where:

$\sigma_{\lambda(t0)}{}^2(\lambda_{S2})$ is the sum of the variances of each of the $S2$ bands at the time $t0$.

$\sigma_{\lambda(t0+1)}{}^2(\lambda_{S2})$ is sum of the variances of each of the $S2$ bands at the time $t0 + 1$.

4.4. Supervised Classification Algorithms

Starting with the $S2$ input dataset and the training set, the supervised classifications were performed in SAGA GIS vers. 8.0.0 and the confusion matrix was computed. Given the characteristics of the $S2$ input dataset and the analyzed alpine territory, the best performing algorithms adopted were the k-nearest neighbors classification-KMC and the minimum distance with pre-segmentation (SNIC) by applying a distance threshold of 50. The k-nearest neighbors (k-NN) is an algorithm used in pattern recognition for the classification of objects based on the characteristics of the objects in close proximity to the ones considered. It is a non-parametric classification method. In both cases, the input is the closest k training example in the feature space. The output depends on whether the k-NN is used for classification or regression. In the k-NN classification, the output is a membership in a class. An object is classified by a plurality vote of its neighbors, with the object assigned to the most common class among its k closest neighbors (k is a positive, typically small, integer). If k = 1, the object is simply assigned to the class of that single closest neighbor. In the k-NN regression, the output is the property value for the object. This value is the average of the closest neighboring k values. On the other hand, the minimum distance classifier is used to classify unknown image data to classes which minimize the distance between the image data and the class in the multi-feature space. The distance is defined as an index of the similarity so that the minimum distance is identical to the maximum similarity. Therefore, the minimum distance technique uses the mean vectors of each endmember and calculates the Euclidean distance from each unknown pixel to the mean vector for each class. All pixels are classified to the nearest class unless a standard deviation or distance threshold is specified, in which case some pixels may be unclassified if they do not meet the selected criteria. The classification was performed following a hierarchical approach described in the analysis section.

4.5. Deep Learning Using the Convolutional Neural Network (CNN)

Deep learning is a type of machine learning that relies on multiple layers of nonlinear processing for feature identification and pattern recognition, described in a model. Deep learning models can be used on different tools or by performing Python codes related to libraries such as PyTorch, Keras, TensorFlow, Onnx, Fats.ai etc. In this case, open-source libraries and Python scripts were integrated with ESRI ArcGIS Pro v.2.8 for object detection, object classification and image classification. In order to extract the building and roads, deep learning techniques using convolutional neural networks (CNNs) were adopted starting with the ortho-rectified images acquired by air flights over the Aosta Valley region. In particular, the two images regarding the AGEA (Agency for Disbursements in Agriculture) 2020 and the Aosta Valley 2018 orthophoto were used.

An inferencing process was performed to extract the roads and buildings. This phase was crucial because the information learned during the deep learning training process was put to work in detecting similar features in the datasets. ESRI ArcGIS Pro uses an external third-party framework and model definition file to run the inference geoprocessing tools. Therefore, the library and dependencies were appropriately installed. In this case, the two models provided by ESRI and edited accordingly considering the alpine areas were adopted. It is worth noting that the model definition files and (.dlpk) packages can be used multiple times as inputs for the geoprocessing tools, allowing for the assessment of multiple images over different locations and time periods using the same trained model.

The main settings adopted to perform CNN deep learning on ArcGIS Pro are reported in Table 7.

Table 7. Deep learning CNN settings in ArcGIS Pro v.2.8.

PARAMETERS	INPUT SETTINGS
Input Raster	Orthophoto.ecw
Output Detected Object	Buildings and Roads
Model Definition:	Edited models from ESRI .dplk
Padding:	32
Batch_size:	16
Threshold	0.9
Filtering threshold	99.999
Return_bboxes	False
Non-Maximum SuppressionOther parameters	CheckedDefault
ENVIRONMENTS	**INPUT SETTINGS**
Processing Extent	Raster extent
Processor Type	GPU
GPU id	Default
Cell size	Raster native GSD
Parallel processing	8

It is worth noting that filtering and threshold is normally not present in the parameter settings. In fact, to avoid deep filtering out features, in this case buildings and roads, with a surface less than 100 m square, a script was realized to include this command and perform this analysis during the building extraction phase.

5. Results and Discussion

The classification was performed following a hierarchical approach. First, a supervised k-nearest neighbor classification-KMC OpenCV considering all classes was performed. The KMC classification was carried out by normalizing the dataset due to the diversity of the input variables to make them homogeneous. The parameters adopted in the k-nearest neighbor classification-KMC (OpenCV) were a number of neighbors equal to 8, a training method classification and a type of Brute Force algorithm [41].

It is worth noting that the water and urban areas classified with SAR and urban deep learning were joined together with the urban and water classes that were mapped with the optical data to improve these classes especially in isolated mountain villages. This improved these classes by performing a semi-automatic GIS procedure. During this joining phase, a minimum mapping unit (mmu) of 100 m was considered. Therefore, only the pixels that have this mmu were mapped as urban while the other pixels that did not intersect with urban (both SAR/deep learning and optic multispectral) with less than 100 m were considered as classified by the optical data.

Since woody crops were particularly complex to discriminate (hereinafter called WC), performing only a KMC classification due to the single multispectral composite input dataset did not permit us to consider the whole phenological active season. A hierarchical classification approach was then implemented to try to overcome this issue. In particular, the developed EO service foresaw the first classification (considering the S2 main input dataset) with all the classes according to the new EAGLE land cover legend and a subsequent one with only WC class. In the end, the two classifications were subjected to a mosaicking process by first applying an overlap for WC. Then, the doubtful areas were corrected manually by photo-interpretation of composite PlanetScope imagery.

Regarding the WC class, a supervised minimum distance classification (MDC) was performed, including the following input datasets: a yearly cloud-shadow masked NDVI stack filtered (Savitzky-Golay) [73–75] and regularized at 10 days times-steps [76] on the GEE, and an annual stack of the NDRE index (normalized difference red-edge index for agriculture) following the same procedure of the NDVI stack [77]:

$$\text{NDRE} = \frac{\text{NIR} - \text{RE}}{\text{NIR} + \text{RE}}$$

NDVI composite Entropy [32,78]

$$H_{\text{NDVI}} = -\sum_{i=0}^{N-1}\sum_{j=0}^{N-1} \text{NDVI}_{i,j}\log\left(\text{NDVI}_{i,j}\right)$$

where $\text{NDVI}_{i,j}$ is the NDVI value at the i-th row and the j-th column in the local square window, measuring N pixels. For this study, a kernel window size of 10×10 pixels was adopted.

Using Rao's Q Diversity index on the S2 NDVI composite [79], Rao's Q is calculated using half the squared Euclidean distance. Therefore, the resulting index is [80]:

$$Q = \sum\sum d_{ij} * p_i * p_j$$

where p_i and p_j are the proportion of the area for each category per the rows and columns in the pairwise distance d_{ij}.

The pattern analysis of the S2 NDVI composite used the following parameters: (a) dominance, (b) diversity, (c) relative richness and (d) fragmentation [81]. Then, the KMC data were mosaicked using as first overlap onto the MDC to refine only the WC class. The same was done considering urban and water masks mapped using the S1 data. As a last step, a simple filter was performed using a radius greater than 20 m. Furthermore, in the final classification, deep learning features considered in the urban and anthropic areas were included and the confusion matrix computed.

The scalable Earth Observation service to map land cover in geomorphological complex areas beyond the Dynamic World developed for the Aosta Valley region are reported in Figure 4.

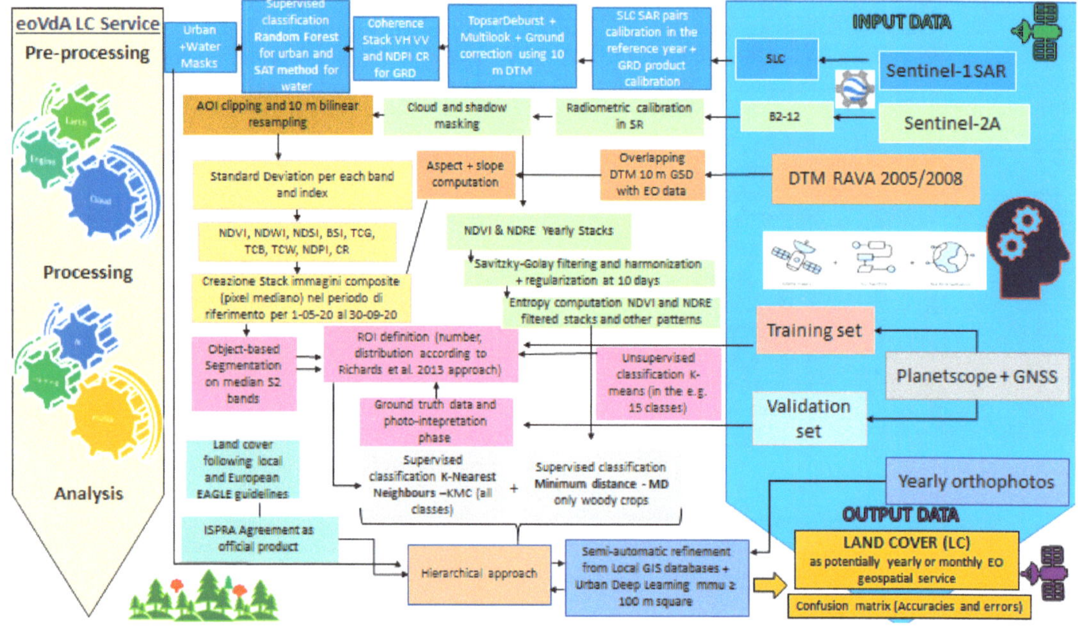

Figure 4. Workflow processing to process the scalable Earth Observation service to map land cover in geomorphological complex areas.

The parameters reported in the confusion matrix [82] were: overall accuracy, errors of commission and omission, user and producer accuracy, sum users and sum producers and unclassified pixels (in this case each pixel were classified). The sum user indicates the number of pixels for each class in each row while the sum producer represents the number of pixels for each class in each column. The overall accuracy is calculated by summing the number of the correctly classified values and dividing it by the total number of values. Finally, the kappa coefficient measures the agreement between the classification and the truth values.

It is worth noting that a comparison with traditional methods was followed to prove the real effectiveness of the suggested approach and the developed EO services. Therefore, the k-coefficients were computed per each approach.

A traditional approach that adopted only the optic multispectral data was followed by performing a unique one-shot classification using KMC. A combined approach adopted a single KMC supervised classification with the optical data, considering all the classes and the two classifications involving only urban and water with random forest and SAT, respectively, including SAR data. Finally, the mixed hierarchical approach with the two optical supervised classifications (KMC + MD), the two SAR classifications (for urban and water respectively) and deep learning was described in this work.

The hierarchical approach improved the quality of the obtained classifications, as shown in Table 8.

Table 8. Accuracies.

Approach	Overall Accuracy	K-Coefficient
Traditional approach	88%	0.88
Combined approach	89%	0.89
Mixed Hierarchical approach	97%	0.97

The developed EO service represented a valuable to map land cover at high temporal and spatial resolutions. The combined application of S1 (only for a couple of classes) and S2 EO data coupling deep learning techniques boosted the classification of land cover components in geomorphological complex areas such as the Alps. It is worth noting that the S1 data were adopted only to better map urban and water areas due to misleading classifications that may occur due to the physical limitations of SAR in mountainous areas. Moreover, the S1 processing especially related to interferometry required high performance computing machines and would not permit a rapid land cover mapping. The developed EO service was considered to be scalable to other morphological complex realities, in particular the mountainous areas. The realized EO local service with free EO data and open-source tools, except from ESRI ArcGIS (that can be replaced by QGIS and Python script for deep learning), represented a possible workflow to perform ongoing territorial planning and management. The present EO service led to an important technology transfer in the Aosta Valley territory answering various requests at different levels (European, national and local). This EO service will streamline the implementation of local policies concerning land cover monitoring and assessment. In this regard, the Aosta Valley, similar to many Italian regions, needs to assign development funds to each municipality every year, which are largely based on the distribution and extension of the land cover components within its borders. The maps developed with the present EO service can be freely downloaded in an ESRI shapefile format or be requested in raster (.tif) from the official Aosta Valley geoportale, reachable at this link (last access 11 November 2022): https://geoportale.regione.vda.it/download/carta-copertura-suolo/. The land cover developed starting with the reference year 2020 is reported in Figure 5 with its confusion matrix in Figure 6.

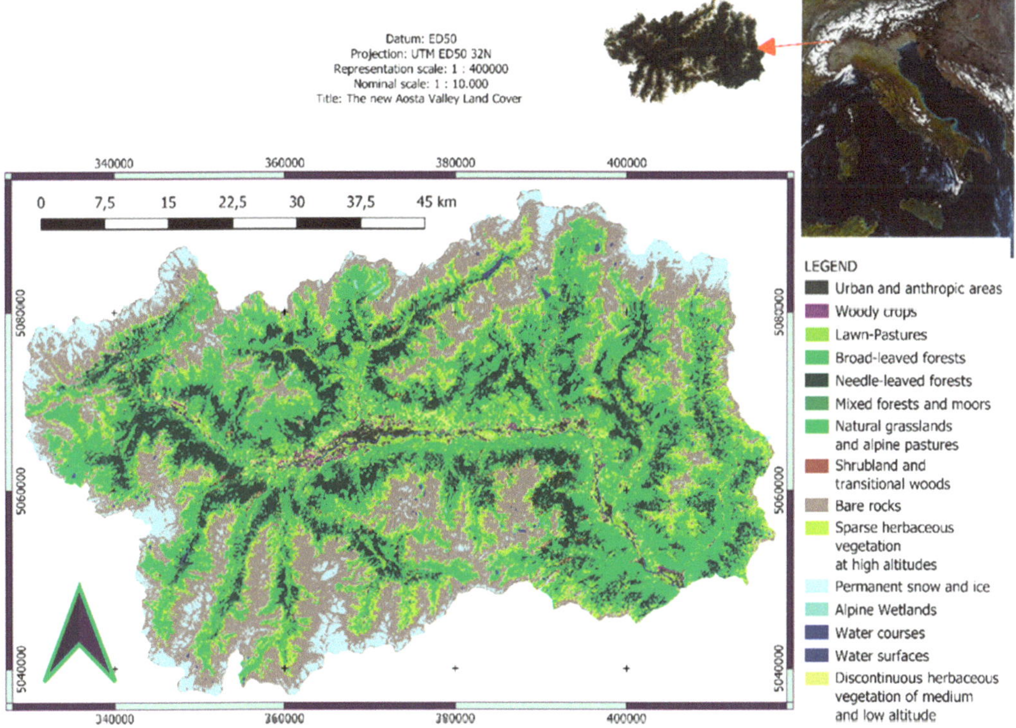

Figure 5. Aosta Valley Land Cover 2020.

Legend classes	ID	Urban and anthropic areas (111)	Permanet snow and ice (335)	Bare rocks (332)	Sparse herbaceous vegetation at high altitudes (333)	Water courses (511)	Water surfaces (512)	Discontinuous herbaceous vegetation of medium and low altitude (909)	Broad-leaved forests (311)	Needle-leaved forests (312)	Mixed forests and moors (313)	Alpine Wetlands (410)	Woody crops (221)	Shrubland and transitional woods (324)	Lawn-pastures (231)	Natural grasslands and alpine pastures (321)	User sum	User Accuracy	Omission error
Urban and anthropic areas	111	1252		195		2		1						1			1451	86.29	0.137
Permanet snow and ice	335		23866	30													23896	99.87	0.001
Bare rocks	332	82	49	9031	2										1	1	9166	98.53	0.015
Sparse herbaceous vegetation at high altitudes	333			64	1857									4		232	2157	86.09	0.139
Water courses	511	33				1496	189					11					1729	86.52	0.135
Water surfaces	512					159	1319					56					1534	85.98	0.140
Discontinuous herbaceous vegetation of medium and low altitude	909	2						2065					207		44		2318	89.09	0.109
Broad-leaved forests	311								2385	11	36		139	37			2608	91.45	0.086
Needle-leaved forests	312			180					14	24159	4			55			24412	98.96	0.010
Mixed forests and moors	313								44		982						1026	95.71	0.043
Alpine Wetlands	410					37	44					1900		24			2005	94.76	0.052
Woody crops	221					88	40	22					3128	307			3585	87.25	0.127
Shrubland and transitional woods	324												107	5483			5590	98.09	0.019
Lawn-pastures	231	6						78					26		14592	401	15103	96.62	0.034
Natural grasslands and alpine pastures	321			196	77				4	100			21	3	223	18132	18756	96.67	0.033
Producer sum		1375	23915	9696	1936	1782	1592	2166	2447	24270	1022	1967	3629	5914	14859	18766	115336		
Unclassified pixel		0	0	0	0	0	0	0	0	0	0	0	0	0	0	0		Overall Accuracy	K Coefficient
Producer Accuracy		91.05	99.80	93.14	95.92	83.95	82.85	95.34	97.47	99.54	96.09	96.59	86.19	92.71	98.20	96.62		0.97	0.97
Commission error		0.089	0.002	0.069	0.041	0.180	0.171	0.047	0.025	0.005	0.039	0.034	0.138	0.073	0.018	0.034			

Figure 6. Aosta valley land cover 2020 confusion matrix.

The final confusion matrix was reported in the caption section below.

6. Conclusions

EO services regarding land cover mapping are crucial to monitor and assess land cover changes and propose useful sustainable management and planning policies. Free Copernicus data, offered by S1 and S2 missions as well as PlanetScope may play a great role in land cover mapping. Nevertheless, the exploitation of these kinds of EO data is well known in literature. However, there is still a lack in the development of robust services to map mountainous areas (such as the Alps) with a high level of accuracy according to the newest EAGLE guidelines. In this regard, this work has successfully explored a possible scalable and repeatable service for mountainous areas that predominantly uses optical data, but also use radar data for some components, aiming to compensate native SAR acquisition mode distortions by adopting a mixed hierarchical approach to map land cover. This geospatial service based on EO data may help with the implementation of European, global and local policies concerning land cover mapping both at high spatial and temporal resolutions to assess land cover changes due to anthropic pressure and climate change and pursue a sustainable development perspective, empowering the technological transfer in mountainous realities with a higher degree of detail beyond the GEE-based Dynamic World.

Author Contributions: Conceptualization, T.O.; methodology, T.O.; software: T.O. and D.C.; validation, T.O. and D.C.; formal analysis, T.O.; investigation, T.O.; resources, T.O.; data curation, T.O.; writing—original draft preparation, T.O.; writing—review and editing, T.O. and E.B.M.; visualization, T.O. and D.C.; supervision, E.B.M.; project administration, T.O.; funding acquisition, T.O. All authors have read and agreed to the published version of the manuscript.

Funding: This research received no external funding.

Institutional Review Board Statement: Not applicable.

Informed Consent Statement: Not applicable.

Data Availability Statement: The findings may be reachable at https://geoportale.regione.vda.it/download/carta-copertura-suolo/ (last accessed 27 December 2022).

Acknowledgments: Thanks to our colleagues at INVA spa and GEO4Agri DISAFA Laboratory and Luigi Perotti, as well as Annalisa Viani for the support in performing the following work. A remarkable thanks to Edoardo Cremonese for the great feedback regarding this product and Fabrizia Joly, both of the Environmental Protection Agency of Aosta Valley, and Luca Congedo, Michele Munafò and Ines Marinosci of the ISPRA Land Unit. A huge thanks to the Regione Autonoma Valle d'Aosta, the head of the INVA spa GIS area Davide Freppaz and the head of the Regional Cartographic

Office Chantal Tréves of the Aosta Valley autonomous region for permitting the realization of this work. A final thanks to all INVA spa GIS colleagues and to Pierre Vuillermoz. Last but not least, thanks to everyone in the Aosta Valley who expressed positive or negative feedbacks about the present work that has pushed us to do better day after day and never give up.

Conflicts of Interest: The authors declare no conflict of interest.

Appendix A

A legend description is described as follows.

RAVA Legend	EAGLE—ISPRA Legend	Description
Urban and anthropic areas (111)	Artificial abiotic surfaces	Surfaces strongly influenced by anthropic activity and characterized by human settlements. These are areas in which structures are built without distinction for the intended use or are under construction, as well as roads, airports, railways, parking lots and any artifact capable of determining a permanent or semi-permanent loss of the soil resource, including caves and mines.
Shrubland and transitional woods (324)	Shrubland	Natural or natural-shaped surfaces. Areas characterized by arboreal species and generally sparse woods near grazing areas or areas with reduced herbaceous vegetation and rocks (such as rubble). These areas indicate the dynamics of the ecological forest succession following the abandonment of grazing areas and consequent expansion of forest areas or following disturbances to natural or anthropogenic disturbances to the forest.
Woody crops (221)	Not defined (considered separately vineyards and orchards)	Surfaces characterized by the presence of various cultivation systems, in particular orchards and vineyards. Surfaces influenced by human activity and agronomic practices.
Water surfaces (512)	Water	Natural or natural-shaped surfaces. Areas characterized by the presence of bodies of water such as natural lakes of fluvial and/or glacial origin, artificial reservoirs and bodies of water in wetlands.
Water courses (511)	Water	Natural or natural-shaped surfaces. Areas characterized by the presence of watercourses such as rivers, streams along runoff lines and slope impluviums.
Needle-leaved forests (312)	Needle-leaved	Natural or natural-shaped surfaces. Wooded areas characterized by a prevalent and widespread presence of coniferous trees on a given surface (larch, spruce, fir, pine, Douglas fir, etc.)
Broad-leaved forests (311)	Broad-leaved forests	Natural or natural-shaped surfaces. Wooded areas characterized by a prevalent presence of broad-leaved trees on a given surface (oak, chestnut, ash, maple, linden, alder, birch, poplars, etc.)
Mixed forests and moors (313)	Not defined	Natural or natural-shaped surfaces. Wooded areas characterized by the presence of both broad-leaved and conifers with no evident prevalence and sometimes shrubs or the presence of heather (*Erica* spp. and *Calluna vulgaris* L.).

RAVA Legend	EAGLE—ISPRA Legend	Description
Permanent snow and ice (335)	Permanent snow and ice	Natural surfaces. Areas characterized by the presence of glaciers and glaciated surfaces such as seracs, icefalls and frozen or snow-covered surfaces such as snowfields in the considered observation period. It should be noted how the measurements carried out fall within the full ablation season and can, therefore, constitute a useful data on the perimeter in this sense. The rock glaciers entirely covered by debris and rocks are not included in this class, preferring to follow a criterion of spectral uniformity based on the typical characteristics of remote sensing with s1-s2 data in the context of the Copernicus Programme that are capable of investigating the surfaces, not the subsoil, as indicated in international scientific literature regarding both optical and sar remote sensing data of these missions. Therefore, we refer to the rock class.
Natural grasslands and alpine pastures (321)	Defined as generic pastures	Natural or natural-shaped surfaces. Areas characterized by natural evolution or by pastoral management conditioning practices. These areas are characterized by the presence of medium-high altitude herbaceous species.
Lawn pastures (231)	Defined as generic pastures	Natural-shaped surfaces. Areas characterized by herbaceous cover conditioned by pastoral and agronomic practices in this case mowing, haymaking and eventual irrigation. The areas can be characterized by both grazing and mowing.
Bare rocks (332)	Consolidated surfaces	Natural surfaces. Areas characterized by the presence of outcropping rocks and coherent non-vegetated soils.
Discontinuous herbaceous vegetation of medium-low altitude (909)	Not defined (only an unconsolidated class is present in a non-vegetated macro-class)	Natural or natural-shaped surfaces. Areas characterized by unconsolidated soils with continuous vegetation cover over time as they have reduced annual vegetation, xeric sparse vegetation or poorly managed grassing with little or no agronomic conditioning practices. This coverage also includes rock jumps provided with vegetation spots with occasional but not very powerful soils and extremely limited or absent vegetation.
Sparse herbaceous vegetation at high altitudes (333)	Herbaceous vegetation permanenet	Natural surfaces. Areas characterized by the presence of scarce but permanent vegetation that is difficult to graze given both the characteristics of the vegetation and, in some cases, the slope. These are high-altitude surfaces near rocks or natural grasslands and woods.
Alpine wetlands (410)	Defined as generic wetlands	Natural surfaces. Areas characterized by the presence of wetlands at different altitudes such as swamps, peat bogs and vegetation typical of these areas. Only the stretches of water in correspondence with these areas return to the water bodies.

References

1. Gorelick, N.; Hancher, M.; Dixon, M.; Ilyushchenko, S.; Thau, D.; Moore, R. Google Earth Engine: Planetary-Scale Geospatial Analysis for Everyone. *Remote Sens. Environ.* **2017**, *202*, 18–27. [CrossRef]
2. Lukacz, P.M. Data Capitalism, Microsoft's Planetary Computer, and the Biodiversity Informatics Community. In Proceedings of the International Conference on Information, Virtual Event, 28 February–4 March 2022; pp. 355–369.
3. Mutanga, O.; Kumar, L. Google Earth Engine Applications. *Remote Sens.* **2019**, *11*, 591. [CrossRef]
4. Highfill, T.C.; MacDonald, A.C. Estimating the United States Space Economy Using Input-Output Frameworks. *Space Policy* **2022**, *60*, 101474. [CrossRef]
5. Miraux, L. Environmental Limits to the Space Sector's Growth. *Sci. Total Environ.* **2022**, *806*, 150862. [CrossRef]

6. Andreatta, D.; Gianelle, D.; Scotton, M.; Vescovo, L.; Dalponte, M. Detection of Grassland Mowing Frequency Using Time Series of Vegetation Indices from Sentinel-2 Imagery. *GISci. Remote Sens.* **2022**, *59*, 481–500. [CrossRef]
7. Orusa, T.; Borgogno Mondino, E. Exploring Short-Term Climate Change Effects on Rangelands and Broad-Leaved Forests by Free Satellite Data in Aosta Valley (Northwest Italy). *Climate* **2021**, *9*, 47. [CrossRef]
8. Orusa, T.; Orusa, R.; Viani, A.; Carella, E.; Borgogno Mondino, E. Geomatics and EO Data to Support Wildlife Diseases Assessment at Landscape Level: A Pilot Experience to Map Infectious Keratoconjunctivitis in Chamois and Phenological Trends in Aosta Valley (NW Italy). *Remote Sens.* **2020**, *12*, 3542. [CrossRef]
9. Feranec, J.; Soukup, T.; Hazeu, G.; Jaffrain, G. *European Landscape Dynamics: CORINE Land Cover Data;* CRC Press: Boca Raton, FL, USA, 2016.
10. Panagos, P.; Jones, A.; Van Liedekerke, M.; Orgiazzi, A.; Lugato, E.; Montanarella, L. *JRC Support to the European Joint Programme for Soil (EJP SOIL);* Technical Report by the Joint Research Centre (JRC), EUR 30450 EN, JRC122248; JRC: Ispra, Italy, 2020.
11. Žlebir, S. Copernicus Earth Observation Programme. *40th COSPAR Sci. Assem.* **2014**, *40*, A0–A1.
12. De Fioravante, P.; Luti, T.; Cavalli, A.; Giuliani, C.; Dichicco, P.; Marchetti, M.; Chirici, G.; Congedo, L.; Munafò, M. Multispectral Sentinel-2 and SAR Sentinel-1 Integration for Automatic Land Cover Classification. *Land* **2021**, *10*, 611. [CrossRef]
13. Anderson, K.; Ryan, B.; Sonntag, W.; Kavvada, A.; Friedl, L. Earth Observation in Service of the 2030 Agenda for Sustainable Development. *Geo-Spat. Inf. Sci.* **2017**, *20*, 77–96. [CrossRef]
14. Wunder, S.; Kaphengst, T.; Frelih-Larsen, A. Implementing Land Degradation Neutrality (SDG 15.3) at National Level: General Approach, Indicator Selection and Experiences from Germany. In *International Yearbook of Soil Law and Policy 2017;* Springer: Berlin/Heidelberg, Germany, 2018; pp. 191–219.
15. Arnold, S.; Kosztra, B.; Banko, G.; Milenov, P.; Smith, G.; Hazeu, G. *Explanatory Content Documentation of the EAGLE Concept 2021;* Version 3.1; EEA: Copenhagen, Denmark, 2021.
16. Gómez, C.; White, J.C.; Wulder, M.A. Optical Remotely Sensed Time Series Data for Land Cover Classification: A Review. *ISPRS J. Photogramm. Remote Sens.* **2016**, *116*, 55–72. [CrossRef]
17. Zhu, Z.; Gallant, A.L.; Woodcock, C.E.; Pengra, B.; Olofsson, P.; Loveland, T.R.; Jin, S.; Dahal, D.; Yang, L.; Auch, R.F. Optimizing Selection of Training and Auxiliary Data for Operational Land Cover Classification for the LCMAP Initiative. *ISPRS J. Photogramm. Remote Sens.* **2016**, *122*, 206–221. [CrossRef]
18. Holloway, J.; Helmstedt, K.J.; Mengersen, K.; Schmidt, M. A Decision Tree Approach for Spatially Interpolating Missing Land Cover Data and Classifying Satellite Images. *Remote Sens.* **2019**, *11*, 1796. [CrossRef]
19. Aune-Lundberg, L.; Strand, G.-H. The Content and Accuracy of the CORINE Land Cover Dataset for Norway. *Int. J. Appl. Earth Obs. Geoinf.* **2021**, *96*, 102266. [CrossRef]
20. Waser, L.T.; Schwarz, M. Comparison of Large-Area Land Cover Products with National Forest Inventories and CORINE Land Cover in the European Alps. *Int. J. Appl. Earth Obs. Geoinf.* **2006**, *8*, 196–207. [CrossRef]
21. De Fioravante, P.; Strollo, A.; Assennato, F.; Marinosci, I.; Congedo, L.; Munafò, M. High Resolution Land Cover Integrating Copernicus Products: A 2012–2020 Map of Italy. *Land* **2021**, *11*, 35. [CrossRef]
22. Congedo, L.; Sallustio, L.; Munafò, M.; Ottaviano, M.; Tonti, D.; Marchetti, M. Copernicus High-Resolution Layers for Land Cover Classification in Italy. *J. Maps* **2016**, *12*, 1195–1205. [CrossRef]
23. Karra, K.; Kontgis, C.; Statman-Weil, Z.; Mazzariello, J.C.; Mathis, M.; Brumby, S.P. Global Land Use/Land Cover with Sentinel 2 and Deep Learning. In Proceedings of the 2021 IEEE International Geoscience and Remote Sensing Symposium IGARSS, Brussels, Belgium, 11–16 July 2021; pp. 4704–4707.
24. Comber, A.; Fisher, P.; Wadsworth, R. What Is Land Cover? *Environ. Plan. B Plan. Des.* **2005**, *32*, 199–209. [CrossRef]
25. Comber, A.J.; Wadsworth, R.; Fisher, P. Using Semantics to Clarify the Conceptual Confusion between Land Cover and Land Use: The Example of "Forest". *J. Land Use Sci.* **2008**, *3*, 185–198. [CrossRef]
26. Vizzari, M. PlanetScope, Sentinel-2, and Sentinel-1 Data Integration for Object-Based Land Cover Classification in Google Earth Engine. *Remote Sens.* **2022**, *14*, 2628. [CrossRef]
27. Velastegui-Montoya, A.; Rivera-Torres, H.; Herrera-Matamoros, V.; Sadeck, L.; Quevedo, R.P. Application of Google Earth Engine for Land Cover Classification in Yasuni National Park, Ecuador. In Proceedings of the IGARSS 2022—2022 IEEE International Geoscience and Remote Sensing Symposium, Kuala Lumpur, Malaysia, 17–22 July 2022; pp. 6376–6379.
28. Huang, K.; Yang, G.; Yuan, Y.; Sun, W.; Meng, X.; Ge, Y. Optical and SAR Images Combined Mangrove Index Based on Multi-Feature Fusion. *Sci. Remote Sens.* **2022**, *5*, 100040. [CrossRef]
29. Meng, X.; Liu, Q.; Shao, F.; Li, S. Spatio–Temporal–Spectral Collaborative Learning for Spatio–Temporal Fusion with Land Cover Changes. *IEEE Trans. Geosci. Remote Sens.* **2022**, *60*, 5704116. [CrossRef]
30. Büttner, G.; Feranec, J.; Jaffrain, G.; Mari, L.; Maucha, G.; Soukup, T. The CORINE Land Cover 2000 Project. *EARSeL eProceedings* **2004**, *3*, 331–346.
31. Büttner, G. CORINE Land Cover and Land Cover Change Products. In *Land Use and Land Cover Mapping in Europe;* Springer: Berlin/Heidelberg, Germany, 2014; pp. 55–74.
32. Carella, E.; Orusa, T.; Viani, A.; Meloni, D.; Borgogno-Mondino, E.; Orusa, R. An Integrated, Tentative Remote-Sensing Approach Based on NDVI Entropy to Model Canine Distemper Virus in Wildlife and to Prompt Science-Based Management Policies. *Animals* **2022**, *12*, 1049. [CrossRef]

33. Orusa, T.; Mondino, E.B. Landsat 8 Thermal Data to Support Urban Management and Planning in the Climate Change Era: A Case Study in Torino Area, NW Italy. In Proceedings of the Remote Sensing Technologies and Applications in Urban Environments IV, International Society for Optics and Photonics, Strasbourg, France, 9–10 September 2019; Volume 11157.

34. Weiß, T.; Fincke, T. SenSARP: A Pipeline to Pre-Process Sentinel-1 SLC Data by Using ESA SNAP Sentinel-1 Toolbox. *J. Open Source Softw.* **2022**, *7*, 3337. [CrossRef]

35. Amani, M.; Ghorbanian, A.; Ahmadi, S.A.; Kakooei, M.; Moghimi, A.; Mirmazloumi, S.M.; Moghaddam, S.H.A.; Mahdavi, S.; Ghahremanloo, M.; Parsian, S.; et al. Google Earth Engine Cloud Computing Platform for Remote Sensing Big Data Applications: A Comprehensive Review. *IEEE J. Sel. Top. Appl. Earth Obs. Remote Sens.* **2020**, *13*, 5326–5350. [CrossRef]

36. Braun, A. Retrieval of Digital Elevation Models from Sentinel-1 Radar Data–Open Applications, Techniques, and Limitations. *Open Geosci.* **2021**, *13*, 532–569. [CrossRef]

37. Richards, J.A. *Remote Sensing with Imaging Radar*; Springer: Berlin/Heidelberg, Germany, 2009; Volume 1.

38. Samuele, D.P.; Filippo, S.; Orusa, T.; Enrico, B.-M. Mapping SAR Geometric Distortions and Their Stability along Time: A New Tool in Google Earth Engine Based on Sentinel-1 Image Time Series. *Int. J. Remote Sens.* **2021**, *42*, 9126–9145. [CrossRef]

39. Grizonnet, M.; Michel, J.; Poughon, V.; Inglada, J.; Savinaud, M.; Cresson, R. Orfeo ToolBox: Open Source Processing of Remote Sensing Images. *Open Geospat. Data Softw. Stand.* **2017**, *2*, 15. [CrossRef]

40. Inglada, J.; Christophe, E. The Orfeo Toolbox Remote Sensing Image Processing Software. In Proceedings of the 2009 IEEE International Geoscience and Remote Sensing Symposium, Cape Town, South Africa, 12–17 July 2009; Volume 4, p. IV-733.

41. Conrad, O.; Bechtel, B.; Bock, M.; Dietrich, H.; Fischer, E.; Gerlitz, L.; Wehberg, J.; Wichmann, V.; Böhner, J. System for Automated Geoscientific Analyses (SAGA) v. 2.1. 4. *Geosci. Model Dev.* **2015**, *8*, 1991–2007. [CrossRef]

42. QGIS Development Team. QGIS Development Team. QGIS Geographic Information System. In *A Free and Open Source Geographic Information System*; QGIS: Rome, Italy, 2018.

43. Neteler, M.; Bowman, M.H.; Landa, M.; Metz, M. GRASS GIS: A Multi-Purpose Open Source GIS. *Environ. Model. Softw.* **2012**, *31*, 124–130. [CrossRef]

44. Neteler, M.; Mitasova, H. *Open Source GIS: A GRASS GIS Approach*; Springer Science & Business Media: Cham, Switzerland, 2013; Volume 689.

45. Deering, D.W. *Rangeland Reflectance Characteristics Measured by Aircraft and Spacecraftsensors*; Texas A&M University: College Station, TX, USA, 1978.

46. Deering, D. Measuring" Forage Production" of Grazing Units from Landsat MSS Data. In Proceedings of the Tenth International Symposium of Remote Sensing of the Envrionment, Ann Arbor, MI, USA, 6–10 October 1975; pp. 1169–1198.

47. Rouse, J., Jr.; Haas, R.H.; Deering, D.; Schell, J.; Harlan, J.C. *Monitoring the Vernal Advancement and Retrogradation (Green Wave Effect) of Natural Vegetation*; NASA: College Station, TX, USA, 1974.

48. Rouse, J.; Haas, R.; Schell, J.; Deering, D. NASA SP-351. Monitoring vegetation systems in the Great Plains with ERTS. In Proceedings of the Third ERTS (Earth Resources Technology Satellite) Symposium, Washington, DC, USA, 10–14 December 1973; Volume 1, pp. 309–317.

49. Tucker, C.J. Red and Photographic Infrared Linear Combinations for Monitoring Vegetation. *Remote Sens. Environ.* **1979**, *8*, 127–150. [CrossRef]

50. Liu, H.; Zhan, Q.; Yang, C.; Wang, J. The Multi-Timescale Temporal Patterns and Dynamics of Land Surface Temperature Using Ensemble Empirical Mode Decomposition. *Sci. Total Environ.* **2019**, *652*, 243–255. [CrossRef] [PubMed]

51. McFeeters, S.K. The Use of the Normalized Difference Water Index (NDWI) in the Delineation of Open Water Features. *Int. J. Remote Sens.* **1996**, *17*, 1425–1432. [CrossRef]

52. Xu, H. Modification of Normalised Difference Water Index (NDWI) to Enhance Open Water Features in Remotely Sensed Imagery. *Int. J. Remote Sens.* **2006**, *27*, 3025–3033. [CrossRef]

53. Valovcin, F. *Snow/Cloud Discrimination*; Air Force Geophysics Laboratory Rep; AFGL-TR-76-0174/ADA 032385; AFRL: Wright-Patterson Air Force Base, OH, USA, 1976.

54. Valovcin, F.R. *Spectral Radiance of Snow and Clouds in the near Infrared Spectral Region*; AFRL: Wright-Patterson Air Force Base, OH, USA, 1978; Volume 78.

55. Kyle, H.; Curran, R.; Barnes, W.; Escoe, D. A Cloud Physics Radiometer. In Proceedings of the 3rd Conference on Atmospheric Radiation, Berkeley, CA, USA, 28–30 June 1978; pp. 107–109.

56. Bunting, J.T. *Improved Cloud Detection Utilizing Defense Meteorological Satellite Program near Infrared Measurements*; AFRL: Wright-Patterson Air Force Base, OH, USA, 1982.

57. Jensen, J. *Introductory Digital Image Processing—A Remote Sensing Perspective*; New Jersey Prentice Hall: Englewood Cliffs, NJ, USA, 1986; 379p.

58. Crist, E.P.; Cicone, R.C. Others Application of the Tasseled Cap Concept to Simulated Thematic Mapper Data. *Photogramm. Eng. Remote Sens.* **1984**, *50*, 343–352.

59. Kauth, R.; Lambeck, P.; Richardson, W.; Thomas, G.; Pentland, A. Feature Extraction Applied to Agricultural Crops as Seen by Landsat. *NASA. Johns. Space Cent. Proc. Tech. Sess.* **1979**, *1–2*, 705–721.

60. Huang, C.; Wylie, B.; Yang, L.; Homer, C.; Zylstra, G. Derivation of a Tasselled Cap Transformation Based on Landsat 7 At-Satellite Reflectance. *Int. J. Remote Sens.* **2002**, *23*, 1741–1748. [CrossRef]

61. Kellndorfer, J.; Cartus, O.; Lavalle, M.; Magnard, C.; Milillo, P.; Oveisgharan, S.; Osmanoglu, B.; Rosen, P.A.; Wegmüller, U. Global Seasonal Sentinel-1 Interferometric Coherence and Backscatter Data Set. *Sci. Data* **2022**, *9*, 73. [CrossRef]
62. Ohki, M.; Abe, T.; Tadono, T.; Shimada, M. Landslide Detection in Mountainous Forest Areas Using Polarimetry and Interferometric Coherence. *Earth Planets Space* **2020**, *72*, 67. [CrossRef]
63. Sica, F.; Pulella, A.; Nannini, M.; Pinheiro, M.; Rizzoli, P. Repeat-Pass SAR Interferometry for Land Cover Classification: A Methodology Using Sentinel-1 Short-Time-Series. *Remote Sens. Environ.* **2019**, *232*, 111277. [CrossRef]
64. Veci, L. *Sentinel-1 Toolbox—TOPS Interferometry Tutorial*; European Space Agency; Array Systems Computing Inc.: Toronto, ON, Canada, 2015.
65. Veci, L.; Interferometry Tutorial. Array Systems. 2015. Available online: http://sentinel1.s3.amazonaws.com/docs/S1TBX%20 Stripmap%20Interferometry%20with%20Sentinel-1%20Tutorial.pdf (accessed on 12 August 2017).
66. Semenzato, A.; Pappalardo, S.E.; Codato, D.; Trivelloni, U.; De Zorzi, S.; Ferrari, S.; De Marchi, M.; Massironi, M. Mapping and Monitoring Urban Environment through Sentinel-1 SAR Data: A Case Study in the Veneto Region (Italy). *ISPRS Int. J. Geo-Inf.* **2020**, *9*, 375. [CrossRef]
67. Filipponi, F. Sentinel-1 GRD Preprocessing Workflow. In Proceedings of the Multidisciplinary Digital Publishing Institute Proceedings, Online Event, 22 May–5 June 2019; Volume 18, p. 11.
68. Knott, E.F.; Schaeffer, J.F.; Tulley, M.T. *Radar Cross Section*; SciTech Publishing: Sunnyvale, CA, USA, 2004.
69. Zhang, W.; Hu, B.; Brown, G.S. Automatic Surface Water Mapping Using Polarimetric SAR Data for Long-Term Change Detection. *Water* **2020**, *12*, 872. [CrossRef]
70. Davis, S.M.; Swain, P.H. *Remote Sensing: The Quantitative Approach*; McGraw-Hill International Book Company: New York, NY, USA, 1978.
71. Hossain, M.D.; Chen, D. Segmentation for Object-Based Image Analysis (OBIA): A Review of Algorithms and Challenges from Remote Sensing Perspective. *ISPRS J. Photogramm. Remote Sens.* **2019**, *150*, 115–134.
72. Stromann, O.; Nascetti, A.; Yousif, O.; Ban, Y. Dimensionality Reduction and Feature Selection for Object-Based Land Cover Classification Based on Sentinel-1 and Sentinel-2 Time Series Using Google Earth Engine. *Remote Sens.* **2020**, *12*, 76. [CrossRef]
73. Press, W.H.; Teukolsky, S.A. Savitzky-Golay Smoothing Filters. *Comput. Phys.* **1990**, *4*, 669–672. [CrossRef]
74. Chen, J.; Jönsson, P.; Tamura, M.; Gu, Z.; Matsushita, B.; Eklundh, L. A Simple Method for Reconstructing a High-Quality NDVI Time-Series Data Set Based on the Savitzky–Golay Filter. *Remote Sens. Environ.* **2004**, *91*, 332–344. [CrossRef]
75. Schafer, R.W. What Is a Savitzky-Golay Filter? *IEEE Signal Process. Mag.* **2011**, *28*, 111–117. [CrossRef]
76. Nguyen, M.D.; Baez-Villanueva, O.M.; Bui, D.D.; Nguyen, P.T.; Ribbe, L. Harmonization of Landsat and Sentinel 2 for Crop Monitoring in Drought Prone Areas: Case Studies of Ninh Thuan (Vietnam) and Bekaa (Lebanon). *Remote Sens.* **2020**, *12*, 281. [CrossRef]
77. Barnes, E.; Clarke, T.; Richards, S.; Colaizzi, P.; Haberland, J.; Kostrzewski, M.; Waller, P.; Choi, C.; Riley, E.; Thompson, T.; et al. Coincident Detection of Crop Water Stress, Nitrogen Status and Canopy Density Using Ground Based Multispectral Data. In Proceedings of the Fifth International Conference on Precision Agriculture, Bloomington, MN, USA, 16–19 July 2000; Volume 1619.
78. De Marinis, P.; De Petris, S.; Sarvia, F.; Manfron, G.; Momo, E.J.; Orusa, T.; Corvino, G.; Sali, G.; Borgogno, E.M. Supporting Pro-Poor Reforms of Agricultural Systems in Eastern DRC (Africa) with Remotely Sensed Data: A Possible Contribution of Spatial Entropy to Interpret Land Management Practices. *Land* **2021**, *10*, 1368. [CrossRef]
79. Rocchini, D.; Marcantonio, M.; Ricotta, C. Measuring Rao's Q Diversity Index from Remote Sensing: An Open Source Solution. *Ecol. Indic.* **2017**, *72*, 234–238. [CrossRef]
80. Pavoine, S. Clarifying and Developing Analyses of Biodiversity: Towards a Generalisation of Current Approaches. *Methods Ecol. Evol.* **2012**, *3*, 509–518. [CrossRef]
81. Borgogno-Mondino, E.; Farbo, A.; Novello, V.; Palma, L. de A Fast Regression-Based Approach to Map Water Status of Pomegranate Orchards with Sentinel 2 Data. *Horticulturae* **2022**, *8*, 759. [CrossRef]
82. Foody, G.M. Status of Land Cover Classification Accuracy Assessment. *Remote Sens. Environ.* **2002**, *80*, 185–201. [CrossRef]

Article

The Geomorphological and Geological Structure of the Samaria Gorge, Crete, Greece—Geological Models Comprehensive Review and the Link with the Geomorphological Evolution

Emmanouil Manoutsoglou, Ilias Lazos *, Emmanouil Steiakakis and Antonios Vafeidis

School of Mineral Resources Engineering, Technical University of Crete, 731 00 Chania, Greece
* Correspondence: ilazos@tuc.gr; Tel.: +30-2821-037652

Abstract: The Samaria Gorge is a dominant geomorphological and geological structure on Crete Island and it is one of the national parks established in Greece. Due to the complex tectonics and the stratigraphic ambiguities imprinted in the geological formations of the area, a comprehensive review of the geological models referring to the geological evolution of the area is essential in order to clarify its geomorphological evolution. In particular, the study area is geologically structured by the Gigilos formation, the Plattenkalk series and the Trypali unit. Regarding lithology, the Gigilos formation predominantly includes phyllites and slates, while the Plattenkalk series and the Trypali unit are mainly structured by metacarbonate rocks; the Plattenkalk series metacarbonate rocks include cherts, while the corresponding ones of the Trypali unit do not. Furthermore, the wider region was subjected to compressional tectonics, resulting in folding occurrences and intense faulting, accompanied by high dip angles of the formations, causing similar differentiations in the relief. Significant lithological differentiations are documented among them, which are further analyzed in relation to stratigraphy, the tectonics, and the erosion rate that changes, due to differentiations of the lithological composition. In addition, the existing hydrological conditions are decisive for further geomorphological evolution.

Keywords: geomorphological and geological features; Samaria Gorge; White Mountains; Crete; Greece

Citation: Manoutsoglou, E.; Lazos, I.; Steiakakis, E.; Vafeidis, A. The Geomorphological and Geological Structure of the Samaria Gorge, Crete, Greece—Geological Models Comprehensive Review and the Link with the Geomorphological Evolution. *Appl. Sci.* **2022**, *12*, 10670. https://doi.org/10.3390/app122010670

Academic Editors: Giovanni Randazzo, Stefania Lanza and Anselme Muzirafuti

Received: 20 September 2022
Accepted: 19 October 2022
Published: 21 October 2022

Publisher's Note: MDPI stays neutral with regard to jurisdictional claims in published maps and institutional affiliations.

1. Introduction

The western part of Crete is predominantly structured by Plattenkalk Group geological formations, which belong to the Hellenides foreland [1]. However, the stratigraphically lowest formations of this group are additionally documented in Central Crete, where the bed inversion occurred [2]. This complicated geological structure, which significantly affects geomorphological evolution, resulted in proposing three different geological models for the wider area [3–7], while the need for implementing cutting-edge techniques was generated, in order to display them accurately; this is achieved by implementing three-dimensional (3D) geological modeling [8,9].

Three-dimensional geological modeling was initiated in the early 60s as a tool for improving mining excavation. In particular, the 3D fixed block model was an initial application, performed for tectonically deformed stratified deposit excavation [10,11], as well as the gridded seam model, which was implemented on tectonically undisturbed ore deposits [12]. Regarding geological simulation, Houlding [13] proposed boundary representation, which is based on geological mapping data recorded during fieldwork; these are imprinted in an artificial environment and subjected to geometrical rules, while representative geological cross-sections and preexisting legends of geological formations are necessary for 3D geological composition.

Considering both the proposed geological models and the 3D geological modeling principles, the objective of this paper is to review the proposed geological models of the Samaria area, in order to link the geological regime, geological formation properties (lithology, tectonics and stratigraphy) and the drainage network to the geomorphological evolution.

2. Geology of Crete Island

Crete Island is located just over the active subduction zone of the African plate beneath the Eurasian one ([14] and references therein), resulting in the gradual uplift of the entirety of Crete [15]), characterized by complicated active tectonics. The Alpine tectonostratigraphic regime indicates complexity as the various units and lithological formations covering the island are controlled by compressional tectonics, such as folding, thrust sheets etc. [16]. In particular, the major tectonic units of Crete (stratigraphically from the lower to the upper one) are the following [17–23]: (i) Plattenkalk Group (parautochthonous), (ii) Phyllite Nappe, (iii) Tripolis Nappe, (iv) Pindos Nappe and (v) Uppermost Nappe (Figure 1).

Various lithological formations, composing these nappes, tectonically overlie different parts of the parautochthonous formations of the Plattenkalk Group, which is predominantly structured by HP/LT-affected metamorphic rocks [18,24]. Furthermore, the Ravdoucha (slate-carbonate) beds tectonically overlie the dynamometamorphic sequence in different sites of Western Crete, constituting the lower part of the Tripolis unit, whilst the upper part includes the carbonate sequence and flysch (Figure 1). The Tripolis unit is approximately aged between the Middle and Upper Triassic, while the corresponding age of the overlying Pindos (or Olonos-Pindos) unit ranges between the Upper Triassic and the Middle Palaeocene [25–29].

It is worth mentioning that the Tripolis and Pindos geotectonic zones, which are part of the External Hellenides, tectonically underlie the allochthonous Internal Hellenides nappes, and they are characterized as the "Uppermost unit"; this unit constitutes complicated lithofacies, a tectonic complex consisting of a nappes pile [2].

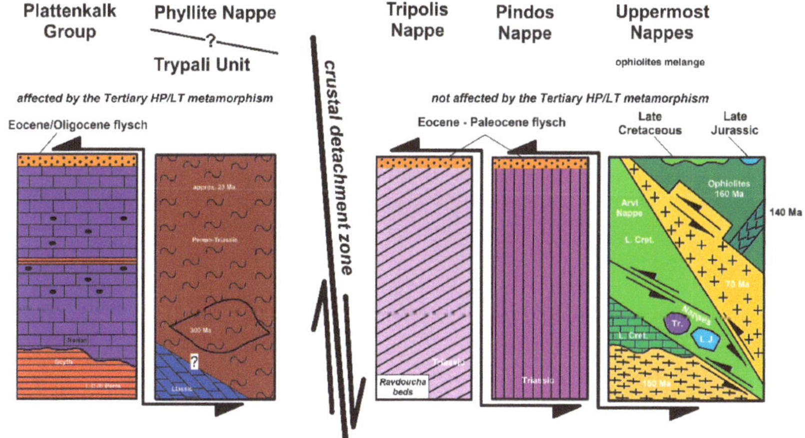

Figure 1. Tectonic emplacement of the tectonostratigraphic units and nappes of Crete (modified from Seidel et al. [23]).

In particular, these different nappes are as follows (from the uppermost to the lowermost): (i) ophiolites nappe (serpentinized peridotites, gabbros, diorites, dolerites and diabases) of Lower Jurassic–Upper Cretaceous age. The vast majority of ophiolites are exposed at the central part of Crete and they are placed at the top of the Uppermost unit [30], extending in areas that vary from a few hundred square meters to several square kilometers, (ii) Asterousia nappe (meta-silt, mica, chlorite, epidotite gneisses and schists, amphibolites and marbles) of Lower Jurassic–Upper Cretaceous age, (iii) Vatos nappe (dark siltstones, limestone beds and sandstones alternations) of Upper Jurassic age, (iv) Arvi nappe (basalts and pillow lavas) of Upper Cretaceous age. Eventually, the Alpine formations which extend in different Crete regions underlie the Neogene and Quaternary units of various thicknesses; these formations are associated with the post-orogenetic processes that occurred in the area. The uppermost Quaternary formations, affecting the geomorphological evolution of the area, unconformably overlie the Neogene formations or the pre-Neogene bedrock [31,32]. They consist of loose or compact terrestrial formations (Pleistocene or more recent), documented as colluvial deposits or alluvial fans.

The Samaria Gorge is located in the White Mountains mass in the southwestern part of Crete (Figure 2), being one of the National Parks of Greece (established in 1962). It is considered as one of the most significant geomorphological structures in the Mediterranean region, and it extends in an N–S direction for approximately 13 km. The global recognition of the Samaria Gorge is documented by numerous international distinctions: (1) UNESCO Man and the Biosphere Reserve, (2) European Diploma of Protected Areas, awarded by the Council of Europe, (3) European Biogenetic Reserve of the Council of Europe, (4) Important Bird Areas by the Birdlife International and (5) NATURA 2000 protected area, under code GR4340014 (Zone of Special Protection: ZSP), and it is considered as one of the most significant geotopes of Greece. Furthermore, the wider area of the White Mountains belongs to the Natura 2000 European Network of Protected Areas, under code GR4340008H, and it is certified as a Place of Universal Importance (PUI).

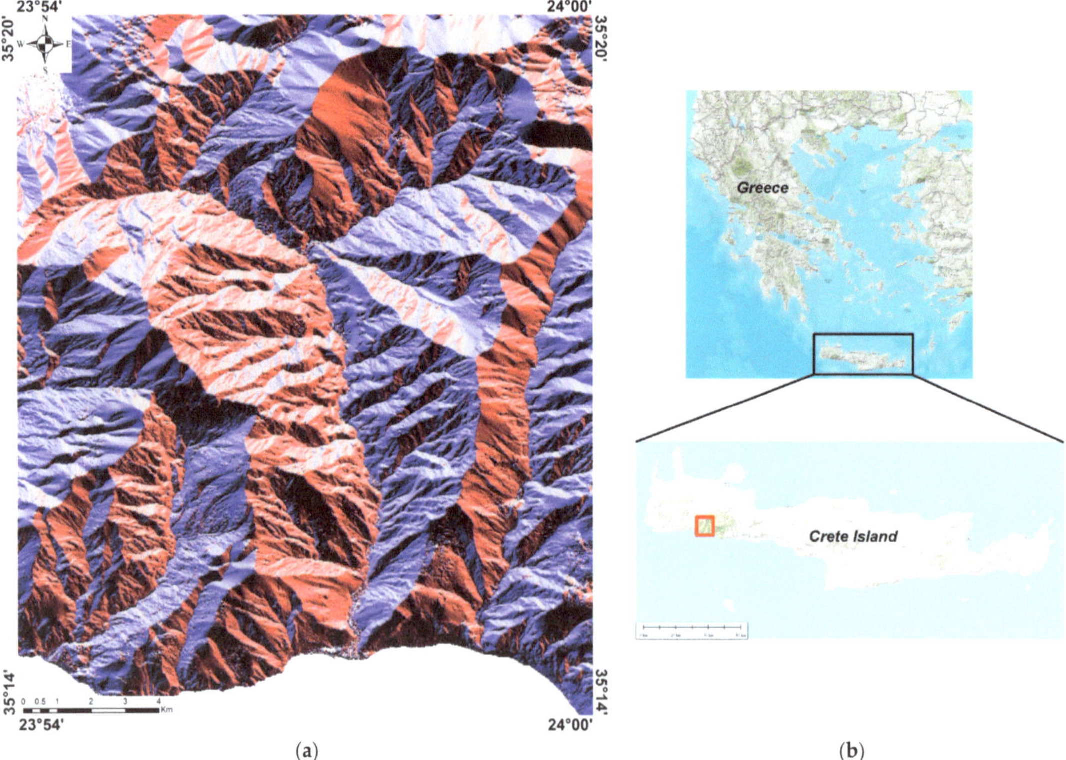

Figure 2. (**a**) Digital Elevation Model (DEM) of Samaria Gorge; (**b**) Location of Samaria Gorge in Greece.

The Samaria Gorge is also characterized by complicated tectonics, affecting the geological and geomorphological features of the area. It represents the geological profile of the region, highlighting the Plattenkalk Group geological formations (Figure 3), which is the lower parautochthonous group of all the sequences, structuring the White Mountains' core, as well as the major mountain masses of Crete (Mt Talea, Mt Psiloritis, Mt Lasithiotika etc.).

It is worth noting that the lower tectonic units, such as the Fodele (including the Galinos beds) and Sisses beds [1] are lacking in the study area.

Figure 3. Contact (yellow line) between the Gigilos and Plattenkalk formations.

Regarding the geological formations of the Plattenkalk Group, they are geomorphologically associated with the sharp topographic relief of the area, as well as with the steep slopes that are presented. Moreover, the drainage network density, and mainly the river crossing the gorge, contributes decisively to the geomorphological configuration of the area. The erosion caused by the river flow is directly related to precipitation; the increased winter precipitation is expressed by significant water flow quantities, recorded in the winter period between May and October. On the contrary, the limited precipitation from June to September (summer period) results in lower flow water quantities and hence a lower erosion rate.

Concerning the northern part of the Samaria Gorge, a different geological formation (Gigilos formation) is documented, which outcrops as a tectonic window structure, while a smoother relief is observed; despite the numerous drainage network branches, the main river, mentioned above, is absent. Considering the elevation variety and the drainage network, the Samaria river catchment area shows significant differentiations between the northern and the southern part, related both to the shape and extension. In addition, dense forest is documented in the northern part of the study area, related to the aforementioned factors [33]. Particularly, nine distinct land ecotopes have been identified in the Samaria Gorge, including 162 plant species, while 36 of them are protected and characterized as endangered [34].

3. Geology of the Study Area

The Samaria Gorge area is geologically structured by Neogene–Quaternary geological formations, as well as by Alpine bedrock. The geological and topographical complexity of the study area has been highlighted by various researchers, who contributed to the theoretical background establishment. In these proposals, the following geological formations/terms are referred to:

The "Plattenkalk" term, which was initially introduced by Chalikiopoulos [35], describes a thin- to medium-bedded crystalline limestone sequence, including chert intercalations or/and nodules, located in Eastern Crete. Numerous geological studies, performed in the Peloponnese and Crete regions, consider that this sequence is predominantly exhumated as a tectonic window throughout these regions. On the contrary, Creutzburg [36] applied a different terminology, introducing the "Madara-Kalke" term in order to describe a thick complex of white to dark recrystallized limestones («mächtiger Komplex»), bedded in the lower part and unsorted in the upper part. The Triassic age was questioned as it underlies the Permian "Phyllites-Quartzites" unit.

Regarding the Western Crete region, the "underlying system of the Plattenkalk" term was introduced by Tataris and Christodoulou [3] to describe a system of approximately 300 m in thickness that consists of phyllites, dolomites, microbreccia limestones, cherts and clays, while numerous Upper Triassic occurrences (with Ostrea, Myophoria and probably Halobia) are documented within the dark, thin-bedded limestones.

In the geological map of Crete (1:200,000 scale) constructed by Creutzburg et al. [37], the aforementioned beds are defined as the "Gigilos formation", which underlies the Plattenkalk" formation. On the contrary, Fytrolakis [38] proposed the existence of a tectonic unconformity between the Gigilos beds and the Plattenkalk formation; the gradual stratigraphic transition is the potential between the two formations, and the estimated age of the Gigilos formation is Liasian–Doggerian. Furthermore, apart from the common clastic sediment alternations (clays, calc-phyllites, sandstones, microbreccia limestones), an additional formation was identified, which resembles flysch and consists of clays with a thin gypsum layer, thin-bedded sandstones and thin-bedded, clastic limestones.

Therefore, according to various researchers [17,31,38–41], the generally accepted geological model highlights an anticlinic megastructure occurrence, which considers Mt Gigilos as the core [6,7]. It should be emphasized that the Trypali unit dolomitic limestones are documented southwest of the Samaria Gorge, while the corresponding underlying system of the Plattenkalk Group is not. The Trypali dolomitic limestones tectonically overlie the Plattenkalk formation, while the in-between contact is accompanied by a tectonic breccia, locally exceeding 2 m thick.

In particular, the Trypali unit and Plattenkalk Group constituting the bedrock of the wider area of Samaria Gorge show different lithological characteristics; intensively brecciated, strongly karstified, predominantly carbonate formations dominate in the Trypali unit, characterized by their cellular texture, resembling rauhwackes. The Plattenkalk Formation consists of thin-bedded marbles, including chert intercalations and/or nodules, while the underlying Gigilos Formation rocks consist of alternations of metaclastic and metacarbonate rocks with cherts.

4. Data and Methodology

The 3D geological model construction was performed by applying software packages, which are widely implemented in geosciences. In particular, the presentation of the first and the second models was carried out using the ArcGIS 10.6.1 software package, while the third one was performed by the SURPAC2000 software package. The methodological process of geological models in the ArcGIS 10.6.1 software package includes the georeferencing of the original geological maps [3,4] and the existing formations' digitalization; moreover, a colored legend was extracted based on the age of each formation. Then, a detailed digital elevation model (DEM) of the study area was implemented, resulting in the three-dimensional configuration of geological formations; for this purpose, the ArcScene application (extension of ArcGIS 10.6.1 software) was applied. Regarding the corresponding process in the SURPAC2000 software package, the principles of digitalization are identical to the ArcGIS 10.6.1 software package ones, while limited technical differentiations are documented.

5. Results—Discussion

5.1. Geological Models Review

Based on the various approaches concerning the geology of the Samaria Gorge, and the theoretical background improvement, different geological models were proposed, highlighting the geological regime of the wider area. Particularly, three geological models have been proposed by: (a) Tataris and Christodoulou [3,4]—1st model, (b) Pavlaki and Perleros [5]—2nd model, (c) Manutsoglu et al. [6,7]—3rd model.

Regarding the 1st model [3,4] shown in Figure 4, the underlying bed system of Plattenkalk (after Chalikiopoulos [35]) is characterized by the following lithological formations (stratigraphically from the upper to the lower one): (a) Phyllites, (b) Gray–white grayish, massive carbonates with sparse chert nodules, (c) Gray limestones with a cellular texture (in some locations), (d) Thin-bedded limestones with marl intercalations, (e) Dark, massive limestones with chert nodules and phyllite intercalations, (f) Calcitic phyllites and dark, crystalline limestones alternations, (g) Crystalline limestones, (h) Calcitic phyllites and dark, crystalline limestone alternations and (i) Calcitic phyllites. According to this model, the Plattenkalk system consists of thin-bedded, crystalline limestones, including nodules and/or thin chert intercalations and thin phyllite intercalations. These limestones are locally documented in a thick-bedded form, without chert occurrences, maintaining the crystallization. Moreover, the transition of these formations to calcitic phyllites is observed. The formations system, overlying Plattenkalk, includes a lower sequence of limestones and dolomites (Madara-Kalke), and an upper sequence consisting of phyllites, quartzites, rauhwackes, gypsum, limestones, eruptive formations and iron ores.

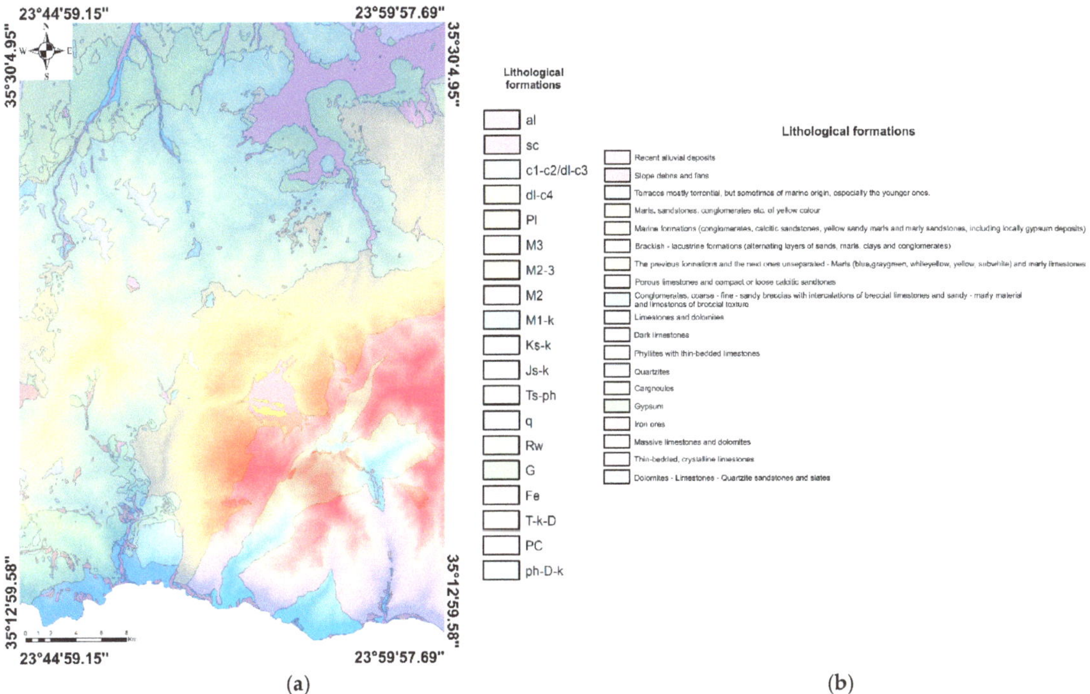

Figure 4. The geological model for the wider Samaria Gorge modified from Tataris and Christodoulou [3] —1st geological model: (**a**) Lithological map of the study area; (**b**) Description of the geological formations.

According to Pavlaki and Perleros [5], the 2nd model (shown in Figure 5) suggests that the underlying system of the Plattenkalk includes: (a) Thin-bedded alternations of clay phyllites and (meta)sandstones with low-grade metamorphism, marls, as well as sparse thin-bedded limestones and chert occurrences. This sequence is also known as "Gigilos beds"; (b) A carbonate system, subjected to different tectonic phases of folding and uplift. It is characterized by significant thickness in the greatest part of the White Mountains region, and it is divided into (a) White-grayish–whitish marbles, locally cracked, showing karstic features, (b) Stromatolitic dolomites and (c) Black dolomites, showing a cellular texture, strongly cracked with karstic features.

Regarding the Plattenkalk series, the upper members consist of thick-bedded carbonate formations, alternating with green, calcitic phyllites, while brown-black slates are locally documented (White Mountains metaflysch). The lower members consist of thin-bedded and strongly recrystallized gray-black limestones and dolomites, forming beds with thin chert intercalations and nodules.

Finally, the overlying Trypali unit, strongly karstified and tectonically affected, includes (meta)carbonate formations, which locally show a conglomerate-breccia formation. In particular, the upper horizons consist of coarse carbonate conglomerate-breccia formations and recrystallized limestones–dolomitic limestones, while the corresponding lower ones include strongly recrystallized, white-grayish, thick-bedded and cracked limestones, as well as dark dolomites.

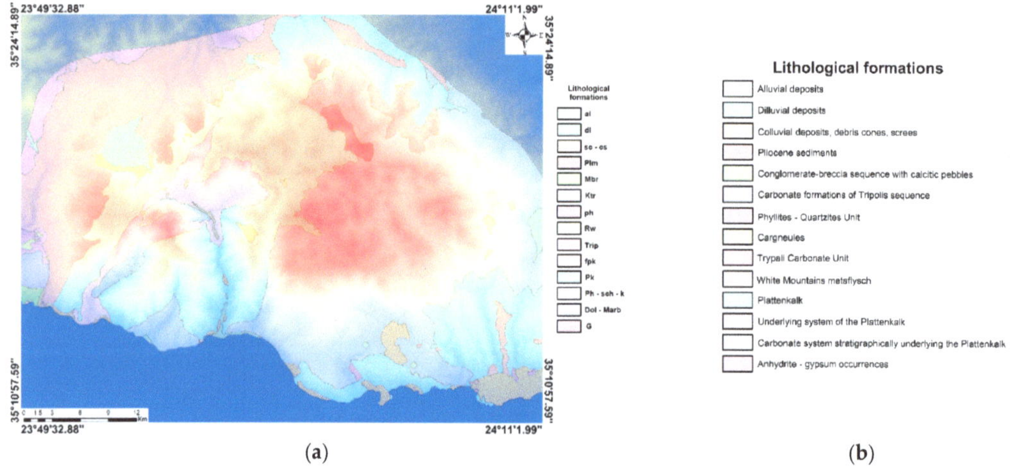

(a)   (b)

Figure 5. The 2nd geological model for the wider Samaria area (modified from Pavlaki and Perleros, 2015 [5]): (**a**) Lithological map of the study area; (**b**) Description of the geological formations.

According to Manutsoglu et al. [6,7], the 3rd model (Figure 6), which considers the study of Soujon et al. [40], the Plattenkalk Group of the White Mountains, consists of the "Mavri" and the "Aloides" formations, which include the following lithologies (from the upper to the lower): (a) Carbonate breccia, (b) Dolomitic marbles with chert nodules, (c) Chert-clay-carbonate sequence, (d) Thin-bedded calcitic marbles with chert intercalations and nodules, (e) Medium-bedded to thin-bedded calcitic marbles with chert nodules and layers and (f) Thin-bedded marbles with red/green calc-silt horizons and cherts. In addition, the upper part includes marls and calc-schists (Kalavros formation), which are considered the White Mountains metaflysch. Eventually, the Trypali unit overlies the Plattenkalk Group sequence.

Considering the aforementioned viewpoints, we correlated the geological-lithological formations of the models highlighting the similarities and differences between them. The results are summarized in Table 1.

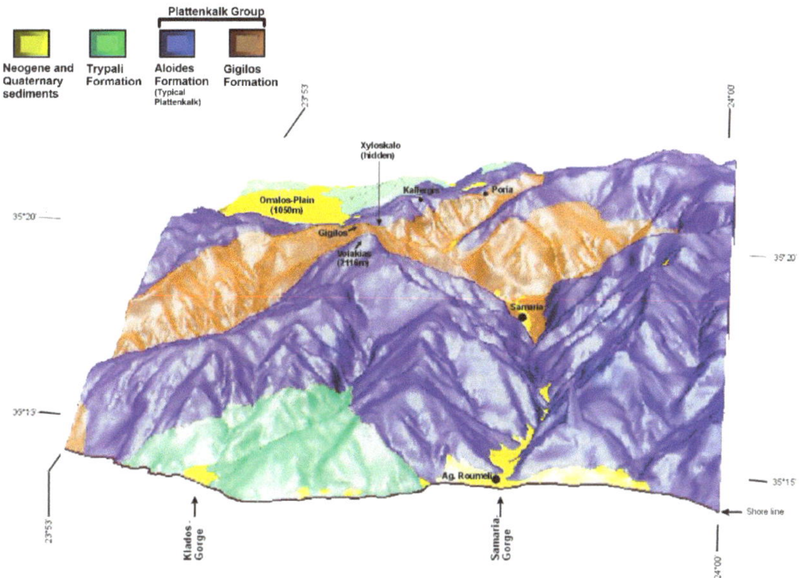

Figure 6. The 3rd geological model for the wider region of Samaria Gorge (modified from Manutsoglu et al., 1999; 2001 [6,7]).

Table 1. Correlation of the proposed geological models for the wider Samaria Gorge area.

1st Model (by Tataris and Christodoulou [3,4])	2nd Model (by Pavlaki and Perleros [5])	3rd Model (by Manutsoglu et al. [6,7])
Formations system, overlying Plattenkalk: a. Upper sequence b. Lower sequence—Madara-Kalke (limestones and dolomites)	Trypali unit	Trypali formation
Plattenkalk system: crystalline, light gray–dark gray limestones with phyllite intercalations. Thin-bedded with chert intercalations and nodules. The uppermost horizons are thick-bedded without cherts, changing locally into calcitic phyllites.	White Mountains metaflysch: Thick-bedded carbonate formations with intercalations of green, calcitic phyllites Plattenkalk: (detailed lithological description in the text)	Aloides formation
Underlying bed system of Plattenkalk: phyllites, dolomites, limestones, quartz sandstones and slates. These formations are documented in the Klados and Trypiti Gorges, as well as within the Gigilos and Poria areas.	System underlying the Plattenkalk: division into metacarbonate and metaclastic formations (detailed lithological description in the text)	Gigilos (Mavri) formation

5.2. Geomorphological Evolution of Samaria Gorge

Sea-level change is a significant factor affecting the geomorphological evolution of a region. In particular, numerous and various sea-level changes have occurred throughout the Mediterranean Sea since the Middle Pleistocene, due to climate change ([42] and references therein), and therefore Crete Island was correspondingly affected.

Especially, the Samaria Gorge's geomorphological evolution is based on the combination of lithology, stratigraphy, tectonics, karstic processes and erosion, resulting in an intense topographic relief configuration, as well as in the formation of a surface and underground drainage network. In general, the Samaria Gorge structure is part of the Samaria River catchment area (Figure 7), composed of geological formations with different vulnerabilities.

Figure 7. Catchment area of Samaria River, including the drainage network and springs outlets.

Particularly, post-Alpine formations (recent alluvial deposits, slope debris and fans, terraces and Quaternary formations) aged between the Quaternary to the present (Figure 8) affect the geomorphological evolution of the area. Specifically:

Recent alluvial deposits, slope debris and fans include sand, dune or torrent depositions formed in different depositional environments. Slope debris and fans are additionally documented, forming variable debris cones. In particular, the slope debris is derived from the older formations weathering, resulting in rock fragment occurrences, which are displaced by gravity, forming loose and compact sediments. Alluvial fans differ from slope debris due to the different clastic parts' distribution, while they display sorting and layering. Furthermore, alluvial fans significantly contribute to different geomorphological evolutions depending on the climate environment [43,44].

Terrace formations have predominantly been formed at the exit of Samaria Gorge, as well as the adjacent gorges, while they have been interpreted as an index of the marine environment; the estimated thickness ranges between one and several meters. The material of their formation derived from alluvial deposits and fans, while the intense tectonic activity, which uplifts the wider area, results in the formation of beach rocks, which are also uplifted, forming terraces.

The Quaternary formations are divided into Lower and Upper parts. Particularly, the Lower Quaternary formations include marls, sandstones and conglomerates, while the Upper Quaternary ones are characterized by torrential origin and consist of sandy marls, clays, sandstones and conglomerates, with considerable thickness. Similarly to the terraces, they are affected by significant tectonic activity.

Figure 8. Representative post-Alpine formations of various thicknesses, documented throughout the Samaria Gorge area.

Furthermore, the Gigilos formation (northern part of the Samaria Gorge), which is part of the Alpine formations, is susceptible to erosion, as, in this formation, phyllites and slates with smooth dip angles are predominantly included (Figure 9a). In particular, the minerals composing these formations, such as moscovite, kaolinite, montmorillonite and illite, are prone to erosional processes. Therefore, these lithological properties affect the catchment area shape, which is significantly widened in the northern part. On the contrary, the Plattenkalk (metacarbonate formations with chert intercalations consisting of calcite/dolomite and cryptocrystalline silicon dioxide, respectively) formation is characterized by steep slopes along the southern part of Samaria Gorge (Figure 9b), resulting in the narrowing of the catchment area throughout this part. Moreover, it should be mentioned that the low (25°–35°) and high angles (75°–85°) of slopes which are observed in the Gigilos and Plattenkalk formations, respectively, resulted from local erosion activity. Moreover, remarkable karstic features are documented, especially in the southern part of the Samaria Gorge (Figure 10).

Regarding groundwater permeability, the Gigilos formation is considered impermeable, as the watertight horizons of chert and schists restrict the downward movement of the water. Therefore, the limited infiltration and the high surface drainage result in a high runoff coefficient, affecting the geomorphological relief. Moreover, the lithological alternation between the metaclastic and metacarbonate rocks, as well as the extended fracture, favors the formation of low-capacity aquifers, represented by low-discharge value springs. However, some of these springs, located within the Gigilos beds, show significant discharge values, such as the Mytatouli (5.9 m^3/h) and Potistiria (5.3 m^3/h) springs [45]. It is worth mentioning that the transmissibility of the Gigilos formation is significantly lower than the corresponding one of the Trypali unit. Regarding the Plattenkalk series, it consists of several hundreds of meters of crystalline limestone and dolomite, while it is interbedded with watertight horizons of chert and schists. These horizons restrict the downward movement of the water and strongly influence its movement and concentration in the overlying carbonate formation of the Trypali unit. The transmissibility of the Trypali carbonate formations ranges between 0.1 to 1 m^2/sec, which is equivalent to permeability between 10^{-3} and 10^{-2} m/sec [46].

(**a**) (**b**)

Figure 9. The main lithological formations outcropping in the Samaria Gorge: (**a**) Gigilos formation, consisting of phyllites and slates. The slopes are characterized by smooth dip angles; (**b**) Plattenkalk formation, consisting of recrystallized limestones with chert intercalations, showing steep slopes.

Figure 10. Extensive karst features predominantly observed in the southern part of Samaria Gorge.

6. Conclusions

Considering the models relating to geological processes and their relationship with geomorphological features, the following concluding remarks arise for the Samaria Gorge area:

1. The entire region is a part of a megastructure, with the core of Mt Gigilos, consisting of unsorted, dolomitic stromatolite limestones. On either side of the mountain, the dip direction of the formations is differentiated, maintaining the same NNE–SSW striking, indicating that Mt Gigilos is a tectonic window.
2. Although the proposed models are detailed and provide different aspects of the geological regime of the study area, we conclude that the 3rd model is more accurate as the Trypali unit's occurrence at the southwestern part of Samaria Gorge suggested; this is verified by field observations.
3. According to the field observations, this megastructure dips in the NE direction, resulting in the geological formations' intense erosion rate. Therefore, the geological formations' inclination, which is directly associated with tectonic processes, occurred in the area, controlling the geomorphological evolution of the region.
4. The lithology constitution, combined with the tectonic evidence, is a significant factor affecting the geomorphological relief. In particular, the phyllite and slate formations (Gigilos formation) with medium dip angles are characterized by smooth relief. On the contrary, the metacarbonate formations of the Plattenkalk Group, accompanied by high (up to vertical) dip angles, show sharp slopes on both sides of the river that crosses the Samaria Gorge and form slopes with heights of up to 300 m in some locations.
5. Overall, the Samaria Gorge constitutes a typical catchment area, in which the surface runoff is favored in the northern part, due to the lithological formations, while the underground one is favored in the southern part, respectively.

Author Contributions: Conceptualization, E.M. and I.L.; methodology, E.M.; software, I.L.; validation, E.M., E.S., and A.V.; formal analysis, E.M.; investigation, E.M., I.L., and E.S.; resources, E.M. and E.S.; data curation, I.L. and E.S.; writing—original draft preparation, I.L.; writing—review and editing, E.M., I.L., E.S., and A.V.; visualization, I.L.; supervision, E.M.; project administration, E.M.; funding acquisition, E.M. All authors have read and agreed to the published version of the manuscript.

Funding: This research was funded by the Applied Research Project "3D geological structure modeling of Samaria Gorge, aiming at the investigation of hydrogeological conditions and water reservoir levels in the core of Samaria National Park (Lefka Ori), Western Crete" (funding number: 82230), which is financially supported by the Green Fund "Forest Protection and Upgrading 2019" under "Other Nationals. Green Fund".

Institutional Review Board Statement: Not applicable.

Informed Consent Statement: Not applicable.

Data Availability Statement: Not applicable.

Conflicts of Interest: The authors declare no conflict of interest.

References

1. Epting, M.; Kudrass, H.-R.; Leppig, U.; Schafer, A. Geologie Der Talea Ori/Kreta. *Neues Jb. Geol. Paläont. Abh.* **1972**, *141*, 259–285.
2. Bonneau, M. Correlation of the Hellenide Nappes in the South-East Aegean and Their Tectonic Reconstruction. *Geol. Soc. Spec. Publ.* **1984**, *17*, 517–527. [CrossRef]
3. Tataris, A.A.; Christodoulou, G. The Geological Structure of Leuca Mountains (Western Crete). *Bull. Geol. Soc. Greece* **1965**, *6*, 319–347.
4. Tataris, A.A.; Christodoulou, G. *Geological Map of Greece, 1:50.000 Scale Map Series, "Alikianos" Sheet*; Institute of Geological and Mining Research (IGME): Athens, Greece, 1969.
5. Pavlaki, A.; Perleros, V. *Updated Geological Map of the White Mountains National Park with Identification and Map of Geotope Locations*; Chania, Greece, 2015.
6. Manutsoglu, E.; Jacobshagen, V.; Spyridonos, E.; Skala, W. Geologische 3D-Modellierung Der Plattenkalk-Gruppe West-Kretas. *Math. Geol.* **1999**, *4*, 73–79.
7. Manutsoglu, E.; Spyridonos, E.; Soujon, A.; Jacobshagen, V. Revision of the Geological Map and 3D Modelling of the Geological Structure of the Samaria Gorge Region, W. Crete. *Bull. Geol. Soc. Greece* **2001**, *34*, 29–36. [CrossRef]
8. Turner, A.K. Challenges and Trends for Geological Modelling and Visualisation. *Bull. Eng. Geol. Environ.* **2006**, *65*, 109–127. [CrossRef]
9. Thornton, J.M.; Mariethoz, G.; Brunner, P. A 3D Geological Model of a Structurally Complex Alpine Region as a Basis for Interdisciplinary Research. *Sci. Data* **2018**, *5*, 180238. [CrossRef]
10. Shurtz, R.F. The Electronic Computer and Statistics for Predicting Ore Recovery. *Min. Eng.* **1959**, *11*, 1035–1044.
11. Zhao, X.; Niu, J. Method of Predicting Ore Dilution Based on a Neural Network and Its Application. *Sustainability* **2020**, *12*, 1550. [CrossRef]
12. Zensus, T. Tagebauplanung Mit Automatischer Daten-Verarbeitung. *Braunkohle Waerme Und Energ.* **1963**, *15*, 253–266.
13. Houlding, S.W. *3D Geoscience Modeling*; Springer: Berlin/Heidelberg, Germany, 1994. [CrossRef]
14. Jolivet, L.; Menant, A.; Clerc, C.; Sternai, P.; Bellahsen, N.; Leroy, S.; Pik, R.; Stab, M.; Faccenna, C.; Gorini, C. Extensional Crustal Tectonics and Crust-Mantle Coupling, a View from the Geological Record. *Earth Sci. Rev.* **2018**, *185*, 1187–1209. [CrossRef]
15. Strobl, M.; Hetzel, R.; Fassoulas, C.; Kubik, P.W. A Long-Term Rock Uplift Rate for Eastern Crete and Geodynamic Implications for the Hellenic Subduction Zone. *J. Geodyn.* **2014**, *78*, 21–31. [CrossRef]
16. Fassoulas, C.; Kilias, A.; Mountrakis, D. Postnappe Stacking Extension and Exhumation of High-Pressure/Low-Temperature Rocks in the Island of Crete, Greece. *Tectonics* **1994**, *13*, 127–138. [CrossRef]
17. Creutzburg, N.; Seidel, E. Zum Stand Der Geologie Des Preaneogens Auf Kreta. *Neu. Jb. Geol. Palaontol. Abh.* **1975**, *149*, 363–383.
18. Seidel, E. *Zur Petrologie Der Phyllit-Quarzit-Serie Kretas*; Technische Universitat: Braunschweig, Germany, 1978.
19. Fassoulas, C. The Structural Evolution of Central Crete: Insight into the Tectonic Evolution of the South Aegean (Greece). *J. Geodyn.* **1998**, *27*, 23–43. [CrossRef]
20. Papanikolaou, D.; Vassilakis, E. Thrust Faults and Extensional Detachment Faults in Cretan Tectono-Stratigraphy: Implications for Middle Miocene Extension. *Tectonophysics* **2010**, *488*, 233–247. [CrossRef]
21. Papanikolaou, D.I. *The Geology of Greece*; Regional Geology Reviews; Springer International Publishing: Cham, Switzerland, 2021; ISBN 978-3-030-60730-2.
22. Robertson, A.H.F. Palaeozoic-Early Mesozoic Transition from Palaeotethys to Neotethys: Synthesis of Data and Interpretations from the Northern Periphery of Gondwana (Central and Western Anatolia, Aegean, Balkans and Sicily). *Earth Sci. Rev.* **2022**, *230*, 104000. [CrossRef]

23. Seidel, E.; Kreuzer, H.; Harre, W. A Late Oligocene/Early Miocene High-Pressure Belt in the External Hellenides. *Geol. Jb.* **1982**, *E23*, 165–206.
24. Robertson, A.H.F.; Dixon, J.E.; Brown, S.; Collins, A.; Morris, A.; Pickett, E.; Sharp, I.; Ustaömer, T. Alternative Tectonic Models for the Late Palaeozoic-Early Tertiary Development of Tethys in the Eastern Mediterranean Region. *Geol. Soc. Spec. Publ.* **1996**, *105*, 239–263. [CrossRef]
25. Robertson, A.H.F.; Clift, P.D.; Degnan, P.J.; Jones, G. Palaeogeographic and Palaeotectonic Evolution of the Eastern Mediterranean Neotethys. *Palaeogeogr Palaeoclim. Palaeoecol.* **1991**, *87*, 289–343. [CrossRef]
26. Klein, T.; Craddock, J.P.; Zulauf, G. Constraints on the Geodynamical Evolution of Crete: Insights from Illite Crystallinity, Raman Spectroscopy and Calcite Twinning above and below the 'Cretan Detachment. *Int. J. Earth Sci.* **2013**, *102*, 139–182. [CrossRef]
27. Zulauf, G.; Dörr, W.; Fisher-Spurlock, S.C.; Gerdes, A.; Chatzaras, V.; Xypolias, P. Closure of the Paleotethys in the External Hellenides: Constraints from U–Pb Ages of Magmatic and Detrital Zircons (Crete). *Gondwana Res.* **2015**, *28*, 642–667. [CrossRef]
28. Zulauf, G.; Dörr, W.; Marko, L.; Krahl, J. The Late Eo-Cimmerian Evolution of the External Hellenides: Constraints from Microfabrics and U–Pb Detrital Zircon Ages of Upper Triassic (Meta)Sediments (Crete, Greece). *Int. J. Earth Sci.* **2018**, *107*, 2859–2894. [CrossRef]
29. Zulauf, G.; Dörr, W.; Xypolias, P.; Gerdes, A.; Kowalczyk, G.; Linckens, J. Triassic Evolution of the Western Neotethys: Constraints from Microfabrics and U–Pb Detrital Zircon Ages of the Plattenkalk Unit (External Hellenides, Greece). *Int. J. Earth Sci.* **2019**, *108*, 2493–2529. [CrossRef]
30. Koepke, J.; Seidel, E.; Kreuzer, H. Ophiolites on the Southern Aegean Islands Crete, Karpathos and Rhodes: Composition, Geochronology and Position within the Ophiolite Belts of the Eastern Mediterranean. *Lithos* **2002**, *65*, 183–203. [CrossRef]
31. Fytrolakis, N. *The Geological Structure of Crete*; National Technical University of Athens: Athens, Greece, 1980.
32. Meulenkamp, J.E.; van der Zwaan, G.J.; van Wamel, W.A. On Late Miocene to Recent Vertical Motions in the Cretan Segment of the Hellenic Arc. *Tectonophysics* **1994**, *234*, 53–72. [CrossRef]
33. Wei, H.; Chen, G.; Chen, X.; Zhao, H. Geographical Distribution of Aralia Elata Characteristics Correlated with Topography and Forest Structure in Heilongjiang and Jilin Provinces, Northeast China. *J. For. Res.* **2021**, *32*, 1115–1125. [CrossRef]
34. Kargiolaki, H.; Manutsoglu, E.K.; Kasiotakis, V.; Rekatsinas, Y. Management Body of a National Park and a Biosphere Reserve: Samaria Gorge and the White Mountains. In Proceedings of the Protection and Restoration of the Environment VI, Skiathos, Greece, 5 July 2002; pp. 595–602.
35. Chalikiopoulos, L. Sitia, Die Osthalbinsel Kreta's. *Veroff. Inst. Meereskd. Berl.* **1903**, *4*, 1–138.
36. Creutzburg, N. Kreta, Leben Und Landschaft. *Zeitschr. Ges. Erdkd. Berl.* **1928**, 16–38.
37. Creutzburg, N.; Drooger, C.W.; Meulenkamp, J.E. *General Geologic Map of Crete 1:200.000*; Institute of Geological and Mining Research (IGME): Athens, Greece, 1977.
38. Fytrolakis, N. Contribution in the Geological Research of Crete. *Bull. Geol. Soc. Greece* **1978**, *13*, 101–115.
39. Xavier, J.-P. *Contribution a l'étude Géologique de l'arc Égéen: La Crète Occidentale, Secteurs d'Omalos et de Kastelli*; Université Pierre et Marie Curie: Paris, France, 1976.
40. Soujon, A.; Jacobshagen, V.; Manutsoglu, E. A Lithostratigraphic Correlation of the Plattenkalk Occurrences of Crete (Greece). *Bull. Geol. Soc. Greece* **1998**, *32*, 41–48.
41. Manutsoglu, E.; Soujon, A.; Jacobshagen, V. Tectonic Structure and Fabric Development of the Plattenkalk Unit around the Samaria Gorge, Western Crete, Greece. *Z. Der Dtsch. Geol. Ges.* **2003**, *154*, 85–100. [CrossRef]
42. Lichter, M.; Zviely, D.; Klein, M.; Sivan, D. Sea-Level Changes in the Mediterranean: Past, Present, and Future—A Review. In *Seaweeds and Their Role in Globally Changing Environments. Cellular Origin, Life in Extreme Habitats and Astrobiology*; Seckbach, J., Einav, R., Israel, A., Eds.; Springer: Dordrecht, The Netherlands, 2010; Volume 15, pp. 3–17.
43. Huang, S.; Tang, L.; Hupy, J.P.; Wang, Y.; Shao, G. A Commentary Review on the Use of Normalized Difference Vegetation Index (NDVI) in the Era of Popular Remote Sensing. *J. For. Res.* **2021**, *32*, 1–6. [CrossRef]
44. Thammanu, S.; Marod, D.; Han, H.; Bhusal, N.; Asanok, L.; Ketdee, P.; Gaewsingha, N.; Lee, S.; Chung, J. The Influence of Environmental Factors on Species Composition and Distribution in a Community Forest in Northern Thailand. *J. For. Res.* **2021**, *32*, 649–662. [CrossRef]
45. Lazos, I.; Grigorakis, E.; Gouskos, Z.; Steiakakis, E.; Vafidis, A.; Manoutsoglou, E. Investigation of the Geological Stucture and Assessment of the Groundwater Potential in Samaria Region (West Crete). In Proceedings of the 12th International Hydrogeological Congress, Nicosia, Cyprus, 20 March 2022; pp. 326–329.
46. Steiakakis, E.; Monopolis, D.; Vavadakis, D.; Manutsoglu, E. Hydrogeological Research in Trypali Carbonate Unit (NW Crete). In *Advances in the Research of Aquatic Environment*; Springer: Berlin/Heidelberg, Germany, 2011; pp. 561–567.

Article

The Influence of Flight Direction and Camera Orientation on the Quality Products of UAV-Based SfM-Photogrammetry

Shaker Ahmed [1,2], **Adel El-Shazly** [1], **Fanar Abed** [3,4] and **Wael Ahmed** [1,*]

1 Faculty of Engineering, Cairo University, Giza 12613, Egypt
2 College of Engineering, University of Anbar, Ramadi 31001, Iraq
3 College of Engineering, University of Baghdad, Baghdad 10001, Iraq
4 Environment and Sustainability Institute, University of Exeter, Penryn Campus, Penryn, Falmouth TR10 9FE, Cornwall, UK
* Correspondence: wael.m.ahmed@cu.edu.eg

Abstract: Unmanned aerial vehicles (UAVs) can provide valuable spatial information products for many projects across a wide range of applications. One of the major challenges in this discipline is the quality of positioning accuracy of the resulting mapping products in professional photogrammetric projects. This is especially true when using low-cost UAV systems equipped with GNSS receivers for navigation. In this study, the influence of UAV flight direction and camera orientation on positioning accuracy in an urban area on the west bank of the Euphrates river in Iraq was investigated. Positioning accuracy was tested in this study with different flight directions and camera orientation settings using a UAV autopilot app (Pix4Dcapture software (Ver. 4.11.0)). The different combinations of these two main parameters (camera orientation and flight direction) resulted in 11 different flight cases for which individual planimetric and vertical accuracies were evaluated. Eleven flight sets of dense point clouds, DEMs, and ortho-imagery were created in this way to compare the achieved positional accuracies. One set was created using the direct georeferencing method (without using GCPs), while the other ten sets were created using the indirect georeferencing approach based on ground truth measurements of five artificially created GCPs. Positional accuracy was found to vary depending on the user-defined flight plan settings, despite an approximately constant flight altitude. However, it was found that the horizontal accuracy achieved was better than the vertical accuracy for all flight sets. This study revealed that combining multiple sets of images with different flight directions and camera orientations can significantly improve the overall positional accuracy to reach several centimeters.

Keywords: unmanned aerial vehicle (UAV); flight parameters; positional accuracy; indirect georeferencing; evaluation; accuracy analysis

Citation: Ahmed, S.; El-Shazly, A.; Abed, F.; Ahmed, W. The Influence of Flight Direction and Camera Orientation on the Quality Products of UAV-Based SfM-Photogrammetry. *Appl. Sci.* **2022**, *12*, 10492. https://doi.org/10.3390/app122010492

Academic Editors: Anselme Muzirafuti, Giovanni Randazzo and Stefania Lanza

Received: 16 September 2022
Accepted: 13 October 2022
Published: 18 October 2022

1. Introduction

The positional accuracy of UAV geospatial products is affected by numerous factors, including the nature of the survey area and its morphological characteristics [1]. The other main factors that affect the positional accuracy of UAV products are the model of the UAVs, the type and accuracy of the navigation and orientation instruments on board the UAVs, the camera specification, the ground sensing distance (GSD) [2], the degree of overlap, the UAV flight altitude, and the number and configuration of ground control points (GCPs) [3]. Several previous studies have analyzed and investigated the effects of these factors on the positional accuracy of UAV products [4]. In general, the positional accuracy of UAV products has been tested and analyzed in some studies and research as follows [5]:

In 2015, Whitehead and Hugenholtz used GCPs and Pix4D software to photogrammetrically map a gravelly river channel and achieved a horizontal accuracy of 0.048 m (RMSEH) and vertical accuracy of 0.035 m (RMSEZ), which were compatible with the accuracy standards developed by the American Society for Photogrammetry and Remote Sensing (ASPRS) for digital geospatial data at the 0.1-m RMSE level [6].

Next, Hugenholtz et al. (2016) compared the spatial resolution of three cases of UAV imagery acquired during two missions to a gravel pit. The first case was a surveillance-level GNSS/RTK receiver (direct georeferencing), the second was a lower-level GPS receiver (direct georeferencing), and the third was a lower-level GPS receiver with GCPs (indirect georeferencing). The horizontal and vertical accuracies were RMSEH = 0.032 m and RMSEZ = 0.120 m in the first case, RMSEH = 0.843 m and RMSEZ = 2.144 m in the second case, and RMSEH = 0.034 m and RMSEZ = 0.063 m in the third case, respectively [7].

Later, Lee and Sung (2018) evaluated and analyzed the positional accuracy of UAV mapping with onboard RTK without GCPs compared with UAV mapping without RTK with different numbers and configurations of GCPs. In the case of the non-RTK method with GCPs, the horizontal and vertical accuracies were 4.8 and 8.2 cm with 5 GCPs, 5.4 and 10.3 cm with 4 GCPs, and 6.2 and 12.0 cm with 3 GCPs, respectively, depending on the number of GCPs. The horizontal and vertical accuracies of the non-RTK method without GCPs deteriorated to about 112.9 and 204.6 cm, respectively. The horizontal and vertical accuracies of the RTK-onboard-UAV method without GCPs were 13.1 and 15.7 cm, respectively [8].

In addition, Yang et al. (2020) investigated the influence of the number of GCPs on the vertical positional accuracy of UAV products. This study was conducted on sandy beaches in China. The authors concluded that the number 11 was the optimal number of GCPs in this study area and had a vertical RMSE of about 15 cm [9].

Further, Štroner et al. (2021) investigated georeferencing of UAV imagery aboard GNSS RTK (without GCPs). The researchers evaluated and analyzed the reasons for the high-altitude error, which is a challenge for this technique. They proposed strategies to reduce these types of errors. Several missions were conducted in two study areas with different flight altitudes and image-capturing axes [10]. This study showed that a combination of two flights at the same altitude but with vertical and oblique image acquisition axes could reduce the systematic vertical error to less than 3 cm. In addition, this study demonstrated for the first time the linear relationship between the systematic vertical error and the variation of the focal length. Finally, this study proved that georeferencing without GCPs is a suitable alternative to using GCPs.

All these studies have produced important results and paved the way for further studies. However, few studies have addressed the effects of camera orientation on positional accuracy. Similarly, there have been no studies that have addressed the effects of the direction of the flight lines and their position within the study area on the positional accuracy values.

This paper fills this gap by investigating the effects of camera orientation and flight direction on positioning accuracy. It also investigates and analyzes the positioning accuracy of UAV products to improve this accuracy, especially vertical accuracy.

There were several limitations related to the steps involved in carrying out work for this research, including those related to the possibility of obtaining a drone and those related to the possibility of obtaining security clearances. This was in addition to the social determinants related to the inhabitants' refusal to be photographed from above, given that there was a kind of intrusion of disturbances, especially at lower elevations. Therefore, we made sure to choose the time of filming when the presence of people was minimal, taking into account the appropriate weather conditions.

2. Importance of This Research

All conventional surveying, whether using a global positioning system (GPS), a total station, or other surveying equipment, requires a great deal of time, effort, and labor. Sometimes the sites where engineering work is performed are harsh and hazardous environments where it is difficult for humans to freely move. To reduce the time, labor, and manpower required for conventional survey work, photogrammetry has been used to survey land areas and locations of engineering projects. The latest findings of researchers and professionals in this field have been related to the use of unmanned aerial vehicles (UAVs) to produce survey maps, digital elevation models (DEMs), orthophotos, and 3D models.

The creation of maps using drones, like other surveying work, differs in its positional accuracy. Therefore, this research was very important because it dealt with the testing, evaluation, and analysis of the positional accuracy of geospatial data products using drones. In this study, positional accuracies without ground control points (direct georeferencing) and with ground control points (indirect georeferencing) were tested using different combinations of image sets acquired with different camera orientations and flight directions.

In reviewing previous studies that have addressed the evaluation and analysis of positional accuracy, it was found that positional accuracy varies from one study to another, as it is influenced by many different factors. It was also noted that the topic of positional accuracy, especially in the vertical, is still in need of renewed study and research, and this was one of the motivations for this study. In this study, UAV photogrammetry products were tested to determine if they are accurate enough to replace current GNSS and total station surveying methods in engineering, cadastral, and topographic surveying [11].

3. Materials and Methods

The work methodology in this study included two main phases: the fieldwork phase and the office work phase. The fieldwork included selecting the study area, defining and establishing ground control points and checkpoints, measuring their coordinates with a global positioning system (GPS), and conducting aerial photography with a UAV. The office work included the preparation of the flight plan, data processing, extraction of map products (dense point cloud, DEMs, and orthophotos), and evaluation of positional accuracies.

In this study, the effects of flight direction and camera orientation on UAV photogrammetry geospatial products were studied and analyzed. For this purpose, photogrammetric flight missions were conducted using low-cost UAVs equipped with consumer-level image sensors and low-quality navigation and attitude sensors. Different combinations of image sets were used to evaluate the effects of some parameters such as the use of GCPs, the image acquisition angle, and the flight direction on the positioning accuracy of the UAV products.

Figure 1 shows the methodology proposed in this study, which included the main phases mentioned above and the steps and tasks performed for each phase. The UAV flights in this study were performed using a Mavic 2 manufactured by the Chinese company DJI-Innovation Technology Co. LTD (Figure 2a). This multi-rotation rotary drone has a diameter of only 30 cm and weighs only 2 kg. The Mavic 2 can fly autonomously and is controlled by route-planning software that also takes off and returns automatically. The autonomous flight time of the drone is about 25 min per battery group. The Mavic 2 is equipped with a Hasselblad L1D-20c camera from the manufacturer. This camera provides an image of 12.825×8.550 mm, a focal length of 10 mm, and an image size of 5472×3648 pixels.

3.1. Study Area

The study area was a rural region which was located on the southern bank of the Euphrates river in Al-Anbar Governorate in the western part of Iraq. It was part of a residential area, as indicated by the yellow line in Figure 2b, with an average built-up area of 150×200 m. This area consisted of several internal roads, blocks of buildings, some undeveloped land, and residential buildings, most of which were two-story buildings. The height of the buildings varied from 4 to 12 m. In this area, there were also scattered plantings such as trees and home gardens.

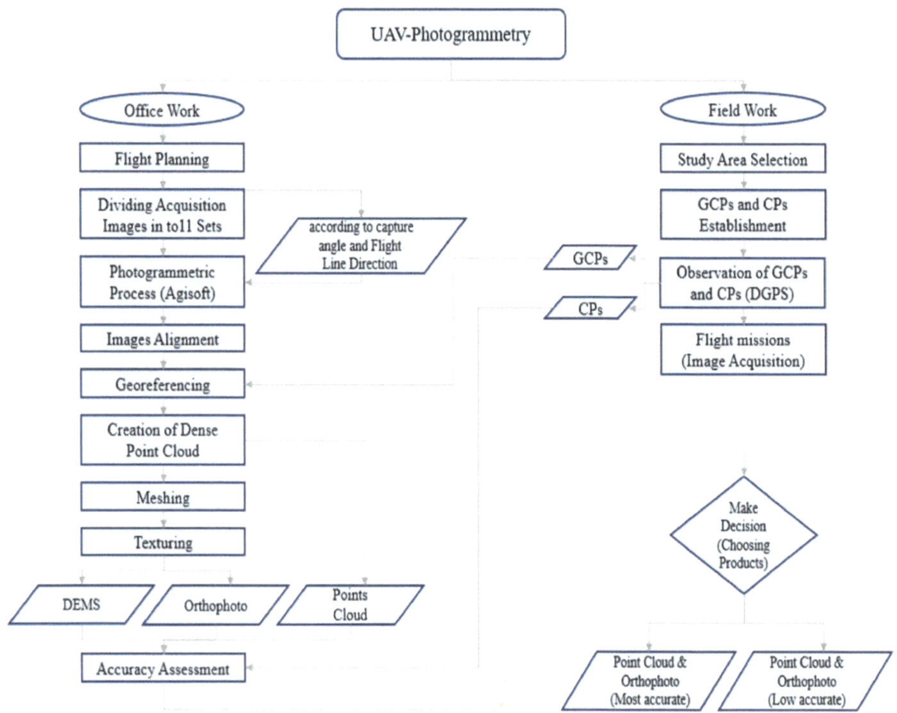

Figure 1. Workflow diagram of the methodology adopted in this paper.

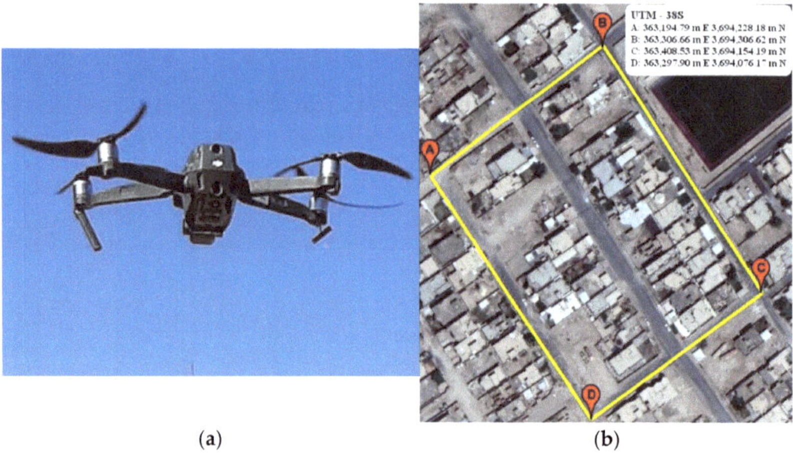

(**a**)   (**b**)

Figure 2. (**a**) DJI Mavic Pro 2 at the study area; (**b**) the study area.

3.2. Flight Planning and Data Acquisition

For any photogrammetric work with UAVs or even with conventional technology, flight planning must first be performed. This step requires some activities, such as obtaining a flight permit, selecting appropriate software, studying the characteristics of the project area, the specifications of the onboard digital camera, and the ground sample distance (GSD) value [12]. Also, the selection of weather conditions suitable for the flight process is one of the most important factors contributing to the success of the study. In this study, the field survey was conducted in a rural region away from vital places between ten to twelve o'clock on 18 November 2022. Shooting in the morning was selected so that the sun's rays could shine without much misdirection, and the wind speed was very low so as to not affect the drone's course.

Three flights were conducted over the study area to verify and evaluate the positional accuracy of the UAV photogrammetry products and the influencing factors. Depending on the flights performed, multiple datasets were created, differing in acquisition angle, flight direction, and the georeferencing method. To compare the positional accuracies, eleven groups of dense point clouds, DEMs, and orthoimages with different parameters were created in this study (see Table 1).

Table 1. The eleven cases of processed images in this study.

No. of Case	Georeferencing Method	Combinations Image Acquisition Angle/Flight Lines Direction	No. of Images
1	Without GCPs	90°/Longitudinal flight lines	113
2	With Five GCPs	90°/Longitudinal flight lines	113
3		90°/Transverse flight lines	120
4		70°/Longitudinal flight lines	126
5		70°/Transverse flight lines	115
6		90°/Longitudinal flight lines + 90°/Transverse flight lines	113 + 120 = 233
7		90°/Longitudinal flight lines + 70°/Longitudinal flight lines	113 + 126 = 239
8		90°/Longitudinal flight lines + 70°/Transverse flight lines	113 + 115 = 228
9		90°/Transverse flight lines + 70°/Longitudinal flight lines	120 + 126 = 246
10		90°/Transverse flight lines + 70°/Transverse flight lines	120 + 115 = 235
11		70°/Longitudinal flight lines + 70°/Transverse flight lines	126 + 115 = 241

Pix4Dcapture software (Ver. 4.11.0) was used to create the flight plan in this study, as shown in Figure 3. The flight altitude was set at 60 m with 75% longitudinal overlap and 70% transversal overlap for all UAV photogrammetry tasks. Taking into account the specific imaging time of the UAV used, three tasks were performed in this study.

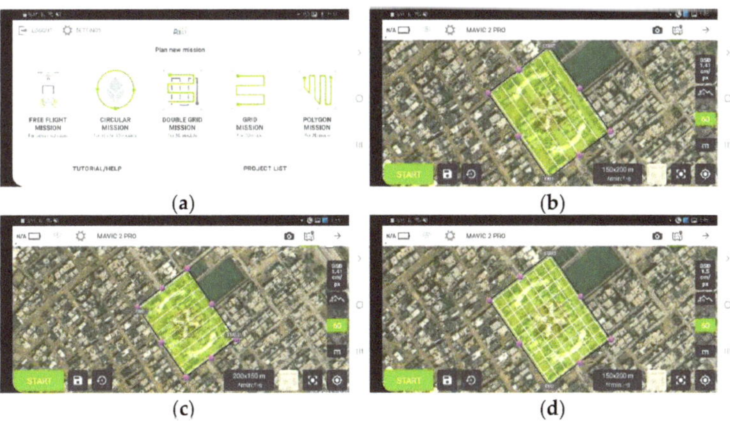

Figure 3. Pix4D capture software: (**a**) the main page, (**b**) first mission, (**c**) second mission, (**d**) third mission.

The first task was to select the GRID MISSION option in the Pix4Dcapture software, set the acquisition angle to 90° (vertical), and align the flight lines along the length of the study area (Figure 3b). These parameters resulted in a GSD value of 1.41 cm/pixel. In this task, 113 images were acquired.

The second task was also selected as a GRID MISSION option in the Pix4Dcapture software, and the acquisition angle was also set to 90° (vertical), but the direction of the flight lines was set in the direction of the width of the study area (Figure 3c). With these parameters, the GSD value was 1.41 cm/pixel, and the number of images in this task was 120 images.

In the third task, the DOUBLE GRID MISSION option of the Pix4Dcapture software was selected and the oblique acquisition angle was set to 70°. With these parameters, the GSD value was 1.50 cm/pixel, and the number of images in this task was 241. The images in this task were divided into two groups: one group with flight lines along the length of the study area, which contained 126 images, and the other group with flight lines along the width of the study area, which contained 115 images. All missions were planned as autonomous flights. Figure 4 shows examples of four consecutive images acquired during this study.

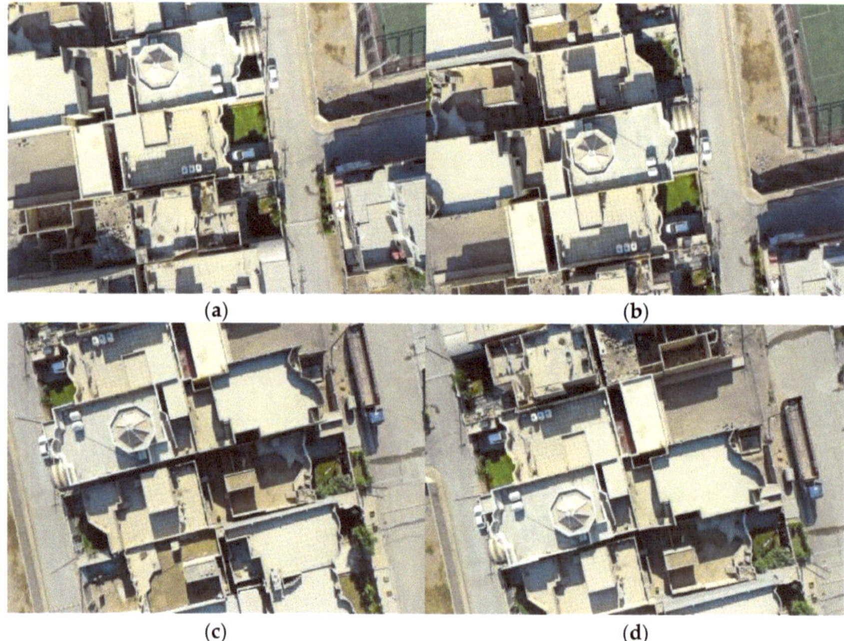

Figure 4. Samples from four adjacent images; the top two images were taken in one flight line and the bottom two images were taken in an adjacent flight line. (**a**) DJI_0891. (**b**) DJI_0892. (**c**) DJI_0908. (**d**) DJI_0909.

3.3. GCP and CP Distribution

The UAV used in this study was not equipped with precise navigation and orientation instruments, so a sufficient number of GCPs in the study area had to be established and measured before the UAV flight for georeferencing [6]. This step had two main objectives:

1. The identification and measurement of ground control points (GCPs) for indirect georeferencing.

2. The identification and measurement of checkpoints (CPs) to determine and evaluate the positional accuracy of the generated dense point cloud, DEMs, and orthophotos [13].

Field measurements before flights included measuring the coordinates of ground control points and checkpoints using a GNSS/RTK base station and rover [7]. The GCP and CP markers were precisely fabricated from square iron sheets with a side length of 50 cm and divided into four identical parts, which were assigned different colors (black and white) to facilitate their visibility and distinction by drone imagery (Figure 5).

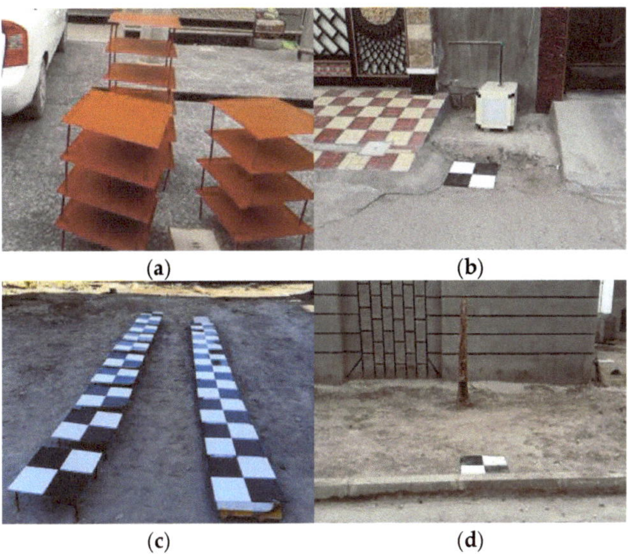

Figure 5. Preparation of GCPs and CPs. (**a**) initial iron sheets. (**c**) final Colored iron sheets. (**b,d**) samples of CPs.

In this study, 18 points were set and measured (numbered from 1 to 18). Point 1 was used as a reference point or base station, 5 points served as ground control points (GCPs) for indirect georeferencing, and 12 points were used as checkpoints (CPs) to calculate the positional accuracy of the work products. All points were placed in appropriate distribution throughout the study area (Figure 6) and were visible in the images. The GNSS survey of the GCPs and CPs was performed in RTK mode. The planimetric coordinate system was the UTM zone 38N with the geodetic reference system WGS84. The altimetric coordinate system was the EGM96 geodetic model for converting ellipsoidal heights to orthometric heights [14].

3.4. Photogrammetric Processing

All images in this study were processed using Agisoft Metashape Professional software, version 1.8.0 [15], and the parameters chosen for the different steps are listed in Table 2. Agisoft Metashape is a software tool based on Structure from Motion (SfM) for bundle adjustment and Multiview Stereo (MVS) for dense image matching. This software was the appropriate choice for the reconstruction of DEMs and orthophotos for the scene. The images were automatically processed in this software, except for some interventions in some details, such as the identification and measurement of ground control points (GCPs) and checkpoints (CPs), which were manually performed by the user.

Figure 6. Distribution of ground control points and checkpoints.

The processing of UAV images in Agisoft Metashape software was performed in the following main steps [2]:

1. Import images, GNSS data, and IMU data.
2. Align images for interior and exterior orientation and generate a sparse point cloud.
3. Import and mark GCPs.
4. Optimize the camera.
5. Build a dense point cloud using image matching methods.
6. Create of a mesh and texture.
7. Generate a DEM.
8. Create an orthophoto.

In this study, eleven sets of the dense point cloud, DEMs, and orthomosaic were created from the combinations of the acquisition images; one set was processed based on the direct georeferencing method (without GCPs), and the other ten sets were processed by indirect georeferencing using five GCPs. In all cases, the dense point cloud was used to create DEMs, and the last one was used to create the orthophoto for each image. Then, the orthophoto images were aligned to create the orthomosaic for the study area.

3.5. Accuracy Assessment

As mentioned earlier, field measurements were made at eighteen fixed points in the study area using GNSS/RTK before the flight and imaging procedure. One of these points was monitored for approximately 5 h as it served as a reference point. Then, the remaining points were monitored and measured relative to this base point. Five of these points were used as ground control points (GCPs) in indirect georeferencing, and the remaining twelve served as checkpoints (CPs) to verify the accuracy of the digital elevation models and the resulting orthophotos.

The checkpoints were used to calculate the positional accuracy of the DEMs and orthophotos produced. The positional accuracy was determined from the value of the calculated errors between the observed coordinates of the CPs using GNSS/RTK measurements and the predicted coordinates of the CPs from the geospatial data products (DEMs and orthophotos). In addition, the root mean square errors (RMSEs) were calculated for the east (X), north (Y), and vertical (Z) coordinates, as well as the total RMSE.

Table 2. Setting parameters of Agisoft Metashape used in this study.

Processing Steps	Parameters	Chosen Value
Align Photos	Accuracy	High
	Key point limit	40,000
	Tie point limit	4000
Optimization		Fit all constants (f, cx, cy, k1–k4, p1, p2, b1, and b2)
Build Dense Cloud	Quality	Medium
	Depth filtering	Moderate
Build Mesh	Source data	Dense cloud
	Surface type	Height field
	Face count	High
	Interpolation	Enabled
	Point classes	All
Build Texture	Texture type	Diffuse map
	Source data	Images
	Mapping mode	Adaptive orthophoto
	Blending mode	Mosaic
	Texture size/count:	4096
Build DEM	Projection	Geographic
	Coordinate system	WGS 84/UTM zone 38N
	Source data	Dense cloud
	Interpolation	Enabled
	Point classes	All
	Resolution	0.0532556 m
Build Orthomosaic	Projection	Geographic
	Coordinate system	WGS 84/UTM zone 38N
	Surface	Mesh
	Blending mode	Mosaic
	Pixel size	0.013 m

4. Results and Discussion

This study tested the positional accuracy of geospatial products created from UAV imagery processed with and without GCPs and examined the effects of camera orientation and flight direction. The positional accuracy results for the eleven cases are shown in Table 3. The results showed that the horizontal and vertical accuracies of the geospatial data mainly depended on the quality of the GNSS and IMU on board the UAV, and whether or not GCPs were used in the photogrammetric processing [16]. Overall, the achieved horizontal accuracy was better than the vertical accuracy in all cases. The RMSEs resulting from direct georeferencing based on the consumer-grade GNSS and IMU showed that the accuracy was severely degraded with an H-RMSE and Z-RMSE of more than 1 and 2 m, respectively.

Table 3. The positional accuracies of the eleven cases. [red color for maximum error, green color for minimum error].

No. of Case	E. errs. (m)	N. errs. (m)	Z. errs. (m)	Tot. err. (m)
1	1.830	1.203	2.885	3.622
2	0.028	0.019	0.054	0.064
3	0.029	0.029	0.112	0.120
4	0.026	0.036	0.366	0.368
5	0.066	0.080	0.059	0.120
6	0.007	0.017	0.048	0.052
7	0.074	0.125	0.107	0.181
8	0.044	0.028	0.022	0.057
9	0.049	0.039	0.043	0.076
10	0.062	0.082	0.071	0.125
11	0.008	0.009	0.028	0.031

From the results of the positional accuracy test, reproduced in Table 3, it is clear that the discrepancy was either on the horizontal or vertical planes. From these results, the effect of flight direction and camera orientation on the level of this accuracy was also evident. It can be seen that the best positional accuracy was obtained in the eleventh case (3 cm), in which the DOUBLE GRID MISSION option of the Pix4Dcapture software was selected to create the flight plan [17]. In this case, the images were acquired with an acquisition angle of 70°, regardless of whether the flight direction was longitudinal or transversal.

Figure 7 shows the plots of X, Y, Z, and the total RMSE values for all eleven cases (from the first to eleventh cases). This figure shows the significant difference in RMSEs between the first and other cases, since the first case was based on direct georeferencing, although the onboard GNSS and IMU were of low quality [18]. For further explanation, Figure 8 shows the values of X, Y, Z, and the total RMSEs for the last ten cases (from the second to eleventh cases).

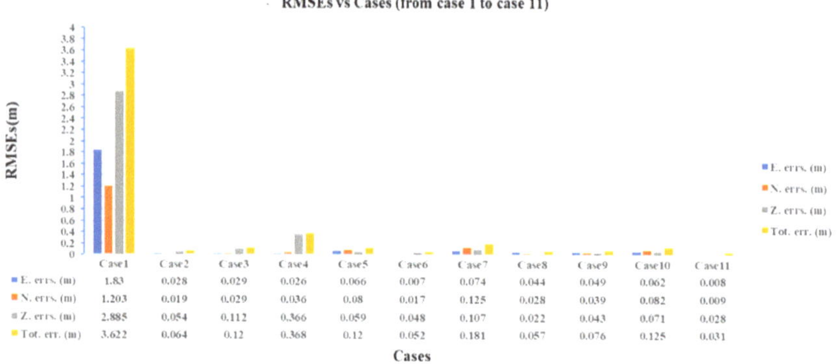

Figure 7. The RMSEs for the eleven cases.

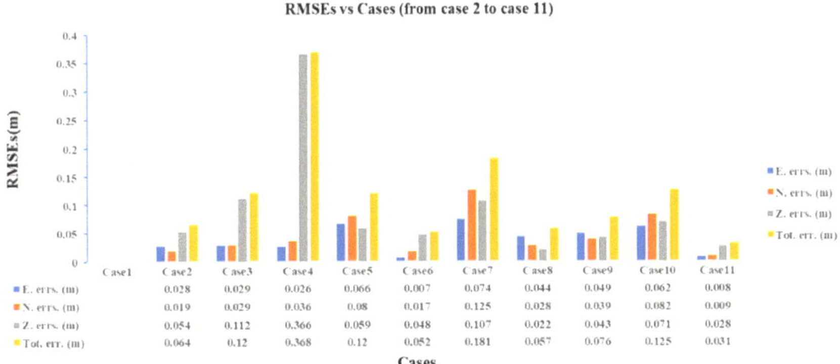

Figure 8. The RMSEs for the last ten cases.

Figures 7 and 8 show that the most influential part of the RMSEs in positional accuracy concerned heights (Z-RMSEs) [19]. From the study of all cases, it can be seen that the Z-RMSEs were strongly influenced by the image acquisition angle and flight direction, especially when the checkpoints (CPs) were at a high elevation while the ground control points (GCPs) were at ground level. In this study, the GCPs used for georeferencing, numbered 2, 7, 12, 15, and 18, were at ground level, and all checkpoints except numbers 4 and 11 were also at ground level. Checkpoints 4 and 11 were located on the roof of the second floor (about 8 m above ground level). Therefore, these checkpoints had large Z errors in all cases, especially in cases with longitudinal flight lines and 70° acquisition angles as in case four.

4.1. Positional Accuracy Test Results Depending on Flight Direction

To investigate the effect of flight direction on positional accuracy, all other parameters were set the same, including the orientation of the camera perpendicular to the ground at a 90° angle; the flight direction relative to the study area was the only variable. This resulted in three cases (two, three, and six) where the direction of the flight was longitudinal, transversal, or a combination of both directions. As shown in Figure 9, the results indicated that when the angle of detection was 90°, the accuracy obtained with the longitudinal flight direction (6 cm) was better than that with the transversal flight direction (12 cm). However, the best accuracy was generally obtained with the combined acquisition of the longitudinal and transversal flight directions (5 cm).

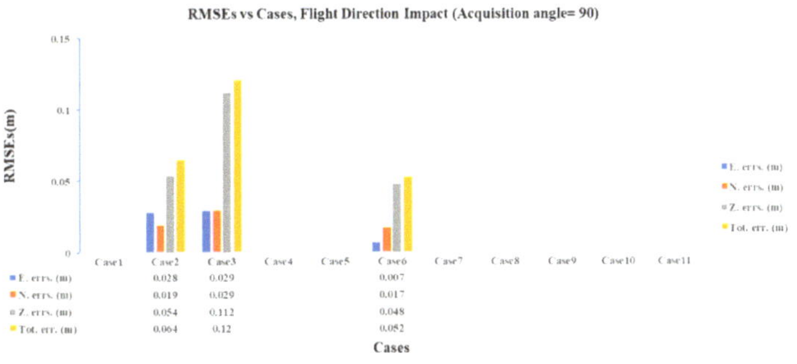

Figure 9. The impact of flight direction (acquisition angle = 90°).

On the other hand, if all parameters were set the same and the camera orientation in this test was set to 70°, and the only variable was the direction of flight, then there were three cases (4, 5, and 11). In this case, the accuracy obtained with the longitudinal flight direction (36 cm) was very poor and the accuracy obtained with the transversal flight direction (12 cm) was the best compared with the first case (Figure 10). This was due to the different detection angles (90° and 70°). In addition, the accuracy was still the best when the longitudinal and transversal flight directions were combined (3 cm).

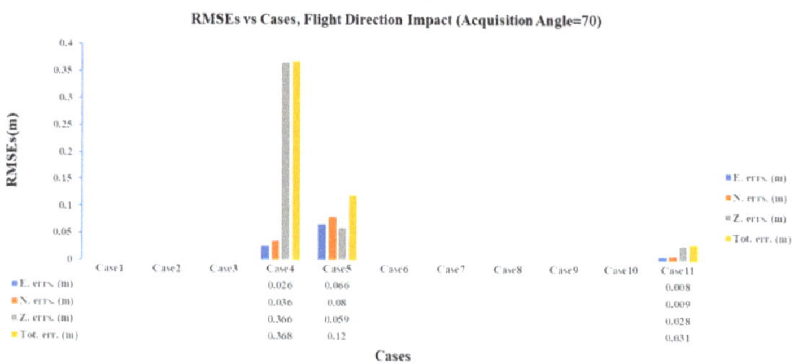

Figure 10. The impact of flight direction (acquisition angle = 70°).

4.2. Positional Accuracy Test Results Depending on Camera Orientation

To analyze the effect of individual camera orientation on positional accuracy, all other parameters were set the same, including the direction of flight longitudinally to the study area. This resulted in three cases (two, four, and seven) where the camera orientation axis was taken perpendicular (90°), oblique (70°), and both together. As shown in Figure 11, the results showed that for the longitudinal flight direction, the positioning accuracy of the vertical acquisition angle (90°), which was 6 cm, was better than that in the other cases. The worst positioning accuracy (36 cm) was obtained with an oblique detection angle (70°). This difference was very reliable because at an oblique acquisition angle (70°), the geometric distortions increased at large distances, as in the case of the longitudinal flight direction.

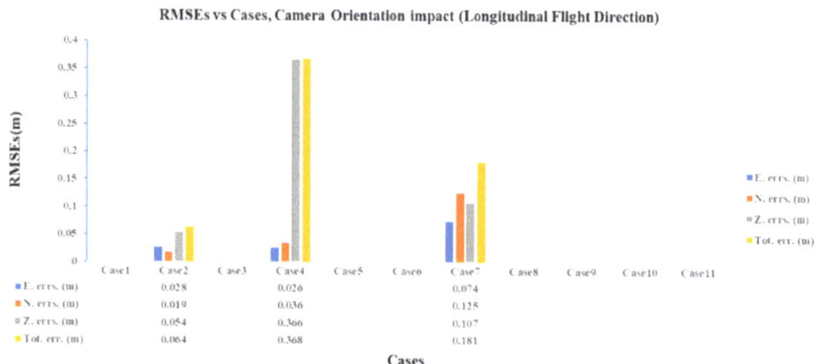

Figure 11. The impact of camera orientation (longitudinal flight direction).

On the other hand, if all parameters were set equal and the direction of flight was chosen as traversing relative to the study area, and the only variable was the camera orientation, then three cases were obtained (three, five, and ten). The results showed that the overall accuracy of the three cases in this test was almost the same (12 cm). The horizontal accuracy was better with a vertical acquisition angle (90°), while the vertical accuracy was better with an oblique acquisition angle (70°) (Figure 12).

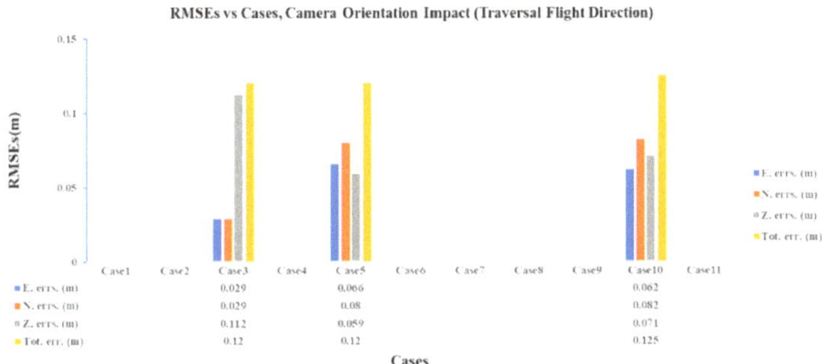

	Case3	Case5	Case10
E. errs. (m)	0.029	0.066	0.062
N. errs. (m)	0.029	0.08	0.082
Z. errs. (m)	0.112	0.059	0.071
Tot. err. (m)	0.12	0.12	0.125

Figure 12. The impact of camera orientation (traversal flight direction).

Figure 13 shows the sparse point cloud, dense point cloud, 3D model, DEM, and orthomosaic of the study area for the eleventh case.

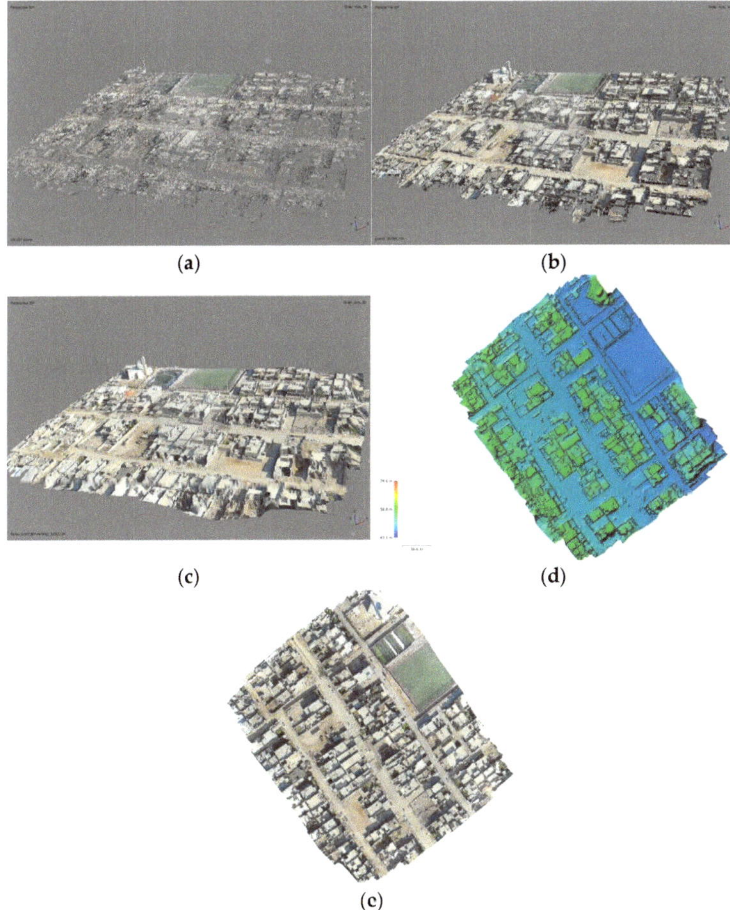

(a) **(b)**

(c) **(d)**

(e)

Figure 13. Geospatial products for the eleventh case: (**a**) sparse point cloud, (**b**) dense point cloud, (**c**) 3D model, (**d**) DEM, (**e**) orthophoto mosaic.

5. Conclusions

We investigated the positional accuracy of UAV geospatial products and analyzed the influences of flight direction and camera orientation on this accuracy. For this purpose, many sets of DSMs and orthophoto mosaics were created by processing UAV images without GCPs and then with five GCPs at different combinations of flight directions and image acquisition angles. In this study, 474 images were acquired from a DJI Mavic 2 multi-rotary UAV equipped with a Hasselblad camera during three flight missions. The workflow in this project was conducted using a case study on the west bank of the Euphrates river in Iraq. The flight altitude was set at 60 m and the overlap in the forward and lateral directions was set at 75% and 70%, respectively. For the first two missions, the images were acquired with a vertical (90°) image acquisition angle, and for the third mission, an oblique (70°) image acquisition angle was set.

The calculated root means square errors (RMSEs) in the horizontal and vertical planes showed that the horizontal accuracy was higher than the vertical accuracy in all cases. This result supports previous studies that have looked at the horizontal and vertical accuracies of UAV geospatial data. In this study, it was shown that processing UAV imagery using a direct georeferencing method for low-cost UAVs resulted in low-accuracy geospatial data products even at low altitudes. The RMSEs (X, Y, Z, and total) for geospatial data products in the first case were 1.830, 1.203, 2.885, and 3.622 m, respectively. The indirect georeferencing approach led to high accuracy, where the RMSEs decreased to several centimeters. On the other hand, this study found that a fusion of two sets of images acquired with different acquisition angles and flight directions could significantly improve positional accuracy. This result was evident from the analysis of the RMSEs (X, Y, Z, and total) of the geospatial products for the last ten cases. The highest positional accuracy for these ten cases came from the eleventh case, where the processed imagery, in this case, combined imagery acquired with a 70° acquisition angle with longitudinal flight direction and imagery acquired with a 70° acquisition angle with a transversal flight direction. The X, Y, Z, and total RMSEs for this case were 0.008, 0.009, 0.028, and 0.031 m, respectively.

Author Contributions: Conceptualization, S.A.; Formal analysis, S.A. and W.A.; Methodology, S.A.; Project administration, A.E.-S.; Supervision, A.E.-S., F.A. and W.A.; Validation, S.A.; Visualization, S.A.; Writing—original draft, S.A.; Writing—review & editing, F.A. and W.A. All authors have read and agreed to the published version of the manuscript.

Funding: This research received no external funding.

Conflicts of Interest: The authors declare no conflict of interest.

References

1. Jeong, E.; Park, J.-Y.; Hwang, C.-S. Assessment of UAV photogrammetric mapping accuracy in the beach environment. *J. Coast. Res.* **2018**, *85*, 176–180. [CrossRef]
2. Rangel, J.M.G.; Gonçalves, G.R.; Pérez, J.A. The impact of number and spatial distribution of GCPs on the positional accuracy of geospatial products derived from low-cost UASs. *Int. J. Remote Sens.* **2018**, *39*, 7154–7171. [CrossRef]
3. Nex, F.; Remondino, F. UAV for 3D mapping applications: A review. *Appl. Geomat.* **2014**, *6*, 1–15. [CrossRef]
4. Muzirafuti, A.; Cascio, M.; Lanza, S.; Randazzo, G. UAV Photogrammetry-based Mapping of the Pocket Beaches of Isola Bella Bay, Taormina (Eastern Sicily). In Proceedings of the 2021 International Workshop on Metrology for the Sea; Learning to Measure Sea Health Parameters (MetroSea), Reggio Calabria, Italy, 4–6 October 2021; pp. 418–422.
5. Ostrowski, S.; Jozkow, G.; Toth, C.; Jagt, B.V. Analysis of point cloud generation from UAS images. *ISPRS Ann. Photogramm. Remote Sens. Spat. Inf. Sci.* **2014**, *2*, 45. [CrossRef]
6. Whitehead, K.; Hugenholtz, C.H. Applying ASPRS accuracy standards to surveys from small unmanned aircraft systems (UAS). *Photogramm. Eng. Remote Sens.* **2015**, *81*, 787–793. [CrossRef]
7. Hugenholtz, C.; Brown, O.; Walker, J.; Barchyn, T.; Nesbit, P.; Kucharczyk, M.; Myshak, S. Spatial accuracy of UAV-derived orthoimagery and topography: Comparing photogrammetric models processed with direct geo-referencing and ground control points. *Geomatica* **2016**, *70*, 21–30. [CrossRef]
8. Lee, J.-O.; Sung, S.-M. Assessment of positioning accuracy of UAV photogrammetry based on RTK-GPS. *J. Korea Acad. Coop. Soc.* **2018**, *19*, 63–68.

9. Yang, H.; Li, H.; Gong, Z.; Dai, W.; Lu, S. Relations between the Number of GCPs and Accuracy of UAV Photogrammetry in the Foreshore of the Sandy Beach. *J. Coast. Res.* **2020**, *95*, 1372–1376. [CrossRef]

10. Štroner, M.; Urban, R.; Seidl, J.; Reindl, T.; Brouček, J. Photogrammetry using UAV-mounted GNSS RTK: Georeferencing strategies without GCPs. *Remote Sens.* **2021**, *13*, 1336. [CrossRef]

11. Barry, P.; Coakley, R. Accuracy of UAV photogrammetry compared with network RTK GPS. *Int. Arch. Photogramm. Remote Sens.* **2013**, *2*, 2731.

12. Koeva, M.; Muneza, M.; Gevaert, C.; Gerke, M.; Nex, F. Using UAVs for map creation and updating. A case study in Rwanda. *Surv. Rev.* **2018**, *50*, 312–325. [CrossRef]

13. Mesas-Carrascosa, F.J.; Rumbao, I.C.; Berrocal, J.A.B.; Porras, A.G.-F. Positional quality assessment of orthophotos obtained from sensors onboard multi-rotor UAV platforms. *Sensors* **2014**, *14*, 22394–22407. [CrossRef] [PubMed]

14. Ekaso, D.; Nex, F.; Kerle, N. Accuracy assessment of real-time kinematics (RTK) measurements on unmanned aerial vehicles (UAV) for direct geo-referencing. *Geo-Spat. Inf. Sci.* **2020**, *23*, 165–181. [CrossRef]

15. "Agisoft Metashape". Available online: https://www.agisoft.com (accessed on 1 October 2022).

16. Shahbazi, M.; Sohn, G.; Théau, J.; Menard, P. Development and evaluation of a UAV-photogrammetry system for precise 3D environmental modeling. *Sensors* **2015**, *15*, 27493–27524. [CrossRef] [PubMed]

17. Chaudhry, M.H.; Ahmad, A.; Gulzar, Q.; Farid, M.S.; Shahabi, H.; Al-Ansari, N. Assessment of DSM based on radiometric transformation of UAV data. *Sensors* **2021**, *21*, 1649. [CrossRef] [PubMed]

18. Forlani, G.; Diotri, F.; di Cella, U.M.; Roncella, R. Indirect UAV strip georeferencing by on-board GNSS data under poor satellite coverage. *Remote Sens.* **2019**, *11*, 1765. [CrossRef]

19. Wiącek, P.; Pyka, K. The test field for uav accuracy assessments. *Int. Arch. Photogramm. Remote Sens. Spat. Inf. Sci.* **2019**, *42*, 67–73. [CrossRef]

Article

Morphology of Dome- and Tepee-like Landforms Generated by Expansive Hydration of Weathering Anhydrite: A Case Study at Dingwall, Nova Scotia, Canada

Adrian Jarzyna [1],*, Maciej Bąbel [1],*, Damian Ługowski [1] and Firouz Vladi [2]

1   Faculty of Geology, University of Warsaw, ul. Żwirki i Wigury 93, PL-02-089 Warsaw, Poland; lugowski.damian@gmail.com
2   Deutsches Gipsmuseum und Karstwanderweg e.V., Duna 9a, D-37520 Osterode, Germany; fvladi@t-online.de
*   Correspondence: a.jarzyna@student.uw.edu.pl (A.J.); m.babel@uw.edu.pl (M.B.)

Abstract: The gypsum-anhydrite rocks in the abandoned quarry at Dingwall (Nova Scotia, Canada) are subjected to physical and chemical weathering, including hydration of the anhydrite, i.e., its transformation into secondary gypsum under the influence of water. This process is known to lead to the localized volume increase of the rock and the formation of spectacular hydration landforms: domes, tepees and ridges. Cavities appearing in the interior of these domes are often unique hydration caves (*Quellungshöhlen* in German). For the first time, this paper gives detailed geomorphometric characteristics of the 77 dome- and tepee-like hydration landforms growing today at Dingwall based on their digital surface models and orthophotomaps, made with the method of photogrammetry integrated with direct measurements. The length of hydration landforms varies from 1.86 to 23.05 m and the relative height varies from 0.33 to 2.09 m. Their approximate shape in a plan view varies from nearly circular, through oval, to elongated with a length-to-width ratio rarely exceeding 5:2. Length, width and relative height are characterized by moderate mutual correlation with proportional relations expressed by linear equations, testifying that the hydration landforms generally preserve the same or very similar shape independent of their sizes. The averaged thickness of the detached rock layer ranges from 6 to 46 cm. The size of the forms seems to depend on this thickness—the forms larger in extent (longer) generally have a thicker detached rock layer. Master (and other) joints and, to a lesser extent, layering in the bedrock influence the development of hydration landforms, particularly by controlling the place where the entrances are open to internal cavities or caves. Three structural types of the bedrock influencing the growth of hydration forms were recognized: with master joints, with layering and with both of them. The latter type of bedrock has the most complex impact on the morphology of hydration landforms because it depends on the number of master joint sets and the mutual orientation of joints and layering, which are changeable across the quarry. The durability of the hydration forms over time depends, among others, on the density of fractures in the detached rock layer.

Keywords: anhydrite; gypsum; hydration landforms; weathering; photogrammetry; geomorphometric analysis; structural analysis

Citation: Jarzyna, A.; Bąbel, M.; Ługowski, D.; Vladi, F. Morphology of Dome- and Tepee-like Landforms Generated by Expansive Hydration of Weathering Anhydrite: A Case Study at Dingwall, Nova Scotia, Canada. *Appl. Sci.* **2022**, *12*, 7374. https://doi.org/10.3390/app12157374

Academic Editors: Giovanni Randazzo, Anselme Muzirafuti and Stefania Lanza

Received: 9 May 2022
Accepted: 14 July 2022
Published: 22 July 2022

Publisher's Note: MDPI stays neutral with regard to jurisdictional claims in published maps and institutional affiliations.

1. Introduction

The hydration forms of relief described in this paper are rare morphological forms appearing in the weathering zone of gypsum-anhydrite rocks [1–9]. They are more or less convex in shape and reach up to 2–3 m in height and over several meters in lateral extension. Their origin is related to the phenomenon of volume increase during hydration of exposed anhydrite rocks under the influence of surface and subsurface waters [10–12] (and further references in [13,14]). During this process, which can be called expansive gypsification of anhydrite [14], the anhydrous calcium sulfate (the mineral anhydrite; $CaSO_4$) transforms into calcium sulfate dihydrate ($CaSO_4 \cdot 2H_2O$; gypsum) according to the reaction: $CaSO_4 + 2H_2O \rightarrow CaSO_4 \cdot 2H_2O$. This reaction can lead to volume increase only if the system is open, i.e., when the water is added from the outside [11].

The term hydration form was first used by Hunt et al. [3] (p. 6) for characteristic small-scale forms of relief "unique to karst developed on anhydrite" when "anhydrite hydrates and is changed into gypsum" and "experiences a 1.557 times increase in volume." They described hydration polygons, domes and crusts in Tripolitania province in Libya. Earlier, hydration domes with internal hydration caves were recognized in the Harz Mts. region in Germany [15].

The majority of the known hydration forms are created as a result of the localized hydration of the anhydrite substrate, leading to volume expansion and local uplift of the surface layer of the rock split of the substrate. These forms commonly show a narrow empty fissure or cavity inside [6,16–18], rarely are they massive [3,19]. Some empty cavities reach sizes large enough to be called caves. They are up to 1.5 [5,7] or even 2.3–2.4 m high [4,20], providing the possibility for an adult man to enter inside. The name hydration caves or swelling caves (*Quellungshölen* in German) was proposed for them [15,21–24]. Caves of this type are very rare in the world [25–27].

The convex hydration forms of the relief were apparently created by expansion of the rock material taking place locally, very close to the surface. According to Breisch and Wefer [17], the shape of the forms is generally a product of eccentrically acting compressive stress within the localized hydration zone. The differential expansion greater at the surface than in the rock interior causes detachment of the surface layer of the rock and its uplift, leading to empty dome- or tepee-like hydration forms, including pressure ridges. The volume expansion of the rock material is led by displacive crystallization of secondary gypsum, exerting crystallization pressure on the surroundings [14]. It has been recognized that these processes and expansions leading to the growth of the forms take place not only within the uplifted layer but also in the nearest vicinity of the forms [4,28].

The hydration forms of relief have a specified origin, shapes and sizes, and occur as discrete objects on the Earth's surface. Thus, they show the basic features defining landforms [29–31], and as such, they can be a subject of geomorphometric analysis [30,32–35].

Hydration landforms may occur in groups, creating a rare type of morphology, which can be called a hydration landscape. The discussed forms (as well as the similar weathering forms of the relief—gypsum tumuli [36,37])—can be considered a unique type of landform characteristic of sulfate karst [38] (pp. 125–126), yet not satisfactorily recognized and known.

Occurrences of hydration forms and caves have been documented in many places around the world, both in natural (e.g., Woodward Co. in Oklahoma, Ottawa Co. in Ohio, Culberson Co. in Texas, and Eddy Co. in New Mexico in the USA, Alberta in Canada, and Tripolitania in Libya) and artificial exposures of anhydrite rocks [3,4,15–17,19,22,39–50]. However, so far, only four places are known with remarkable amounts of these forms creating spectacular hydration landscape: most notably at environs of Walkenried at the south margin of Harz Mts. in Germany [2,4,6,15,47,49,51–53], in the abandoned quarry at Pisky in Ukraine [13,18,54–60], on the small Alebastrovyye Islands in Russia [5] and at the abandoned quarry at Dingwall in Canada [6–9,14,48,61]. Among them, the site at Dingwall in Canada (Figures 1 and 2a–e), studied by the authors, is the largest in number and size of hydration forms and caves. This permits the investigation of their geomorphometric features in a statistical way.

During the growth of these landforms, the rock creates a generally convex shape, described as dome-like, but also tepee-like, triangular in cross-section (tents in [48]) [6,61]. The other authors called them gypsum bubbles due to their predominantly dome-like shape [6,17,39,48], mounds [5,16], and blisters [42,48], but because of their similarity to the characteristic low wall-less army tents—also tents or A-tents [48,62]. Based on the observation of changes in the form shape over time (Figure 2e,f) [7,9], it has been proposed that the shape of the forms from Dingwall will change from a dome (bubble) to a tepee (tent [48]), similar to the case of the hydration form called Waldschmiede in Germany [4,49].

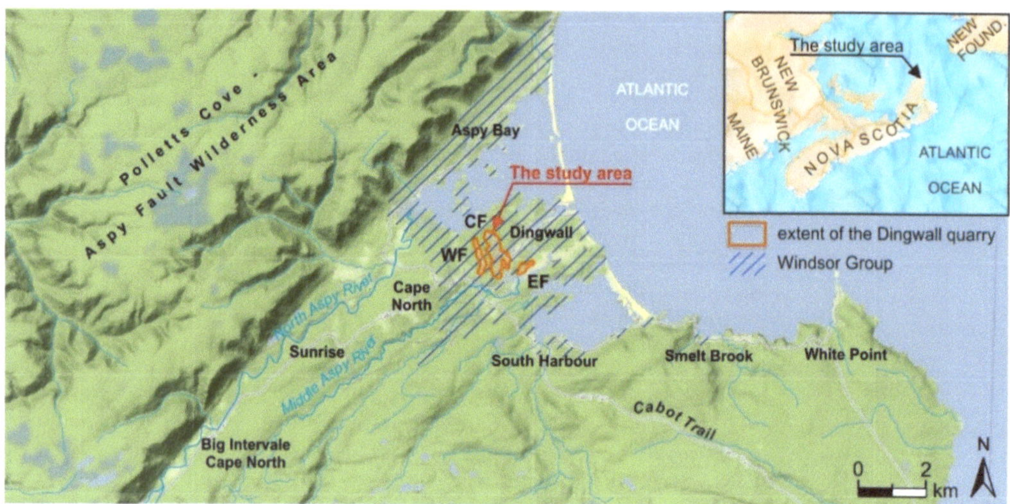

Figure 1. Location of the abandoned gypsum quarry at Dingwall, eastern Canada, along with Windsor Group. Quarry is divided into the western field (WF), the central field (CF) and the eastern field (EF). Source of maps: ArcGIS program (Esri, Redlands, CA, USA) and http://maps.stamen.com (accessed on 25 June 2022).

Figure 2. *Cont.*

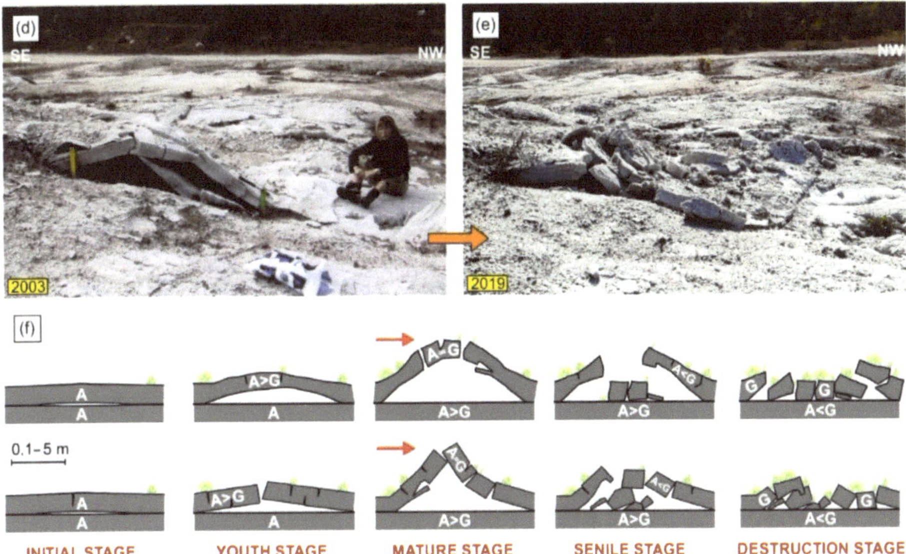

Figure 2. Hydration landscape and landforms at Dingwall in photographs taken with the use of UAV (**a**–**c**) and digital camera (**d**,**e**) and interpretation of their development (**f**); (**a**) northern part of the CF (Figure 1); (**b**) view of the CF showing form no. 15 with hydration cave named Personal Grotto (see Section 4.1) with the author (D.Ł.), (**a**) and (**b**) photographs were taken on 28 August 2019; (**c**) the largest hydration cave, Ramesh Cave seen from above (the authors F.V., A.J., D.Ł. are on the photograph taken on 21 September 2018); (**d**,**e**) sixteen-year morphological evolution of exemplary hydration dome from mature to destruction stage; note the development of weathering debris from massive anhydrite bedrock, such as one on which Susanne Vladi is sitting; form without identification number located 10 m east of the Ramesh Cave (see Section 4.1); (**f**)—evolution stages of domed (upper row) and tepee-like (lower row) anhydrite hydration form associated with changes in mineralogical composition (A–anhydrite, G–gypsum); maximum relative height is indicated by red arrow [57].

An intermediate form similar to both was called the rounded tent [48]. Deformations formed during gypsification can also take a more diverse, irregular and difficult to define shape, such as a bridge [18], or elongated forms, such as long ridges, called pressure ridges [7].

The studied tepee-like hydration landforms represent the morphological equivalent of the well-known tepee structures commonly recorded in carbonate and evaporite rocks [63] but differ from them in not forming polygonal ridges but creating isolated forms rounded in a plan view. Only the elongated hydration pressure ridges co-occurring with the studied forms are equivalents of the typical tepee structures. Many tepee-like hydration landforms are similar to granite A-tents or "pop-ups" landforms [62,64–66]. On the other hand, the dome-like hydration forms are very similar to those of the gypsum tumuli [36,37,67].

Hydration forms usually develop on the anhydrite-dominated bedrock, characterized by varying degrees of coverage by fractures but also with massive structures [4,5,14]. The fractures are not only the migration path of the fluids, causing the transition of anhydrite to gypsum and generating the development of forms. They also influence the shape and orientation of the growing forms. Stenson [48] found at Dingwall that when fractures are present in the bedrock, "tents" with sharp crests and triangular vertical cross-sections develop, while in the absence of them, the rock deforms in a more "plastic" way, generating more rounded forms, such as bubbles or blisters. Near Walkenried, the dependence of the position and elongation of hydration domes and the accompanying depressions on the system of joints have been documented by Reimann [4]. In turn, the elongation of hydration forms from the Alebastrovyye Islands was found to be generally parallel to the strike of the anhydrite layers [5].

Lateral sizes of the discussed forms reach values of more than 10 m (Table 1) [2,5,8,18], in extreme cases of pressure ridges, even 25 m [61]. The largest documented hydration landform was the Great Dome at Pisky, which has dimensions of 18.1 × 15.5 × 3.0 m, with an inner cave of 9.5 × 7.8 × 1.2 m [18,58]. The other large hydration form found in the forest near Walkenried was Waldschmiede (Forest Forge). Its dimensions in 1933 were 10.0 × 7.0 × 1.9 m [68] and its inner cave measured a few years earlier was 8.0 × 7.5 × 2.0 m in size [53]. The height of this growing cave was 2.3 m, although their lateral sizes shrunk [1,4,58]. Another hydration form of extraordinary size is the Ramesh Cave in the Dingwall quarry, with dimensions of 9.95 × 4.10 × 1.32 m [8]. Documented sizes of hydration caves are often changeable and grow or decrease with time [4,9,49]. A detached rock layer is commonly from a few cm to several tens of cm thick [5,6,18,51,53,61], and in extreme cases, in side parts of forms 200 cm thick [9]. The thicker layers, as a rule, occur above the largest hydration caves [45].

Table 1. Morphometric data of anhydrite hydration landforms (including pressure ridges at Dingwall) and hydration caves at Pisky, Walkenried, the Alebastrovyye Islands and Dingwall.

Site Name/ Province/ Country	Pisky/ Lviv Oblast/ Ukraine	Walkeried/ Lower Saxony/ Germany	Alebastrovyye Islands/ Novaya Zemlya/Russia	Dingwall/ Nova Scotia/ Canada
References	[7,13,18,20,28,54–60]	[1,2,4,6,7,49,51,68]	[5]	[6–9,14,48,61]
Number of documented hydration forms	99	26	unknown number	69 *
Area with hydration relief	200 × 300 m	1300 × 600 m	800 × 150 m	1800 × 1400 m
Range of dimensions of hydration forms	length: 0.58–18.1 m width: 0.32–15.5 m height: 0.05–3.0 m	length: <10 m width: 7.0 m height: <3 m	length: 3–15 m width: 1–10 m height: 0.5–1.5 m	length: <25 m width: 0.8–8.2 m height: 0.1–2.4 m
Range of dimensions of inner hydration caves	length: <9.5 m. width: <7.8 m height: <2.4 m	length: <10 m width: <7.5 m height: <2.3 m	length: unknown width: unknown height: <1.5 m	length: <10.7 m width: <6.6 m height: <1.32 m

*—data after Stenson [48].

Until now, the most frequently used methods in the study of hydration forms were standard geological and geomorphological observations, drawing and photographing. Measurements of the most important dimensions, such as length, width and height, were made with a ruler, but also a laser rangefinder. Moreover, the azimuth of elongation was measured with a geological compass and location was noted by a GPS device [18,48]. Terrestrial photogrammetry was recently used at Pisky, which, in combination with ArcGIS (Esri, Redlands, CA, USA) and Photoscan (Agisoft LLC, St. Petersburg, Russia) computer programs, was the basis for supplementary morphometric measurements [57,60]. To measure the change in the distance between the two points in time, the benchmark method with metal bolts was used [6,60]. At Dingwall, Stenson [48] applied statistical analysis of measured geomorphometric parameters of 69 hydration landforms (Table 1), calculating, inter alia, the average or the minimum and maximum value, and also calculated unknown parameters on the basis of already determined dimensions. However, Dingwall does not yet have detailed and comprehensive documentation of aerial and terrestrial photogrammetry and geomorphometric analysis based on these data.

Morphometric three-dimensional features, sizes and spacings are crucial for characterizing any landform [29] and also for remote recognition of them both on Earth and the other planets [30,69]. Hydration forms of relief analogous to terrestrial ones can potentially be present on Mars [70].

The aim of this paper is to give the geomorphometric characteristics of the anhydrite hydration landforms from Dingwall. The shapes, sizes and other distinctive morphological features of these poorly known forms are described quantitatively with the use of photogrammetric and statistical methods of analysis for the first time. Our study is based on the examination of a relatively large number of forms and supplements to previous information on this subject. We also attempted to recognize the influence of the structural elements of the weathering anhydrite bedrocks, such as joints and layering, on the morphology of the hydration forms and their evolution. The hydration forms from Dingwall are compared with similar hydration forms and other landforms found elsewhere on the substrate of weathering anhydrite and gypsum rocks.

2. Study Area

The studied quarry is located on the north side of Cape Breton Island in Victoria Co., on the edge of both Dingwall and Cape North villages (Figure 1). The mining of gypsum in the excavation was carried out by the National Gypsum Company in 1933–1955 [71,72].

The bedrock of the study area consists of Precambrian igneous-metamorphic rocks intruded by Devonian magmatic bodies with a predominantly granite composition [71,73]. The bedrock is covered with the Carboniferous evaporitic-clastic deposits [71], which include the Horton Group (Tournaisian) [74] and the Windsor Group (Visean) [71], the latter exposed at Dingwall quarry (Figure 1). The next younger rocks are Quaternary in age and are related to the activity of the Wisconsinan ice sheet, lasting from 75 or 65 to 11–10 ka [75,76]. It left the layers of tills lying on the eroded Carboniferous and older rocks.

The gypsum-anhydrite layers in the studied quarry show a variable arrangement. The data collected in the field and from the available maps [71] indicate that the predominant dip is in the eastern direction and its value mostly oscillates between 15° and 30°, but also is horizontal (see Section 4.6.3) [14].

The outcrop is characterized by several meters deep, vertically elongated and narrow in shape Schlottenkarren (densely packed, funnel-shaped sinkholes draining into vertical cylindrical shafts [76,77]). According to Moseley [76], they can be either exhumed forms originating before the ice sheet comes or post-glacial in age. Currently, the Dingwall area is subjected to intensive karst solution, including the formation of gypsum karren [48,78].

The climate of Dingwall is temperate and cold [14]. According to ClimateCharts. net [79,80] in the quarry area (34 m a.s.l., based on data from 1901–2019), the precipitation reaches 1242.7 mm/year with monthly values between 85.8 mm (in July) and 136.6 mm (in November). The mean temperature is 4.7 °C. The coldest month is February, with an average temperature of −6.7 °C, and the warmest one is August, with an average temperature of 16.4 °C.

3. Materials and Methods
3.1. Field and Laboratory Photogrammetric Works

Fieldworks were led during two expeditions in 2018 (10–26 September) and 2019 (15–29 August). We measured all the largest dome- and tepee-like hydration forms with internal caves and the largest hydration chambers using a laser range finger and tape measure. Smaller hydration domes and tepee-shaped forms of a few to several tens of centimeters in height and less than ca. 1.8 m in diameter in a plan view are very abundant at the site (e.g., see Section 4.1) (Figure 7 in [9]). These as well as badly damaged forms (Figure 2d,e) and some pressure ridges up to 1 m or more in height were omitted from our quantitative documentation. We also did not measure irregular detachments of the rock layers (see Section 4.1) and extensive, almost flat uplifts of these layers, with an extension of several meters or more.

Our investigations focused on convex hydration landforms showing significant sizes—those that attain more than about 2 m in diameter. The basic identification criteria used for their recognition were as follows: (1) pronounced convex shape, (2) the presence of a cavity under the rock surface; when this cavity was invisible, then its presence was detected by the rumble sound during hitting the ground with a hammer.

The hydration forms were recorded by direct measurements and photogrammetric methods using Unmanned Aerial Vehicle (UAV) and digital cameras. Photogrammetry was also used to visualize the bottom of the quarry and the individual forms. During field work, the form was identified with the assignation of a documentation number, its most important morphometric parameters (Table 2; Figure 3) and the location. The delimitation of the forms in the plan view was outlined on the basemap of the orthophotomap with the use of a portable mobile device (Lenovo Yoga Tab 3, Hong Kong, China). Joints and fractures were measured with a geological compass and distances with the laser rangefinder and the tape measure.

Table 2. Geomorphometric parameters measured and computed for 74 hydration forms, including symbol, unit, formula, maximum, minimum and average value, references and frequency distribution diagrams (horizontal axis has the same unit as the parameter, vertical axis is the number of measurements).

Name of Parameter	Symbol, Unit and Formula (If Applied)	Maximum–Minimum; Average Value	Number of Measurements	References	Simplified Frequency Distribution Diagrams
Length *	a (m)	1.86–23.05; 5.25	74	[36] *modified*; [67] *modified*; [81] *modified*; [82,83]	
Width	b (m)	0.92–9.01; 8.38	74	[36] *modified*; [67] *modified*; [81] *modified*; [82,83]	
Relative height	h_r (m)	0.33–2.09; 0.83	74	[84] *modified*	
Thickness of detached layer	e (m)	0.06–0.46; 0.21	54	[18,36,64]	
Coefficient of circularity	C (–) $(a - b)/a$	0.04–0.73; 0.33	74	[18,36,67]	
Azimuth of elongation orientation	Az (°)	0–176 ****; 84	68	New	
Bulge degree	B (–) b/h_r	1.82–8.58; 4.28	74	[84]	
Fracture density	F_D (m/m^2) $\Sigma Ls/P$ ***	0.04–3.38; 0.85	69	[85,86] *modified*	
Length of the entrance line **	— (m)	0.30–4.52; 1.21	98	New	
Azimuth of the entrance line **	— (°)	0–177 ****; 92	98	New	
Azimuth of entrance **	— (°)	0–350 ****; 92	92	[18]	
Direction of maximum lateral expansion	— (°)	3–67, 80–115, 130–176 ****; 83	66	[66] *modified*	

*—measurements do not involve small hydration forms with a length less than 1.8 m, **—parameter concerns the entrance to hydration cavity or cave inside the form, ***—explanations of symbols in the text (see Section 3.2.2), ****—range of the values.

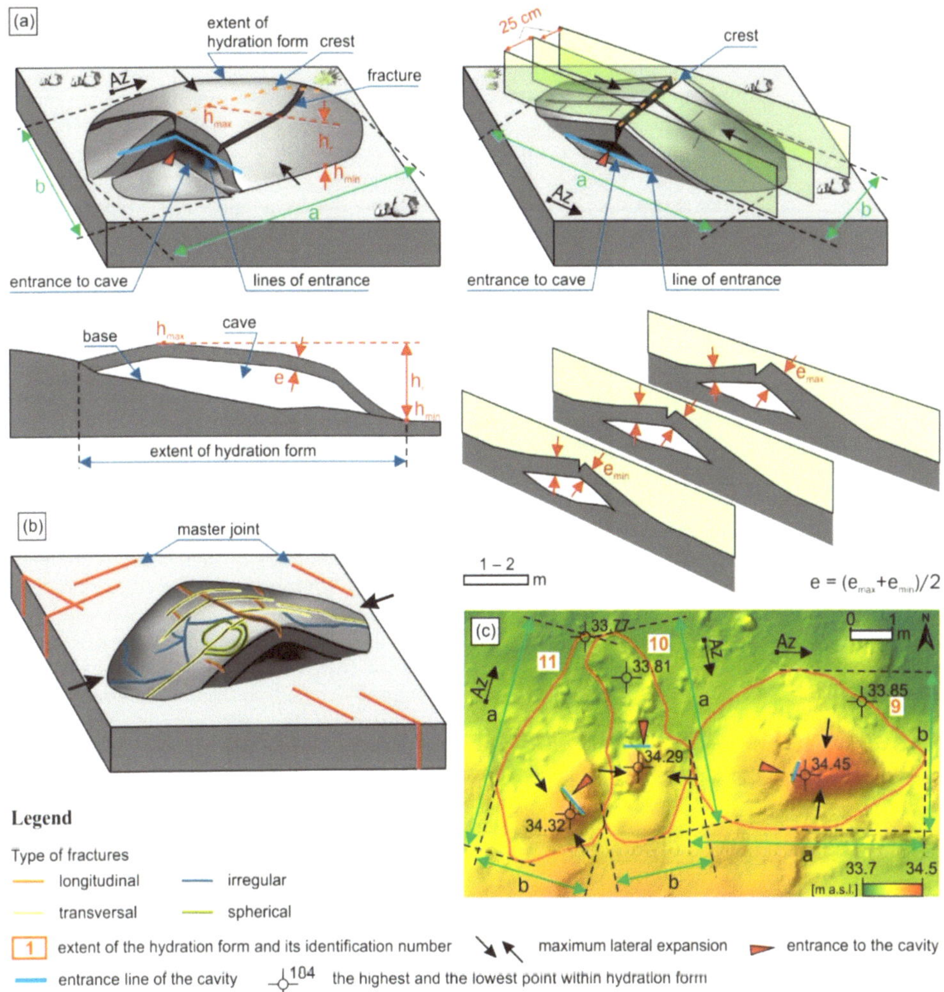

Figure 3. Morphometric parameters of domed (top left) and tepee-like hydration forms (top right) with method of measuring and calculation of the thickness of detached rock layer using cross sections through *.las* data, presented both on schematic drawings (**a**) and exemplary DMSs (**c**), and types of fractures recorded within hydration landforms including those inherited from master joints (**b**); a—length of the form in plan view; b—width of the form in plan view; Az—azimuth of elongation orientation; e—thickness of detached rock layer; e_{min}—the smallest, measured thickness; e_{max}—the largest, measured thickness; h_r—relative height; h_{max}—the largest value of elevation within extent of the form; h_{min}—the smallest value of elevation within extent of the form. Further explanations in the text.

The applied methodology of photogrammetric documentation is described in the previous work of the authors [7], the other papers [87–89], the instructions for the program used [90] and in the supplementary materials (Figure S1). These works mainly involved the use of the Photoscan program (Agisoft LLC, St. Petersburg, Russia) based on the Structure-From-Motion method [89]. The usefulness and effectiveness of the applied photogrammetric methodology in geospatial analysis have been demonstrated in many papers (e.g., [91]).

3.2. Methodology of Dataset Analysis

3.2.1. Analysis of Morphology and Morphometry

An orthophotomap and a Digital Surface Model (DSM) were created for every studied hydration landform involving its surroundings. The forms are most commonly domed (e.g., Figure 2b) or tepee-like in shape (e.g., Figure 2c). The basis for distinguishing the tepee-shaped forms is the presence of at least two rock slabs with semi-flat or concave morphology, raised in place of their contact and leaning against each other, "pressing" each other from opposite directions, occupying the dominant part of the form, usually in the middle of it. The dome, on the other hand, has a dominant rounded (convex) shape without a sharp crest.

The geomorphometric parameters for the characteristics of the studied landforms were adopted from descriptions of the analogous domed forms of relief: hydration forms [18,48], gypsum tumuli [36,37,67], drumlins [82,83,92], and granite A-tents [64,66]. Some parameters were modified and the new ones were also introduced as described below. The geomorphometric description of the landforms was carried out during both field and office work and was adapted to their convex morphology.

During the office work, photogrammetric data and data obtained as a result of their further processing, such as 3D and 2.5D models [93], orthophotomaps, point clouds with extension *.las* and DSMs were collected and elaborated. These works mainly involved ArcGIS, Photoscan, Excel and Corel Draw X9 programs.

Work in the ArcGIS program (Esri, Redlands, CA, USA) began with the transfer of the extent line for each of the 74 identified hydration forms to the orthophotomap of the area (3 forms were excluded from the analysis because of the lack of photogrammetric data; see Table S1). Based on the extent of the form, lines of the longest axis (length of the form, a—longer axis; Figure 3) were drawn and measured on the map similarly to Spagnolo et al. [82] and Maclachlan and Eyles [83], as the longest straight line possible within the mapped forms area. Then, the width of the form (b—short axis; Figure 3) perpendicular to the a axis was drawn and measured [82,83]. The crest lines through the points of maximum curvature of the majority of contours were also marked (Figure 3) [92].

An orientation of the axis a, i.e., elongation of the form, was computed to acquire the azimuth of elongation orientation Az of every form and, on the basis of these measurements, to generate a rose diagram. On the basis of the DSM, the Zonal Statistic as Table tool [94] (pp. 327, 349) was used to measure the values of elevations (in m a.s.l.) marked by h_{max} and h_{min} (Figure 3). Value h_{max} is the peak at the highest elevation point (usually located in the middle of the form) while h_{min} is the lowest elevation point of the hydration form within its extent. The relative height of form h_r is understood as the form altitude range and is the result of subtraction h_{min} from h_{max} (see [84]). This value does not take into account the hillslope of the terrain surface, which at Dingwall is generally minimal. When the forms were overgrown with vegetation, the covered part was excluded from the calculation of relative height.

As part of the statistical analysis realized in the Excel program for all the calculated parameters (Table 2), a simple linear regression analysis was performed to reveal the possible regular relationships between these parameters [95]. The regression lines were generated and the Pearson correlation coefficient was calculated. For the values: a, b, h_r, the regression line was fixed at a point (0,0) in the coordinate system to obtain a more realistic equation of their mutual dependence. All the values (a, h_r, B, C; see below) were used to create their frequency distribution diagrams (histograms). They were further analyzed in order to select or distinguish the characteristic groups that best reflect the real state.

The length and width of the hydration forms were used to calculate the coefficient of circularity (C), characterizing the degree of elongation according to the formula [18,36]:

$$C = (a - b)/a \tag{1}$$

It was calculated in a similar way as the so-called flattening in case of elliptical objects [67]. For its value closest to 0, the form is the most round; for the value closest to 1, the form is the most elongated. The coefficient of circularity is a parameter analogous to the elongation (degree of elongation, E), calculated as the ratio a/b often used in the description of similar forms of relief [81,84]. The relationship between both parameters is described by the formula:

$$C = 1 - 1/E \qquad (2)$$

The ratio called bulge degree (B) was calculated as in the case of drumlins [84] by the formula:

$$B = b/h_r \qquad (3)$$

It was used in order to determine the convexity of hydration forms quantitatively. The increasing value of the ratio b/h_r informs about a higher flattening.

A detached rock layer in the studied forms is characterized by a thickness (e in Figure 3) measured perpendicularly to the surfaces of this layer, as in the case of the A-tent landforms [64] or gypsum tumuli [18,36,67]. Based on the field measurements and detailed cross-sections, its thickness was determined, and when *.las* data were not available, only field measurements were taken into account. Cross-sections were made thanks to the *.las* format data generated in Photoscan (Agisoft LLC, St. Petersburg, Russia) and analyzed in the ArcGIS program (Esri, Redlands, CA, USA). The cross-sections were drawn, as shown in Figure 3, in the ArcGIS program using the Stack Profile tool. Where the rock layer thickens considerably toward the margin of the form, the thickness measured in these places was not considered.

Additionally, the orientation of entrances to an internal space or cave of hydration forms was analyzed. The entrance is defined as breaking the continuity of the detached rock layer, revealing the interior of the form. Some of the internal spaces of the forms are large enough for direct adult human exploration and therefore represent hydration caves [27,96], or, strictly speaking, proper caves sensu Curl [96,97]. On the orthophotomap, the upper edge of the entrances was commonly seen as a more or less long and straight line (entrance line; Figure 3). These lines were marked and their orientations were determined to detect some regularities of their orientations and a possible influence of the structure of bedrock on entrance formation. Only the entrance lines that show the minimal length of 0.30 m or more were marked. In cases where the lines were curved, they were approximated to the straight lines, while in the case of an entrance creating two distinct lines, the azimuths of both lines were measured separately (see Figure 3). The influence of the bedrock structure on the development of entrances was examined by tracing the fractures and layering visible in the quarry and comparing their orientation with the orientation of the entrance lines. Entrance lines were used to mark the azimuth of the entrance to the cave or cavity as generally normal to these lines. The azimuths directed out of the internal space of the forms were calculated and analyzed statistically.

3.2.2. Analysis of Structural Elements

The characteristics of the structural elements included the mentioned layering and fractures. They were documented in order to recognize their relationship with the morphology of the hydration forms (their elongation, orientation of entrances).

The strike and dip of the gypsum-anhydrite layers were measured directly in the quarry, and the strike of the layers visible in the bedrock was traced and documented as lineaments on the orthophotomaps in the ArcGIS program (Esri, Redlands, CA, USA).

Similarly, the fractures were both measured directly and traced using the orthophotomap in the ArcGIS program (Esri, Redlands, CA, USA). Attention was paid to fractures showing the features of systematic joints and master joints [98]. Traces of such fractures or joints were recognized on the orthophotomap and marked on areas of the quarry where they were best seen. The tilt value of these structures was assumed to be vertical or nearly vertical, based on the direct field measurements and observations. Joints and fractures measured in the quarry were analyzed on Angelier diagrams, whereas joint traces on the quarry bottom were characterized on rose diagrams [99]. The rose diagrams (in the ArcGIS program using polar plots), unlike the traditional ones, took into account the length of the fractures, which translated into a more realistic analysis. The description of the fractures was based on the terms and classifications presented by Peacock et al. [100].

To characterize the deformation a rock has undergone to give convex hydration forms, we attempted to determine the direction of the maximum lateral (horizontal) expansion (or extension) within this surface layer (Figure 3) in a similar way as Ericson and Olvmo (Figure 9 in [66]) designated the direction of extension in A-tens (see also [65]). The maximum lateral expansion (MLE) was marked as normal to the line of the crest of the form. The wavy-shaped crests were approximated to straight lines and MLE was assumed to be normal to this line. In many cases where precise determination of the direction of the MLE was difficult or impossible, it was marked arbitrary. The relations of MLE to elongation of the forms were investigated by determining the angles between both directions.

Attention was paid to the fractures occurring on the surface and in the vicinity of the hydration forms. Their orientations were measured to reveal the influence of master joints on the development of fractures within hydration forms and in relation to the direction of MLE [85]. The intersections of fractures were studied to determine the order of their generation [100,101]. In addition, the surface density of fracture traces [102] in 3D space, F_D, was calculated for every form according to the modified formula [85] and similar to the trace density formula [86]:

$$F_D = \frac{\Sigma Ls}{P} \tag{4}$$

where:

Σls—sum of the length of the fracture traces within the hydration form surface measured in 3D (m)
P—surface area of hydration form (m^2)

4. Results

The generated photogrammetric models and their survey, taking into account field observations and statistical analysis of the collected data, provided the basis for morphological and structural analyses of the hydration forms. Photogrammetric data include several good-quality 3D and 2.5D models of hydration landforms (see Figure S1 for more details). A map of the quarry with marked traces of fractures and layers, as well as the extent of hydration forms, is a main result of the work (Figure S8). The complete documentary and statistical data are presented in Supplementary Materials (Table S1, Figures S1–S9), as well as on video materials available on Youtube.com (profile name: hydration caves), which present a comprehensive view of the hydration landscape. The photographic and cartographic documentation of the studied hydration forms is presented both in Supplementary Materials and on the website [103].

4.1. Morphology of the Quarry Bottom

The quarry, divided into three parts: WF, CF and EF (Figures 1 and 4), occupies about 78 ha at an altitude of 0–53 m a.s.l. with the lowest terrain in the EF and south of the CF. The rocks are very well exposed and only in places covered with low sparse vegetation. Young trees up to 1.5 m high grow in the oldest parts of the exploitation fields, containing more weathering debris. Small ponds occur in several places and the EF is partially covered by periodic wetlands (Figures 4 and S3).

At the end of mining operations in 1955, the bottom of the quarry, mostly cut in the anhydrite bedrocks, was left generally flatted and remained exposed to weathering. Hydration domes and ridges ("pressure blisters" and "pressure-ridges") growing at the quarry bottom, as well as already destroyed ones, were observed as early as 1969, testifying to their earlier development [104]. A present-day rough, uneven relief of the rocky bottom is mainly a product of expansive hydration of anhydrite due to the reaction with waters of meteoric derivation. This process has been acting since 1955 and from the 1930s at the earliest—in places where the first gypsum mining was started. The relief produced by present-day dissolution (karren, [78]) is of minor importance. The created hydration landscape dominates in many parts of the quarry (Figures 2, 4 and 5a,b). The large hydration domes and tepee-shaped forms, with a relative height of above 0.5–1.0 m and an extension in a plan view up to several and a dozen meters, are the most distinctive products of expansive hydration. Seventy-seven of these forms were documented, of which three were only partly measured and did not have photogrammetric records (Figure 4, Table S1). Within the examined forms and in their vicinities, numerous fractures are present, including fractures belonging to the master joint system. The formation of new fractures, evidently younger than the master joints, was documented by photographs (e.g., Figure 4 in [8]; Figure 5 in [9]). They have apparently been formed due to present-day weathering, including anhydrite hydration [14].

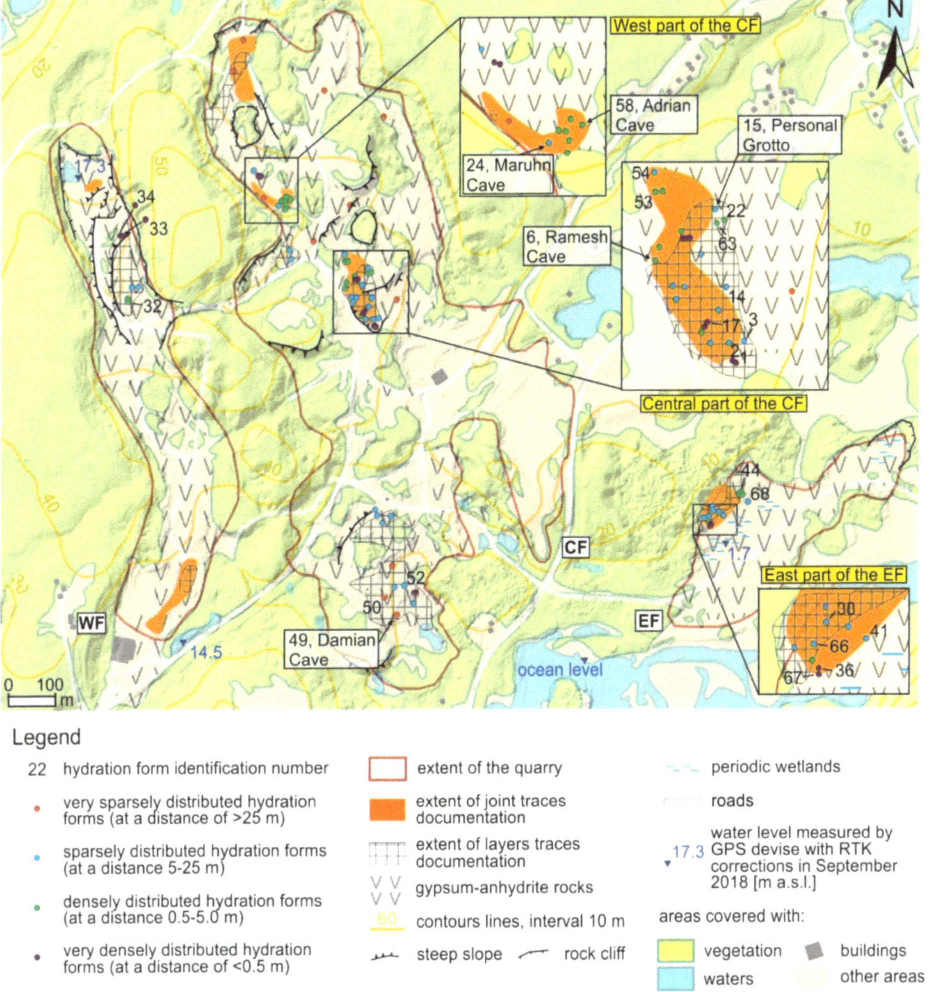

Figure 4. The map of 77 studied anhydrite hydration landforms at Dingwall (source of base maps: ArcGIS program (Esri, Redlands, CA, USA) and nsgi.novascotia.ca); forms mentioned in the text are numbered (full numeration is shown on the maps in Figure S3).

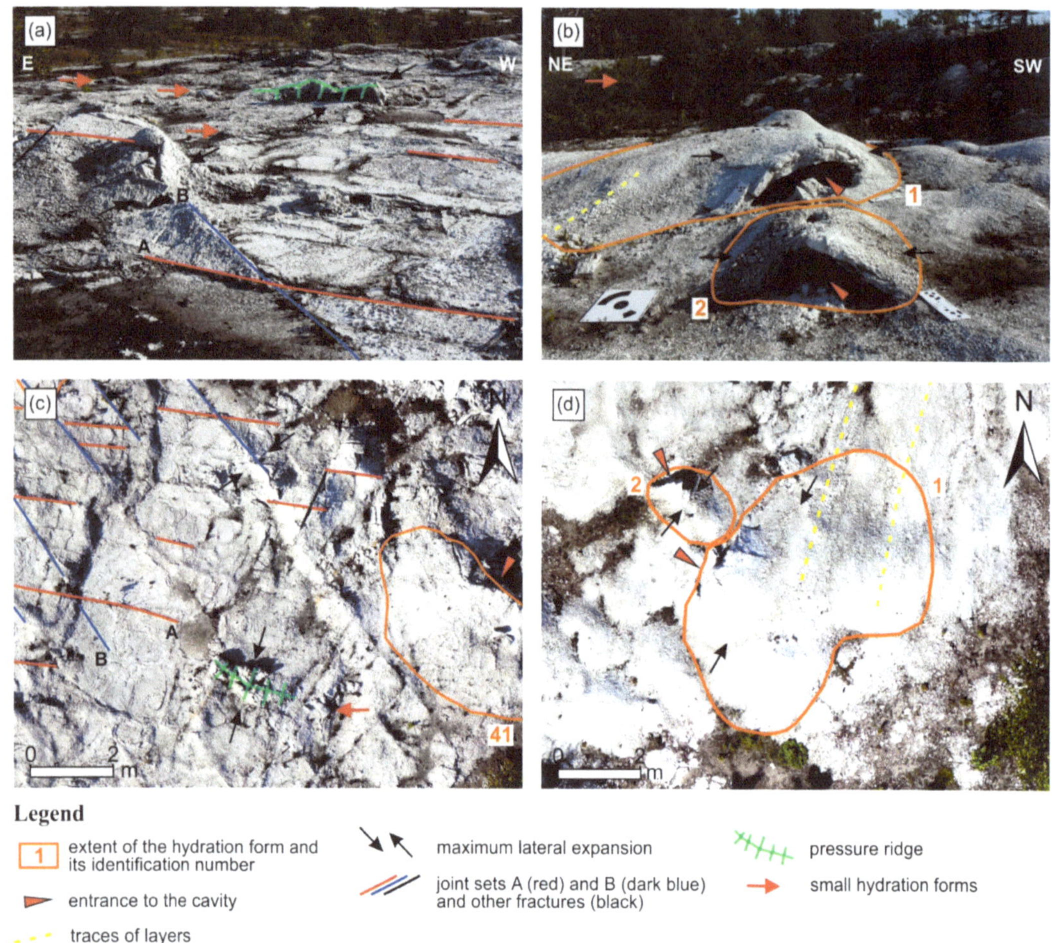

Legend

1	extent of the hydration form and its identification number

entrance to the cavity

traces of layers

maximum lateral expansion

joint sets A (red) and B (dark blue) and other fractures (black)

pressure ridge

small hydration forms

Figure 5. Hydration relief of the Dingwall site; (**a,c**) part of the quarry bottom with visible bulges, fractures and a pressure ridge; note unnumbered forms not measured in this study. (**b,d**) Dome-like hydration forms with entrances to the inner cave (no. 1) and smaller cavity (no. 2); see Figure 4 for the location of the forms.

The rock surface of hydration forms is solid but also covered with loose gypsum-anhydrite weathering debris composed of cm-mm-sized sharp-edged rock fragments [8]. This debris develops relatively rapidly with time, which is very well seen in the photographs taken several years apart (Figure 2d,e). In the oldest parts of the quarry, initial soil and vegetation have developed on the debris, forming a several cm thick weathering mantle. In some parts of the quarry, this weathering mantle (regolith) has been cut by a network of erosional furrows (Figure S8). Because of the presence of the weathering mantle, debris and vegetation, the structure of the bedrock is entirely invisible in vast areas of the quarry.

4.2. Distribution of Hydration Landforms

The hydration landforms were recorded only in a few parts of the quarry and were presented on a topographic map (Figure 4) and a satellite map (Figure S3). The distances between the forms ranged from less than 0.5 to more than 25 m. Considering the following three limits of distances—0.5 m, 5 m, and 25 m—four groups were distinguished according to the range of a distance between the forms: very sparsely (14%), sparsely (40%), densely (26%) and very densely (19%) distributed forms (Figure 4). A group is a set of at least two forms. Two close to each other forms from the CF (no. 1 and 2, Figures 4 and S2) are an example of a very densely distributed hydration form (Figure 5b,d). The highest number of forms (31) occurs in the middle of the CF, but they vary, from very sparsely to very densely distributed (Figure 4).

4.3. Shape and Morphometry of Hydration Landforms

The hydration forms show a diversified shape in a plan view. Most of them are quite regular and more or less rounded, but some of them are irregular in shape (Table 2). Taking into account the coefficient of circularity (Table S1, Figure S5), the more regular forms are divided into round, slightly oval, oval, slightly elongated, elongated and strongly elongated (Figure 6). The oval and slightly elongated forms constitute 42% of all examined objects, and 8% of the forms are irregular in shape (7 forms, e.g., Figure 7i). The latter group is the least numerous (Figure 7b) and has an uncommon geometry without any clear elongation orientation (Figure 7i, Table S1).

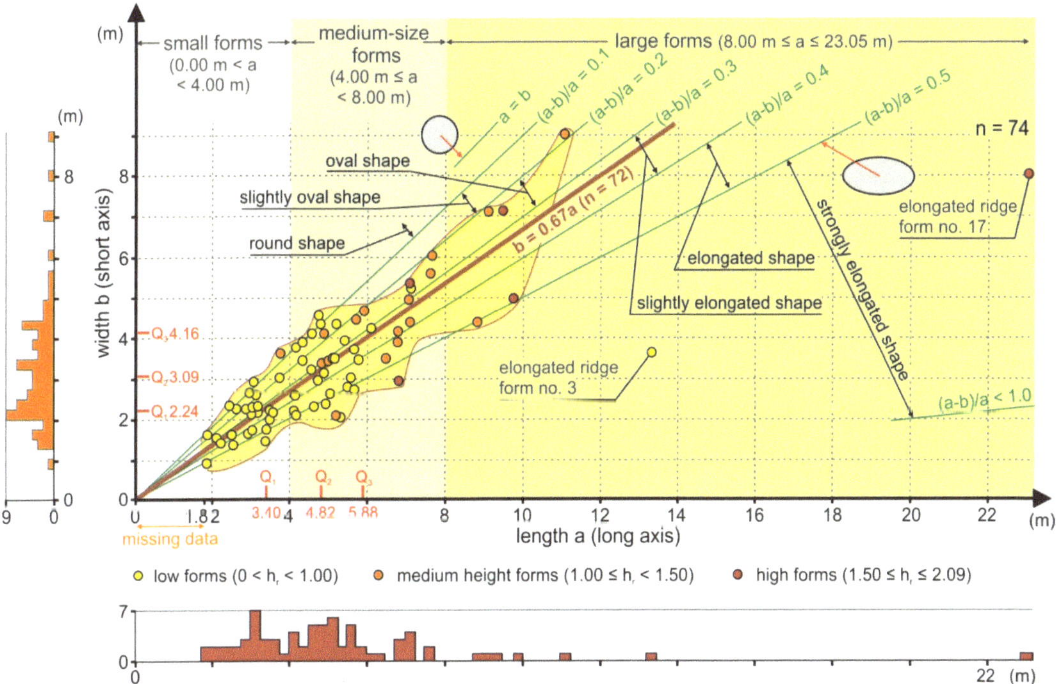

Figure 6. Diagram showing relation of width to length in 74 anhydrite hydration landforms divided according to h_r value and a value, showing the diagram areas for different values of the coefficient of circularity $(a - b)/a$, first (Q_1), second (Q_2) and third (Q_3) quartile of a and b value, as well as showing regression line of the width and length relation for 72 hydration forms (yellow area), excluding two elongated ridges, supplemented by frequency distribution diagrams of the width and length. Further explanations in the text.

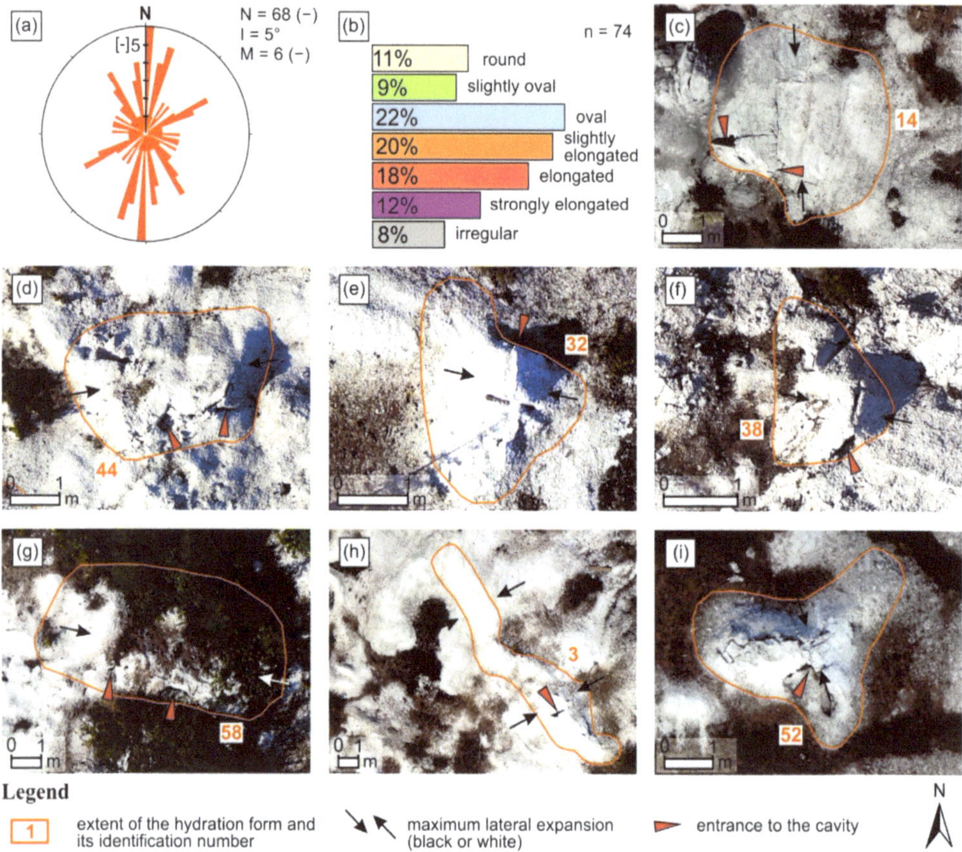

Legend

Figure 7. Characteristic of the elongation of the hydration forms at Dingwall; (**a**) rose diagram of azimuths of form elongation (Figure 3; N—total number of the forms, I—azimuth interval, M—number of forms represented by a circle radius); (**b**) frequency distribution of forms divided into groups according to coefficient of circularity (Table 2), including irregular forms (Figure 6); (**c–i**) examples of variously elongated forms and irregular form presented on orthophotomaps: (**c**) round form; (**d**) slightly oval form; (**e**) oval form; (**f**) slightly elongated form; (**g**) elongated form; (**h**) strongly elongated form; (**i**) irregular form.

The most dominant azimuths of the landform elongation have values of 0–5°, 15–20°, 65–75°, and 140–150° (Figure 7a). These dominant orientations poorly coincide with the orientations of master joints, although these joints create as many as five distinct sets (see Section 4.6.1). Additionally, direct field observations indicate that, despite the presence of the characteristic-oriented master joints in the particular parts of the quarry, the forms occurring there show rather random elongation (Figure S8).

Elongation of the landforms shows oriented relations with direction of MLE and is perpendicular to it, with an accuracy of ±5°, in 46% of forms, and parallel to it, with the same accuracy, in 29% of forms.

The relationship between landform elongation and the strike of layers is unclear. There is a certain convergence of dominant orientations of elongation and layering in several hydration forms, e.g., in the southern part of CF. In places, the traces of strike are perpendicular or at some acute angle to the elongation of the forms (Figure S8).

The measured lengths of the landforms range from 1.86 to 23.05 m (Table 2, Figure 6). The measurements clearly showed that the longer the forms, the less frequently they occur (Figure 6). From the length frequency distribution, the limits of 4 and 8 m were established, situated in intervals of occurrence of low frequency length values, as the most closed to natural limits for dividing the forms according to sizes into the following groups: the small, the medium-size, and the large forms (constituting 34%, 57% and 9% of all the objects, respectively; Figures 6 and S5). Among the 74 measured hydration forms, the largest (longest) form is 23.05 m in length and up to 8.02 m wide (no. 17, Tables 2 and S1, Figures 4 and S2). It stands out from the others with its strongly elongated shape, the morphology of a pressure ridge, a tepee-like shape in cross section, and an outstanding relative height of 2.08 m. On the other hand, the shortest documented form is 1.86 m long and has almost the lowest relative height of 0.35 m. It is a form with a distinct tepee shape and characteristic concave bending of detached layers (no. 66, Tables 2 and S1, Figures 4 and S2).

According to general shape, 52 dome- and 25 tepee-like forms were distinguished, representing 67.5% and 32.5% of all the studied objects (Table S1).

The relative heights of the landforms range from 0.33 to 2.09 m (Table 2), and half of the recorded values range from 0.56 to 1.05 m. Again, using the frequency distribution of the relative heights, two limits of 1.0 and 1.5 m were determined within the low-frequency intervals on the histogram, and considering these limits, the forms were classified into three groups: the low forms (72%), the medium height forms (22%), and the high forms (7%) (Figures 6 and S5). The highest hydration form shows a relative height of 2.09 m and is characterized by a domed shape (no. 49, Figures 4 and S2, Table S1). Such a large height of that form corresponds to the largest area of the internal cave, 29.4 m^2 (the Damian Cave [61]).

The width of the landforms appears to change proportionally to their length, except for the most elongated forms represented by the two pressure ridges (no. 3 and no. 17, Figure 6). Similarly, the relative height seems to change proportionally to both the length and width of the forms (Figure S6). The linear correlation between the discussed parameters can be expressed by the following equations, characterized by relatively large values of correlation (Pearson) coefficients (r), indicating moderate correlation (Figures 6 and 8) [95] (p. 529).

$$b = 0.67a \tag{5}$$

$$h_r = 0.16a \tag{6}$$

$$h_r = 0.23b \tag{7}$$

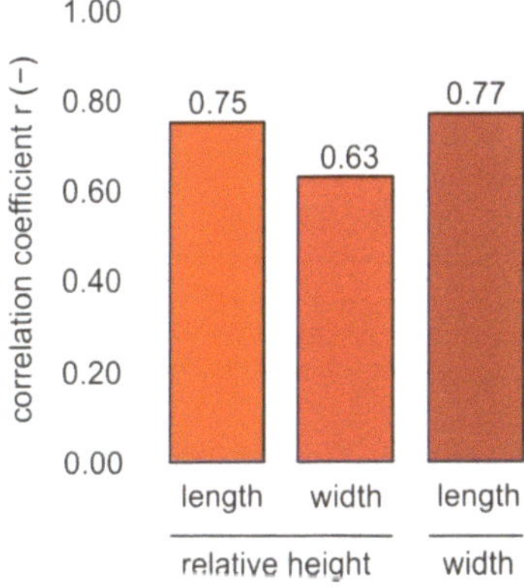

Figure 8. Correlation coefficient r for relations of relative height with length and width and also for the relationship between length and width computed for 74 anhydrite hydration forms.

Stenson [48] obtained the best correlation between the height and the width of the forms (expressed by the ratio height/width $\approx$ 0.16, n = 69). Our data suggest the best correlation between the length and the width (Figures 6 and 8).

The frequency distribution of b/h_r ratio shows that the average bulge degree of the studied forms is 4.28 (Table 2, see Figure S5). This parameter has an exceptionally high value of 8.58, indicating the least convex form, with a relative height ca. eight times less than its width (no. 68, Figure 9a, Table S1). In spite of its flatness, this form has an internal hydration cave, which is a proper cave with a proper entrance sensu Curl [96,97,105], i.e., an open cavity large enough to crawl inside by a man. On the other hand, the lowest value of bulge degree 1.82, indicating the greatest convexity, is recorded in the typical tepee-shaped form with a relative height of only ca. two times less than its width and created by two rock slabs leaning against each other (no. 15, Figure 9b and Figure S2, Table S1). This strongly bulged form also currently has an internal cave (in 2019), reaching 1.27 m in height (the Personal Grotto, Figure 9b). The height of this cave continues to rise (Figure 13 in [9]).

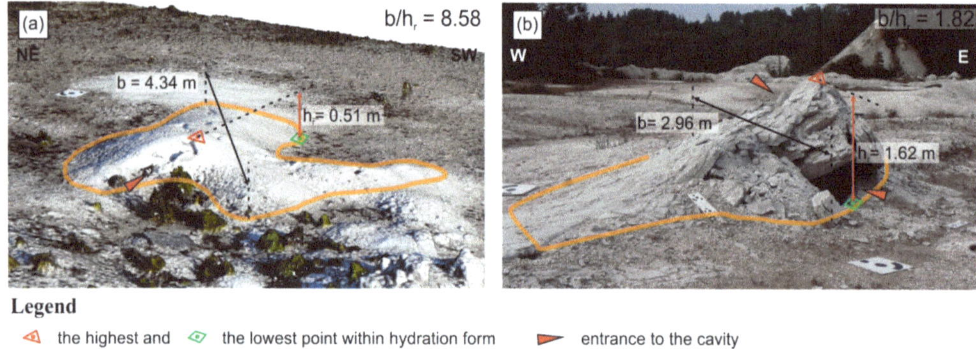

Figure 9. Anhydrite hydration landforms with extreme values of bulge degree (b/h_r ratio); (**a**) form no. 68 with the highest value of bulge degree visible on the 3D model; (**b**) form no. 15 with the lowest value of bulge degree. See Figure 4 or Figure S3 for the locations of these forms.

Generally, it can be noticed that, on average, the tepee-shaped forms have a slightly higher convexity (average b/h_r is 3.69) than dome-shaped forms (average b/h_r is 4.51; see Figure S5). The degree of bulge does not correlate with the length of the forms in a linear manner (Figure 10a). Such a correlation is also weak between bulge degree and width (Figure 10b) and bulge degree and coefficient of circularity (Figure 10c). Generally, it is seen that the higher convexity is reached by forms with a smaller length (Figure 10a), and the lower bulge degree by forms with a smaller width (Figure 10b).

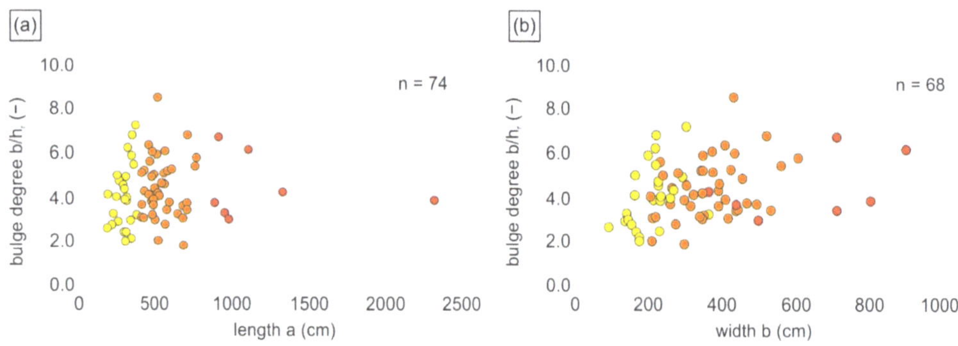

Figure 10. *Cont.*

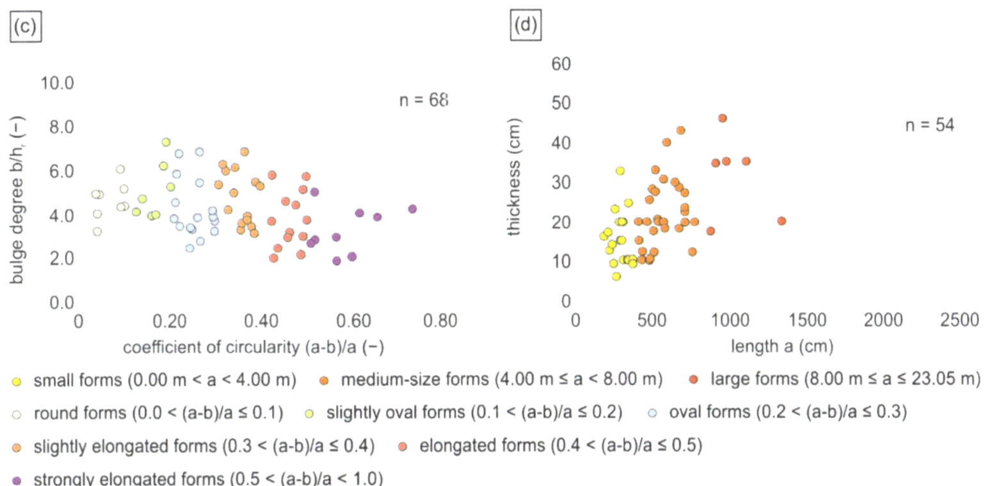

Figure 10. (**a**–**c**) Dependence of bulge degree on length (**a**), width (**b**) and coefficient of circularity (**c**) of the anhydrite hydration landforms (Figure 3); (**d**) dependence of thickness of detached layer on length of the hydration forms. Diagrams (**a**,**b**,**d**) take into account the size groups of the forms according to length of the long axis a and diagram (**c**) takes into account the groups of the forms according to coefficient of circularity.

4.4. Detached Layer

The averaged thickness of the detached and uplifted gypsum-anhydrite layer for 54 hydration landforms ranges from 6 to 46 cm, with an average value of 21 cm (Tables 2 and S1, Figure S4). It is generally observed that the greater the length of the forms, the greater the thickness of detached layers (Figure 10d). There is no visible difference between the thickness of the detached layer in the tepee- and dome-shaped forms.

4.5. Inner Cavity

Documented internal cavities of the studied 74 hydration landforms involve 48 proper caves, including 43 caves with a proper entrance [96,97]. The other cavities are too small to be a proper cave or cannot be verified whether they are such caves because of restricted access to their interior. The lengths of the documented 47 caves and cavities are from 0.70 to 8.87 m, and the range of the heights is from 0.28 to 1.35 m (data from 2019, in 2008 the length of the Ramesh Cave was larger—its floor space was ca. 10.7 × 6.6 m and it was 1.10 m high, which makes this cave the largest known hydration cave in the world) [8,61]. The largest cavities were identified in domed forms; the tepee-shaped forms had a relatively smaller inner space. The ceiling is generally flat, with a slightly rugged surface, but it also shows fractures and thin protruding rolled rock layers. Larger rock fragments fell from the ceiling, crystalline gravel derived from the till and leaves cover the bottom. Caves are predominantly opened only from one side; rarely, they have two entrances (Figure S2). Inlets leading to several caves are too narrow for an adult man to crawl inside.

The entrance lines (with the minimal length ≥0.30 m) were determined in 69 of the 74 hydration landforms (Figure 11). The length of the entrance line marked on the map (in six cases—two distinct lines) ranges from 0.30 to 4.52 m (Table 2). A single entrance documented by the entrance line occurs in 48 forms, two entrances appear in 20 forms and three entrances in one form. They lead both to proper caves, enabling humans to enter, and to smaller cavities.

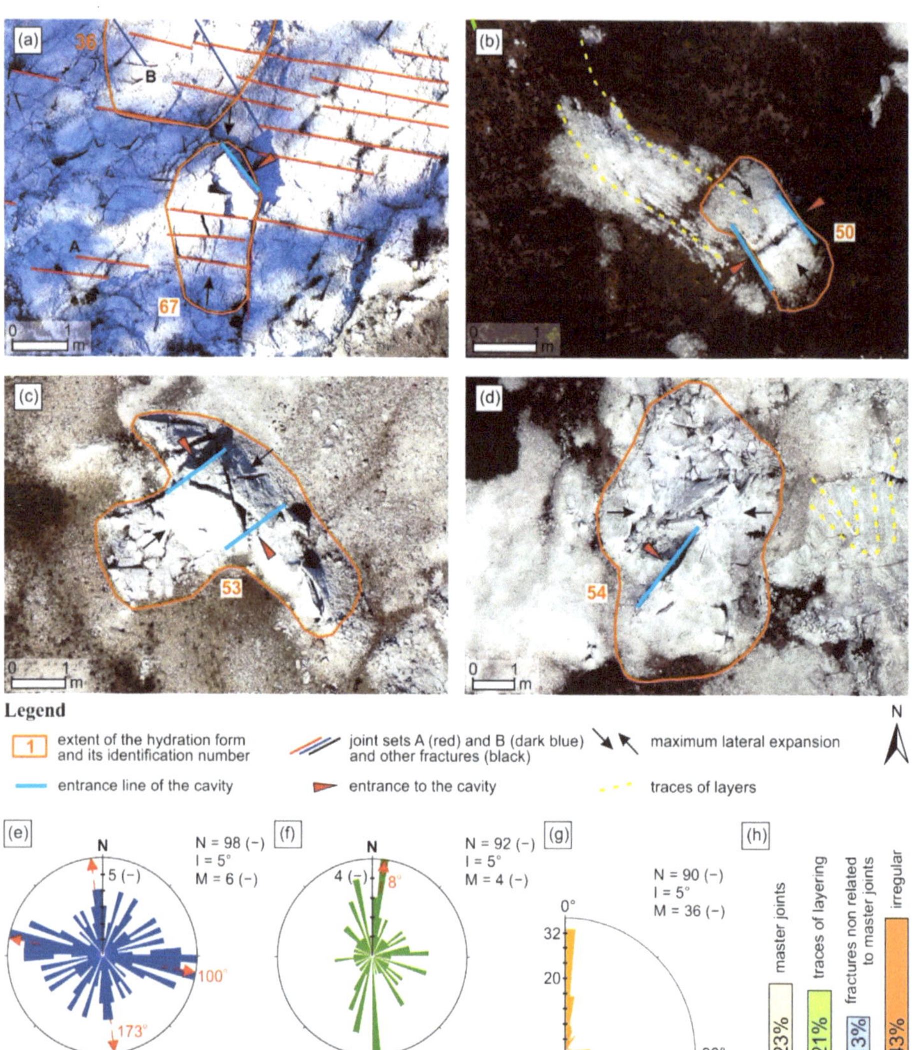

Figure 11. Morphometric features of entrances to hydration caves and cavities and their relation to master joints and layering; (**a**–**d**) exemplary anhydrite hydration landforms showing parallelism of the entrance lines (Figure 3) with master joint system (**a**), layering (**b**), and fractures within form related neither to master joints nor layering (**c**) and the form with lack of any regular relations between entrances and above features (**d**); (**e**,**f**) rose diagrams of the orientation of the entrance lines (**e**) and directions of the entrance (from inside the form to the outside) (**f**) with red arrow symbolizing dominating azimuths (N—total number of the forms, I—azimuth interval, M—number of forms represented by a circle radius); (**g**) rose diagram of the values of angles between entrance lines and directions of MLE (Figure 3); (**h**)—frequency distribution of entrances according to coincidence (more or less common orientation) of entrance lines with listed linear features and to lack of any visible relations with these features (irregular). The master joints and layering in the vicinity of forms illustrated in (**c**,**d**) are seen in Figure S8.

In 23% of cases, the entrance line coincides with some master joints in the vicinity of the form and in 21% with layering (Figure 11h). However, entrance lines are also more or less parallel to the other fractures, probably generated mostly by expansive hydration (Figure 11h). The rose diagram of entrance lines azimuths points out two dominant orientations: more common 100° and less common 173° (Figure 11e). The former orientation appears to coincide with the master joints set A (Figure 12a–c,e,f) and the latter with the common strike orientation (Figure 11d,f). Many entrance lines are irregularly oriented (Figure 11d,h). Azimuths of the directions of entrances are variously spread out, but the north direction is the most numerous (Figure 11f).

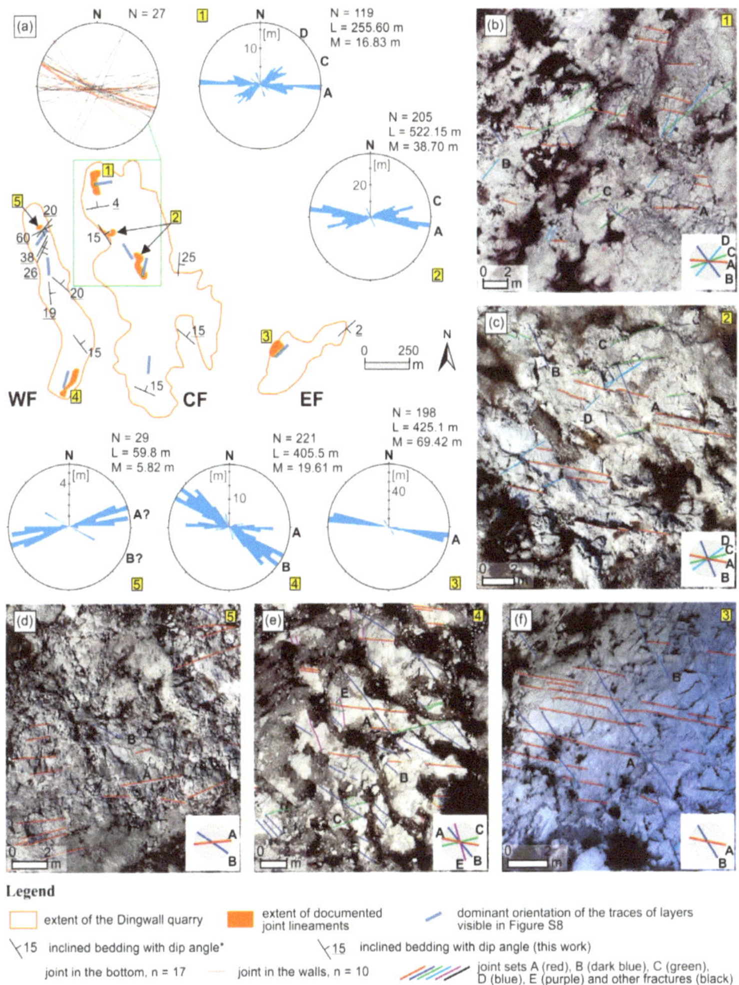

Figure 12. System of master joints recorded in the quarry at Dingwall; (**a**) joints measured in bottom and walls of the quarry within area marked by green rectangle (shown on Angelier diagram, upper hemisphere; upper left, N—number of measurements) and joints traces documented in particular parts of the quarry (shown on rose diagrams numbered 1–5; N—total number of fractures, L—total length of fractures, M—maximum length of a circle radius) together with dominant direction of layering; (**b–f**) fragments of the orthophotomap showing characteristic sets of joints A–E documented in particular parts of the quarry marked on figure (**a**); each set is marked with the same characteristic color on the map and in circles at bottom right. Note the rotation of main joint set A across the quarry. In all the figures (**a–f**), north is in the same orientation as in figure (**a**); *—data after Neale [71].

Most commonly, the entrances have opened with entrance lines parallel to the MLE, which means that they are simultaneously perpendicular to the crests of the forms (Figure 7e). In some cases, the entrance line is oblique or rarely perpendicular to that direction (Figure 11g).

4.6. Fractures and Layers

4.6.1. System of Master Joints and Fractures

Within the entire quarry, more than 754 fractures and traces of fractures were documented as lineaments and most of them represent joints (Figure S8). Well-developed systematic joints with an approximate vertical orientation were identified in six areas (Figure 4). These joints appear at a distance from ca. 0.5–2.0 m up to 10.0 or more meters (Figures 5a,c, 11a and 12). Twenty-seven joints and fractures were measured directly in the quarry (Figure 12). Azimuths of joint traces show significant similarity in orientation, permitting recognition of three main sets A, B and C and two less pronounced sets D and E (Figure 12). From one to four of these sets were observed to occur together in one place (Figure 12b,c,e). Joint set A (red in Figure 12) is commonest and has changeable azimuths and an average acute angle in relation to set B (dark blue line). The fractures measured in the north of the CF (Figure 12a, upper left) are mainly equivalents of joint sets A and C recognized in this part of the quarry (areas 1 and 2; Figure 12a). The joint system tends to cut the layering at angles of 40–90°, except in the northernmost part of the quarry (Figure 12a).

4.6.2. Fractures of Hydration Forms

Fractures in the detached rock layer cut it at right and oblique angles or parallel to its top or base. They represent both extension and shear types occurring with the opening between fracture walls and without any space-separating walls. The landform surface is dominated mainly by transverse fractures parallel to the MLE direction, as well as longitudinal, equally frequent ones perpendicular to this direction (Figures 3b and 7c,e,f). Irregular fractures (oblique to the MLE direction, Figure 3b) occur in a smaller amount than the abovementioned ones, but they can still be found in 34 of the 74 investigated forms. Spherical (Figures 3b and 5a,c) and radial fractures (form no. 22, Figure S2) are by far the least common. Spherical fractures are associated with the effects of spheroidal weathering of anhydrite blocks subjected to hydration [14], described elsewhere as "cannonball" structures [106].

The presence of joints cutting the forms and continuing outside of them representing the above-described sets A–E are characteristic of 28 of the 74 examined forms. Such joints are excellently visible in the exemplary form in EF (Figures 11a and S8).

The surface density of fractures in the detached rock layer (recorded in 69 forms; five forms were without visible fractures) ranges from 0.04 to 3.38 m/m^2 (Table 2, Figures 13 and 14b). The areas of the quarry with the large impact of the master joints are characterized by a higher F_D value compared to other areas (Figure 13). For all the hydration landforms, regardless of whether they are connected with master joints or not, the density of fractures is only slightly larger for the tepee-shaped forms than for the dome-shaped ones (Figure 13). The highest average F_D is recorded in the SW part of the EF influenced by the master joint system, where one form is characterized by a plentiful amount of fractures and the largest recorded F_D of 3.38 m/m^2 (no. 67, Figures 4, 13 and S8; Table S1).

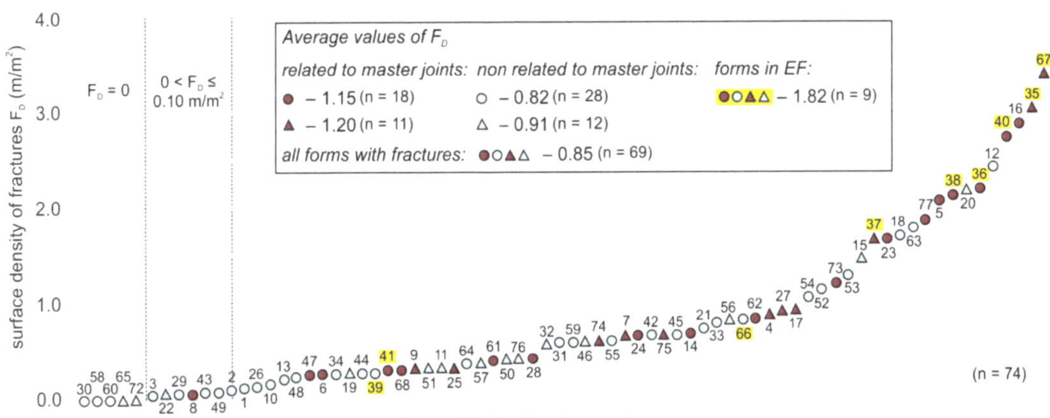

Figure 13. Fracture surface density F_D for 74 anhydrite hydration landforms taking into account dome-like shape (circle symbol) and tepee-like shape (triangle symbol), relation to master joints (claret color), lack of relation to master joints (blue color, further explanations in the text) and location in the EF of the quarry (yellow color). Average values of F_D are presented for both particular groups of forms and for all forms (n—number of measurements).

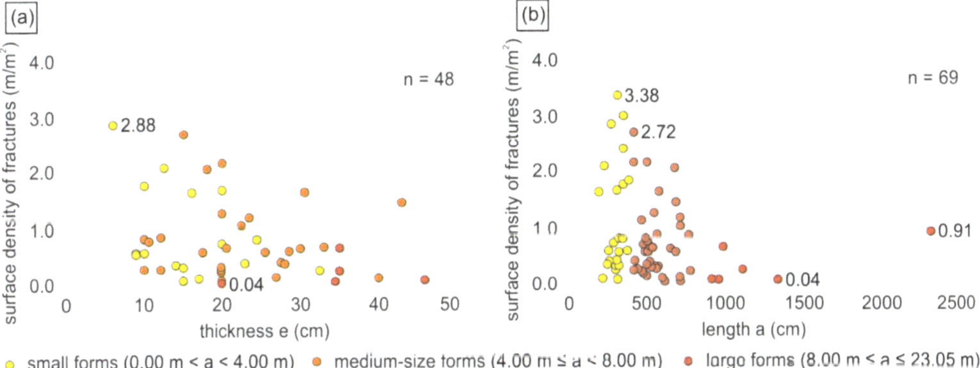

Figure 14. Dependence of surface density of fracture traces on thickness of detached rock layer (**a**) and on length of form (**b**), taking into account size classification of landforms according to length of axis a (Figures 3 and 6).

For 48 hydration landforms with measured thickness of detached layer, the density of fractures is variable, both very large and modest (Figure 14a). The lower F_D values are usually connected to greater thickness. For a thickness lower than or equal to 20 cm, the surface density can attain as much as almost 3 m/m^2, but if the thickness is higher than 20 cm, then this density does not overpass 1.66 m/m^2. Additionally, the relationship between F_D and the length of forms indicates that the smaller forms attain a higher fracture density (Figure 14b).

4.6.3. Strike and Dip of Layers

Traces of layering seen on the surface of the bedrock were documented as lineaments on the orthophotomap of the quarry bottom. Because the surface of the bottom is approximately horizontal, these traces represent strikes of the layers quite well. In some places of the bedrock, the layers are not visible because of the cover of weathering debris (e.g., in surroundings of forms no. 24–28, 58 and 61 in NW part of the CF). Where seen, these traces are often oriented in one generalized azimuth but also run in a wavy manner, rapidly changing the orientation or even forming closed, oval structures (Figure 11d) [14,57]. The rose diagrams of the azimuths of the strike lines (lineaments) were created separately for eight parts of the quarry (Figure S8). The diagrams demonstrate that the dominant azimuths of strike are in the ranges of 16°–48° and 147°–172°. The dominant dip is to the east (Figure 12) [71].

4.7. Morphological Relationships and Structural Characteristics of Hydration Forms
4.7.1. Distribution

The expansive anhydrite hydration causes numerous and variable rock deformations at the bottom of the quarry. The accomplished observations demonstrate that the formation of the most distinctive convex hydration forms analyzed in this paper takes place rather in a non-uniform and irregular manner throughout the area, which is the opposite way of many ordered natural landforms, such as drumlins, patterned grounds, dunes or ripples showing regular spacing (e.g., [29,107]). In contrast to the ordered forms, the distribution of hydration forms is very variable; they occur with different densities throughout the quarry. Such a way of distribution, also noted in the other occurrences of dome- and tepee-like hydration landforms (Pisky, Walkenried), is not easy to explain. It seems that most of all, it is related to the features of anhydrite rocks controlling the anhydrite gypsification, which was discussed in a separate publication [14] but also to the topographic condition influencing the water migration paths.

4.7.2. Sizes

The values of the basic morphometric parameters for the 74 forms (length, width, relative height) show a relatively good ("moderate" [95]) mutual correlation for the relations: $h_r - a$, $h_r - b$, and $a - b$, characterized by simple Equations (5)–(7) (see also [48]). Such correlations suggest that the studied landforms, irrespective of their size, show fixed proportions of the mentioned parameters. Therefore, as in the case of the gypsum tumuli [36], their morphological evolution appears to follow the same or similar simple rules. In terms of shape, they are thus relatively regular landforms "printed" in the landscape, with nearly the same fixed proportions of the basic morphometric parameters (compare [84] p. 189).

The recorded variability of the measured basic dimensions can be explained by asynchronous initiation of the form growth, i.e., different ages of the forms [48]. Stenson [48] also supposed that the size of the forms depends on the depth of anhydrite hydration. Many of the studied objects represent so-called non-equilibrium landforms, i.e., landforms that are still growing or decaying [29]. Such forms usually show broader size ranges than more "stable" equilibrium forms, such as bedforms.

The dome- and tepee-like hydration landforms from Dingwall show specified variations in size and shape. They are limited in size, which is a characteristic feature of any landform defined as a discrete object [29,30]. We do not have quantitative data for the determination of their lower size limit, but it seems to be a few tens of centimeters or less (Figure 7 in [9]) [48]. Their maximum length attains ca. 23 m, a maximum width—ca. 9 m and a maximum relative height—slightly over 2 m. The large forms are, however, distinctly relatively rare, which is a typical feature of the majority of landforms [29]. Small- and medium-size forms with lengths in the range of 1.8 to 8.0 m are the most common and constitute 90% of the studied forms. A similar frequency distribution of lengths of hydration forms, with evidently rare larger forms, was documented at Pisky [57].

The basic dimensions of the hydration landforms at Pisky, Walkenried, Dingwall and the Alebastrovyye Islands sites are very similar. As evidenced by the ranges of their lengths, they reached a maximum value of more than 10 and even 20 m (Table 1). Heights are also on the same scale, reaching 2–3 m. In this regard, dome- and tepee-like hydration landforms are evidently globally scale-specific [29], although this conclusion is based on limited data from only fourth sites on Earth. A worldwide similarity of these hydration forms in sizes is also present within the thickness range of the detached rock layer, which is from a few cm to nearly 50 cm at Dingwall and Walkenried [2,4], also in the Alebastrovyye Islands, as indicated by the drawings of forms [5], but at Pisky the thickness exceeds 50 cm in some cases reaching over 100 cm [18,58]. The slight differences in size parameters between the forms from the four discussed sites (Table 1) presumably reflect the environmental differences between the particular localities. Thus, the hydration forms at these sites are, to some extent, regionally or locally scale-specific [29].

4.7.3. Shape

During the development of hydration landforms, the rock layer rising on the bottom of the quarry creates a variety of shapes in a plan view, which were approximated to round, oval and elongated. The irregular shape formation in a plan view is manifested in a curved or even winding course of the erected rock and the formation of three arms, the crests of which do not give an unambiguous orientation to elongation.

Among the studied landforms, dome-like shapes, characterizing 2/3 of all the forms, predominate over the tepee-shapes. Stenson [48] distinguished blisters (domes), rounded-tents and A-tents (tepees) at Dingwall. We have found, however, that it is difficult to distinguish between the blisters and the rounded-tents and have classified both forms as dome-like forms according to the introduced criteria. Nevertheless, our almost 20 years of study of the site have revealed changes in the domed to tepee-like shape, as Stenson did [48]. However, we also noticed the constant domed shapes of many hydration forms during the same period of evolution.

The values of bulge degree (b/h_r) are variable. On the basis of the realized research, it appears that the tepee-shaped forms are characterized by the lower value of b/h_r, being more privileged to create steep surfaces and lift the top up. This, in turn, may indicate that the tepee-shaped forms could be created as a result of the action of larger expansive gypsification deforming the rock or deeper hydration, as supposed by Stenson [48].

Orientation of the crest line and direction of the MLE supply information on the way of creation of the particular hydration landforms and, among others, on the formation of their shape (reflected by the mentioned morphometric parameters). They provide information on the method of deformation or strain, which is mainly dependent on the place and course of anhydrite hydration and thus on the petrological structure of the bedrock [14]. In the later stages of the discussed landforms' development, the hydration and volume increase have apparently been concentrated in the vicinity of the erected forms within which, at that time, the hydration processes have slowed down [4] (Figure 15 in [9]) [28,49]. In the case of the studied landforms, particularly those with the tepee shape (Figure 3, right), the centers of accelerated volume expansion were located near the forms, somewhere on their opposite sides, indicated by the MLE direction. Such forms have risen by pushing the detached rock slabs against each other just from these opposite sides in a way similar to buckling [98]. The centripetal displacements of the rock slabs are well documented in the case of such tepee-like forms (Figure 43 in [4]) (Figure 6c–e in [9]) [49] proving this interpretation. Restoration of the course of the deformational processes is, however, not a subject of this paper devoted to the morphometric characteristics of hydration landforms.

4.7.4. Entrance to Internal Caves or Cavities

Entrances to the internal cavity of elevated hydration landforms are quite a common feature. They allow access to the interior of hydration forms and, in the case of 43 cavities being proper caves [96,105], entry and direct examination. The entrance within a detached rock layer was opened by the collapse or uplift of a part of the rock, commonly along with the flat fractures visible as a straight line in a plan view (Figures 2d and 11d). Many of these fractures existed before the uplift of the layer and represent some joint sets developed in the bedrock. This is clearly evident from the parallel orientation of many entrance lines and such joints (like set A) and also directly from the continuity of the fracture forming the entrance line with the joint in the bedrock near the form. These older fractures have easily been widened or opened during the growth of the form, particularly due to the differential movements of the disconnected fracture walls and blocks on opposite sides of the fractures.

Moreover, the same orientation of the entrance lines and layering suggest that the fractures responsible for the creation of the entrances easily opened along with the bedding of the bedrock. Indeed, some newest fractures opening recently in the weathering anhydrite rocks are parallel to the layering (Figure 5b–d in [9]) and minor fractures parallel to the bedding are also noted (Figure 5i,l in [14]). However, it is remarkable that one-third of the entrances have been created in an irregular manner. Such entrances prove it difficult to recognize factors influencing their orientation and origin. Thus, many entrances have been created without a clear connection with the bedrock structure. Simultaneously, the entrance line is predominantly parallel to the MLE direction, which informs about the expected orientation of the inlets to the inner cave or cavity (Figure 11g).

The formation of single or two entrances to the cavity progresses from both the marginal and central parts of the forms, as in the case of gypsum tumuli landforms in the Sorbas region in Spain, where erosion is an important factor in broadening small inlets to the size of a large entrance [36,37].

4.7.5. Relation to Gypsum Tumuli

The origin of gypsum tumuli is not related to anhydrite dissolution and succeeding gypsum crystallization, but solely to the crystallization of secondary gypsum after gypsum dissolution within weathering gypsum rocks taking place in an arid or semi-arid climate or microclimate [36,37,108]. Morphologically, gypsum tumuli are generally domes similar to many anhydrite hydration landforms; however, they form tepee-shaped forms either, although much less frequently [37]. Researchers of gypsum tumuli in the Spanish region of Sorbas have recognized their maximum length up to 11.7 m [67], whereas in Sicily, up to 11 m [109] or 15 m [108], which is similar in value to the size of hydration forms. Similar to hydration landforms, the largest gypsum tumuli are less common. On the other hand, both hydration forms and gypsum tumuli have similar small dimensions in a plan view—a few tens of centimeters [37].

The significant morphometric difference between the hydration landforms and gypsum tumuli, as noted earlier [36,48], is a different height, determined as a maximum of 1.30 m for gypsum tumuli, which is significantly less than the 2.09 m recognized for the forms at Dingwall, ca. 3 m at Pisky [18,60] and 3 m for the Waldschmiede in Walkenried [4]. The difference between the maximum height of the tumuli and the hydration landforms is even larger, taking into account that the reported "height" of gypsum tumuli is not a real relative height but the distance measured from the bottom of the internal cavity to the top of the form [67]. The other morphometric difference is the higher convexity reached by hydration landforms, reflected by the ratio of elevation (relative height) to their lateral dimensions [36] (p. 927). There are also some differences in the shape of both forms related to different mechanisms of volume expansion and potentially higher volume increase and more complicated deformations in case of anhydrite hydration (resulting, e.g., in a higher steepness of side parts of the mature hydration forms such as Waldschmiede, the Great Tepee from Pisky, or some forms from Dingwall) [1,4] (Figure 6b,d in [9]) [18,49,61]. Some other structural differences between anhydrite hydration landforms and gypsum tumuli were noted by Calaforra and Pulido-Bosch [36].

4.7.6. Structural Characteristic

Stenson [48] recorded at Dingwall large hydration landforms (>2.0 m) with thicknesses of detached layer <10 cm and also small forms (<1.0 m) showing a thickness of >35 cm and concluded that the thickness depends neither on the size nor on the type (blister or tent) of hydration forms. Our data suggest, however, that the size of the forms depends to some extent on this thickness—the landforms larger in extent (longer) have a remarkably thicker detached rock layer (Figure 10d). The same rule was recognized very clearly in the gypsum tumuli [36,67]. The larger thickness of the detached layer may be related to deeper localized hydration. Herrmann [45] believed that the greater the thickness of the detached layer, the greater the hydration cave.

The surface of hydration forms is commonly strongly fractured, and the majority of fractures are evidently the result of expansive anhydrite hydration. They formed during the growth of the landforms, when rock slabs rise up, deform, bend, break and push against each other. However, as proved above, some fractures have been evidently inherited from the joint system in the bedrock, occurring in the place where the hydration form is now. These joints commonly widened during the rise of the detached rock layer. The master joints significantly contribute to a greater density of fractures, recorded, e.g., in the EF (see Figure 12). It was demonstrated that with decreasing fracturing, the dimensions of forms increase, both in length and height (see Figure S3). This feature is related to the strength of the rock, which is significant when the rock is not fractured and weakens as soon as it is more fractured. The hydration forms densely covered with fractures, as rocks documented worldwide [110] are more prone to erosion and rapid transition into the destruction stage.

In addition, the density of fractures generally decreases when the thickness of the raised layer increases (Figure 14a), just as in the case of joints in sedimentary rocks, where the joint spacing increases with the rise in thickness of the rock layers [111,112]. In addition, contrary to Stenson's [48] opinion, fractures have been formed just as easily within both domed and tepee-shaped forms because both of them have a similar number of fractures (Figure 13).

4.8. Bedrock Structure Impact for Hydration Forms

Our observations suggest that the morphology of hydration forms at Dingwall is dependent on the structure of the anhydrite bedrock (Figure S9). In particular, the location of the entrances to the inner cavities and the way they have been opened are strongly dependent on the structural features of the rocks, such as master (and other) joints and layering. These features, the presence or absence of joints and layering, and surface fracture density appear to be the important factors controlling the development and structure of the hydration landforms at Dingwall. In general, three structural types of bedrocks can be distinguished, involving both master joints and pronounced layering, or one of them (Figure 15a–c).

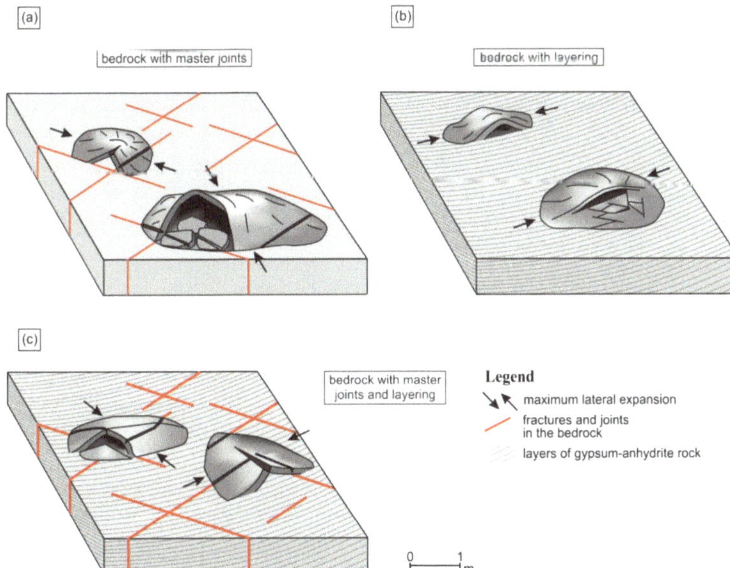

Figure 15. (a–c) Three structural types of bedrock influencing the development of hydration landforms. Note that the location of the entrances to hydration cavities and caves is determined by the presence of master joints and/or layering in the bedrock. Further explanations in the text.

4.8.1. Master Joints

These joints were present in the rock layer before they underwent detachment and uplift. The joints forming particular sets cross-cut, causing that the bottom of the quarry is usually divided into segments with triangular, rhombic or polygonal contours (e.g., Figure 11a) (Figure 1e in [14]). The presence of these blocks significantly influences the structure of the developing hydration landforms, fracture density and fracture orientations. Rock within the range of the form at the beginning of its uplift inherits older fractures or joints (Figure 3b). They affect its strength and cause distortions of the layer separated from the bedrock. Such distortions are mainly manifested by the collapse of fragments of the elevated rock, along with inherited fractures and the formation of entrances to the internal cavity. The collapse may occur along two fractures (joint sets) intersecting at a certain angle and creating segments (Figure 15a,c). Movements along the fracture surface may also lead to an uplift of the rock layer on one side of the fracture; on the other side, the rock may remain lower.

New fractures are also generated by expansive anhydrite hydration connected inter alia with the direction of MLE, affecting the development of landforms and their morphology (Figure 3b). These fractures—not related to earlier masters and other joints—are also the reason for weakening the strength of the rock layer. They equally favor the direction and location of rock detachment and uprising, and they are also a place for the formation of entrances to the inner cavity (Figures 11d and 15a).

4.8.2. Strike of Layers

Another impact on the morphology of hydration landforms is the strike of the rock layers. Observing its azimuth, it could be supposed that a similar orientation has some entrances to the internal cavities of the growing landforms (entrance lines; Figure 3) and also fractures within them. Moreover, in some cases, the orientation of the layers seems to control the azimuth of the elongation of the discussed forms. These observations lead to the conclusion that the strike of rock layers guides both the shape and orientation of the hydration landforms (direction of their elongation), as in the case of the site at the Alebastrovyye Islands [5], but in the study area that impact is less pronounced (Figure 15b).

4.8.3. Master Joints and Strike of Layers

The occurrence of both master joints and prominent layering in the bedrocks influences the development of hydration landforms even more (Figure 15c). The joints increase the fracture density, determine the structure of the growing form and affect the orientation of the entrance to the cavity. The strike of the layers can also influence the direction of elongation of the form. However, the influence of these factors is more complex and hardly predictable because it depends on the number of master joint sets (from 1 to 4), as well as the mutual orientation of joints and layering, which are changeable across the quarry.

5. Conclusions and Final Remarks

It was demonstrated that expansive hydration of anhydrite during the weathering of these rocks leads to the creation of peculiar convex landforms characterized by defined sizes and shapes. Such landforms studied in the abandoned gypsum quarry at Dingwall have developed for more than 65 years, counting since the time of anhydrite bedrock exposure, i.e., longer than the growth of hydration forms at the Pisky site and much shorter than at the Walkenried site, where the hydration continues for more than 300 years [52]. At present, the majority of the studied landforms (estimated at about 91%) seem to be in the mature or senile stage of development (Figure 2f). 62% of them have internal proper caves [96,97], being a unique speleological object known as hydration or swelling caves. Among them is one of the largest hydration caves, measuring $10.7 \times 6.6 \times 1.10$ m.

The landforms from Dingwall, similar to the other mentioned sites, are characterized by defined sizes and shapes. The two basic shapes include dome-like, dominating and tepee-like, less frequent; the latter is defined by a sharp crest. In a plan view, the landforms vary from circular through oval to elongated in shape, with length (a) to width (b) ratio rarely exceeding 5:2 (Figure 6). These shapes are an idealized approximation of the real shapes, which are commonly irregular and complex and difficult to define and classify, particularly in the case of "senile" forms undergoing destruction.

The sizes of the studied landforms are limited. Most commonly, they are 1.8–8.0 m in length (90% of forms). The lower limit of length was not determined precisely, but presumably is a few tens of centimeters. The upper limit is 23 m, and as a rule, the longer the form, the less frequently it occurs. The relative height ranges from 0.33 to 2.09 m and is 0.83 m on average. The frequency distribution of sizes (length, width, relative height) confirms the general rule observed in the dimensions of many landforms characterized by "large numbers of small values and smaller numbers of large ones" [29] (p. 62). All of the obtained histograms appear to show right-skewed distributions, although the smallest sizes were not documented (Figure 6, Table 2).

The size of the hydration landforms depends mainly on the thickness of the detached rock layer—the thicker the layer, the larger the hydration form.

The length (a), width (b) and relative height (h_r) for the majority of forms overall show proportional relations, testifying that the hydration landforms preserve generally the same or very similar shape independent of their sizes. The relations between these basic parameters can be expressed by the following equations: $b = 0.67a$, $h_r = 0.16a$, $h_r = 0.23b$; with correlation (Pearson) coefficients (r) of 0.77, 0.75, and 0.63, respectively, indicating moderate correlation.

The convexity of hydration landforms was characterized by a b/h_r ratio—bulge degree. It ranges from 1.82 for the most convex forms to 8.58 for the least convex forms. According to the average value of this parameter, the domed forms are slightly less convex than the tepee-shaped ones.

Maximum lateral expansion, measured normally to the approximated crest line of the landforms, provides information on the strain responsible for the creation of their convex shape. Simplifying informs us about the direction of dominating "forces" pushing up or buckling the detached layer. Although MLE was most commonly normal or nearly normal to the elongation of the forms (46% of them), quite often, it was parallel to the elongation (29% of forms; Figure 7d,g). It can be concluded that the centers of accelerated greater volume expansion (crystallization of the secondary gypsum) were presumably located at sites near the erected forms, situated as a rule normally to their elongation or, less commonly, along this direction.

The development of hydration landforms depends on the structure of the bedrock on which they occur. The joint system present in the bedrock is inherited by the growing landforms—master joints are recognizable on almost half of the hydration forms. These earlier fractures determined the way of opening the entrance to the inner cavities or caves and weakened the stability of the lifted layers.

The layering in the massive bedrock without joints may control the orientation of both the landform elongation and the entrance to the internal cavity. The bedrock with both mentioned features has a more complex impact on the growing hydration forms, which depends on the mutual orientation of the layering and joints. Such landforms develop in a more complicated and unpredictable manner.

The number of fractures on the surface of hydration forms varies significantly. Both landforms without fractures and those with densely distributed fractures (with their total length up to 33.5 m per 10 m^2) have been documented. Fractures are most common in the hydration forms developed on the bedrock disturbed by the extensive master joint system and are the least common in forms growing on the substrate without any visible joints or fractures.

Fractures on the landform surfaces are represented by the most common transverse, longitudinal and irregular ones, and the uncommon ones, such as spherical and radial. The durability of landforms over time depends on the number of fractures associated with them.

The authors hope that the presented geomorphometric documentation and characteristics of the hydration landforms will help in remote recognition of such forms and in distinguishing them from other similar objects, as well as in a better understanding of their not entirely clear origin.

The photogrammetric models and maps of the zone of the weathering anhydrite will help in monitoring, further analysis and study of the ongoing morphological evolution of the Dingwall quarry area.

In closing, it should be emphasized that expansive anhydrite hydration is still not a well-recognized process that requires further study. This process, which leads to swelling of the ground surface, can be very destructive. It currently takes place with highly catastrophic results in several urban and industrial areas around the world [14] (with references therein) [113–115]. In spite of many studies, engineering knowledge on how to prevent, control and stop this process is still insufficient. The Dingwall site is an excellent place where many aspects of the anhydrite hydration process can be directly investigated, and we hope that further research will contribute to addressing many of the unsolved anhydrite hydration problems.

Supplementary Materials: The following supporting information can be downloaded at: https://www.mdpi.com/article/10.3390/app12157374/s1, Table S1: Inventory of hydration landforms with basic morphometric parameters and GPS locations; Figure S1: Applied photogrammetric method—equipment, course of work and results; Figure S2: Hydration landforms on the orthophotomaps and digital surface models; Figure S3: Location of hydration landforms on the base map of satellite images with characteristics of their sizes; Figure S4: Location of hydration landforms with measured thickness of the detached layer on the base map of the satellite image; Figure S5: Frequency distribution diagrams of morphometric parameters characterizing hydration landforms; Figure S6: Dependence of relative height of hydration landforms on length, width, bulge degree and coefficient of circularity; Figure S7: Pattern of fractures within hydration forms; Figure S8: Traces of fractures and rock layers on the orthophotomaps of the quarry bottom; Figure S9: Exemplary hydration landforms influenced by structural type of bedrock.

Author Contributions: Conceptualization, A.J. and M.B.; methodology, A.J.; software, A.J.; validation, A.J. and M.B.; formal analysis, A.J.; investigation, A.J., F.V. and D.Ł.; resources, A.J.; data curation, A.J.; writing—original draft preparation, A.J. and M.B.; writing—review and editing, A.J. and M.B.; visualization, A.J.; supervision, M.B.; project administration, A.J.; funding acquisition, A.J. All authors have read and agreed to the published version of the manuscript.

Funding: This research was funded by the Ministry of Higher Education and Science project within the "Diamond Grant" program, grant no. 0002/DIA/2017/46.

Institutional Review Board Statement: Not applicable.

Informed Consent Statement: Not applicable.

Data Availability Statement: Not applicable.

Acknowledgments: The authors acknowledge the owner of the Dingwall quarry area, Gary Burchell for enabling the research work at the site, and Robert Ryan from the Nova Scotia Department of Energy and Mines for supplying the valuable geological information. Many thanks to Janet Connor and John Connor for providing the accommodation and support during expeditions to the Dingwall quarry. The authors also acknowledge Canada Transport for permission to launch the UAV in the quarry. The authors also thank the five reviewers and Editors of the volume for their contributions to improving the paper.

Conflicts of Interest: The authors declare no conflict of interest.

References

1. Reinboth, F. Die Waldschmiede bei Walkenried eingestüzt. *Mitt. Verb. dt. Höhlen-u. Karstf.* **1967**, *13*, 68.
2. Reinboth, F. Die Zwerglöcher bei Walkenried am Südharz–Bemerkungen zur Frage der Quellungshöhlen. *Die Höhle Z. für Karst–und Höhlenkunde* **1997**, *48*, 1–13.
3. Hunt, C.O.; Gale, S.J.; Gilbertson, D.D. The UNESCO Libyan Valley Survey IX: Anhydrite and limestone karst of the Tripolita-nian pre-desert. *Libyan Stud.* **1985**, *16*, 1–13. [CrossRef]
4. Reimann, M. Geologisch–lagerstättenkundliche und mineralogische Untersuchungen zur Vergipsung und Volumenzunahme der Anhydrite verschiedener geologischer Formationen unter natürlichen und labormäßigen Bedingungen. *Geol. Jahrb.* **1991**, *D97*, 21–125.
5. Yushkin, N.P. Supergenesis of Carboniferous anhydrites of Novaya Zemlya. *Polar. Geogr. Geol.* **1994**, *18*, 33–43. [CrossRef]
6. Reimann, M.; Vladi, F. Zur Entwicklung der sog. Zwergenkirche am Sachsenstein bei Walkenried, Landkreis Osterode am Harz, Niedersachsen und vergleichende Beobachtungen zur rezenten Entstehung von Quellungshöhlen in einem aufgelassenen Gipssteinbruch bei Dingwall, Nova Scotia, Kanada. *Mitt. Verb. dt. Höhlen-u. Karstforscher* **2003**, *49*, 75–77.
7. Jarzyna, A.; Bąbel, M.; Ługowski, D.; Vladi, F.; Yatsyshyn, A.; Olszewska-Nejbert, D.; Nejbert, K.; Bogucki, A. Unique hydration caves and recommended photogrammetric methods for their documentation. *Geoheritage* **2020**, *12*, 27. [CrossRef]
8. Vladi, F.; Bąbel, M. Recent growth and decay of the hydration (swelling) caves in the former gypsum quarry of Dingwall in Cape Breton, Nova Scotia, Canada. In *The Weathering of Anhydrite and Gypsum Rocks*; Bąbel, M., Olszewska-Nejbert, D., Nejbert, K., Eds.; GIMPO Agencja Wydawniczo-Poligraficzna: Warsaw, Poland, 2020; pp. 223–232. [CrossRef]
9. Vladi, F.; Bąbel, M.; Jarzyna, A. Wachstum und Zerfall rezenter Quellungshöhlen im ehemaligen Gipssteinbruch von Dingwall in Cape Breton, Nova Scotia, Kanada, nebst Anmerkungen zum dortigen Sulfatkarst und dem Forschungsstand. *Abh. Zur Karst-und Höhlenkunde* **2022**, *40*, 119–134.
10. Mossop, G.D.; Shearman, D.J. Origin of secondary gypsum rocks. *Trans. Inst. Min. Metall. Sec. B Appl. Earth Sci.* **1973**, *82*, B147–B154.
11. Zanbak, C.; Arthur, R.C. Geochemical and engineering aspects of anhydrite/gypsum phase transition. *Bull. Assoc. Eng. Geol.* **1986**, *23*, 419–433. [CrossRef]
12. James, A.N. *Soluble Materials in Civil Engineering*; Ellis Horwood: New York, NY, USA, 1992; pp. 1–434.
13. Bąbel, M.; Olszewska-Nejbert, D.; Ługowski, D.; Nejbert, K.; Jacyszyn, A. Petrogenesis of the zone of present day anhydrite weathering at Pisky near Lviv. In *The Weathering of Anhydrite and Gypsum Rocks*; Bąbel, M., Olszewska-Nejbert, D., Nejbert, K., Eds.; GIMPO Agencja Wydawniczo-Poligraficzna: Warsaw, Poland, 2020; pp. 145–213. (In Polish with English Summary) [CrossRef]
14. Jarzyna, A.; Bąbel, M.; Ługowski, D.; Vladi, F. Petrographic record and conditions of expansive hydration of anhydrite in the recent weathering zone at the abandoned Dingwall gypsum quarry, Nova Scotia, Canada. *Minerals* **2022**, *12*, 58. [CrossRef]
15. Biese, W. Über Höhlenbildung, Teil 1. Entstehung der Gipshöhlen am südlichen Harzrand und am Kyffhäuser. *Abh. Preußischen Geol. Landesanst. Neue Folge* **1931**, *137*, 1–71.
16. Myers, A.J. Geology of the Alabaster Cavern area. In *Guide to Alabaster Cavern and Woodward County, Oklahoma*; Guidebook 15; Myers, A.J., Gibson, A.M., Glass, B.P., Patrick, C.R., Eds.; Oklahoma Geological Survey: Norman, OK, USA, 1969; pp. 6–16.
17. Breisch, R.L.; Wefer, F.L. The shape of gypsum bubbles. In *Proceedings of the 8th International Congress of Speleology*; Beck, B.F., Ed.; National Speleological Society: Huntsville, AL, USA, 1981; Volume 2, pp. 757–759.
18. Jarzyna, A.; Ługowski, D.; Olszewska-Nejbert, D.; Jacyszyn, A.; Bąbel, M. Inventory of the dome-shaped hydration landforms from the area of weathering anhydrite rocks at Pisky near Lviv. In *The Weathering of Anhydrite and Gypsum Rocks*; Bąbel, M., Olszewska-Nejbert, D., Nejbert, K., Eds.; GIMPO Agencja Wydawniczo-Poligraficzna: Warsaw, Poland, 2020; pp. 47–77. (In Polish with English Summary) [CrossRef]
19. Tsui, P.C.; Cruden, D.M. Deformation associated with gypsum karst in the Salt River Escarpment, northeastern Alberta. *Can. J. Earth Sci.* **1984**, *21*, 949–959. [CrossRef]
20. Bąbel, M.; Yatsyshyn, A.; Ługowski, D.; Nejbert, K.; Olszewska-Nejbert, D.; Bogucki, A.; Kremer, B. Swelling caves from the weathering zone of anhydrite rocks in western Ukraine. In *Sedimentology at the Crossroads of New Frontiers, Proceedings of the 19th International Sedimentological Congress, Geneva, Switzerland, 18–22 August 2014, Abstracts Book*; Université de Genève: Geneva, Switzerland, 2014; p. 35.
21. Kraus, E.H. Hydration caves. *Sci. New Ser.* **1905**, *22*, 502–503.
22. Gorbunova, K.A. Caves of hydration. *Peshchery* **1978**, *17*, 61–63. (In Russian)
23. Bögli, A. *Karst Hydrology and Physical Speleology*; German Edition, Karsthydrographie und Physische Speläologie, 1978; Schmid, J.C., Translator; Springer: Berlin, Germany, 1980; pp. 1–270. [CrossRef]
24. Kempe, S. Gypsum karst of Germany. *Int. J. Speleol.* **1996**, *25*, 209–224. [CrossRef]
25. Bella, P. *Genetické Typy Jaskýň*; Speleologia Slovaca 2; Verbum: Ružomberok, Slovakia, 2011; pp. 1–220. (In Slovak with English Summary)
26. Bella, P.; Gaál, L. Genetic types of non-solution caves. In *Proceedings of the 16th International Congress of Speleology, Brno, Czech Republic, 21–28 July 2013*; Filippi, M., Bosák, P., Eds.; Speleological Society: Praha, Czech Republic, 2013; Volume 3, pp. 237–242.

27. White, W.B.; Culver, D.C. Cave, definition of. In *Encyclopedia of Caves*, 3rd ed.; White, W.B., Culver, D.C., Pipan, T., Eds.; Elsevier: Amsterdam, The Netherlands, 2019; pp. 255–259. [CrossRef]
28. Ługowski, D.; Bąbel, M. Distribution of gypsum content in anhydrite hydration dome from Pisky near Lviv and its development. In *The Weathering of Anhydrite and Gypsum Rocks*; Bąbel, M., Olszewska-Nejbert, D., Nejbert, K., Eds.; GIMPO Agencja Wydawniczo-Poligraficzna: Warsaw, Poland, 2020; pp. 135–144, (In Polish with English Summary). [CrossRef]
29. Evans, I.S. Scale-specific landforms and aspects of the land surface. In *Concepts and Modelling in Geomorphology: International Perspectives*; Evans, I.S., Dikau, R., Tokunaga, E., Ohmori, H., Hirano, M., Eds.; TERRAPUB: Tokyo, Japan, 2003; pp. 61–84.
30. MacMillan, R.A.; Shary, P.A. Landforms and landform elements in geomorphometry. *Dev. Soil Sci.* **2009**, *33*, 227–254. [CrossRef]
31. Mokarram, M.; Sathyamoorthy, D. A review of landform classification methods. *Spat. Inf. Res.* **2018**, *26*, 647–660. [CrossRef]
32. Mark, D.M. Geomorphometric parameters: A review and evaluation. *Geogr. Ann.* **1975**, *57*, 165–177. [CrossRef]
33. Pike, R.J. Geomorphometry—Diversity in quantitative surface analysis. *Prog. Phys. Geogr.* **2000**, *24*, 1–20. [CrossRef]
34. Pike, R.J.; Evans, I.; Hengl, T. Geomorphometry: A brief guide. *Dev. Soil Sci.* **2009**, *33*, 227–254. [CrossRef]
35. Muzirafuti, A.; Boualoul, M.; Barreca, G.; Allaoui, A.; Bouikbane, H.; Lanza, S.; Crupi, A.; Randazzo, G. Fusion of remote sensing and applied geophysics for sinkholes identification in Tabular Middle Atlas of Morocco (the Causse of El Hajeb): Impact on the protection of water resource. *Resources* **2020**, *9*, 51. [CrossRef]
36. Calaforra, J.M.; Pulido-Bosch, A. Genesis and evolution of gypsum tumuli. *Earth Surf. Proc. Land.* **1999**, *24*, 919–930. [CrossRef]
37. Jarzyna, A.; Bąbel, M.; Ługowski, D. Morphological diversity and form evolution of the gypsum tumuli from Sorbas region in Spain. In *The Weathering of Anhydrite and Gypsum Rocks*; Bąbel, M., Olszewska-Nejbert, D., Nejbert, K., Eds.; GIMPO Agencja Wydawniczo-Poligraficzna: Warsaw, Poland, 2020; pp. 233–269, (In Polish with English Summary). [CrossRef]
38. Gutiérrez, F.; Gutiérrez, M. *Landforms of the Earth: An Illustrated Guide*; Springer: Berlin/Heidelberg, Germany, 2016; pp. 1–270. [CrossRef]
39. Breisch, R.L. The truth about gypsum caves. *Natl. Speleol. Soc. News* **1978**, *36*, 183–185.
40. Kraus, E.H. On the origin of the caves of the island of Put-in-Bay, Lake Erie. *Am. Geol.* **1905**, *35*, 167–171.
41. Cottingham, K. The origin of the caves at Put-in-Bay, Ohio. *Ohio J. Sci.* **1919**, *20*, 38–42.
42. Muir, J.L. Anhydrite–gypsum problem of Blaine Formation, Oklahoma. *Am. Assoc. Petr. Geol. B.* **1934**, *18*, 1310–1311, The text discussion by Griley. [CrossRef]
43. Verber, J.L.; Stansbury, D.H. Caves in Lake Erie Islands. *Ohio J. Sci.* **1953**, *53*, 358–362.
44. Kes, A.S. Gypsum domes in a desert. *Priroda* **1961**, *2*, 114–115. (In Russian)
45. Herrmann, A. Vergipsung und oberflächenformung im gipskarst. In *Proceedings of the Third International Congress of Speleology, Vienna, Austria, 1963*; Trimmel, H., Ed.; Verband Österreichischer Höhlenforscher: Vienna, Austria, 1966; Volume 5, pp. 99–108. Available online: http://uis-speleo.org/wp-content/uploads/2020/07/3rd_Kongress_fuer_Spelaeologie_Band_V.pdf (accessed on 13 July 2022).
46. Gorbunova, K.A. Exogenetic karst tectonics. In *Proceedings of the 7th International Speleological Congress, Sheffield, 1977*; Ford, T.D., Ed.; BCRA: Bridgwater, UK, 1977; pp. 222–223. Available online: http://uis-speleo.org/wp-content/uploads/2020/0 8/Proceedings-of-The-7th-International-Speleological-Congress-Sheffield-England-September-1977.pdf (accessed on 13 July 2022).
47. Völker, C.; Völker, R. Gipskuppen und Gipsbuckel—Elemente der Sulfatkarstlandschaft. *Mitt. Karstmus. Heimkehle* **1988**, *19*, 1–19.
48. Stenson, R.E. The Morphometry and Spatial Distribution of Surface Depressions in Gypsum, with Examples from Nova Scotia, Newfoundland and Manitoba. Master's Thesis, McMaster University, Hamilton, ON, Canada, 1990; pp. 1–134.
49. Reimann, M. Quellungshöhlen am südharz, landschaftsprägende auswirkung der vergipsung von anhydritstein. In Proceedings of the Zechstein 1987—International Symposium—Abstracts, Posters, Program, Kassel, Hannover, Germany, 28 April–9 May 1987; Bernd, K., Schröder, B., Eds.; Poster, MS of the Summary. p. 9.
50. Stafford, K.W.; Ehrhart, J.; Majzoub, A.; Shields, J.; Brown, W. Unconfined hypogene evaporite karst: West Texas and southeastern New Mexico, USA. *Int. J. Speleol.* **2018**, *47*, 293–305. [CrossRef]
51. Beyrich, E. *Erläuterungen zur Geologischen Specialkarte von Preussen und den Thüringischen Staaten*; No. 255, Blatt Ehrlich, Bande 6, Blatt 2; Verlag von J.H. Neumann: Berlin, Germany, 1870; pp. 1–18.
52. Behrens, G.H. *Hercynia Curiosa Oder Curiöser Hartz-Wald*; Verlegts Carl Chriftian Neuenhahn: Nordhausen, Germany, 1703; pp. 1–216. Available online: http://www.deutschestextarchiv.de/book/show/behrens_hercynia_1703 (accessed on 6 May 2022).
53. Stolberg, F. *Die Höhlen des Harzes, Bd. 1. Einleitung und Südharzer Zechsteinhöhlen. Sonderausgabe der illustrierten Monatsschrift "Der Harz", Heft 2*; Eilers-Verlag G.m.b.H.: Magdeburg, Germany, 1926; pp. 1–40.
54. Bąbel, M.; Jacyszyn, A.; Olszewska-Nejbert, D.; Nejbert, K.; Bogucki, A.; Maksymiw, I.; Mik, W.; Bermes, A.; Ługowski, D.; Kacprzak, K.; et al. Jaskinie z pęcznienia (ang. swelling caves) w strefie współczesnego wietrzenia anhydrytów w kamieniołomie Pisky w okolicach Lwowa. In *Od Czarnohory po Góry Świętokrzyskie—Geologiczne Peregrynacje, Polsko–Ukraińska Sesja Naukowa, Warszawa, Bocheniec, 15–19 Października 2013*; Bąbel, M., Dzierżek, J., Olszewska-Nejbert, D., Eds.; Instytut Geologii Podstawowej WG UW: Warsaw, Poland, 2013; pp. 19–26. [CrossRef]

55. Bąbel, M.; Yatsyshyn, A.; Bogucki, A.; Jarzyna, A.; Ługowski, D.; Olszewska-Nejbert, D.; Nejbert, K.; Kotowski, J.; Przybylik, G.; Bermes, A.; et al. Development of the unique landforms in the zone of weathering gypsum-anhydrite rocks at Pisky (Shchyrka river valley, Dnister basin). In *Scientific Principles of Conservation Management of Ecosystems in the Dniester Canyon Area, Proceedings of the Second International Scientific and Practical Conference Dedicated to the 170th Anniversary of Publication of Rudolf Kner's Work Which Marked the Beginning of the Profound Paleontological Investigations in the Dniester Canyon, Zalishchyky, Ternopil, Ukraine, 14–15 September 2017*; Skilsky, I.V., Vikyrchak, O.K., Eds.; Druk Art: Chernivtsi, Ukraine, 2017; pp. 23–25. [CrossRef]
56. Ługowski, D.; Jarzyna, A. The map of the site of weathering anhydrites and hydration caves at Pisky at environs of Lviv. In *The Weathering of Anhydrite and Gypsum Rocks*; Bąbel, M., Olszewska-Nejbert, D., Nejbert, K., Eds.; GIMPO Agencja Wydawniczo-Poligraficzna: Warsaw, Poland, 2020; pp. 21–48. (In Polish with English Summary) [CrossRef]
57. Jarzyna, A. Geological Setting and Morphology of the Site of the Weathering Miocene Anhydrites at Pisky near Lviv. Master's Thesis, University of Warsaw, Warsaw, Poland, 2021; pp. 1–131. (In Polish with English Summary)
58. Bąbel, M.; Bogucki, A.; Jacyszyn, A.; Ługowski, D.; Olszewska-Nejbert, D.; Nejbert, K.; Jarzyna, A.; Bermes, A.; Przybylik, G.; Tomeniuk, O. Weathering anhydrites at Pisky quarry, Part I. General characteristic. In *Weathering of Gypsum and Anhydrite Rocks, Proceedings of the Polish-Ukrainian Scientific Seminar, Warsaw, Poland, 19–21 January 2017*; Bąbel, M., Olszewska-Nejbert, D., Nejbert, K., Eds.; IGP WG UW: Warszaw, Poland, 2017; pp. 18–22. (In Polish) [CrossRef]
59. Maksymiw, I.P. Geomorphological Effects of the Hydration Processes in the Gypsum Layer of the "Pisky" Section. Master's Thesis, Faculty of Geography, University of Lviv, Lviv, Ukraine, 2013; pp. 1–70. (In Ukrainian)
60. Ługowski, D.; Jarzyna, A.; Bąbel, M.; Nejbert, K. Data collecting methods used in the field study of weathering anhydrites at Pisky near Lviv. *Biul. Państwowego Inst. Geol.* **2016**, *466*, 201–214. [CrossRef]
61. Jarzyna, A.; Bąbel, M.; Ługowski, D.; Vladi, F. Preliminary morphological analysis of the anhydrite hydration forms at Dingwall (Canada, Nova Scotia) on the base of photogrammetric documentation. In *Forum GIS UW. GIS na Uniwersytecie Warszawskim. Materiały Pokonferencyjne z 5. i 6. Forum GIS na UW*; Chyla, J., Lechnio, J., Stępień, M., Zaszewski, D., Eds.; UW WG, WGiSR, IAWH: Warsaw, Poland, 2019; pp. 92–102. (In Polish with English Summary) [CrossRef]
62. Twidale, C.R.; Bourne, J.A. On the origin of A-tents (pop-ups), sheet structures, and associated forms. *Prog. Phys. Geogr.* **2009**, *33*, 147–162. [CrossRef]
63. Demicco, R.V.; Hardie, L.A. *Sedimentary Structures and Early Diagenetic Features of Shallow Marine Carbonate Deposits*; SEPM Atlas Series; SEPM: Tulsa, OK, USA, 1994; pp. 1–255. [CrossRef]
64. Jennings, J.N.; Twidale, C.R. Origin and implications of the A-tent, a minor granite landform. *Aust. Geogr. Stud.* **1971**, *9*, 41–53. [CrossRef]
65. Folk, R.L.; Begle Patton, E. Buttressed expansion of granite and development of grus in Central Texas. *Z. Geomorphol. Neue Folge* **1982**, *26*, 17–32.
66. Ericson, K.; Olvmo, M. A-tents in the Central Sierra Nevada, California: A geomorphological indicator of tectonic stress. *Phys. Geogr.* **2004**, *25*, 291–312. [CrossRef]
67. Pulido-Bosch, A. Le karst dans le gypses de Sorbas (Almeria). Aspects morphologiques et hydrogéologiques. In *Karst et Cavités d'Andalousie, Cordillères Bétiques Centrales et Occidentales*; Karstologia, Mémoires, 1; Association Française de Karstologie: Nîmes, France, 1986; pp. 27–35.
68. Von Gaertner, H.-R. Petrographie und paläogeographische Stellung der Gipse vom Südrande des Harzes. *Jahrb. Preußischen Geol. Landesanst. Berl.* **1933**, *53*, 655–694.
69. Hargitai, H.; Kereszturi, A. *Encyclopedia of Planetary Landforms*; Springer: New York, NY, USA, 2015; pp. 1–2460. [CrossRef]
70. Bishop, J.L.; Yeşilbaş, M.; Hinman, N.W.; Burton, Z.F.M.; Englert, P.A.J.; Toner, J.D.; McEwen, A.S.; Gulick, V.C.; Gibson, E.K.; Koeberl, C. Martian subsurface cryosalt expansion and collapse as trigger for landslides. *Sci. Adv.* **2021**, *7*, eabe4459. [CrossRef]
71. Neale, E.R.W. *Geology, Dingwall, Nova Scotia, Map 1124A*; Geological Survey of Canada, Department of Mines and Technical Surveys: Ottawa, ON, Canada, 1963.
72. Adams, G.C. Gypsum and anhydrite resources in Nova Scotia. *Econ. Geol. Ser.* **1991**, *91*, 1–293.
73. Wiebe, R.A. Igneous and tectonic events in northeastern Cape Breton Island, Nova Scotia. *Can. J. Earth Sci.* **1972**, *9*, 1262–1277. [CrossRef]
74. Lynch, G.; Tremblay, C. Late Devonian–Carboniferous detachment faulting and extensional tectonics in western Cape Breton Island, Nova Scotia, Canada. *Tectonophysics* **1994**, *238*, 55–69. [CrossRef]
75. Grant, D.R. Quaternary geology, Cape Breton Island, Nova Scotia. *Geol. Surv. Can.* **1994**, *482*, 1–159. [CrossRef]
76. Moseley, M. Genesis of schlottenkarren on the Avon Peninsula of Nova Scotia (Canada) with implications for the geochronology of evaporate karsts and caves of Atlantic Canada. *Int. J. Speleol.* **2017**, *46*, 267–276. [CrossRef]
77. Ford, D. Principal features of evaporite karst in Canada. *Carbonate Evaporite* **1997**, *12*, 15–23. [CrossRef]
78. Stenson, R.E.; Ford, D.C. Rillenkarren on gypsum in Nova Scotia. *Geogr. Phys. Quatern.* **1993**, *47*, 239–243. [CrossRef]
79. Climate-Data.org. Available online: https://en.climate-data.org/north-america/canada/nova-scotia/dalem-lake-98946/ (accessed on 18 September 2021).
80. Zepner, L.; Karrasch, P.; Wiemann, F.; Bernard, L. ClimateCharts.net—An interactive climate analysis web platform. *Int. J. Digit. Earth* **2021**, *14*, 338–356. [CrossRef]

81. Clark, C.D.; Hughes, A.L.C.; Greenwood, S.L.; Spagnolo, M.; Ng, F.S.L. Size and shape characteristics of drumlins, derived from a large sample, and associated scaling laws. *Quatern. Sci. Rev.* **2009**, *28*, 677–692. [CrossRef]
82. Spagnolo, M.; Clark, C.D.; Hughes, A.L.C.; Dunlop, P.; Stokes, C.R. The planar shape of drumlins. *Sediment. Geol.* **2010**, *232*, 119–129. [CrossRef]
83. Maclachlan, J.C.; Eyles, C.H. Quantitave geomorphological analysis of drumlins in the Peterborough Drumlin Field, Ontario, Canada. *Geogr. Ann. Ser. A Phys. Geogr.* **2013**, *95*, 125–144. [CrossRef]
84. Spagnolo, M.; Clark, C.D.; Hughes, A.L.C. Drumlin relief. *Geomorphology* **2012**, *153–154*, 179–191. [CrossRef]
85. Dadlez, R.; Jaroszewski, W. *Tektonika*; Wydawnictwo Naukowe PWN: Warsaw, Poland, 1994; pp. 1–743.
86. Mathumaniraja, C.K.; Anbazhagan, S.; Jothibasu, A.; Chinnamuthu, M. Remote sensing and fuzzy logic approach for artificial recharge studies in hard rock terrain of South India. In *GIS and Geostatistical Techniques for Groundwater Science*; Senapathi, V., Viswanathan, P.M., Chung, S.Y., Eds.; Elsevier: Amsterdam, The Netherlands, 2019; pp. 91–112. [CrossRef]
87. Triggs, B.; Mclauchlan, P.; Hartley, R.; Fitzgibbon, A. Bundle adjustment—A modern synthesis. In *Vision Algorithms: Theory and Practice*; IWVA 1999, Lecture Notes in Computer Science; Triggs, B., Zisserman, A., Szeliski, R., Eds.; Springer: Berlin/Heidelberg, Germany, 2000; Volume 1883, pp. 1–71. [CrossRef]
88. Snavely, K.N. Scene Reconstruction and Visualization from Internet Photo Collections. Ph.D. Thesis, University of Washington, Seattle, WA, USA, 2008; pp. 1–192.
89. Westoby, M.J.; Brasington, J.; Glasser, N.F.; Hambrey, M.J.; Reynolds, J.M. 'Structure-from-motion' photogrammetry: A low-cost, effective tool for geoscience applications. *Geomorphology* **2012**, *179*, 300–314. [CrossRef]
90. Agisoft. *Agisoft PhotoScan User Manual Professional Edition, Version 1.2*; Agisoft LLC: St. Petersburg, Russia, 2016; pp. 1–97.
91. Randazzo, G.; Italiano, F.; Micallef, A.; Tomasello, A.; Cassetti, F.P.; Zammit, A.; D'Amico, S.; Saliba, O.; Cascio, M.; Cavallaro, F.; et al. WebGIS Implementation for dynamic mapping and visualization of coastal geospatial data: A case study of BESS project. *Appl. Sci.* **2021**, *11*, 8233. [CrossRef]
92. Rose, J.; Letzer, J.M. Drumlin measurements: A test of the reliability of data derived from 1:25,000 scale topographic maps. *Geol. Mag.* **1975**, *112*, 361–371. [CrossRef]
93. Turner, K. What's the Difference among 2-D, 2.5-D, 3-D and 4-D? Applied Geoscience Forum, GIS World Article. 1997. Available online: http://dusk.geo.orst.edu/gis/gis_world_article.pdf (accessed on 13 July 2022).
94. Price, M.H. *Mastering ArcGIS*, 7th ed.; McGraw-Hill Education: New York, NY, USA, 2016; pp. 1–606.
95. Devore, J.L. *Probability and Statistics for Engineering and the Sciences*, 9th ed.; Cengage Learning: Boston, MA, USA, 2016; pp. 1–715.
96. Curl, R.L. On the definition of a cave. *NSS Bull.* **1964**, *26*, 1–6.
97. Curl, R.L. Entranceless and fractal caves revisited. In *Karst Modeling, Proceedings of the Symposium, Charlottesville, VA, USA, 24–27 February 1999*; Karst Waters Institute Special Publication 5; Palmer, A.N., Palmer, M.V., Sasowsky, I.D., Eds.; Karst Waters Institute Inc.: Charles Town, WV, USA, 1999; pp. 183–185.
98. Twiss, R.J.; Moores, E.M. *Structural Geology*, 2nd ed.; W.H. Freeman and Company: New York, NY, USA, 2007; pp. 1–736. [CrossRef]
99. Potter, P.E.; Pettijohn, F.J. *Paleocurrents and Basin Analysis*, 2nd ed.; Springer: Berlin/Heidelberg, Germany, 1977; pp. 1–425. [CrossRef]
100. Peacock, D.C.P.; Sanderson, D.J.; Rotevatn, A. Relationship between fractures. *J. Struct. Geol.* **2017**, *106*, 41–53. [CrossRef]
101. Hancock, P.L. Brittle microtectonics: Principles and practice. *J. Struct. Geol.* **1985**, *7*, 437–457. [CrossRef]
102. Laslett, G.M. Censoring and edge effects in areal and line transect sampling of rock joint traces. *Math. Geol.* **1982**, *14*, 125–140. [CrossRef]
103. Hydration Caves. Available online: http://hydrationcaves.com/en/77-hydration-forms/ (accessed on 8 May 2022).
104. Beales, F.W.; Oldershaw, A.E. Evaporite-solution brecciation and Devonian carbonate reservoir porosity in Western Canada. *Am. Assoc. Petr. Geol. B.* **1969**, *53*, 503–512. [CrossRef]
105. Curl, R.L. Caves as a measure of karst. *J. Geol.* **1966**, *74*, 798–830. [CrossRef]
106. Webb, T.C. Geology and economic development of early Carboniferous marine evaporates, southeastern New Brunswick. In *Lands, Minerals and Petroleum Division*; Field Guide No. 6; Lands, Minerals and Petroleum Division, Department of Natural Resources: Tracadie-Sheila, NB, Canada, 2010; pp. 1–71.
107. Ball, P. *The Self-Made Tapestry: Pattern Formation in Nature*; Oxford University Press: Oxford, UK, 2001; pp. 1–312.
108. Ferrarese, F.; Macaluso, T.; Madonia, G.; Palmeri, A.; Sauro, U. Solution and recrystallisation processes and associated landforms in gypsum outcrops of Sicily. *Geomorphology* **2002**, *49*, 25–43. [CrossRef]
109. Macaluso, T.; Sauro, U. Aspects of weathering and landforms evolution on gypsum slopes and ridges of Sicily. *Suppl. Geogr. Fis. Dinam. Quat.* **1998**, *3*, 91–99.
110. Scott, D.N.; Wohl, E.E. Bedrock fracture influence on geomorphic process and form across process domains and scales. *Earth Surf. Proc. Land.* **2018**, *44*, 27–45. [CrossRef]
111. Narr, W.; Suppe, J. Joint spacing in sedimentary rocks. *J. Struct. Geol.* **1991**, *13*, 1037–1048. [CrossRef]
112. Shaocheng, J.; Zheming, Z.; Zichao, W. Relationship between joint spacing and bed thickness in sedimentary rocks: Effects of interbed slip. *Geol. Mag.* **1998**, *135*, 637–655. [CrossRef]
113. Butscher, C.; Mutschler, T.; Blum, P. Swelling of clay-sulfate rocks: A review of processes and controls. *Rock Mech. Rock Eng.* **2016**, *49*, 1533–1549. [CrossRef]

114. Fleuchaus, P.; Blum, P. Damage event analysis of vertical ground source heat pump systems in Germany. *Geotherm. Energy* **2017**, 5, 1–15. [CrossRef]
115. Hou, Z.; Wu, J. A practical swelling constitutive model of anhydrite and its application on tunnel engineering. *Res. Sq.* **2022**. [CrossRef]

125

Article

The Influence of the Quality of Digital Elevation Data on the Modelling of Terrain Vehicle Movement

Marian Rybansky [1,*] and Josef Rada [2]

1 Faculty of Military Technology, University of Defence, Kounicova 65, 662 10 Brno, Czech Republic
2 NATO HQ SHAPE J2 GEOMETOC, Rue Grande, 7010 Mons, Belgium; josef.rada@army.cz
* Correspondence: marian.rybansky@unob.cz

Abstract: This study investigated digital terrain models and options for their evaluation and effective usage. The most important result of this study was the introduction of the slope reduction method for low-detail elevation models. It enabled accurate results of passability analyses by performing adjustments of slopes. In addition, the goal was to determine the strengths and weaknesses of selected data for use in cross-country mobility analyses, followed by recommendations on how to use these databases efficiently to obtain accurate results. The selection of elevation databases (1 m, 5 m, 10 m, 30 m) was determined by the focus of data development projects of NATO and current scientific research projects of the Ministry of Defence of the Czech Republic. Key findings showed potential for use in practise for all tested elevation models. Efficient usage of low-detail models in CCM analyses is limited; nevertheless, they can be augmented with additional vector data or automated remote-sensing technologies.

Keywords: DEMs; cross-country mobility; slope analysis; terrain vehicles; geospatial support in NATO

check for updates

Citation: Rybansky, M.; Rada, J. The Influence of the Quality of Digital Elevation Data on the Modelling of Terrain Vehicle Movement. *Appl. Sci.* **2022**, *12*, 6178. https://doi.org/10.3390/app12126178

Academic Editors: Giovanni Randazzo, Stefania Lanza and Anselme Muzirafuti

Received: 13 May 2022
Accepted: 13 June 2022
Published: 17 June 2022

Publisher's Note: MDPI stays neutral with regard to jurisdictional claims in published maps and institutional affiliations.

1. Introduction

Terrain passability analysis as an integral part of battlefield intelligence plays a key role in military operations. Digital elevation models (DEMs) are the most detailed and up-to-date source of landscape information in comparison with other types of data (soil, forests, roads, settlements, etc.). The terrain relief is also the most stable part of the landscape. From the military perspective, DEMs are one of the only reliable sources of information for detailed, accurate and rapid terrain analyses. On the other hand, the quality of elevation models varies substantially depending on an area of the world. North Atlantic Treaty Organization (NATO) nations' territory is covered with data with a high resolution and accuracy. The availability of high-quality data beyond European territory might be very limited (e.g., processed ready-to-use 1 m terrain models). Therefore, any output analysis from the data is constrained as well. It is essential to know what effect the detail of individual terrain models and slope models derived from them has on the results of terrain passability analyses. Besides the military perspective, there are other domains that have analysed DEMs for the extraction of landscape information. Study [1] referred to the impact on the protection of water resources, and photogrammetry-based mapping of microrelief forms was studied in [2,3], which used tri-stereo Pleiades images for the morphometric measurement.

The basis of this article was a study of the currently most used DEMs in the Czech Republic and NATO. To include different aspects of landscapes abroad, available global databases were included in the study as well. The results and recommendations of the article were aimed mainly at the NATO environment, i.e., NATO Command Structure, NATO Force Structure and NATO nations' use cases. The analysis was performed by evaluating slopes from DEMs with the method of raster analysis, including implementing the influence of soils and the quality of terrain surfaces. Values of slopes and results of cross-country movement (CCM) analysis were compared for each terrain model in selected areas.

The key approach of the article, i.e., identifying the influence of slope accuracy on the results of CCM analyses, was based on both national and international studies. The basic methodology of analysis of elevation models was given in [4]. The study compared the accuracy of heights of terrain models with measured values in the field and assigned them specific output statistics. These statistics showed the specific advantage of 5 m × 5 m DEMs. It combined both assets in terrain detail and data availability across the world that are needed for practical use, such as visibility or passability analysis. Despite some problems with accuracy, especially in areas of significant microrelief shapes, DEMs comply with the declared accuracy and they are a major contribution to the field of geoinformatics and terrain analysis. The conducted tests confirmed the declared accuracy of all tested Czech models (used also in this article). Analysis of terrain measurements of the accuracy of elevation models, carried out by the University of Defence, Brno, and a comparison of the possibilities of using available elevation models with different levels of accuracy were the main prerequisites for a new, as yet unpublished methodology aimed at evaluating the influence of microrelief shapes generated from various elevation models on military mobility.

A stereo-vision-based terrain passability estimation method for off-road mobile robots is presented in [5]. The method models surrounding terrain using either sloped planes or a digital elevation model, based on the availability of suitable input data. This combination of two surface-modelling techniques increases the range and information content of the resulting terrain map. As defined in [6], the greatest asset to global elevation models are the TanDEM-X and ALOS projects. The TanDEM-X satellite mission has now mapped the Earth's global topography with a spatial resolution of 0.4 arc-sec (about 12 m). While TanDEM-X is a commercial mission, a down-sampled elevation model (WorldDEM) with a 3 arc-sec resolution (commensurate to SRTM) and global coverage, the ALOS system is a new 3D model of the Earth's surface of up to a 0.15 arc-sec (5 m) resolution. It will be generated from optical stereoscopy carried out aboard the ALOS satellite. The option of comparing various terrain models can be studied from [7]. It assesses uncertainties in a derived slope and aspects from a grid of DEMs. A quantitative methodology was developed for objective and data-independent assessment of errors generated from the algorithms that extract morphologic parameters from grid-based digital elevation model (DEM). The generic approach is to use artificial surfaces that can be described by mathematical models; therefore the 'true' output value can be pre-determined to avoid uncertainty caused by uncontrollable data errors. A prospective development and usage of a new surface model, namely, TanDEM-X High-Resolution Elevation Data Exchange Program (TREx), was introduced in [8]. Using this model to replace insufficient low detailed elevation models seems to be the only option to cover the information gap in some locations. One of the most important impacts of TREx is that it brings accurate results of the terrain analysis from a global perspective. The most useful analyses that can be gathered from it are visibility, propagation of radio signals, searching for areas suitable for air landing and route planning. Data can be procured in its raw form, but users may encounter issues with its extensive processing [9]. Global models can also be substituted for more detailed models in the case of the availability of local LIDAR data [10,11].

The CCM model used in NATO is called New NATO Reference Mobility Model (NRMM); see [12] for a detailed description. The analyses of the model and its gaps are in [13,14]. The NRMM was originally used to facilitate a comparison between vehicle design candidates by assessing the mobility of existing vehicles under specific terrain scenarios but has subsequently and most recently found expanded use in support of complex decision analyses associated with vehicle acquisition and operational planning support. A study [15] verified the usability of NRMM in new, so far untested conditions. It is based on a comparison of the empirically based NRMM with the physics-based Nepean

Tracked Vehicle Performance Model (NTVPM) for assessing the cross-country performance of military tracked vehicles. The NRMM can be used strictly for the tactical level of planning. Such detailed data is not available across the whole Supreme Commanders of Allied Powers in Europe's (SACEURs) area of interest (AOI). The AOI is an extended area beyond the territory of NATO nations where NATO can operate [16,17]. Global availability of high-detail elevation models is limited; therefore, other simplified CCM models must be considered. Future requirements of NATO towards CCM analyses are defined in [18], emphasising the ongoing important position of passability analyses in military planning.

The research of the CCM model in the Czech Republic at the University of Defence at the Department of Military Geography and Meteorology focuses on creating a system of coefficients representing various land features and conditions. Their analyses and implementations defining soil impact are in [19,20], refs. [21–23] focused on forest passability, refs. [24,25] describing problematics of microrelief obstacles and [26–29] consist of all environmental elements in one complex study. This approach was also partially included in the evaluation of data in the article. It is mainly in a form that considers which level of detail is required for data usability in the Czech military CCM model. Talhofer analysed the spatial database quality influence on the modelling of movement of vehicles in terrain in [30]. This general assessment of the quality of geospatial data can serve as a template for how to compare various datasets, including elevation models.

A similar approach as that found in the Czech Republic can be found among authors from other nations. Raster analysis is the main method used in the Polish CCM cartographic model [31,32]. This form was also assumed for the results of the article. The research introduced in [33] was specifically taken into consideration. It defines variances between elevation models with a focus on high-detail models (0.5 m, 1.0 m, 2.5 m, 5.0 m). It shows that the passable area of a smaller tested location is larger with higher detail terrain models. However, when studying large areas in a global aspect, where only lower-detail elevation models are available (5 m, 10 m, 30 m), the passable area is larger with lower-detail models. Processing and adjustment of these elevation models are necessary to obtain reliable CCM analysis results that correspond to the real terrain conditions. The other national models and studies are based on a vector line tactical level CCM analysis [34,35]. They do not consider the unavailability of detailed data; therefore, they do not fit within the focus of this article. Their focus was on automated navigation in the terrain. From the global point of view, this may be the only option before any global high-resolution elevation database is developed.

Elevation models are an important part of the complex analysis of the movement of vehicles in terrain; however, they cannot be considered in analyses without including other data and aspects. The data focus should be put on soil databases. The analysis of the Czech environment is introduced in [36]. The study considered the modelling of geographic and meteorological effects on vehicle movement, focusing on soil conditions and penetrometry; see also [37]. It consisted of an analysis of the characteristics of soil databases. The situation was similar to that of elevation models. National soil databases are relatively detailed enough to support digital terrain models with sufficient accuracy, but the global models are overly generalised and they cannot be used in detailed CCM analyses. Global data can be improved with methods for automatic refinement of soil databases [38] using an algorithm for refining soil data via comparison with relief models in a test grid with a cell of 100 m $\times$ 100 m.

Apart from soil databases, vector databases comprising features of the terrain should be taken into account for accurate passability analyses. One of the most used vector databases in NATO is the Multinational Geospatial Co-Production Program (MGCP) [39]. It is not completely a global database, though it has full coverage of the AOI. Since updating vector databases at the global scale can be substantially challenging, a new trend of using open-source data is rising. The OpenStreetMap is an example of a rapidly developing universal vector database [40]. NATO will most probably be using this kind of data more often in the near future. In general, contemporary vector databases do not consist of all features that are needed to enable accurate tactical-level CCM analyses.

Currently, there are no DEMs that would meet the requirement of high detail and global coverage at the same time in order to fulfil the needs of NATO and cooperating civilian organisations. The aim of this research was to verify the possibilities of usage of lower detail DEMs in practical tasks, predominantly in CCM analyses. The main asset of this study rested in a proposal for processing elevation models for their efficient evaluation in passability analyses done in larger areas.

2. Materials and Methods

2.1. Utilised Methods

The research in the article focused preferentially on elevation models to define the strengths and weaknesses of the models without any external influences. The comparison of the DEMs was based on preparing a suitable representative sample of data and determining a proper method of evaluation. Data from the Czech Republic was selected to test the accuracy of the slopes that were calculated from DEMs: Digital Terrain Model 3rd generation, 4th generation and 5th generation (DTM 3, DTM 4, DTM 5) and Digital Terrain Elevation Data 2 (DTED 2). The selection of the models was based on their occurrence in different parts of the globe. Whilst models with a resolution of up to 1 m predominate in Europe, areas of potential foreign operations are covered predominantly at most with elevation models with a maximum resolution of 30 m. Naturally, more accurate models might be locally available as well, e.g., laser scanning data. Nonetheless, from a military operation perspective, short notice ready-to-use terrain models with global coverage are the ultimate goal in the military environment. This excludes potential surface models that have other disadvantages for accurate CCM analyses. The challenge for elevation databases and their versatile usage is that any additional extensive processing of data prevents it from being operationally usable.

Although the key part of this research was based on the proper selection of elevation data, its usability in CCM analyses was guaranteed only with a suitable processing method. This article introduced its own approach to elevation data evaluation. This method was based on generating random points and general statistical comparisons of their slope values. The second part of the study was a comparison of the reliability of data, which was achieved by evaluating passability with a raster analysis for selected military vehicles. Five operationally representative areas were selected for the tests.

The analyses in the study were undertaken with the software ArcGIS 10.4.1 with its Toolbox functionalities; for points statistics, Extract values to points was used; and for areal statistics, Slope and Zonal statistics were used. Microsoft Excel 2019 was utilised for generating and evaluating statistical functions, e.g., mean slope and standard deviation. A major method used was the comparative method of map results or statistical results. The overview of the methodology and steps used in this research is in Figure 1.

2.2. Selected Digital Elevation Models

The digital terrain model (DTM) approximates the terrain surface or part of it by using a system of points in 2D space with specified height values. These points are predominantly organised into a regular grid or, in more detailed models, a triangular irregular network (TIN) [41]. Both these variants were included in the study. DEMs are the most important basis for evaluating the terrain relief. It varies according to spatial accuracy, where the height accuracy is lower than the position accuracy. In comparison with other types of data (e.g., vegetation, roads), the terrain does not change much in terms of extent and speed of changes, and thus, can be considered the most reliable. An overview of the elevation models considered in this study is in Table 1.

<table>
<tr><th>Stage of Proposed Study of DEMs</th><th></th><th>Description</th></tr>
<tr><td>Data Selection
DEMs</td><td>
<table>
<tr><th>Terrain Models</th><th>Resolution (m)</th><th>Vertical Accuracy (m)</th></tr>
<tr><td colspan="3">Global Models</td></tr>
<tr><td>SRTM 1</td><td>30 m × 30 m</td><td>16–20 m</td></tr>
<tr><td>SRTM 3</td><td>90 m × 90 m</td><td>16–20 m</td></tr>
<tr><td>DTED 1</td><td>90 m × 90 m</td><td>3–20 m</td></tr>
<tr><td>DTED 2</td><td>30 m × 30 m</td><td>3–15 m</td></tr>
<tr><td>TREx</td><td>12 m × 12 m</td><td>2–10 m</td></tr>
<tr><td colspan="3">Czech National Models</td></tr>
<tr><td>DTM 3</td><td>10 m × 10 m</td><td>1–7 m</td></tr>
<tr><td>DTM 4</td><td>5 m × 5 m</td><td>0.3–1 m</td></tr>
<tr><td>DTM 5</td><td>TIN—min. 1 m × 1 m</td><td>0.18–0.3 m</td></tr>
<tr><td colspan="3">Local Models</td></tr>
<tr><td>LIDAR data</td><td>TIN—min. 1 m × 1 m</td><td>0.3 m</td></tr>
</table>
</td><td>Selection and study of Czech and global DEMs.</td></tr>
<tr><td>Data Processing
DEMs</td><td>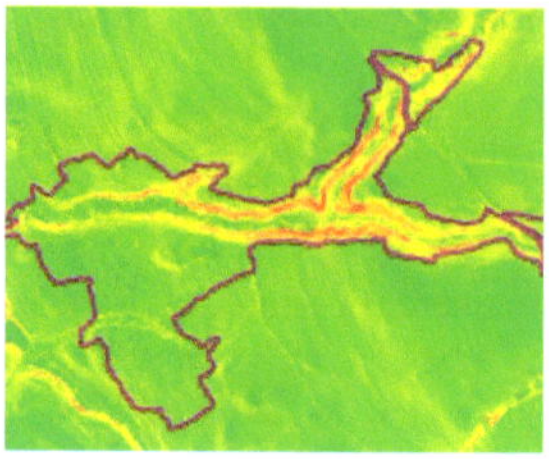</td><td>Description of the processing of DEMs for the needs of passability analysis, e.g., in slope maps.</td></tr>
<tr><td>Data Analysis
Altitude</td><td>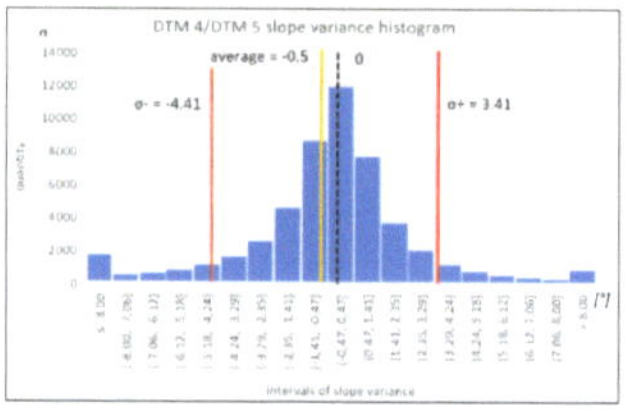
</td><td>Analysis of the accuracy of altitude in DEMs using histograms and graphs (comparison with DTM 5).</td></tr>
<tr><td>Data Analysis
Slope</td><td>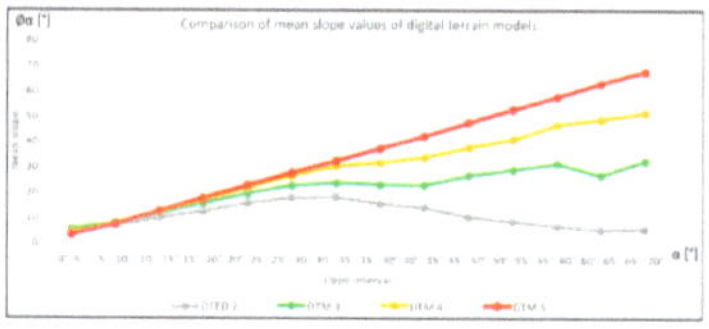</td><td>Analysis of the accuracy of slopes derived from DEMs, evaluation of histograms and graphs (comparison with DTM 5).</td></tr>
<tr><td>Analysis of Passability
Analysis of details</td><td>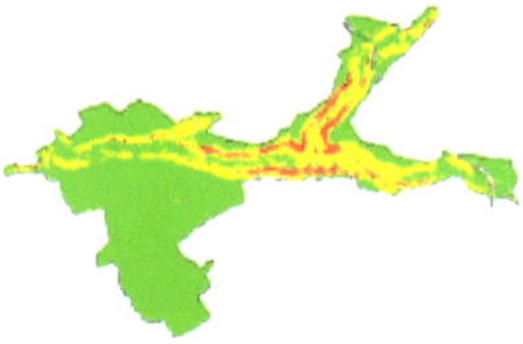</td><td>Comparison of details of the results of the passability analysis using various data and vehicles using the method of raster analysis in a selected area.</td></tr>
<tr><td>Analysis of Passability
Analysis of large areas</td><td>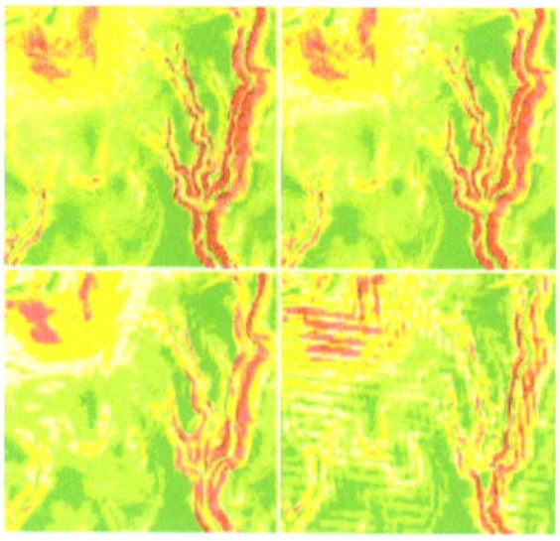</td><td>Comparison of the results of the passability analysis using different data and vehicles in larger areas.</td></tr>
</table>

Figure 1. Scheme of the procedures within the study of DEMs and results of the passability analysis.

Table 1. An overview of global elevation data and data from the territory of the Czech Republic and their altitude accuracy [4]. Unlike the rest of the models, DTM 5 is formed by an irregular grid of points (TIN).

Terrain Models	Resolution (m)	Vertical Accuracy (m)
Global Models		
SRTM 1	30 m × 30 m	16–20 m
SRTM 3	90 m × 90 m	16–20 m
DTED 1	90 m × 90 m	3–20 m
DTED 2	30 m × 30 m	3–15 m
TREx	12 m × 12 m	2–10 m
Czech National Models		
DTM 3	10 m × 10 m	1–7 m
DTM 4	5 m × 5 m	0.3–1 m
DTM 5	TIN—min. 1 m × 1 m	0.18–0.3 m
Local Models		
LIDAR data	TIN—min. 1 m × 1 m	0.3 m

Global data is represented by Shuttle Radar Topography Mission (SRTM) radar data with an average detail of 30 m × 30 m [42]. Furthermore, some areas of the world are covered with DTED models with data levels 0, 1 and 2 [43], which are derived from other various methods of collecting altitude data (for example from SRTM). In 2015, a project of the new accurate global altitude model TREx was established. It is a joint project of NATO nations and non-member partners of the Alliance. It is based on the method of laser scanning. Because of the demanding character in terms of its time and capacity consumption, the project is particularly oriented towards specific locations of interest, primarily areas of deployment of troops of NATO nations.

This project was based on a similar principle as the creation of the global vector database MGCP [8]. TREx is a surface model and, therefore, also contains objects on the terrain (e.g., buildings, forests). Although a digital terrain model is not derived from the TREx data, a new initiative for the automatic generation of a 12 m model was scrutinised. TREx raw data in a combination with DTED 2 data can be used to interpolate a digital terrain model with a resolution of 12 m × 12 m, although the resulting accuracy is low. The main usage of such a model is currently for the MGCP topographic map concept. The contour lines are so far generated from DTED 2, which has limitations regarding use cases of terrain analyses.

The other option for detailed elevation data is local laser-scanning-based data. In areas of military operations, an accurate altitude model created from light detection and ranging (LIDAR) data is often the only detailed source [10]. It was the most accurate available source of altitude data in the International Security Assistance Force (ISAF) operation [11]. Its usability is versatile but obtaining this kind of data is time consuming. In the case of a crisis, the only option is ready-to-use lower-detail global elevation models.

From the elevation data perspective, a representative territory among NATO nations is the Czech Republic. Local sources in the Czech Republic are represented by various models: DTM 3 (10 m × 10 m), DTM 4, (5 m × 5 m) and DTM 5 (1 m × 1 m). Their creation and updating are based on a joint laser-scanning project of the Czech Office for Surveying, Mapping and Cadastre (ČÚZK) and the Office of Military Geography and Hydrometeorology (VGHMÚř). The most detailed model is the DTM 5. Since 2016, it has fully covered the territory of the Czech Republic and a local update of the model is subsequently underway [44].

DTM 5 was taken as a template and comparative database for analyses within this study. The detail of DTED 2 is not sufficient in comparison with DTM 4 and DTM 5; nonetheless, it is the only available detailed global terrain model for a large part of the foreign territory beyond the borders of NATO countries (in the AOI). DTED 1 is also the utilised global model; however, its basic 100 m × 100 m grid is completely insufficient for passability analyses. Large area CCM analyses use Global World Soil Database as a source of information about types of soils. Its scale is 1:500k and its detail is very insufficient. For this reason, small-scale CCM map outputs can use DTED 1 as an input elevation model for an overall picture.

Currently, a global elevation model created within the TREx project with a basic grid dimension of 12 m × 12 m is being developed [8]. This dimension is very close to DTM 3. Therefore, this model was selected for testing in order to determine its relationship with more detailed models and with DTED 2. TREx as a surface model is generally not appropriate for direct implementation in the passability analysis. On the other hand, it can be used without much loss of accuracy, for example, in the flat regions of Afghanistan or Mali.

2.3. Selection of Areas for Testing

To ensure the complexity of analyses, suitable areas representing the landscape of potential NATO foreign operations were selected. The selection represents possible types of reliefs across the AOI. Five different areas with a range of 3 km × 3 km are described in Table 2. The advantage was that each location represents a different type of terrain. This meant that areas with prevailing plains, hills and mountains were included. In terms of the morphometric division, the predominant plains and hills of selected locations represented the type of territory where military operations are conducted most often.

Table 2. Characteristics of the 5 selected locations (3 km × 3 km) for testing the elevation models.

Area	Mean Slope	Mean Altitude (Above Sea Level)	Coordinates (WGS84)
Dobruška	6°	420 m	50.31° N, 16.22° E
Horní Cerekev	5°	645 m	49.32° N, 15.25° E
Znojmo	3°	211 m	48.83° N, 16.16° E
Kdyně	9°	576 m	49.39° N, 13.11° E
Dolní Morava	15°	670 m	50.11° N, 16.86° E

2.4. Selection of Vehicles for Testing

The impact of different elevation models must be tested not only in various territories but also with CCM parameters of different military vehicles. The aim of selecting suitable military vehicles was to create a balanced sample with a basic division into wheeled and tracked vehicles. The following vehicles were selected for data analysis:

- Off-road light vehicle—Land Rover DEFENDER 110 (LRD 110);
- Wheeled truck—TATRA T815 6 × 6 (T815);
- Infantry armoured tracked vehicle—Bojové vozidlo pěchoty 2 (BVP-2).

The selected vehicles are among the main currently used ones in the armed forces of the Czech Republic and partially also in NATO. Each of these vehicles has its own specifics, either in terms of technical parameters or driving characteristics; see the information given in [45]. Vehicle testing did not take place physically in the terrain and is based solely on theoretical research. This took the form of a site survey of local conditions and their possible impact on vehicles.

3. Results

The testing of the quality of digital terrain models was based on the utilisation of two methods. The first method was involved analysing the accuracy of derived slopes from the models. The second method was a comparison of the results of a CCM raster analysis for three selected vehicles.

3.1. Comparison of the Accuracy of Slopes Derived from Digital Terrain Models

The comparison of the accuracy of slopes was based on the usage of the tool *Create random points* in ArcMap. With this tool, 50,000 randomly placed points in the selected areas were created (five locations with 3 km × 3 km areas). A minimum distance between points of three meters was set (points were 3 m or more apart) to ensure a more even distribution. Values of slopes were assigned from the prepared slope maps of individual data sets to the generated points using the tool Extract values to points. The points were then divided into classes with a step of inclination of 5° and comparative statistics were calculated for each class. Mean values and standard deviations were selected as test statistics. The resulting set of values of the most important statistics is shown in Table 3. The average slope was calculated as the mean value of the slope within one class. The average difference was given by subtracting slopes of a selected data set from the DTM 5. The standard deviation of the difference from the DTM 5 indicated the extent to which slope values were spread within one class.

Table 3. A comparison of slopes in random points generated from elevation models and their statistics.

Slope Classes		Mean Slope (°)				Mean Slope Difference from DTM 5 (°)		
Quantity	Slope	DTM 5	DTM 4	DTM 3	DTED 2	DTM 4	DTM 3	DTED 2
10,134	0–5°	3.14	4.09	5.21	5.17	0.95	2.07	2.02
12,389	5–10°	7.17	7.28	7.54	6.82	0.11	0.37	−0.35
6946	10–15°	12.37	12.07	11.74	9.94	−0.30	−0.62	−2.43
5933	15–20°	17.45	16.59	15.42	12.59	−0.87	−2.03	−4.86
5696	20–25°	22.49	21.41	19.11	15.85	−1.08	−3.39	−6.64
4987	25–30°	27.38	26.05	22.37	18.09	−1.33	−5.01	−9.29
2679	30–35°	32.10	29.81	23.37	18.33	−2.29	−8.72	−13.77
740	35–40°	36.95	31.24	22.80	15.75	−5.72	−14.15	−21.20
191	40–45°	42.00	33.36	22.64	14.23	−8.64	−19.36	−27.77
96	45–50°	47.41	37.25	26.36	10.48	−10.16	−21.05	−36.93
54	50–55°	52.53	40.48	28.51	8.73	−12.05	−24.02	−43.80
55	55–60°	57.48	46.31	30.92	6.90	−11.17	−26.56	−50.58
41	60–65°	62.82	48.48	26.30	5.42	−14.34	−36.51	−57.39
35	65–70°	67.43	51.05	31.97	5.80	−16.38	−35.46	−61.63
50,000	Total	14.27	13.77	12.64	10.54	−0.50	−1.63	−3.73

Figures 2–4 show histograms of all values of the standard deviations of slopes without division into classes, unlike in Table 3, where the values were divided into slope classes. The goal was to verify the trends in the accuracy of the data with two different methods. The histograms showed a deviation of the average value from the zero value for all elevation models, and at the same time, the range of standard deviations was compared. The lower the detail of the model, the higher the variance.

The comparison of the accuracy of slopes derived from elevation models produced the following results.

The number of points in each class did not represent an even distribution but fit more suitably to real conditions. Slopes up to 15° were 59% of the points and slopes above 30° were 8% of the points. The key threshold slopes for the assessment of passability, from 15° to 30°, represented 33% of the total number of determined slopes. This was sufficient due to the higher number of testing points. Graphs evaluating the characteristics of elevation models with individual classes of 5° inclination are shown in Figures 5–7.

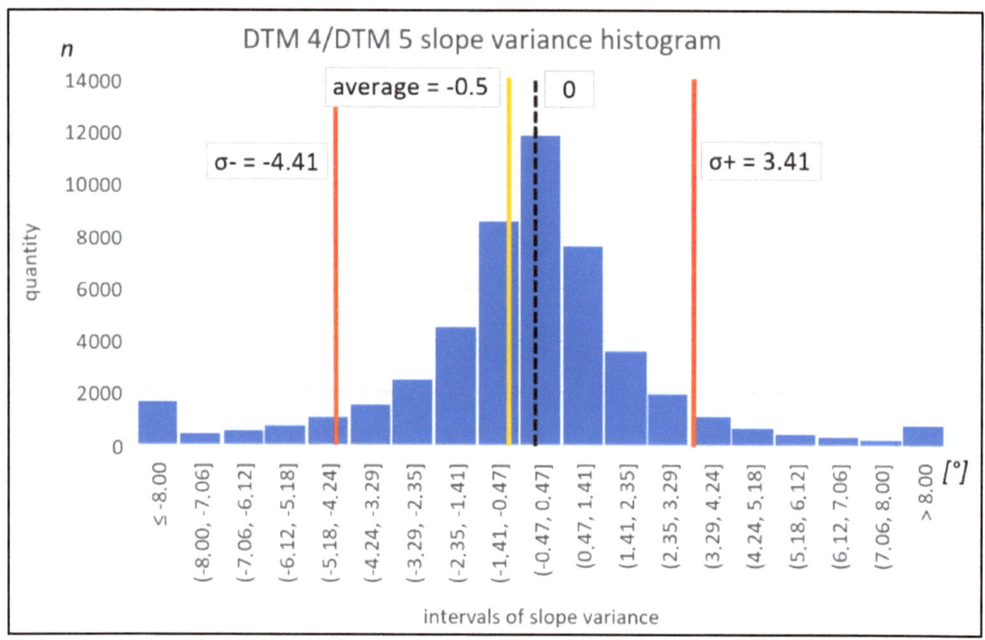

Figure 2. Histogram of standard deviations of the comparison between DTM 4 and DTM 5 for slopes at 50,000 points.

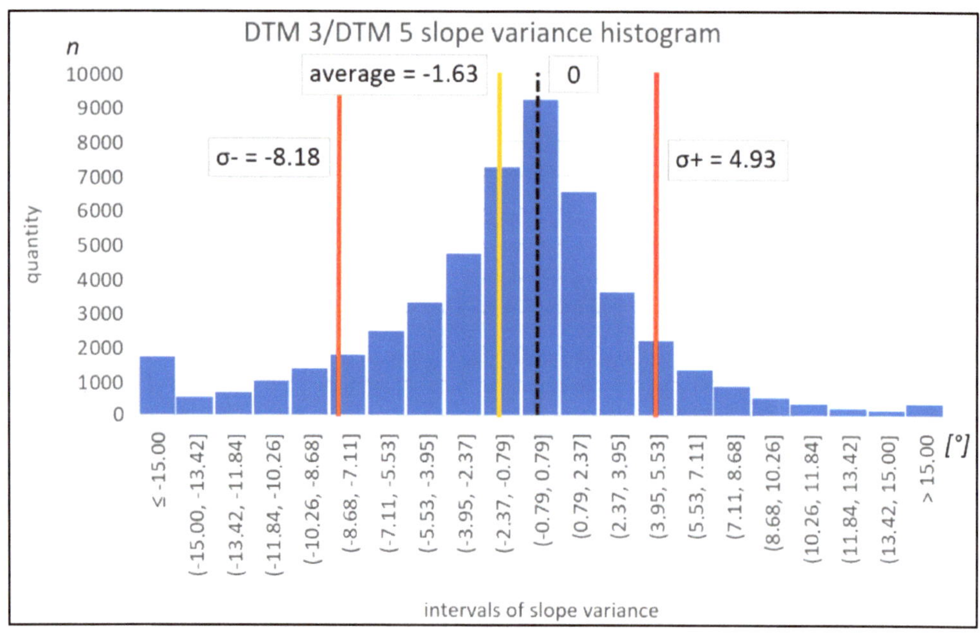

Figure 3. Histogram of standard deviations of the comparison between DTM 3 and DTM 5 for slopes at 50,000 points.

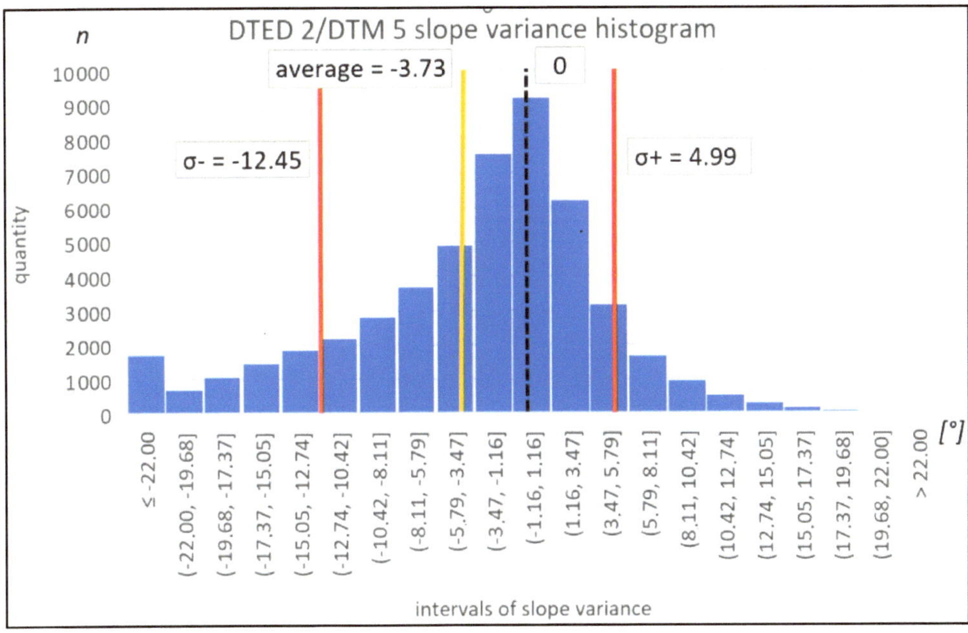

Figure 4. Histogram of standard deviations of comparison between DTED 2 with DTM 5 for slopes at 50,000 points.

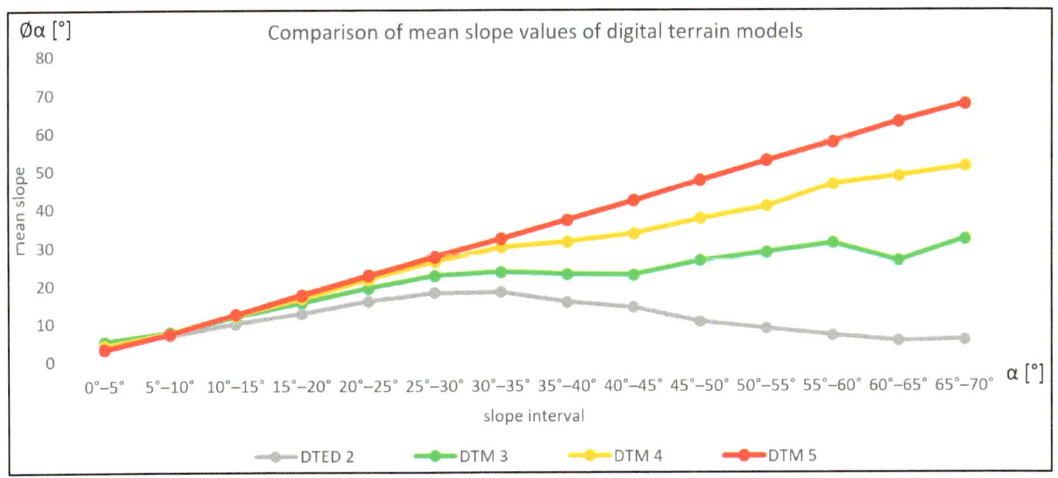

Figure 5. Graph comparing mean slope values of different terrain models (DTM 5, DTM 4, DTM 3, DTED 2) at identical points. Slope classes were determined from DTM 5.

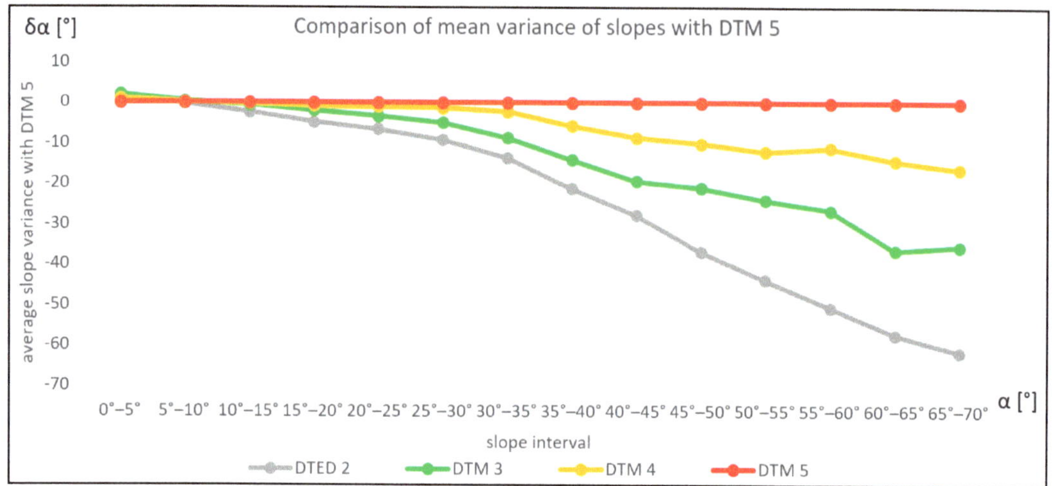

Figure 6. Graph comparing the mean variance of slopes of different terrain models (DTM 5, DTM 4, DTM 3, DTED 2) at identical points. Slope classes were determined from DTM 5.

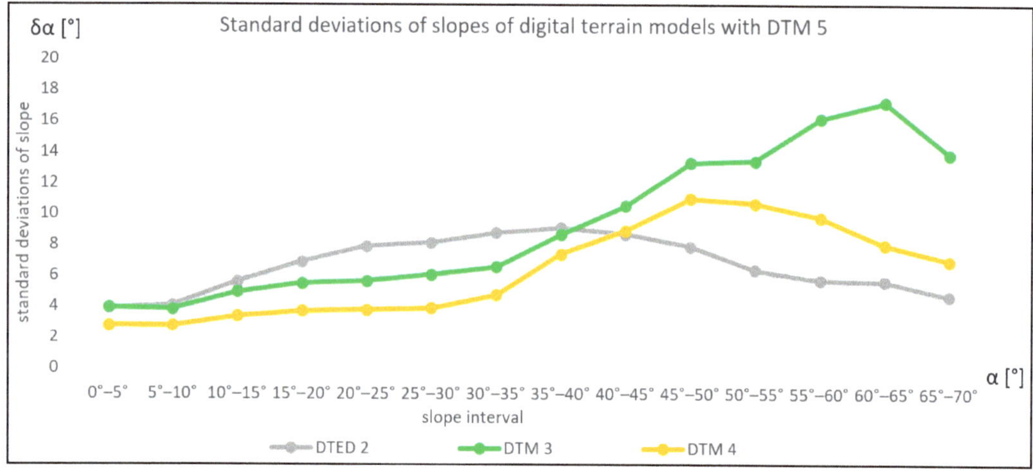

Figure 7. Graph comparing standard deviations of slopes of different terrain models (DTM 5, DTM 4, DTM 3, DTED 2) at identical points. Slope classes were determined from DTM 5.

The assessment of the specific slope values and digital terrain models resulting from the analysis of Figures 2–7 is given below.

Evaluation of specific slope values (slope classes):

- Low-detail elevation models did not differ from high-detail ones for lower slope values (slope values were very similar). The basic threshold was up to 10° for lower-detailed models (10–30 m) and up to 30° for medium-detailed models (5 m).
- Slopes up to 5°—The average slope value was higher in all tested models than in DTM 5. More detailed terrain models (DTM 5) had a more gradual slope in the lowest class. This inversed effect was specifically found in the flat areas only.
- Slopes 5–10°—The average values of slopes were very close for all models and the standard deviations were not very high either. It is possible to use less-detailed models in terrain analyses for slopes up to 10°.

- Slopes 10–30°—Almost a linear increase in slope could be observed depending on the resolution of a model. The average deviations with DTM 5 also had a linear increase. The more detailed the model, the greater the proportion of higher slopes. This slope class was still relatively reliable in all models (except DTED 2).
- Slopes above 30°—The comparative graphs deviated from the trend of the curves due to a smaller number of values, especially above 40° (see Figure 4).

Evaluation of digital terrain models (a comparison with DTM 5—1 m × 1 m):

- The less detailed the model, the bigger the deviation of the average slope and the bigger the variance in the standard deviation.
- The 5 m × 5 m models (DTM 4)—Achieved small deviations below 1° for slopes up to 20°. For bigger slopes, especially above 40°, the slope was almost less than one-third lower than in DTM 5. DTM 4 was suitable for slopes up to 20°, the use is not recommended for slopes above 40°.
- The 10 m × 10 m models (DTM 3)—Achieved small deviations below 1° for slopes up to 15°, and at 30°, the slope exceeded the limit of one-third of the slope difference compared to DTM 5. DTM 3 was suitable for slopes up to 15°; for slopes above 30°, it is not recommended.
- The 30 m × 30 m models (DTED 2)—They differed from DTM 5 by 1° already after 10° of slope inclination; from this value, the difference in inclination compared to DTM 5 was lower by more than one-third. Areas with inclinations above 50° in DTM 5 had less than 10° in DTED 2. The detail of the DTED 2 network of 30 m × 30 m points did not allow for identifying a more fragmented terrain relief. DTED 2 could only be used to determine slopes up to 10°.

3.2. Comparison of the Accuracy of Digital Terrain Models Using Raster Passability Analysis

Properties of digital terrain models can also be studied from the view of their use in the decision-making processes of commanders and staff. Data properties are evaluated from the area difference in the calculated passability. A difference was given by comparing selected models with the most detailed data (e.g., the difference between DTM 3 and DTM 5). Each pixel was assigned one of three passability values (GO, SLOW GO, NO GO) according to the parameters calculated from the traction curves of selected vehicles. For the calculation of the parameters, see [16]. This was based on the findings of the evaluation of dynamics of vehicle movement in the terrain of [46,47]. A total area of all pixels was then converted to percentages of passable, hardly passable, and impassable terrain. Map comparisons of the results of CCM analysis for the T815 vehicle and all tested terrain models are shown in Figure 8. A summary of passable area deviations between lower detail models and DTM 5 is displayed in Table 4.

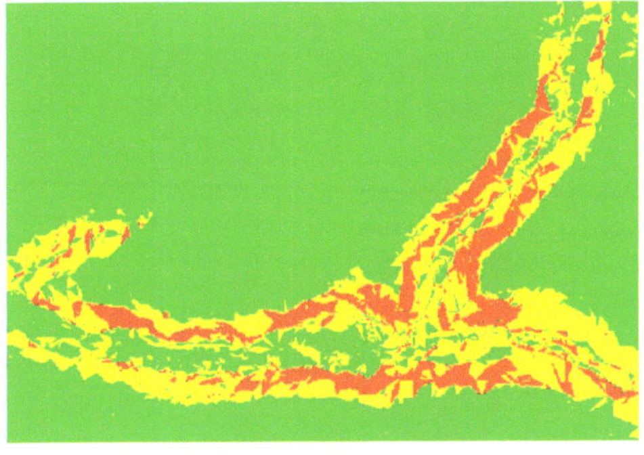

DTM 5 (1 m × 1 m)

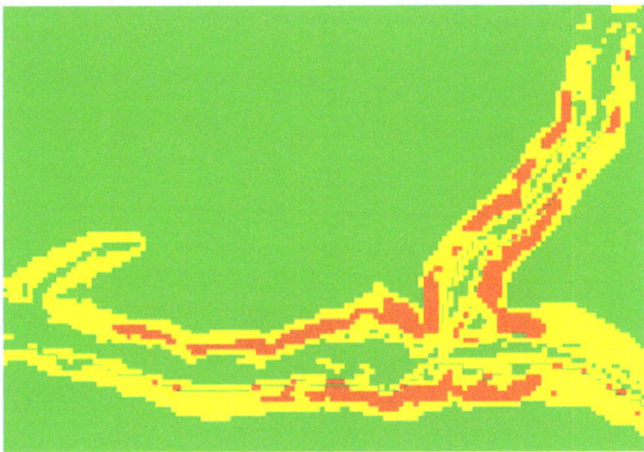

DTM 4 (5 m × 5 m)

Figure 8. *Cont.*

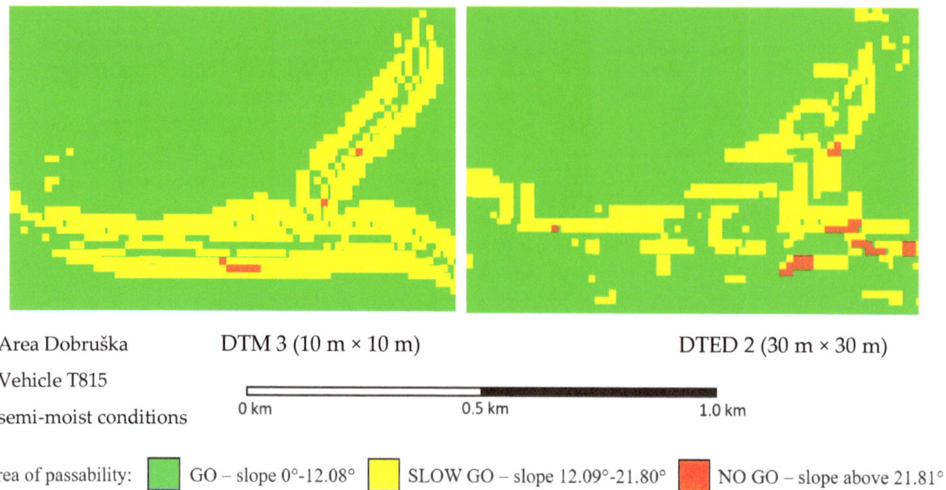

Figure 8. The result of the cross-country mobility analysis of a selected locality for the vehicle T815. A detailed look at a selected valley in the Dobruška locality. Testing was done with the method of raster analysis of the evaluation of pixels for various DEMs (DTM 5, DTM 4, DTM 3, DTED 2).

Table 4. Average deviations of the passable area calculated from elevation models (DTM 4, DTM 3, DTED 2) in comparison with DTM 5 with different parameters (vehicle type, humidity); results are mean values for five 3 km × 3 km areas, max–maximum available slope for the respective conditions, 1% deviation of the area means 9 hectares out of 900 hectares of one area and area deviations of elevation models with DTM 5 are always positive (larger passable areas).

CCM Parameters	Average Deviations of Passable Area in Comparison with DTM 5 (%)		
	DTM 4	DTM 3	DTED 2
T815 GO moist (max slope = 6.25°)	1.32	2.71	4.39
T815 GO semi-moist (max slope = 12.08°)	1.31	2.54	3.32
LRD GO dry (max slope = 16.17°)	1.11	2.32	3.45
LRD GO + SLOW GO semi-moist (max slope = 21.82°)	0.77	1.83	2.37
BVP2 GO + SLOW GO semi-moist (max slope = 28.81°)	0.33	0.98	1.08

The raster analysis of passability produce the following results:

1. Deviations in passable area (GO):

   - The more detailed the model of terrain, the smaller the passable area.
   - The passable area of 851 ha (out of a total of 900 ha of one 3 km × 3 km area) in DTM 5 represented

     ○ 860 ha in DTM 4 (1% larger passable area);
     ○ 870 ha in DTM 3 (2% larger passable area);
     ○ 880 ha in DTED 2 (3% larger passable area).

   - These ratios may significantly vary in different types of landscape (mountains) or surface conditions (impassable soils) but have the same trend.

2. Deviations in the hardly passable area (SLOW GO):

   - The position and structure of hardly passable area remained unchanged in all models.

3. Deviations in the impassable area (NO GO):

   - The total area of impassable territory increased with the detail of a used model.
   - The impassable area in DTM 5 (15 ha) represented

○   75% of the area in DTM 4 (12 ha);
○   33% of the area in DTM 3 (5 ha);
○   20% of the area in DTED 2 (3 ha).

4.   Influence of other factors on the deviation of the passable area:

- The better the passability conditions, the smaller the area deviation when using less accurate relief models (see Table 4); this applied to the following conditions:

○   lower soil moisture;
○   more suitable soil types (clayey-sandy);
○   more powerful vehicles (suitable for cross-country movement).

5.   Evaluation of digital terrain models:

- The 1 m × 1 m models (DTM 5):

○   The accurate model was suitable for detailed CCM analysis.

- The 5 m × 5 m models (DTM 4):

○   The deviation of the area of the passable terrain was small;
○   DTM 5 was more suitable for fragmented terrain in detailed CCM analyses.

- The 10 m × 10 m models (DTM 3):

○   The impassable area was not usually displayed in flat territories in these models (steeper slopes with shorter lengths);
○   The area of the impassable territory was close to DTM 5 in mountainous areas with long slopes;
○   The model could be used for less detailed CCM analyses.

- The 30 m × 30 m models (DTED 2):

○   The impassable area was not usually displayed in flat areas in these models;
○   The impassable area was close to DTM 5 in mountainous areas with long slopes;
○   Reliable results of CCM analysis could not be achieved with DTED 2.

3.3. Efficiency Improvements of the Elevation Models

Before any wider global data improvements were introduced, the only option for more accurate CCM analyses was to implement a methodology regarding how to efficiently use low-detail elevation models. The usability of these models can be achieved by adjusting the maximum slope limits for a passable terrain. A hardly passable terrain can be considered impassable in low-detail models. To obtain similar results for the area of passability as high-detail elevation models, adjustments in Table 5 had to be performed. These adjustments need to be applied to elevation models based on the traction parameters of a vehicle (maximum reachable slope), which were calculated from traction curves or the DMA (Defence Mapping Agency) model; for details, see [16]. The values in Table 5 were calculated from the mean slope values of elevation models for each slope class (see Table 3). Afterwards, these values had to be modified according to the results of CCM map analyses, displayed in Tables 4 and 6 and as map results in Figures 8–10.

Table 5. Adjustments of maximum slope value reachable by a vehicle for passable (GO) and hardly passable terrain (SLOW GO) that were needed to achieve more accurate results for CCM analyses. Modifications calculated for DTM 4 (5 m × 5 m), DTM 3 (10 m × 10 m) and DTED 2 (30 m × 30 m).

| GO or SLOW GO | Reduced Slope Values (°) | | |
Slope Value (°)	DTM 4	DTM 3	DTED 2
10	9.9	9.6	9.3
15	14.5	13.4	12.8
20	19.0	17.3	16.4
25	23.5	21.2	19.8
30	27.8	25.1	23.6

Table 5. *Cont.*

GO or SLOW GO Slope Value (°)	Reduced Slope Values (°)		
	DTM 4	DTM 3	DTED 2
35	32.0	29.0	27.0
40	36.0	32.8	30.5
45	40.0	36.5	34.0

Table 6. The comparison of the extent of passable, hardly passable, and impassable areas with different elevation models (DTM 5, DTM 4, DTM 3, DTED 2) without slope corrections and with slope corrections. Location: Kdyně, 6 km × 6 km, moist conditions, GO max slope = 10°, SLOW GO max slope = 20°. The main improvements in the results are displayed as bold values.

		DTM 5 Area (%)	DTM 4 Area (%)	DTM 3 Area (%)	DTED 2 Area (%)
No slope corrections	GO	72.01	73.17	74.44	73.04
	SLOW GO	24.71	24.70	24.40	26.07
	NO GO	3.28	**2.13**	**1.15**	**0.90**
With slope corrections	GO	72.01	72.65	72.06	69.19
	SLOW GO	24.71	24.35	24.77	26.59
	NO GO	3.28	**3.00**	**3.17**	**4.23**

If a calculated maximum reachable slope for a hardly passable terrain (SLOW GO) for a vehicle was, for example, 25°, then by using the 10 m × 10 m elevation model, the maximum reachable slope was only 21.2° and the CCM analysis map should be adjusted. DTM 5 served as the most accurate model database to calculate the slope deviations of other elevation models from the real terrain. DTM 5 is a very accurate model (1 m × 1 m) but still deviates from the real terrain. Given the trends of lower-detail models displayed in Table 5, the estimated maximum reachable slope value reductions for DTM 5 would be 0.2° for 20° slopes, 0.5° for 30° slopes and 1.5° for 45° slopes. The accuracy of slope reductions was tested for a different extent and conditions. The tests were performed in the form of CCM map analysis in a 3 km × 3 km area in Dolní Morava (arid and moist conditions), a 6 km × 6 km area in Kdyně (moist conditions) and five combined selected areas of 5 km × 5 km (moist conditions). The main comparative characteristics were variations of passable/impassable area between corrected models and models not corrected. The other characteristic was a comparison of the position of a hardly passable and impassable terrain. The results with the calculated slope reductions (Table 5) were more accurate and were markedly closer in position to the most detailed elevation model, namely, DTM 5, than results without slope reductions (see Tables 6–9).

Table 7. The comparison of the extent of passable, hardly passable, and impassable areas with different elevation models without slope corrections and with slope corrections. Location: Dolní Morava, 3 km × 3 km, moist conditions, GO max slope = 10°, SLOW GO max slope = 20°. The main improvements in the results are displayed as bold values.

		DTM 5 Area (%)	DTM 4 Area (%)	DTM 3 Area (%)	DTED 2 Area (%)
No slope corrections	GO	43.67	44.25	43.77	43.72
	SLOW GO	36.36	37.37	40.79	42.53
	NO GO	19.97	**18.38**	**15.44**	**13.76**
With slope corrections	GO	43.67	43.61	41.12	38.89
	SLOW GO	36.36	35.49	35.16	37.03
	NO GO	19.97	**20.90**	**23.72**	**24.08**

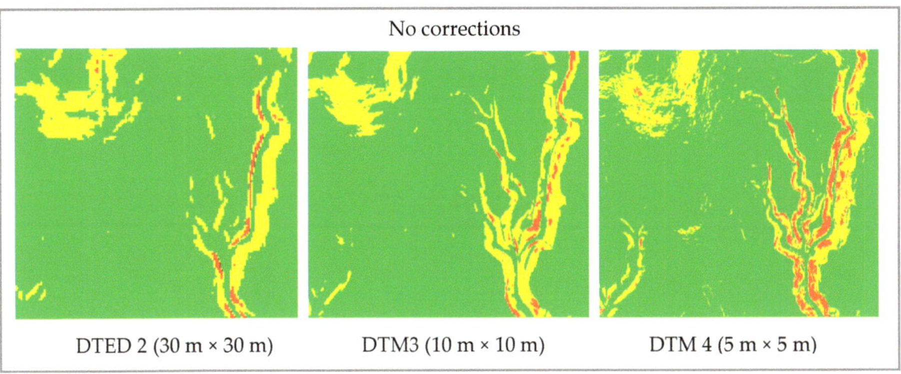

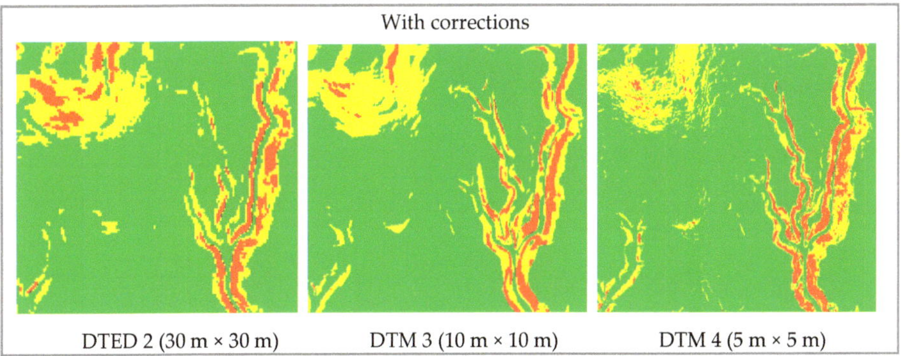

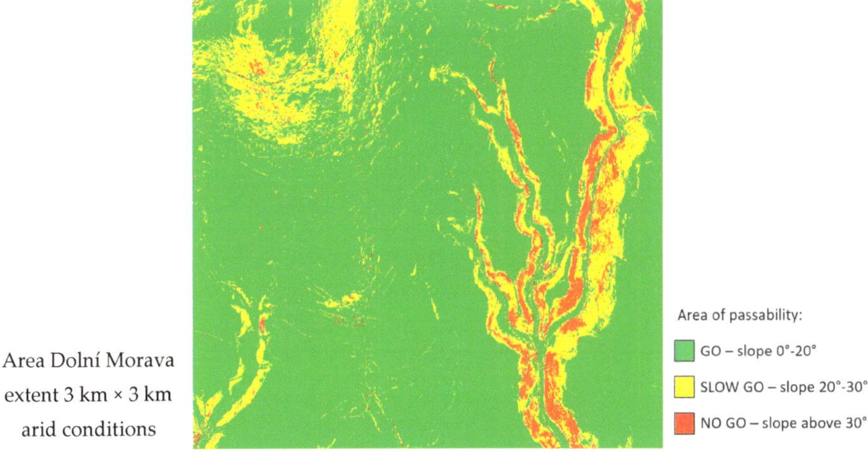

Figure 9. The map comparison of the CCM analysis with different elevation models (DTM 5, DTM 4, DTM 3, DTED 2) without slope corrections and with slope corrections. The slope corrections enabled accurate results of CCM analysis for lower-detail models. DTM 5 (1 m × 1 m) served as the reference model.

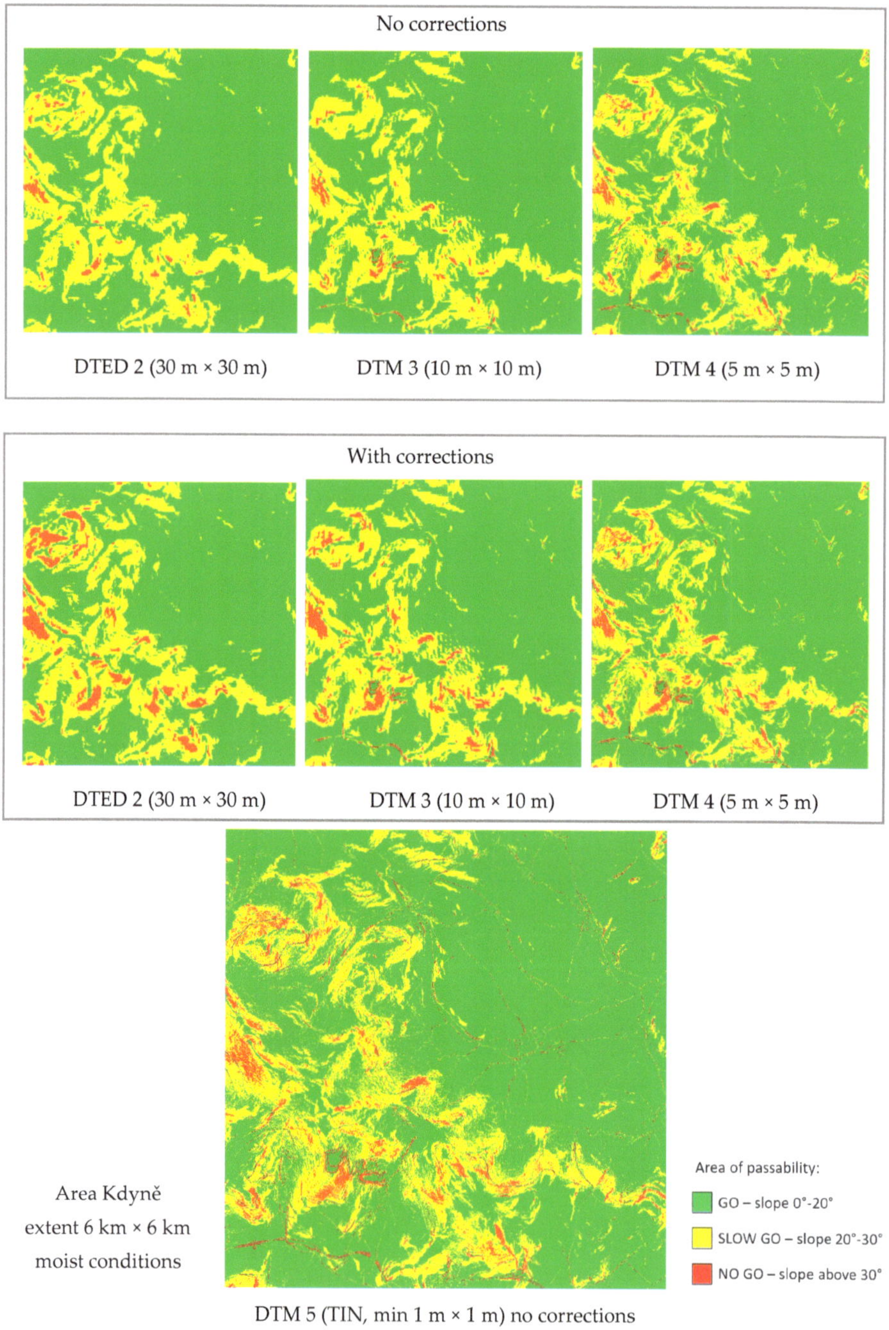

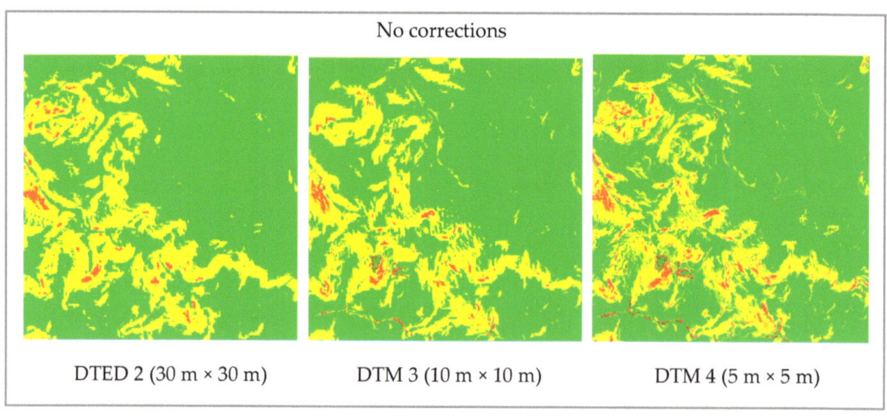

Figure 10. A map comparison of the CCM analysis with different elevation models (DTM 5, DTM 4, DTM 3, DTED 2) without slope corrections and with slope corrections. The slope corrections enabled more accurate results of CCM analysis for lower-detailed models. DTM 5 (1 m × 1 m) served as the reference model Choosing a larger area (6 km × 6 km) confirmed the correctness of the slope-reduction method.

Table 8. The comparison of the extent of passable, hardly passable, and impassable areas with different elevation models without slope corrections and with slope corrections. Location: Dolní Morava, 3 km × 3 km, arid conditions, GO max slope = 20°, SLOW GO max slope = 30°. The main improvements in the results are displayed as bold values.

		DTM 5 Area (%)	DTM 4 Area (%)	DTM 3 Area (%)	DTED 2 Area (%)
No slope corrections	GO	80.03	81.62	84.56	86.24
	SLOW GO	16.12	15.29	14.38	12.89
	NO GO	3.85	**3.09**	**1.06**	**0.87**
With slope corrections	GO	80.03	79.10	76.28	75.92
	SLOW GO	16.12	15.50	18.70	17.81
	NO GO	3.85	**5.40**	**5.02**	**6.27**

Table 9. The comparison of the extent of passable, hardly passable, and impassable areas with different elevation models without slope corrections and with slope corrections. Combined 5 areas, 3 km × 3 km, moist conditions, GO max slope = 10°, SLOW GO max slope = 20°. The main improvements in the results are displayed as bold values.

		DTM 5 Area (%)	DTM 4 Area (%)	DTM 3 Area (%)	DTED 2 Area (%)
No slope corrections	GO	77.34	78.55	79.28	80.54
	SLOW GO	16.64	16.35	16.72	16.27
	NO GO	6.02	**5.09**	**4.00**	**3.19**
With slope corrections	GO	77.34	78.24	77.91	76.82
	SLOW GO	16.64	15.86	15.61	15.55
	NO GO	6.02	**5.90**	**6.48**	**7.63**

The accuracy of the selected method of slope corrections could be observed from an extent of impassable terrain calculated from the selected elevation models. The impassable area was significantly more accurate with implemented slope corrections. More important is the fact that these slope adjustments allowed for distinguishing hardly passable and impassable terrain with noticeably higher precision.

On the right side of the map (Figure 9) is a valley that represents an impassable obstacle. Without corrections, the valley was evaluated as hardly passable when using DTM 3 and DTED 2. With implemented corrections, the valley was correctly displayed as impassable also in these lower detail models. Furthermore, shapes of other major terrain features representing hardly passable terrain were displayed more accurately (e.g., a small valley in the lower-left corner of the map). The detail of DTM 5 allowed for displaying a fragmented terrain (e.g., top left corner).

These locations are usually joined as one area in lower detail models, especially with details that were 10 m × 10 m and above. This was the reason for the higher ratio of impassable terrain with corrected slopes. Due to the fragmented terrain, the mentioned location was, in the end, completely impassable. This meant that the lower-detailed models were displaying the correct situation. The results in Figure 9 were confirmed in Figure 10 for a different location.

The extent of the impassable area is given with the corrections adjusted to a more real coverage and position. It shows that the determination of impassable and hardly passable areas when using a lower-detail elevation model can be very variable. The method of slope correction was the right way to improve a current state, which was verified by better coverage results in comparison with DTM 5.

4. Discussion

The results of analyses of the digital terrain models DTM 5, DTM 4, DTM 3 and DTED 2 showed different qualities based on the resolution of the network of points. Each elevation model had its possible usage in practise: from the most detailed 1 m $\times$ 1 m model, namely, DTM 5, with a universal usage to the low detail model 30 m $\times$ 30 m with limited use cases, mostly for small-scale overview maps. Currently, the lower detail elevation models are not used efficiently. Efficient use can be achieved by adjusting the accuracy to high-detail elevation models. These adjustments would increase the usability of all elevation models. Before implementing any modifications to digital terrain models, other types of data need to be considered as well.

Individual parts of the landscape, i.e., terrain relief and soil cover, are represented by a different quality of data. Terrain relief is covered with detailed and accurate elevation models in the territory of European states. They are suitable for planning military operations down to the tactical level. Soil and surface conditions data is less reliable for detailed CCM analyses. This data has little detail or inaccurate determination of soil type boundaries. It is necessary to consider the shortcomings of the data used to properly interpret the results of passability analyses.

4.1. Evaluation of Digital Terrain Models

Digital terrain models are not globally available in adequate detail (minimum 10 m $\times$ 10 m) and area coverage to support any possible location for a foreign operation at short notice. Currently, it is not possible to perform accurate and detailed CCM analyses in missions abroad without the use of mobile mapping devices (satellite systems, drones), which use various types of sensors and technologies (photogrammetry, laser scanning, radar data, etc.). The biggest disadvantage of this mobile sensors approach is the time delay (months) in collecting and processing data to cover the entire area of a recently established operation. The solution lies in creating a universal database with complete coverage and sufficient detail. This coverage should include the entire NATO area of interest (AOI), which reflects NATO nations' territory and the NATO area of responsibility (AOR) covering adjacent territories in Europe, the western part of Asia and the northern part of Africa. NATO does not have sufficient resources to create or purchase such data. Nevertheless, it could utilise the capabilities of national geographic services, the Multinational Geospatial Support Group (MN GSG), the NATO Communication and Information Agency (NCIA) or a type of direct support of foreign operations, which was done using a Resolute Support Reach-Back Afghanistan Cell (RS RAC). After some time, a military operation is supplied with detailed data, which corresponds to the production in the territories of NATO nations [48]. Nevertheless, sufficient data that would cover immediate needs in the case of a crisis operation are not available, either in the required extent or quality.

Models with a basic resolution of at least 1 m $\times$ 1 m (e.g., DTM 5) are the most detailed terrain models of NATO member countries. However, models with a resolution of 5 m $\times$ 5 m (e.g., DTM 4) are also sufficient in their detail to achieve accurate results of passability analyses. This type of model can be less demanding in terms of data flow for calculations via a network. DTM 3 corresponds in detail with the new worldwide terrain surface model TREx. At this level of detail (10 m $\times$ 10 m), the model cannot be used for detailed CCM analyses. Nonetheless, data can be used for less fragmented terrain without vegetation cover. This is the case for some foreign operations, such as in Mali. These areas are already mostly covered by the TREx model. DTED 2 is a low-detail elevation model that allows for only general terrain analyses. The optimal terrain model with a balanced ratio of detail and accuracy of the results of passability analyses when considering the complexity of its acquisition is a model with a resolution of 5 m $\times$ 5 m. The goal of geographical development within NATO should be to ensure comprehensive coverage with enhanced data in the NATO AOI.

4.2. Usability of Soil Databases in CCM Analyses

Soil databases with sufficient quality are not available in the areas of foreign military operations. Despite a lower accuracy in comparison with elevation models, soil databases can be used for terrain analyses. However, to achieve the required detail of data, a more accurate mapping of soil type boundaries would be necessary. For example, in the territory of the Czech Republic, the Digital Soil Map 1:50,000 need to be updated [20]. Since DTM 4 was identified as a sufficiently accurate terrain model, the optimal density of measurement points for soil characteristics is 5 m × 5 m at the boundaries between soil types. Apart from geometric accuracy improvement, other information needs to be added to the assessment of soil types, such as an actual soil moisture level and an assessment of the quality of the surface itself. Vector databases that contain surface roughness and vegetation coverage parameters with sufficient detail and area coverage do not exist. Accurate soil databases would not be much use without this additional surface information.

From the global perspective, universal soil data is represented by the Harmonized World Soil Database (HWSD), which combines various national soil databases. Its detail is very low and represents a scale of 1:500,000. It can only be used for lower-accuracy estimates of passability analysis for the operational and up to the strategic level of military planning. A detailed study of improvements of the HWSD and its usability in national systems was given in [49]. One of the possible options to replace low-detailed surface data is the Destination Earth initiative. It is a project of the European Commission with the aim of collecting and evaluating satellite images. One of the important use cases of this project for the CCM analysis domain is the evaluation of surface temperatures to determine soil and surface types. Even though this method is not accurate enough, it can improve data availability in territories of potential conflicts across the world [50].

A study [51] showed that if the terrain has arid or semi-moist conditions, then the influence of soil types in the CCM can be disregarded. The only significant exception was the alluvial soil type, which occurs in some humid valleys. Except for arid conditions, these valleys are impassable for most types of vehicles (track or wheeled). The other exception is a sandy surface, which substantially limits the movement of wheeled vehicles. A vast majority of soil types are different from the alluvial and sandy types. Therefore, the focus should be put on distinguishing correct humidity conditions of the surface and properly processing a utilised elevation model.

4.3. Recommendations for Improvements of Elevation Models

The introduced method of slope reduction for the usability of low detail elevation models in CCM analyses was affirmed as correct. All results showed an improvement in the accuracy of passable areas, both in extent and position (as shown by the Figure 10 map comparisons and Tables 6–9 numbers in bold). However, the size and number of testing areas were limited. Verifying this concept at a global scale would require extensive data testing in large areas across Europe, Asia and Africa for various elevation data sources. The key element is also a focus on specific restricted areas for passability. These are hilly and mountain areas, where a key slope limit of approximately 30° occurs more often than in other types of terrain. Additional testing sites should make the slope reduction values presented in Table 5 more accurate. Further slope adjustments to the values in Table 5 would probably only be minor.

Global detailed data readiness is something that NATO currently lacks. The usability of digital terrain models is not limited to only high-detail models. NATO can use low-resolution models in foreign operations in a time frame before any detailed elevation data is procured, e.g., LIDAR data. Low-detail elevation models can fulfil this aim via the additional processing of data introduced in this chapter using the method of slope reduction. Accurate CCM analyses cannot be achieved without including microrelief forms in the analysis. DTM 5 (1 m × 1 m) consists of all microrelief objects, for example, embankments. The other elevation models mostly do not consist of these objects. Remedy can be achieved with vector data that consists of roads and other line or point objects that are important for passability. These objects could be incorporated into analyses with low-detail elevation models as impassable obstacles. The current ongoing project of the global vector database MGCP can serve as a source of information for objects on the terrain. Nowadays, expectations for improvements in global elevation models are focused on the

TREx project. TREx is a surface model; therefore, its usage for accurate CCM analyses is limited.

Nevertheless, new methods for processing TREx data into the form of a digital terrain model are currently in development. Although this might make a new global digital elevation model with a resolution of 12 m × 12 m available, the accuracy of altitude of areas formerly consisting of forest and built-up areas is low. The reason for this is that data from these locations are interpolated from lower detail models, e.g., DTED 2. TREx is also available in its raw unprocessed form as a cloud of points. Using this data can allow for rapid reaction analyses for any crisis region across the world. The disadvantage is in more complicated processing, mostly because of the overly large data volume.

So far, no global elevation model has met all requirements needed for accurate CCM analyses. Global coverage with ready-to-use data is the highest priority. The second priority is the detail of an elevation model, which ideally should be 5 m × 5 m. The surface model, such as TREx, is not a sufficient replacement for a regular digital terrain model. For the moment, the only possible solution for this would be the use of automated systems fitted on military vehicles that evaluate passability parameters in real time. Even though the capability of automated navigation of military vehicles is not widespread in NATO, the systems will become the future of CCM modelling.

To compare the influence of the use of different elevation models on the determination of terrain passability, different statistical methods can be used. The use of the parametric ANOVA method would be possible under the assumption of a normal distribution of data. However, this distribution does not occur in the event of extreme changes in the terrain, e.g., changes in adhesion due to precipitation. For non-parametric analysis of the variance of GO, SLOW GO and NO GO areas, it is possible to use, for example, the Kruskal–Wallis test or multiple comparisons with the adjustment of the significance level using the Bonferonni method. However, the validity of these methods must be verified via physical testing of off-road vehicles in different types of terrain.

The main conclusion of the study is that under the current conditions, the usability of the selected elevation databases for accurate and detailed comprehensive CCM analyses is limited. A future focus should be to find a clear solution that allows for utilising existing databases efficiently; see also [52–55]. The essential part is also adopting selective measures for data collection for CCM analyses in real time using means of remote detection of passability parameters. That includes monitoring necessary methods and technologies of an automated vehicle driving in space and making active steps to implement them into practise.

5. Conclusions

This research studied the possibilities for raster analyses of elevation models and other options for their evaluation and effective usage. One of the most important results of this study was the introduction of the method of slope reduction for low-detail elevation models. Furthermore, the goal was to determine the strengths and weaknesses of selected data for use in cross-country mobility analyses, supplemented by recommendations on how to use these databases efficiently. The selection of databases was determined by the focus of data development projects of NATO and current scientific research projects of the Ministry of Defence of the Czech Republic. The presented methodology of data assessment for the purposes of military geographical analysis of the terrain can be further used in military practice, e.g., in foreign missions where high-detail data may not be present. Key findings of the elevation models analysis show the potential of their usage in practise. DTM 5, which is a model with a minimum resolution of a grid of points of 1 m × 1 m, is the most detailed model in terms of the evaluation of slopes and passability. It can be replaced by using DTM 4 (5 m × 5 m) in terrain analyses, for example, the slope reduction for 30° slopes is 2.2°. The elevation model with lower detail DTM 3 (10 m × 10 m) had a 4.9° slope reduction for 30° slopes and DTED 2 (30 m × 30 m) had a 6.4° reduction. These models can be used for general analyses of less fragmented terrain.

The selected areas, where analyses were performed, were a balanced sample of the landscape. It allowed for the assessment of the most important terrain parameters that have a fundamental influence on the conduct of military operations. Nevertheless, it is necessary to perform a complete validation of the slope reduction method from other various representative regions across NATO AOR (territory of NATO nations) and NATO AOI (territory beyond NATO nations). A larger number of locations would enable a more accurate determination of the reliability of databases when used in the cross-country mobility analysis. The challenge NATO has is a limited global short-notice availability of detailed digital terrain models. Without these models, no detailed and accurate CCM analysis can be performed. The only option is to process available elevation models into a form that allows for acceptable results, such as the slope reduction method.

Author Contributions: Conceptualisation and methodology, J.R. and M.R.; model programming, J.R.; validation, J.R. and M.R.; writing the paper, J.R. and M.R.; original draft preparation, M.R. All authors have read and agreed to the published version of the manuscript.

Funding: This research received no external funding.

Institutional Review Board Statement: Not applicable.

Informed Consent Statement: Not applicable.

Data Availability Statement: The study data were provided by the Military Geographic and Hydrometeorologic Office, Dobruska.

Acknowledgments: This paper is a particular result of the defence research project DZRO VAROPS managed by the University of Defence in Brno, NATO—STO Support Project (CZE-AVT-2019) and specific research project 2021-23 at the department K-210 managed by the University of Defence, Brno.

Conflicts of Interest: The authors declare no conflict of interest. The funders had no role in the design of the study; in the collection, analyses, or interpretation of data; in the writing of the manuscript; or in the decision to publish the results.

References

1. Muzirafuti, A.; Boualoul, M.; Barreca, G.; Allaoui, A.; Bouikbane, H.; Lanza, S.; Crupi, A.; Randazzo, G. Fusion of Remote Sensing and Applied Geophysics for Sinkholes Identification in Tabular Middle Atlas of Morocco (the Causse of El Hajeb): Impact on the Protection of Water Resource. *Resources* **2020**, *9*, 51. [CrossRef]
2. Muzirafuti, A.; Cascio, M.; Lanza, S. UAV Photogrammetry-based Mapping the Pocket Beach of Isola Bella, Taormina (Northeastern Sicily). In Proceedings of the 2021 IEEE International Workshop on Metrology for the Sea (MetroSea 2021), Reggio Calabria, Italy, 4–6 October 2021.
3. Melis, M.T.; Pisani, L.; De Waele, J. On the Use of Tri-Stereo Pleiades Images for the Morphometric Measurement of Dolines in the Basaltic Plateau of Azrou (Middle Atlas, Morocco). *Remote Sens.* **2021**, *13*, 4087. [CrossRef]
4. Hubacek, M.; Kratochvil, V.; Zerzan, P.; Ceplova, L.; Brenova, M. Accuracy of the new generation elevation models. In Proceedings of the International Conference on Military Technologies (ICMT), IEEE, Brno, Czech Republic, 19–21 May 2015; pp. 1–6.
5. Braun, T.; Bitsch, H.; Berns, K. Visual terrain traversability estimation using a combined slope/elevation model. In Proceedings of the IEEE/RSJ International Conference on Intelligent Robots and System, Berlin, Germany, 22–26 September 2008; pp. 177–184.
6. Hirt, C. Digital terrain models. In *Encyclopedia of Geodesy*; Springer: Berlin/Heidelberg, Germany, 2016. [CrossRef]
7. Qiming, Z.; Xuejun, L. Assessing uncertainties in derived slope and aspect from a grid DEM. In *Advances in Digital Terrain Analysis*; Springer: Berlin/Heidelberg, Germany, 2008; pp. 279–306.
8. Belka, L. TREx—New international project for creation of elevation data. *Mil. Geogr. Rev.* **2015**, *58*, 9–11.
9. DLR. *Global 3D Elevation Model of the Mission TanDEM-X Widely Available*; Deutsches Zentrum für Luft und Raumfahrt: Kölle, Germany, 2018. Available online: https://www.dlr.de/content/de/artikel/news/2018/4/20181008_3d-hoehenmodell-tandem-x-mission.html (accessed on 20 December 2021).
10. Zhang, K.; Chen, S.C.; Whitman, D.; Shyu, M.L.; Yan, J.; Zhang, C. A progressive morphological filter for removing nonground measurements from airborne LIDAR data. *IEEE Trans. Geosci. Remote Sens.* **2003**, *41*, 872–882. [CrossRef]
11. Bortl, D. Looking Back on the Action of Geospatial Military Specialists in the Province Reconstruction Team. *Mil. Geogr. Rev.* **2014**, *1*, 3–8.
12. Bradbury, M.; Dasch, J.; Gonzalez-Sanchez, R.; Hodges, H.; Iagnemma, K.; Jain, A.; Jayakumar, P.; Letherwood, M.; McCullough, M.; Priddy, J.; et al. *Next-Generation NATO Reference Mobility Model (NRMM) Development (Développement de la Nouvelle Génération du Modèle de Mobilité de Référence de l'OTAN (NRMM))*; NATO Science and Technology Organization: Neuilly-Sur-Seine, France, 2018.

13. McCullough, M.; Jayakumar, P.; Dasch, J.; Gorsich, D. The next generation NATO reference mobility model development. *J. Terramech.* **2017**, *73*, 49–60. [CrossRef]

14. Gorsich, D.; Gerth, R.; Bradley, S.; Letherwood, M. An Overview of the Next-Generation NATO Reference Mobility Model (NG-NRMM) Cooperative Demonstration of Technology (CDT). In Proceedings of the 2019 NDIA Ground Vehicle Systems Engineering and Technology Symposium, Novi, MI, USA, 13–15 August 2019.

15. Wong, J.Y.; Jayakumar, P.; Toma, E.; Preston-Thomas, J. Comparison of simulation models NRMM and NTVPM for assessing military tracked vehicle cross-country performance. *J. Terramech.* **2018**, *80*, 31–48. [CrossRef]

16. Rada, J. Analysis of Geospatial Data Used in Cross-Country Mobility Modelling. Dissertation Thesis, University of Defence, Brno, Czech Republic, 2021; p. 206.

17. NATO. The Area of Responsibility. NATO Declassified. 23 February 2013. Available online: https://www.nato.int/ebookshop/video/declassified/en/encyclopedia (accessed on 20 December 2021).

18. NSO. *AJP-3.17 Allied Joint Doctrine For Geospatial Support Edition B Version 1 WD 1.0*; NATO Standardization Office: Brussels, Belgium, 2022.

19. Hubacek, M.; Almasiova, L.; Brenova, M.; Bures, M.; Mertova, E. Assessing quality of soil maps and possibilities of their use for computing vehicle mobility. In Proceedings of the 23rd Central European Conference, Brno, Czech Republic, 8–9 January 2016; pp. 99–110.

20. Hubacek, M.; Kovarik, V.; Talhofer, V.; Rybansky, M.; Hofmann, A.; Brenova, M.; Ceplova, L. Modelling of geographic and meteorological effects on vehicle movement in the open terrain. *Cent. Eur. Area View Curr. Geogr.* **2016**, *11*, 149–159.

21. Rybansky, M. Determination the ability of military vehicles to override vegetation. *J. Terramech.* **2020**, *91*, 129–138. [CrossRef]

22. Capek, J.; Zerzan, P.; Simkova, K. Influence of tree spacing on vehicle manoeuvers in forests. In Proceedings of the 7th International Conference on Military Technologies, ICMT 2019, Brno, Czech Republic, 30–31 May 2019.

23. Rada, J.; Rybansky, M.; Dohnal, F. Influence of Quality of Remote Sensing Data on Vegetation Passability by Terrain Vehicles. *ISPRS Int. J. Geo-Inf.* **2020**, *9*, 684. [CrossRef]

24. Dohnal, F.; Hubacek, M.; Sturcova, M.; Bures, M.; Simkova, K. Identification of microrelief shapes along the line objects over DEM data and assessing their impact on the vehicle movement. In Proceedings of the 2017 International Conference on Military Technologies (ICMT), Brno, Czech Republic, 31 May–2 June 2017; pp. 262–267. [CrossRef]

25. Dohnal, F.; Hubacek, M.; Simkova, K. Detection of microrelief objects to impede the movement of vehicles in terrain. *ISPRS Int. J. Geo Inf.* **2019**, *8*, 101. [CrossRef]

26. Talhofer, V.; Bures, M. The Solution of the Mobility Model Using the Database of Roads and Terrain Relief. In *GIS Ostrava 2018 GIS for Support of Security and Crisis Management*; Technical University of Ostrava: Ostrava, Czech Republic, 2018; ISBN 978-80-248-4166-3.

27. Talhofer, V.; Rybansky, M.; Bureš, M.; Šimková, K. The Influence of Spatial Database Quality on Modelling of Vehicle Movement in Terrain. In Proceedings of the 19th International & 14th European-African Regional Conference of the International Society for Terrain-Vehicle Systems, Budapest, Hungary, 25–27 September 2017; pp. 1–15; ISBN 978-1-942112-49-5.

28. Rybansky, M. *The Cross-Country Movement—The Impact and Evaluation of Geographic Factors*; CERM: Brno, Czech Republic, 2009; p. 113; ISBN 978-80-7204-661-4.

29. Rybansky, M.; Hofmann, A.; Hubacek, M.; Kovarik, V.; Talhofer, V. Modelling of cross-country transport in raster format. *Environ. Earth Sci.* **2015**, *74*, 7049–7058. [CrossRef]

30. Talhofer, V.; Hoskova-Mayerova, S.; Hofmann, A. Quality of Spatial Data in Command and Control System. In *Studies in Systems, Decision and Control*; Springer International Publishing: Warszawa, Poland, 2019.

31. Pokonieczny, K. Automatic Military Traverseability Map Generation System. In Proceedings of the 2017 International Conference on Military Technologies (ICMT), IEEE, Brno, Czech Republic, 31 May–2 June 2017; pp. 285–292.

32. Pokonieczny, K. Methods of Using Self-Organising Maps for Terrain Classification, Using an Example of Developing a Military Traverseability Map. Dynamics in GIscience. In *Lecture Notes in Geoinformation and Cartography*; Ivan, I., Horák, J., Inspektor, T., Eds.; Springer International Publishing: Cham, Switzerland, 2018; pp. 359–371.

33. Dawid, W.; Pokonieczny, K. Analysis of the Possibilities of Using Different Resolution DEMs in the Study of Microrelief on the Example of Terrain Passability. *Remote Sens.* **2020**, *12*, 4146. [CrossRef]

34. Bellone, M.; Bellone, M.; Spedicato, L.; Giannoccaro, N.I. 3D traversability awareness for rough terrain mobile robots. *Sens. Rev.* **2014**, *34*, 220–232. [CrossRef]

35. Ivanisevic, V.V.; Lozynskyy, A.; Demkiv, L.; Klos, S. A Foundation for Realtime Tire Mobility Estimation and Control. In Proceedings of the 19th International & 14th European-African Regional Conference of the ISTVS, Budapest, Hungary, 25–27 September 2017.

36. Hubacek, M.; Mertova, E. The Influence of Weather on Soil Properties and on Terrain Traverseability. In *GIS Ostrava 2018: GIS for Supporting Security and Crisis Management*; VSB Technical University of Ostrava: Ostrava, Czech Republic, 2018; pp. 1–6; ISBN 978-80-248-4166-3.

37. Hubacek, M.; Rybansky, M.; Brenova, M.; Ceplova, L. The Soil Traficability Measurement in the Czech Republic for Military and Civil Use. In Proceedings of the 18th International Conference of the ISTVS, Seoul, Korea, 22–25 September 2014; p. 8; ISBN 978-1-942112-45-7.

38. Padarian, J.; Minasny, B.; McBratney, A.B. Using deep learning for digital soil mapping. *SOIL* **2019**, *26*, 79–89. [CrossRef]

39. Wildmann, R.; Bělka, L.; Kotlář, V. Production of foreign territory maps: MGCP derived graphics. *ArcRevue* **2009**, *3*, 6–9.
40. OSM. Project OpenStreetMap. 2021. Available online: https://www.openstreetmap.org/about (accessed on 13 December 2021).
41. Weibel, R.; Heller, M. *Digital Terrain Modelling. Geographical Information Systems: Principles and Applications*; Longman Scientific & Technical: London, UK, 1991.
42. USGS. *Shuttle Radar Topography Mission (SRTM)*; U. S. Geological Survey: Reston, VI, USA, 2021. Available online: https://www.usgs.gov/centers/eros/science/usgs-eros-archive-digital-elevation-shuttle-radar-topography-mission-srtm-1?qt-science_center_objects=0 (accessed on 10 December 2021).
43. NGA. *Digital Terrain Elevation Data*; National Geospatial-Intelligence Agency: Springfield, VA, USA, 2021. Available online: https://www.nga.mil/ProductsServices/TopographicalTerrestrial/Pages/DigitalTerrainElevationData.aspx (accessed on 20 December 2021).
44. ČÚZK. Digital Elevation Model of the Czech Republic of the 5th Generation (DMR5G). 2021. Available online: https://geoportal.cuzk.cz/(S(xg531xtrv3wnozbfp2hdiitu))/Default.aspx?mode=TextMeta&side=vyskopis&metadataID=CZCUZKDMR5GV&head_tab=sekce02gp&menu=30 (accessed on 12 December 2021).
45. MOČR. *Weapons and Equipment of the Armed Forces of the Czech Republic*; Ministry of Defence of the Czech Republic: Praha, Czech Republic, 2021. Available online: https://www.acr.army.cz/technika/default.htm (accessed on 4 December 2021).
46. Hlavacek, V. Evaluation of Dynamics of Vehicle Movement in the Terrain. Diploma Thesis, University of Defence, Brno, Czech Republic, 2017.
47. Vala, M.; Zalud, Z.; Neumann, V. *Theory and Construction of Military and Special Vehicles, Chapter III. Safety and Testing of Vehicles*; Textbook; Faculty of Military Technology, University of Defence: Brno, Czech Republic, 2017; ISBN 978-80-7582-023-5.
48. Rada, J. Smart defence: Joint Geospatial Support in NATO. *GeoScape* **2019**, *13*, 98–105. [CrossRef]
49. McGuire, P.C.; Vidale, P.L.; Best, M.; Case, D.H.; Duran Rojas, C.; Dharssi, I.; Hatcher, R.S.; Lister, G.M.; Martinez-de la Torre, A.; Montzka, C.; et al. Improving the global modeling of soils in JULES and the Unified Model: Updating from UM/HWSD to SoilGrids soil properties and from the Brooks & Corey to the van Genuchten soil-hydraulics model. 2020. In *AGU Fall Meeting Abstracts*; American Geophysical Union: Washington, DC, USA, 2020; Volume 2020, p. H199-0014.
50. Poggio, L.; Gimona, A. Assimilation of optical and radar remote sensing data in 3D mapping of soil properties over large areas. *Sci. Total Environ.* **2017**, *579*, 1094–1110. [CrossRef]
51. Rada, J.; Rybansky, M.; Dohnal, F. The Impact of the Accuracy of Terrain Surface Data on the Navigation of Off-Road Vehicles. *ISPRS Int. J. Geo-Inf.* **2021**, *10*, 106. [CrossRef]
52. Štroner, M.; Urban, R.; Línková, L. A New Method for UAV Lidar Precision Testing Used for the Evaluation of an Affordable DJI ZENMUSE L1 Scanner. *Remote Sens.* **2021**, *13*, 4811. [CrossRef]
53. Surový, P.; Kuželka, K. Acquisition of Forest Attributes for Decision Support at the Forest Enterprise Level Using Remote-Sensing Techniques—A Review. *Forests* **2019**, *10*, 273. [CrossRef]
54. Krůček, M.; Král, K.; Cushman, K.; Missarov, A.; Kellner, J.R. Supervised Segmentation of Ultra-High-Density Drone Lidar for Large-Area Mapping of Individual Trees. *Remote Sens.* **2020**, *12*, 3260. [CrossRef]
55. Rybansky, M. Determination of Forest Structure from Remote Sensing Data for Modeling the Navigation of Rescue Vehicles. *Appl. Sci.* **2022**, *12*, 3939. [CrossRef]

applied sciences

Article

Enhancing 3D Reconstruction Model by Deep Learning and Its Application in Building Damage Assessment after Earthquake

Zhonghua Hong [1], Yahui Yang [1], Jun Liu [2,3,*], Shenlu Jiang [4,*], Haiyan Pan [1], Ruyan Zhou [1], Yun Zhang [1], Yanling Han [1], Jing Wang [1], Shuhu Yang [1] and Changyue Zhong [3]

[1] College of Information Technology, Shanghai Ocean University, Shanghai 201306, China
[2] National Earthquake Response Support Service, Beijing 100049, China
[3] College of Civil Engineering and Architecture, Guizhou Minzu University, Guiyang 550025, China
[4] School of Computer Science and Engineering, Faculty of Innovation Technology, Macau University of Science and Technology, Avenida Wai Long, Taipa, Macau SAR, China
* Correspondence: liujun_eq@sina.com (J.L.); shenlujiang@must.edu.mo (S.J.)

Abstract: A timely and accurate damage assessment of buildings after an earthquake is critical for the safety of people and property. Most of the existing methods based on classification and segmentation use two-dimensional information to determine the damage level of the buildings, which cannot provide the multi-view information of the damaged building, resulting in inaccurate assessment results. According to the knowledge of the authors, there is no related research using the deep-learning-based 3D reconstruction method for the evaluation of building damage. In this paper, we first applied the deep-learning-based MVS model to reconstruct the 3D model of the buildings after an earthquake using multi-view UAV images, to assist the building damage assessment task. The method contains three main steps. Firstly, the camera parameters are calculated. Then, 3D reconstruction is conducted based on CasMVSNet. Finally, a building damage assessment is performed based on the 3D reconstruction result. To evaluate the effectiveness of the proposed method, the method was tested in multi-view UAV aerial images of Yangbi County, Yunnan Province. The results indicate that: (1) the time efficiency of CasMVSNet is significantly higher than that of other deep learning models, which can meet the timeliness requirement of post-earthquake rescue and damage assessment. In addition, the memory consumption of CasMVSNet is the lowest; (2) CasMVSNet exhibits the best 3D reconstruction result in both high and small buildings; (3) the proposed method can provide detail and multi-view information of damaged buildings, which can be used to assist the building damage assessment task. The results of the building damage assessment are very similar to the results of the field survey.

Keywords: multi-view UAV images; deep learning; CasMVSNet; building damage classification

Citation: Hong, Z.; Yang, Y.; Liu, J.; Jiang, S.; Pan, H.; Zhou, R.; Zhang, Y.; Han, Y.; Wang, J.; Yang, S.; et al. Enhancing 3D Reconstruction Model by Deep Learning and Its Application in Building Damage Assessment after Earthquake. *Appl. Sci.* **2022**, *12*, 9790. https://doi.org/10.3390/app12199790

Academic Editors: Anselme Muzirafuti, Giovanni Randazzo and Stefania Lanza

Received: 1 September 2022
Accepted: 25 September 2022
Published: 28 September 2022

Publisher's Note: MDPI stays neutral with regard to jurisdictional claims in published maps and institutional affiliations.

check for **updates**

1. Introduction

Earthquakes are one of the most serious natural disasters affecting humans. They cause many houses to be damaged and collapse, severely affecting the safety of both people and property. One of the key issues after an earthquake is the assessment of the damage of buildings. The results of the assessment can provide important information for disaster relief work. The timely and accurate assessment of damaged buildings is critical for rescues and consequential loss assessment.

Traditionally, post-earthquake building damage is evaluated and counted via manual field surveys, but this method is often time-consuming and laborious. Yamazaki et al. used the QuickBird satellite image after the Ms6.8 earthquake on the Mediterranean coast of Algeria and classified damaged buildings into five grades using the visual interpretation image method [1]. However, atmospheric conditions, such as cloud cover, will affect the image quality and lead to inaccurate evaluation.

With the development of artificial intelligence, machine-learning-related technologies have been gradually applied to the post-earthquake building damage assessment. Li et al. used remote sensing data before and after the earthquake through the decision tree method, in which the damaged buildings were divided into four grades [2]. The neural network of the genetic algorithm (GA) and the neural network composed of multi-layer perceptron (MLP) are used to predict the risk level of damage to reinforced-concrete (RC) structures [3,4]. The method achieves detailed investigation and inspection of buildings before the earthquake, reducing the loss of life and property. SMART SKY EYE (smart building safety assessment system using UAV) evaluates building wall cracks by analyzing natural factors based on machine learning methods such as random forest and support vector machine (SVM) [5]. However, these generalization capabilities are poor and the performance of the model is affected when the study area changes.

Compared with machine learning, convolutional neural networks (CNNs) based on deep learning have strong image processing abilities, strong feature learning and visual recognition abilities, and are widely used in building damage assessment. The dual-temporal methods use CNN to extract information on the characteristics of the images before and after the earthquake to determine the degree of damage to the building [6]. Ci et al. used deep-learning-based automatic detection and classification methods to evaluate and classify the loss levels of buildings in Ludian earthquake aerial images [7]. However, these methods can only achieve good performance when there are few categories, which cannot meet the needs for post-earthquake housing damage assessment. Ji et al. also combined machine learning and deep learning methods to evaluate five types of damage to buildings and improve the evaluation performance of damaged buildings using a combination of texture information from random forests and deep features extracted by CNN [8].

The above methods use two-dimensional semantic information to complete the damage assessment of the building, but they only contain damage information on one side of the building. Therefore, there is a big difference between the assessment results and the actual damage. On the contrary, three-dimensional semantic stereo information can provide structural features and height information of buildings. It is helpful to evaluate the damage grade of buildings after an earthquake. Mustafa et al. extracted the damaged information of buildings based on the differences in elevation between images before and after the earthquake [9]. However, the applicability of this method is limited due to the difficulty in obtaining pre-disaster and post-disaster digital elevation models (DEM). On the contrary, the 3D model efficiently reconstructed using the UAV can describe more detailed information of walls, beams, columns, and roofs from multiple angles [10,11]. Stepinac et al. used a laser scanner and a drone to generate 3D point clouds, after which the damage assessment of the building was performed by analyzing the three-dimensional structure of the building [12]. The scheme is expensive and is not suitable for large-scale 3D reconstruction. SMART SKY EYE used commercial software for 3D reconstruction and found defects in building structures using 3D models to complete the damage assessment of buildings [5]. However, the commercial software was developed from conventional methods [13–15], such as pix4d [16], smart 3d [17], and PhotoScan [18]. They improve the quality of 3D model, but the efficiency cannot meet the urgent needs of post-earthquake assessment.

In recent years, some multi-view stereo (MVS) networks based on deep learning have been widely used in 3D-reconstruction-related research [19]. The basic principle of MVS based on deep learning is to calculate the depth map of all images to complete the 3D reconstruction of the whole scene [20]. Based on the *DTU* dataset [21], MVSNet [20] completed the 3D reconstruction end-to-end for the first time. RMVSNet [22] used the GRU structure [23] to improve its regularization method, making large-scale 3D reconstruction possible. Subsequent improvements proposed by D^2HC-MVSNet [24], AA-RMVSNet [25], and CasMVSNet [26] have further improved network performance. Among them, CasMVSNet uses a multi-layer cascading method to compute the coarse-to-fine depth information [27,28], with higher computational efficiency and reconstruction quality.

However, all of the models mentioned above were tested in public datasets, which are mainly used to validate and evaluate different improved MVS networks. In addition, to the knowledge of the authors, there is little research that has used deep-learning-based MVS models to complete the damage assessment of buildings. The generalization ability of the model is the most important ability to move towards engineering applications, so it is necessary to use real post-earthquake image data analysis to evaluate the performance of the method. At the same time, these data also fill the gaps in the application of this method to the damage assessment of buildings after earthquakes [29,30]. Therefore, the objective of the study is to propose a deep-learning-based MVS method that is suitable for assessing building damage after an earthquake. Most importantly, the applicability of different 3D reconstruction models for post-earthquake building damage assessment is explored from the point view of time efficiency and the construction performance.

The remainder of this paper is organized as follows. Section 2 introduces the dataset used in this study. The methodology is presented in Section 3. In Section 4, the experimental details and results are shown. In Section 5, the performance of different methods is discussed. Finally, the conclusion is shown in Section 6.

2. Datasets

In the experiment, all MVS networks based on deep learning were trained on the public *DTU* dataset [21]. The data contained a wide range of scenarios, including housing models. The training data used 119 scenes, each containing 49 different view images with a pixel resolution of 640 × 512, with seven different intensity illuminations added to all images. The dataset was shot and calibrated by industrial manipulators, which can obtain high-precision camera parameters and improve the training of MVS networks. The dataset was downloaded from https://github.com/YoYo000/MVSNet (accessed on 29 August 2022) [20].

The dataset used in this study contained post-earthquake images of Yangbi County, Dali Prefecture, Yunnan Province, which occurred with an earthquake of magnitude 6.4 on 21 May 2021 with a focal depth of 8 km and an epicenter at 25.67 degrees north latitude and 99.87 degrees east longitude. As shown in Figure 1, a total of 411 UAV images with a pixel resolution of 5472 × 3648 (W × H) were obtained and 153 houses in the area were studied.

Figure 1. Multi-view UAV images of the earthquake area.

3. Methodology

The flow chart of the proposed method is shown in Figure 2, consisting of 3 steps: (1) calculation of camera parameters, (2) 3D reconstruction of MVS using deep learning method, (3) building damage assessment based on the result of the 3D reconstruction.

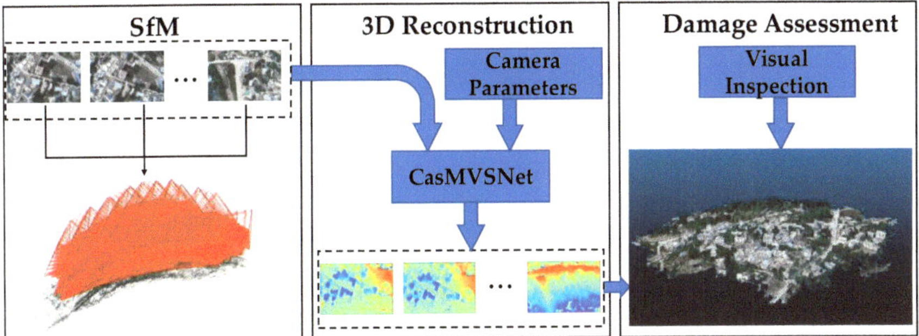

Figure 2. Workflow of building damage classification based on 3D model.

3.1. Calculation of Camera Parameters

When using the deep-learning-based MVS method to reconstruct the UAV image in the earthquake area, the COLMAP based on incremental SfM (structure-from-motion) [31] technology is used to complete the sparse reconstruction part to calculate camera parameters [20]. Incremental SfM is a processing method for sequential iterative reconstruction of 3D scenes. As shown in Figure 3, it usually starts with feature extraction and feature matching and then generates 3D models of scenes through geometric verification iteration. Next, the selected two-view is used as the basis of the model, and the registration of the new image is gradually added and the reconstruction is refined by triangulation and bundle adjustment (BA). After sparse reconstruction, the camera parameters of each image and the horizontal depth range of the model from the camera are output.

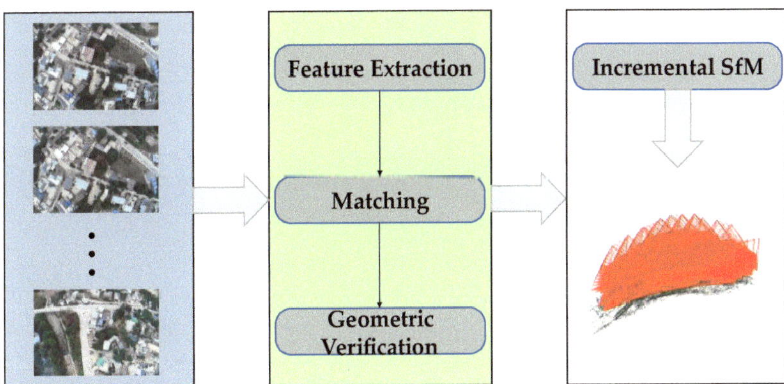

Figure 3. Principle of COLMAP [31].

The processed data are calibrated by the image and camera parameters, and there will be a slight deviation from the original image. In order to adapt to the input of each multi-view semantic stereo network, the image processed by COLMAP is preprocessed to a uniform size, and the image is restored to the original size. According to the image scaling ratio, the same scaling is performed on the camera internal parameters, including the main point offset coordinates and the camera focal length.

$$\frac{W}{W_0} = \frac{H}{H_0} = \frac{u}{u_0} = \frac{v}{v_0} = \frac{f}{f_0} \tag{1}$$

In Formula (1), W is the width of the image, H is the height of the image, u and v are the coordinates of the principal point of the image, and f is the focal length of the camera. Others are scaled corresponding parameters.

3.2. 3D Reconstruction Based on CasMVSNet

Considering the time cost and hardware equipment requirements for the three-dimensional modeling of UAV images, this study uses CasMVSNet [26] for 3D reconstruction due to its faster processing speed and higher accuracy. CasMVSNet extends the regularization of cost volume using 3D CNN [32] in MVSNet, which can better capture spatial feature information and use a multi-layer cascade method from coarse to fine. The multi-scale feature maps extracted from the feature extraction part are matched to construct cost volume with different resolutions. In the previous stage, the rough depth information was obtained via the calculation of the small-resolution feature map, and this is used to adaptively narrow the range of depth calculated by the higher resolution feature map in the next stage.

As shown in Figure 4, to obtain the multi-scale feature map and calculate the depth at different stages, the feature extraction part uses the feature pyramid network (FPN) method to extract the feature map with three scale resolutions, and their size is reduced by [33] times compared to the original image. Based on the above multi-scale feature map, the cost volume is constructed in stages from small to large. As shown in the green line in Figure 4, the depth information calculated by the cost volume with smaller resolution in the previous stage restricts the depth range of the homography transformation in the next stage to the depth value of this calculation. In fact, the depth interval of the previous stage is refined, and a smaller cost volume is constructed, which not only reduces the amount of calculation but also consumes explicit memory when calculating more accurate depth maps.

$$H_i^{(k+1)}\left(d^{(k)} + \Delta^{(k+1)}\right) = \left(d^{(k)} + \Delta^{(k+1)}\right) K_i T_i T_0^{-1} \qquad (2)$$

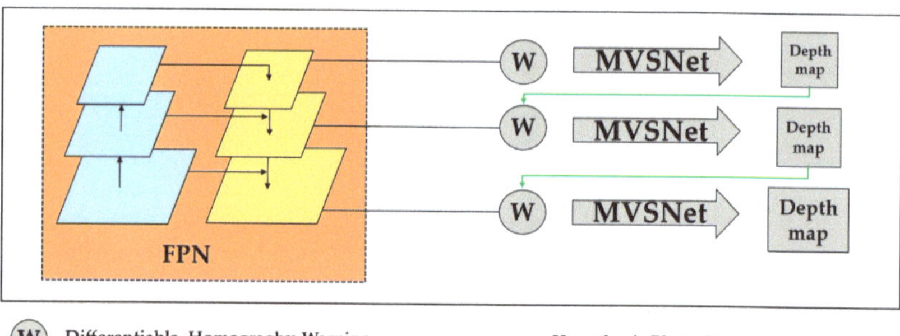

Figure 4. Coarse-to-fine computational deep MVS network.

The differentiable homography [20] of CasMVSNet in different stages is as follows: The depth value of different stages is modified to $d = d^{(k)} + \Delta^{(k+1)}$, where k refers to the number of stages. The depth range for the next stage is calculated by adding the depth result $d^{(k)}$, calculated in the previous stage, to the residual depth $\Delta^{(k+1)}$. As shown in part W of Figure 4, with the change in the feature map, the intrinsic parameters of the camera are scaled equally. Using these parameters, the feature maps of the auxiliary view are warped into the reference image view space, and the cost volume is constructed by aggregating the cost matching the auxiliary view feature maps and the reference image feature maps.

3.3. Assessment of Damaged Buildings

AeDES [34] provides five grades for damaged buildings. Each grade lists detailed close-range pictures of houses. The examples include detailed parameters, such as the width of the wall cracks and the internal structure of the damaged building. Therefore, it is more suitable for field survey. The European Macro-Earthquake Magnitude in 1998 (EMS-98) [35] also classifies damaged buildings into five grade levels in detail, and each level provides a schematic diagram of macroscopic structural damage. Therefore, this sub-standard is more suitable for visual inspection of the 3D model reconstructed using the above method to obtain a post-earthquake damage assessment of buildings. Table 1 shows the detailed descriptions and examples of building damage classifications. Then, according to the EMS-98 standard and the result of 3D reconstruction, a visual interpretation was conducted to determine the damage level of the building.

Table 1. Detailed description of buildings with different damage level.

Reinforced Concrete	Masonry Buildings	3D Model	Classification of Damage
			Grade0: Negligible to slight damage (no structural damage, slight non-structural damage)
			Grade1: Moderate damage (slight structural damage, moderate non-structural damage)
			Grade2: Substantial to heavy damage (moderate structural damage, heavy non-structural damage)
			Grade3: Very heavy damage (heavy structural damage, very heavy non-structural damage)
			Grade4: Destruction (very heavy structural damage)

4. Experiment and Results

4.1. Experimental Details

The generalization ability of the model is the most important engineering application ability. The model trained on the high-precision camera parameter *DTU* dataset [21] was directly used to test the UAV image data of Jinniu Village in Yangbi, Yunnan. The experiment was carried out on a computer with an Intel Xeon (R) W-2295 3.00 GHz * 36 processor and a 24 GB GeForce RTX3090/PCIe/SSE2 graphics processor.

The network was trained for 16 epochs, with an initial learning rate of 0.001, which was reduced by a factor of 2 after 10, 14, and 16 epochs. The pixel resolution of the input image in the network was fine-tuned according to the original public test parameters. The *stage* of CasMVSNet was set to 3, *numdepth* was set to 192, *ndepths* was set to corresponded to {48, 32, 8} and *interval_scale* was set to 1.06.

4.2. Results

The efficiency of 3D reconstruction and the quality of the model are critical for damage assessment of buildings after an earthquake and also take into account the hardware requirements for implementing the work. On the basis of the above experiments, we quantitatively analyzed and compared the results of three indicators of statistical time consumption, video memory consumption, and visual modelling of different methods. For

AA-RMVSNet and D^2HC-MVSNet, *numdepth* was set to 192 and *interval_scale* was set to 1.06.

4.2.1. Time Consumption

Table 2 is the time comparison of different methods. We divided the reconstruction time into two parts: the time for calculation of camera parameters and the time of 3D reconstruction. As can be seen from the table, camera parameters were calculated for deep learning methods based on COLMAP, so the time for the first part was always 61 min. In terms of 3D reconstruction, we set up two sets of resolution experiments to compare the time of 3D reconstruction. When the resolution of the input image was 1184 × 800 (pixel), the time consumption of AA-RMVSNet and D^2HC-MVSNet was approximately 10 times and 7.8 times that of CasMVSNet, respectively. When the resolution of the input image increased to 2160 × 1440 (pixel), the time consumption of CasMVSNet was also significantly lower than the other methods. As indicated in the table, the 3D reconstruction time of AA-RMVSNet and D^2HC-MVSNet was about 9.6 times and 8.6 times that of CasMVSNet, respectively. The CasMVSNet method took the shortest time and had the highest efficiency.

Table 2. Time consumption of various methods.

Meth	Calculating Camera Parameters	3D Reconstruction (min)	
		1184 × 800 Pixel	2160 × 1440 Pixel
AA-RMVSNet	61	163	499
D^2HC-MVSNet	61	124	447
CasMVSNet	61	16	52

The reason for the time difference is that both D^2HC-MVSNet and AA-RMVSNet use a recurrent neural network to regularize the cost map at each depth in the cost aggregation part. Compared with D^2HC-MVSNet, AA-RMVSNet adds an Inter-view AA module to the feature extraction part for multi-scale feature fusion and adds the Inter-view AA module before aggregating cost volume, which takes the longest time. The CascMVSNet model uses a cascading approach to calculate the pixel depth from coarse to fine. The depth range is roughly divided at the initial stage, and the calculation result is then discretized into a depth range for the next stage. Given that the sum of its depth intervals at each stage is much smaller than the depth interval values of the above two methods, and CascMVSNet uses a 3D CNN method that is faster than the recurrent neural network for cost aggregation, the method is the shortest and most efficient.

4.2.2. Memory Consumption

Table 3 shows the results of the memory consumption of different deep-learning-based MVS networks. As can be seen from the table, the memory consumption of CasMVSNet was the lowest. When the resolution of the input image was 1184 × 800, the memory consumption of CasMVSNet was 9232 MiB and 500 MiB lower than that of AA-RMVSNet and D^2HC-RMVSNet, respectively. When the resolution of the input image increased to 2160 × 1440, the memory consumption between different models was more distinct. Compared with the memory of 9955 MiB of CasMVSNet, the memory consumption of D^2HC-RMVSNet was 4400 MiB higher. The memory consumption of AA-RMVSNet was the largest, which was about 2.3 times that of CasMVSNet.

Table 3. Memory consumption of various deep learning methods in different sizes of images.

Meth	1184 × 800 (MiB)	2160 × 1440 (MiB)
AA-RMVSNet	13659	22933
D^2HC-MVSNet	4907	14395
CasMVSNet	4407	9995

The reason for this phenomenon is that CasMVSNet narrows the depth range as the resolution of the feature map increases, building a smaller cost volume at each stage to lower the memory consumption compared to the other two deep learning methods. The D^2HC-MVSNet network using a recurrent neural network instead of 3D CNN for cost aggregation decomposes the whole cost volume into a cost map at each depth. The memory consumption is slightly higher than that of CasMVSNet in low-resolution performance, but with the increase of input image resolution, its consumption will be significantly higher than that of the CasMVSNet model. The AA-RMVSNet model is improved on the basis of the D^2HC-MVSNet model. The Inter-view AA module added to the model performs pixel-level weighted aggregation on the cost volume constructed from multiple perspectives, so the model is higher than the other two models in memory consumption.

4.2.3. Result of 3D Reconstruction

As shown in Figure 5, the visualization results of different 3D reconstruction methods are displayed in CloudCompare [36]. As can be seen from the figures, all the methods exhibited relatively good results and were able to reflect the multidimensional details of the damaged building. Compared to the higher part of the building, the UAV obtained more detailed building information and could obtain more detailed stereo semantic information in the modeling process to complete stereo matching. Therefore, all of the methods performed well in this type of building. The Intra-view AA added in the feature extraction part of AA-RMVSNet maintained the correlation between the original geometric features of the image and the Inter-view AA added in the cost aggregation part on the basis of fusing multi-scale features. Compared to D^2HC-MVSNet, both improved the accuracy of the 3D model. Unlike AA-RMVSNet and D^2HC-MVSNet, CasMVSNet uses a multi-layer cascaded and gradually refined depth calculation method, which can better reflect the advantages of finer division and calculation of depth information when dealing with UAV image reconstruction. The building wall information in the 3D model reconstructed by this method was complete and delicate.

4.2.4. Result of the Evaluation

In this study, the point cloud generated by the 3D reconstruction was converted to 3DTiles format and imported into a seismic information visualization system, which was developed based on Cesium. In Figure 6, the model is marked with a Web page as a carrier to visually view the results of the assessment of the disaster situation. Address at www.peteralbus.com:8085 (accessed on 29 August 2022).

Based on the results of the above comparison, in this experiment, the reconstruction results of the CasMVSNet network were selected to evaluate the damage grade of 153 houses in the Jinniu Village area according to the EMS-98 standard. Furthermore, the proportion of the number of damaged houses at each grade to the total number of houses was calculated and compared with the evaluation results of other methods in this area. In total, two comparison methods were involved. The first was the result obtained by Zhang et al. In this research, visual interpretation was conducted on orthophotos of UAV images of the old street near the Yunlong Bridge in Yangbi County [37]. Another method was visual interpretation using spliced orthophotos. The comparison results are shown in Table 4. As can be seen in the table, our results are very close to those of the other two methods. Compared to the results of the field survey, the relative error was 1.3%, 1.0%, 0.6%, 1%, and 0.6% for G0, G1, G2, G3, and G4, respectively, which indicates the effectiveness of the proposed method.

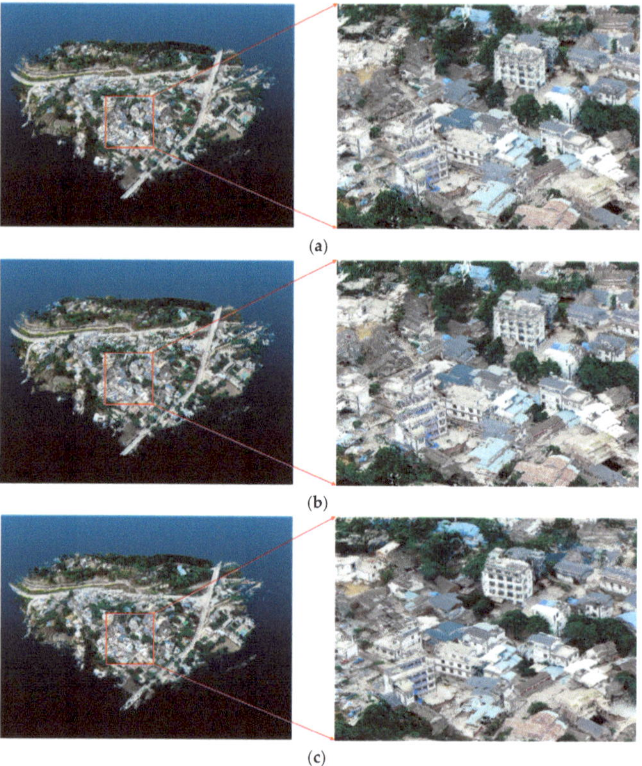

Figure 5. Visualization results of various 3D reconstruction methods. (**a**) D^2HC-MVSNet; (**b**) AA-RMVSNet; (**c**) CasMVSNet.

Figure 6. Results of the damaged assessment in Jinniu Village, Yangbi.

Table 4. Comparison of the results of the 2D and 3D assessment ratio and field survey.

Assessment of Damage	Orthophoto	Field Survey	3D Models
G0	22.7%	20.3%	21.6%
G1	37.7%	37.7%	36.7%
G2	21.6%	26.1%	25.5%
G3	13.2%	10.1%	11.1%
G4	4.8%	5.8%	5.2%
Total	100%	100%	100%

5. Discussion

Timeliness is the most important factor for the damage assessment of buildings after an earthquake. Therefore, the applicability of different MVS models for post-earthquake building damage assessment is discussed from the point of view of the network structures and the robustness. As shown in Figure 7, similar to CasMVSNet, deep-learning-based multi-view stereo networks, such as D^2HC-MVSNet and AA-RMVSNet, have mostly been improved on the basis of MVSNet. The initial MVSNet uses the 3D CNN method to regularize the cost volume and generate the probability. The soft argmin [38] operation calculates the depth value for each pixel in a winner-take-all manner and estimates the initial pixel-level depth map. Finally, the reference image is used to refine the depth map to improve the accuracy of the boundary region and complete refined pixel-level depth map estimation. However, as the input image size increases, the parameters of the model increase exponentially, so the method requires higher memory consumption.

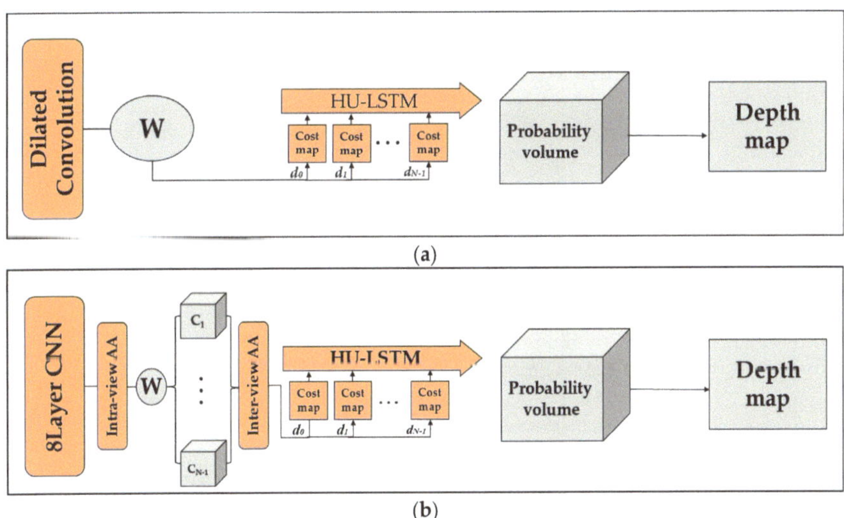

Figure 7. Schematic diagram of various deep learning MVS networks. (a) D^2HC-MVSNet; (b) AA-RMVSNet.

5.1. Network Structure

D^2HC-MVSNet improves the R-MVSNet GRU gating to some extent, with more powerful loop convolution units and dynamic consistency checking strategies. Firstly, the network uses dilated convolution [39] to obtain a larger range of feature information in the 2D feature extraction part, connects the feature output via different convolutional layers, and aggregates context feature information without losing resolution. D^2HC-MVSNet introduces a cyclic encoding–decoding HU-LSTM structure [24] to regularize the cost volume along the depth direction in the cost aggregation part, and realizes the connection memory of the same size features along the depth direction on the feature map of each scale in the regularization process. This method not only aggregates the spatial geometric context information of the cost map but also preserves the cost aggregation output of the original resolution size with lower memory consumption. However, this method decomposes the cost volume into cost maps at each depth and processes them sequentially using a recurrent neural network, resulting in a slower computational efficiency of the network. When used to assist in assessing the damage level of houses in post-earthquake areas, it cannot meet the urgent needs of this work.

AA-RMVSNet is improved on the basis of D^2HC-MVSNet. The feature extraction part uses the common CNN to obtain the high-dimensional information of the image. The last two downsamplings output the feature maps of the original image with 1/4 and 1/16 size 32 channels, respectively. One of the innovations of this method is that an Intra-view AA module composed of deformable convolution [40,41] is added to the feature extraction part, which is used to adaptively aggregate the features of different scales and regions with different texture richness. This module processes the feature maps of the last three layers through deformable convolution and then upsamples the output of the last two layers after processing, and integrates with the previous layer as the final output of the feature extraction part, maintaining the geometric shape of the object in the image to the greatest extent when extracting features. In addition, when the network aggregates the cost volume of multiple perspectives, the Inter-view AA module is added to suppress the mismatched pixels by pixel-level weighting, and the pixels with higher matching correlations are given greater weight, rather than the matching results of all perspectives being treated equally. In short, the network fully retains the original geometric information of the object in the image and pays attention to the correlation problem after matching each perspective, thus improving the quality of 3D reconstruction. However, this method also uses the recurrent neural network, joining the above two optimization methods. Therefore, the quality of 3D reconstruction is guaranteed but the time efficiency is not well balanced when used in the post-earthquake building damage assessment work, and the added Inter-view AA module makes the network's memory consumption higher, resulting in higher GPU hardware requirements when carrying out this work.

5.2. Robustness

Through the MVS network modeling channel based on deep learning, the post-earthquake housing damage assessment work is completed—that is, from the public close-range experimental data migration to the real image of the UAV in the post-earthquake area and reconstruction of the 3D model—so whether the model has good robustness is extremely important. MVS network modeling based on deep learning completes 3D modeling in the form of calculated depth maps, so higher-resolution depth maps can reconstruct finer 3D models. The dilated convolution used by D^2HC-RMVSNet takes into account the features of a larger field of view, and AA-RMVSNet refines feature extraction and filter matching results, both of which improve when rebuilding 3D models. However, when used for high-resolution post-earthquake regional UAV image reconstruction, the series of models obtain higher-resolution depth maps by fine-tuning the *max_w* and *max_h* of the input image. However, from the experimental results, both methods have large fluctuations in time consumption and memory consumption. However, the CasMVSNet method constructs a smaller cost volume, so it has better stability at this time.

The MVS series networks based on deep learning calculate the pixel depth information based on the assumption of discrete depth intervals. Therefore, when using the UAV image data of large scenes to reconstruct the 3D model of the post-earthquake region, the single-stage networks of D^2HC-RMVSNet and AA-RMVSNet can fine-tune and refine the discrete depth intervals. However, the above two networks use the RNN method, so it will increase the workload of the recurrent neural network when fine-tuning the assumed discrete intervals. CasMVSNet can not only improve the robustness of 3D reconstruction by refining the discrete depth intervals of each stage of the network, but also refine the depth map output in the final stage by increasing the level of the network. In this way,

CasMVSNet realizes the 3D reconstruction of UAV image data in large scenes after the earthquake. Whether the number of stages of the network is adjusted or the discrete depth interval is refined, it will refine the calculation of depth information and provide better stability.

6. Conclusions

In this work, a multi-view stereo (MVS) method based deep learning was first applied to assist in assessing the damage level of buildings in post-earthquake areas. The method was tested in aerial UAV images from Yangbi County, Yunnan Province. The time consumption, memory consumption, and the performance of 3D reconstruction of different models were compared. In addition, the applicability of different 3D reconstruction models was discussed. A number of conclusions can be made as follows: (1) the time efficiency of CasMVSNet is significantly higher than that of other deep learning models, and the memory consumption is the lowest, which can meet the timeliness requirement of post-earthquake rescue and damage assessment; (2) CasMVSNet exhibited the best 3D reconstruction result in both high and small buildings; (3) the deep-learning-based 3D reconstruction method can provide the detail and multi-view information of damaged buildings, which can be used to assist the building damage assessment task. The assessment results were very close to the results of the field survey.

The main contributions of our study can be summarized as follows: (1) we first attempted to use deep-learning-based MVS to UAV aerial 3D reconstruction and building damage assessment research, which can provide 3D information for post-earthquake rescue and loss assessment; (2) the applicability of different MVS models for 3D reconstruction of UAV images have been analyzed and discussed.

The limitation of the proposed method is that the damage level of the buildings is determined by visual interpretation. In the future, we will devote efforts to construct a network to realize the automatic classification of buildings, based on the results of the 3D reconstruction.

Author Contributions: Conceptualization, Z.H., J.L. and S.J.; data curation, Y.Y., Z.H., J.L. and C.Z.; methodology, Z.H., Y.Y., S.J. and J.L.; validation, Y.Y., H.P., R.Z., Y.Z., Y.H., J.W. and S.Y.; formal analysis, Z.H. and Y.Y.; investigation, Y.Y. and C.Z.; writing—original draft preparation, Z.H., Y.Y. and H.P.; writing—review and editing, Z.H., Y.Y., H.P., J.L. and S.J.; supervision, J.L. and S.J. All authors have read and agreed to the published version of the manuscript.

Funding: This work was funded by the National Key R&D Program of China, grant number 2017YFC1500906; the National Natural Science Foundation of China, grant number 41871325, 42061073; and the Natural Science and Technology Foundation of Guizhou Province under Grant [2020]1Z056

Institutional Review Board Statement: Not applicable.

Informed Consent Statement: Not applicable.

Data Availability Statement: The *DTU* datasets are freely available online, and can be found at https://github.com/YoYo000/MVSNet (accessed on 29 August 2022).

Acknowledgments: We thank our team's graduate Master Jinmeng Gao for providing UAV images so that our work could be successfully completed.

Conflicts of Interest: The authors declare no conflict of interest.

Abbreviations

UVA	unmanned air vehicle
GA	genetic algorithm
MLP	multi-layer perceptron
RC	reinforced-concrete
SMART SKY EYE	smart building safety assessment system using UAV
CNN	convolutional neural networks
DEM	digital elevation models
MVS	multi-view stereo
SfM	structure-from-motion
BA	bundle adjustment
FPN	feature pyramid networks
EMS-98	European Macro-Earthquake Magnitude in 1998

References

1. Yamazaki, F.; Kouchi, K.I.; Matsuoka, M.; Kohiyama, M.; Muraoka, N. Damage detection from high-resolution satellite images for the 2003 Boumerdes, Algeria earthquake. In Proceedings of the 13th World Conference on Earthquake Engineering, International Association for Earthquake Engineering, Vancouver, BC, Canada, 1–6 August 2004; p. 13.
2. Li, S.; Tang, H. Classification of Building Damage Triggered by Earthquakes Using Decision Tree. *Math. Probl. Eng.* **2020**, *2020*, 1–15. [CrossRef]
3. Bülbül, M.A.; Harirchian, E.; Işık, M.F.; Aghakouchaki Hosseini, S.E.; Işık, E. A Hybrid ANN-GA Model for an Automated Rapid Vulnerability Assessment of Existing RC Buildings. *Appl. Sci.* **2022**, *12*, 5138. [CrossRef]
4. Harirchian, E.; Jadhav, K.; Kumari, V.; Lahmer, T. ML-EHSAPP: A prototype for machine learning-based earthquake hazard safety assessment of structures by using a smartphone app. *Eur. J. Environ. Civ. Eng.* **2021**, *26*, 5279–5299. [CrossRef]
5. Bae, J.; Lee, J.; Jang, A.; Ju, Y.K.; Park, M.J. SMART SKY Eye system for preliminary structural safety assessment of buildings using unmanned aerial vehicles. *Sensors* **2022**, *22*, 2762. [CrossRef] [PubMed]
6. Zheng, Z.; Zhong, Y.; Wang, J.; Ma, A.; Zhang, L. Building damage assessment for rapid disaster response with a deep object-based semantic change detection framework: From natural disasters to man-made disasters. *Remote Sens. Environ.* **2021**, *265*, 112636. [CrossRef]
7. Ci, T.; Liu, Z.; Wang, Y. Assessment of the degree of building damage caused by disaster using convolutional neural networks in combination with ordinal regression. *Remote Sens.* **2019**, *11*, 2858. [CrossRef]
8. Ji, M.; Liu, L.; Du, R.; Buchroithner, M.F. A comparative study of texture and convolutional neural network features for detecting collapsed buildings after earthquakes using pre-and post-event satellite imagery. *Remote Sens.* **2019**, *11*, 1202. [CrossRef]
9. Turker, M.; Cetinkaya, B. Automatic detection of earthquake-damaged buildings using DEMs created from pre-and post-earthquake stereo aerial photographs. *Int. J. Remote Sens.* **2005**, *26*, 823–832. [CrossRef]
10. Muzirafuti, A.; Cascio, M.; Lanza, S.; Randazzo, G. UAV Photogrammetry-based Mapping of the Pocket Beaches of Isola Bella Bay, Taormina (Eastern Sicily). In Proceedings of the 2021 International Workshop on Metrology for the Sea, Learning to Measure Sea Health Parameters (MetroSea), Reggio Calabria, Italy, 4–6 October 2021; pp. 418–422.
11. Randazzo, G.; Italiano, F.; Micallef, A.; Tomasello, A.; Cassetti, F.P.; Zammit, A.; D'Amico, S.; Saliba, O.; Cascio, M.; Cavallaro, F. WebGIS Implementation for Dynamic Mapping and Visualization of Coastal Geospatial Data: A Case Study of BESS Project. *Appl. Sci.* **2021**, *11*, 8233. [CrossRef]
12. Stepinac, M.; Lulić, L.; Ožić, K. The Role of UAV and Laser Scanners in the Post-earthquake Assessment of Heritage Buildings After the 2020 Earthquakes in Croatia. In *Advanced Nondestructive and Structural Techniques for Diagnosis, Redesign and Health Monitoring for the Preservation of Cultural Heritage*; Springer: Berlin/Heidelberg, Germany, 2022; pp. 167–177.
13. Bleyer, M.; Rhemann, C.; Rother, C. Patchmatch stereo-stereo matching with slanted support windows. In Proceedings of the Bmvc, Dundee, UK, 29 August–2 September 2011; pp. 1–11.
14. Hirschmuller, H. Stereo processing by semiglobal matching and mutual information. *IEEE Trans. Pattern Anal. Mach. Intell.* **2007**, *30*, 328–341. [CrossRef] [PubMed]
15. Tola, E.; Strecha, C.; Fua, P. Efficient large-scale multi-view stereo for ultra high-resolution image sets. *Mach. Vis. Appl.* **2012**, *23*, 903–920. [CrossRef]
16. Pix4D. Available online: https://www.pix4d.com/ (accessed on 29 August 2022).
17. ContextCapture. Available online: https://www.bentley.com/en/products/brands/contextcapture (accessed on 29 August 2022).
18. Agisoft. Available online: http://www.agisoft.com (accessed on 29 August 2022).
19. Zhu, Q.; Min, C.; Wei, Z.; Chen, Y.; Wang, G. Deep Learning for Multi-View Stereo via Plane Sweep: A Survey. *arXiv* **2021**, arXiv:2106.15328.
20. Yao, Y.; Luo, Z.; Li, S.; Fang, T.; Quan, L. Mvsnet: Depth inference for unstructured multi-view stereo. In Proceedings of the European Conference on Computer Vision (ECCV), Munich, Germany, 8–14 September 2018; pp. 767–783.

21. Aanæs, H.; Jensen, R.R.; Vogiatzis, G.; Tola, E.; Dahl, A.B. Large-scale data for multiple-view stereopsis. *Int. J. Comput. Vis.* **2016**, *120*, 153–168. [CrossRef]
22. Yao, Y.; Luo, Z.; Li, S.; Shen, T.; Fang, T.; Quan, L. Recurrent mvsnet for high-resolution multi-view stereo depth inference. In Proceedings of the IEEE/CVF Conference on Computer Vision and Pattern Recognition, Long Beach, CA, USA, 15–20 June 2019; pp. 5525–5534.
23. Cho, K.; Van Merriënboer, B.; Gulcehre, C.; Bahdanau, D.; Bougares, F.; Schwenk, H.; Bengio, Y. Learning phrase representations using RNN encoder-decoder for statistical machine translation. *arXiv* **2014**, arXiv:1406.1078.
24. Yan, J.; Wei, Z.; Yi, H.; Ding, M.; Zhang, R.; Chen, Y.; Wang, G.; Tai, Y.-W. Dense hybrid recurrent multi-view stereo net with dynamic consistency checking. In Proceedings of the European Conference on Computer Vision, Glasgow, UK, 23–28 August 2020; pp. 674–689.
25. Wei, Z.; Zhu, Q.; Min, C.; Chen, Y.; Wang, G. Aa-rmvsnet: Adaptive aggregation recurrent multi-view stereo network. In Proceedings of the IEEE/CVF International Conference on Computer Vision, Montreal, QC, Canada, 10–17 October 2021; pp. 6187–6196.
26. Gu, X.; Fan, Z.; Zhu, S.; Dai, Z.; Tan, F.; Tan, P. Cascade cost volume for high-resolution multi-view stereo and stereo matching. In Proceedings of the IEEE/CVF Conference on Computer Vision and Pattern Recognition, Seattle, WA, USA, 20–25 June 2020; pp. 2495–2504.
27. Tonioni, A.; Tosi, F.; Poggi, M.; Mattoccia, S.; Stefano, L.D. Real-time self-adaptive deep stereo. In Proceedings of the IEEE/CVF Conference on Computer Vision and Pattern Recognition, Long Beach, CA, USA, 15–20 June 2019; pp. 195–204.
28. Yin, Z.; Darrell, T.; Yu, F. Hierarchical discrete distribution decomposition for match density estimation. In Proceedings of the Proceedings of the IEEE/CVF Conference on Computer Vision and Pattern Recognition, Long Beach, CA, USA, 15–20 June 2019; pp. 6044–6053.
29. Guptha, G.C.; Swain, S.; Al-Ansari, N.; Taloor, A.K.; Dayal, D. Evaluation of an urban drainage system and its resilience using remote sensing and GIS. *Remote Sens. Appl. Soc. Environ.* **2021**, *23*, 100601. [CrossRef]
30. Kazemian, I.; Torabi, S.A.; Zobel, C.W.; Li, Y.; Baghersad, M. A multi-attribute supply chain network resilience assessment framework based on SNA-inspired indicators. *Oper. Res.* **2022**, *22*, 1853–1883.
31. Schonberger, J.L.; Frahm, J.-M. Structure-from-motion revisited. In *Proceedings of the Proceedings of the IEEE Conference on Computer Vision and Pattern Recognition, Las Vegas, NV, USA, 27–30 June 2016*; pp. 4104–4113.
32. Ji, S.; Xu, W.; Yang, M.; Yu, K. 3D convolutional neural networks for human action recognition. *IEEE Trans. Pattern Anal. Mach. Intell.* **2012**, *35*, 221–231. [CrossRef] [PubMed]
33. Duarte, D.; Nex, F.; Kerle, N.; Vosselman, G. Multi-resolution feature fusion for image classification of building damages with convolutional neural networks. *Remote Sens.* **2018**, *10*, 1636. [CrossRef]
34. Baggio, C.; Bernardini, A.; Colozza, R.; Corazza, L.; Della Bella, M.; Di Pasquale, G.; Dolce, M.; Goretti, A.; Martinelli, A.; Orsini, G. *Field Manual for Post-Earthquake Damage and Safety Assessment and Short Term Countermeasures (AeDES)*; European Commission—Joint Research Centre—Institute for the Protection and Security of the Citizen: Ispra, Italy, 2007; pp. 1–100.
35. Grünthal, G. *European Macroseismic Scale 1998 (EMS-98)*; Conseil De L'europe: Strasbourg, France, 1998.
36. CloudCompare. Available online: http://www.danielgm.net/cc (accessed on 29 August 2022).
37. Zhang, L.; He, F.; Yan, J.; Du, H.; Zhou, Z.; Wang, Y. Quantitative Assessment of Building Damage of the Yangbi Earthquake Based on UAV Images. *South China J. Seismol.* **2021**, *41*, 76–81.
38. Kendall, A.; Martirosyan, H.; Dasgupta, S.; Henry, P.; Kennedy, R.; Bachrach, A.; Bry, A. End-to-end learning of geometry and context for deep stereo regression. In Proceedings of the IEEE International Conference on Computer Vision, Venice, Italy, 22–29 October 2017; pp. 66–75.
39. Yu, F.; Koltun, V. Multi-scale context aggregation by dilated convolutions. *arXiv* **2015**, arXiv:1511.07122.
40. Dai, J.; Qi, H.; Xiong, Y.; Li, Y.; Zhang, G.; Hu, H.; Wei, Y. Deformable convolutional networks. In Proceedings of the IEEE International Conference on Computer Vision, Venice, Italy, 22–29 October 2017; pp. 764–773.
41. Zhu, X.; Hu, H.; Lin, S.; Dai, J. Deformable convnets v2: More deformable, better results. In Proceedings of the IEEE/CVF Conference on Computer Vision and Pattern Recognition, Long Beach, CA, USA, 15–20 June 2019; pp. 9308–9316.

Article

Characterisation of Coastal Sediment Properties from Spectral Reflectance Data

Jasper Knight [1,*] and Mohamed A. M. Abd Elbasit [2]

[1] School of Geography, Archaeology & Environmental Studies, University of the Witwatersrand, Johannesburg 2050, South Africa

[2] School of Natural and Applied Sciences, Sol Plaatje University, Kimberley 8301, South Africa; mohamed.ahmed@spu.ac.za

* Correspondence: jasper.knight@wits.ac.za

Abstract: Remote sensing of coastal sediments for the purpose of automated mapping of their physical properties (grain size, mineralogy and carbonate content) across space has not been widely applied globally or in South Africa. This paper describes a baseline study towards achieving this aim by examining the spectral reflectance signatures of field sediment samples from a beach–dune system at Oyster Bay, Eastern Cape, South Africa. Laboratory measurements of grain size and carbonate content of field samples ($n = 134$) were compared to laboratory measurements of the spectral signature of these samples using an analytical spectral device (ASD), and the results interrogated using different statistical methods. These results show that the proportion of fine sand, $CaCO_3$ content and the distributional range of sediment grain sizes within a sample (here termed *span*) are the parameters with greatest statistical significance—and thus greatest potential interpretive value—with respect to their spectral signatures measured by the ASD. These parameters are also statistically associated with specific wavebands in the visible and near infrared, and the shortwave infrared parts of the spectrum. These results show the potential of spectral reflectance data for discriminating elements of grain size properties of coastal sediments, and thus can provide the baseline towards achieving automated spatial mapping of sediment properties across coastal beach–dune environments using hyperspectral remote sensing techniques.

Keywords: grain size analysis; coastal sediments; analytical spectral device; hyperspectral data

Citation: Knight, J.; Abd Elbasit, M.A.M. Characterisation of Coastal Sediment Properties from Spectral Reflectance Data. *Appl. Sci.* **2022**, *12*, 6826. https://doi.org/10.3390/app12136826

Academic Editors: Giovanni Randazzo, Stefania Lanza and Anselme Muzirafuti

Received: 9 May 2022
Accepted: 3 June 2022
Published: 5 July 2022

Publisher's Note: MDPI stays neutral with regard to jurisdictional claims in published maps and institutional affiliations.

1. Introduction

Sediment properties of sandy beaches and sand dunes, including grain size, carbonate content, moisture content, organic content, magnetic susceptibility and grain mineralogy, are most commonly measured and quantified based on field observations or field sampling, and then laboratory analysis of these samples using different analytical equipment [1–4]. Following this, a range of statistical techniques (e.g., calculation of moment measures, multivariate analyses) can be used on the grain size data in particular, in order to characterise sediment properties and to interpret depositional processes and environments and their changes over time and space, e.g., [5–11]. This standard methodology has been undertaken on many beaches and dunes worldwide, resulting in an understanding of spatial patterns of different sediment properties (based mainly on grain size) across different coastal depositional environments, e.g., [12–15]. The main problem of such a field-based approach is that it provides only a limited view of local-scale coastal sediment properties and dynamics, which is often strongly affected by the specific spatial and temporal context of field sampling at individual sites. In addition, studies also use different sampling strategies and methods of data analysis, which means that results from these individual studies may not be comparable. By contrast, remote sensing methods using a variety of platforms have potential to consistently map and quantify spatial patterns of sediments and landforms across beach–dune systems, and this has been undertaken in several studies e.g., [16–21]. There are fewer studies, however, that have examined spectral data on sediment properties and stratigraphy. Sediment cores have been examined using different hyperspectral imaging techniques, mainly in the shortwave infrared (SWIR) wavebands, in order to identify stratigraphic variations in sediment grain size and mineralogy [22–24]. These studies have been used to produce spectral time series maps that represent variations in sediment properties through the cores, rather than identify individual spectra that represent certain sediment endmembers. There are only a few studies that have examined the spectral properties of sediments in coastal environments, and these have considered the role of variations in water content and mineralogy as key factors influencing their spectral signatures [18,25]. Mineral compositions can then be used to derive endmembers for spatial modelling.

Most work on spectral signatures of sediment has been done on river depositional environments [26,27], and work on coastal sediments can be informed by these previous studies. For example, river and coastal sediment samples in NE Italy were evaluated by Ciampalini et al. [28] using an analytical spectral device (ASD) in order to derive a spectral library representing sample grain size and mineralogy, which was then compared to laboratory results. Principal component analysis was then used to identify sediment provenance endmembers. The same research approach was used by Ibrahim et al. [19] along the Belgian coast. Grain size properties along beaches were examined using Landsat visible, near infrared (VNIR) and thermal infrared bands in SE India [29], but these bands may have been influenced by a high concentration of heavy minerals (50–80%) at this site. Using IKONOS imagery, Park et al. [30] showed that all spectral bands have a good correlation (>0.8) with grain size, and Williams and Greeley [31] showed that different spectral bands from synthetic aperture radar imagery are affected by surface moisture. Thus, there are several studies that have analysed the spectral properties of beach sand but these have tended to focus on the role of local environmental factors rather than the application of different techniques or methodologies. A key question is how location-specific measurements can be applied to similar depositional settings elsewhere [27,32] or how patterns of (for example) grain size, calcium carbonate ($CaCO_3$), organic carbon or biomass content can be mapped across space using automated remote sensing techniques [21,33–35].

Although field and laboratory hyperspectral devices have been used to derive data on coastal sediment properties [16,28,36], there have been hitherto no published studies using the spectral properties of sediments from coastal settings in South Africa. This study uses laboratory hyperspectral measurements of sediment samples collected from a beach–dune system on the coast of South Africa, focusing on relationships between selected properties of the field samples (including grain size and carbonate content) and their associated spectral signatures. The aims of this study are to describe the nature of beach–dune samples in terms of their spectral signatures and to examine these relationships using statistical methods. This can be considered as a first step towards developing a robust methodology for automated mapping of sediment properties across beach–dune environments applicable globally.

2. Study Area and Methods

The study area examined, from which surface sediments were sampled, is at Oyster Bay, Eastern Cape Province, South Africa (Figure 1). Prevailing winds in this region are towards the northeast (in summer) and the west/northwest (in winter). Tidal range is high microtidal/low mesotidal and swell waves from the Southern Ocean have a significant wave height of >5 m [37]. Oyster Bay is an asymmetrical zeta-shape embayment [38] with an extensive sandy beach that is 6.1 km in total length and with a variable beach width of 30 to 290 m at low tide. Bedrock headlands to the east and west define the overall shape of the bay. An extensive supratidal zone is present, containing parallel-aligned transverse dunes with crests that are 40–50 m apart, similar to those found elsewhere along the South African coast [39]. Dune migration periodically blocks off the mouth of the incoming Klipdrift River. The landward boundary of the supratidal dunes at the back of the beach is marked by dune migration into a zone of highly vegetated and variably cemented linear palaeodunes that extend for ~40 km along this coastline. These ridges broadly correspond to the Nahoon Formation of the late Pleistocene Algoa Group, covering the period of marine isotope stages 5 to 2 inclusively [40–42]. Holocene-age dunes in this region, fronting the eroded older dunes, correspond to largely unvegetated foredunes of the Schelm Hoek Formation and are composed of unconsolidated calcareous aeolian sand [43].

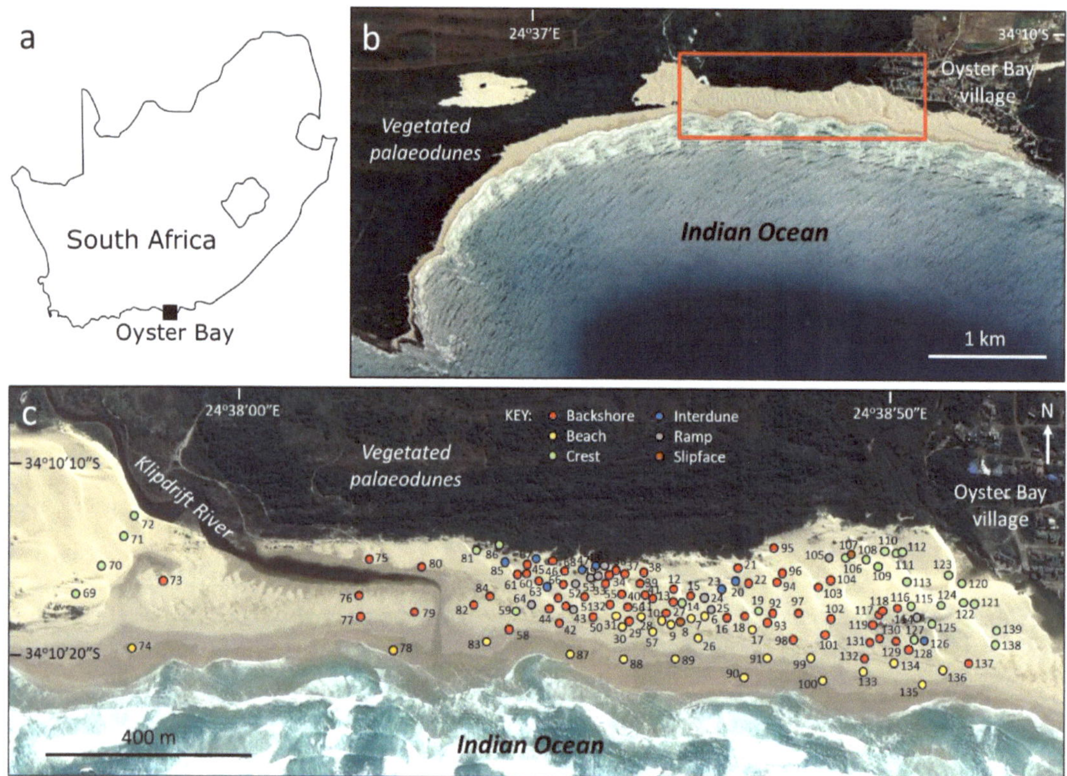

Figure 1. (a) Location of the study area at Oyster Bay, Eastern Cape, South Africa; (b) large-scale geomorphic setting of Oyster Bay with the sampling region (panel **c**) shown in the red box; (c) distribution of sediment sampling points 006–139 (background image in (**c**) from Google Earth, image date 25 August 2013, which is the latest available image before the sampling period).

Surficial (top 5 cm) sediment samples (~400 g each, *n* = 134, labelled 6–139) were collected in the field across the beach–supratidal dune system in the centre of the Oyster Bay embayment (Figure 1c). A random sampling approach was used but covering the full width of the beach including the intertidal zone. These samples were bagged, labelled, and sampling locations and their geomorphic settings marked using a Garmin Etrex 20 handheld GPS (*x y* accuracy ± 3 m). In addition, shells of different species (that were broken and did not contain organisms) were also collected from the intertidal zone. Samples were of the Cape brooding oyster (*Ostrea atherstoni*, sample 1), Brown mussel (*Perna perna*, sample 2), Agulhas ridged nut clam (*Lembulus belcheri*, sample 3), Southern cuttlefish (*Sepia australis*, sample 4), and a mixed shell sample combining these and other shell species found within the intertidal zone (sample 5) (Figure 2). In the laboratory, shell samples (*n* = 5) were dried and crushed using a pestle and mortar to generate broken fragments >2 mm diameter. Sediment samples were dried, sieved to remove the >2 mm fraction, and a subsample (~50 g) evaluated for $CaCO_3$ content using the loss on ignition method. In this method, the subsample was weighed, combusted in a muffle furnace for 5 h at 950 °C, and reweighed. Combustible $CaCO_3$ content (% of sample mass) was then calculated. Three replicates were undertaken for each sample, and the results averaged. Variation between the replicates was commonly <0.1%. The grain size distribution for each sample was measured using a Mastersizer 3000 Hydro EV for the size range 0.01–2000 µm with a subsample size of ~5 g. Each subsample was sonificated for 20 s prior to measurement, and five individual grain size distribution patterns were measured using the Mastersizer, and the average taken. The key grain size distribution parameters (D_{10}, D_{50}, D_{90}, kurtosis, skewness, standard deviation and mean) generated by the Mastersizer software were used for analysis. Additionally, a derived parameter herein called *span*, which describes the width of the particle size distribution, was calculated as

$$Span = \frac{(D_{90} - D_{10})}{D_{50}} \tag{1}$$

where D_{90} and D_{10} are the 90th and 10th percentile values of the grain size distribution, and D_{50} is the median grain size.

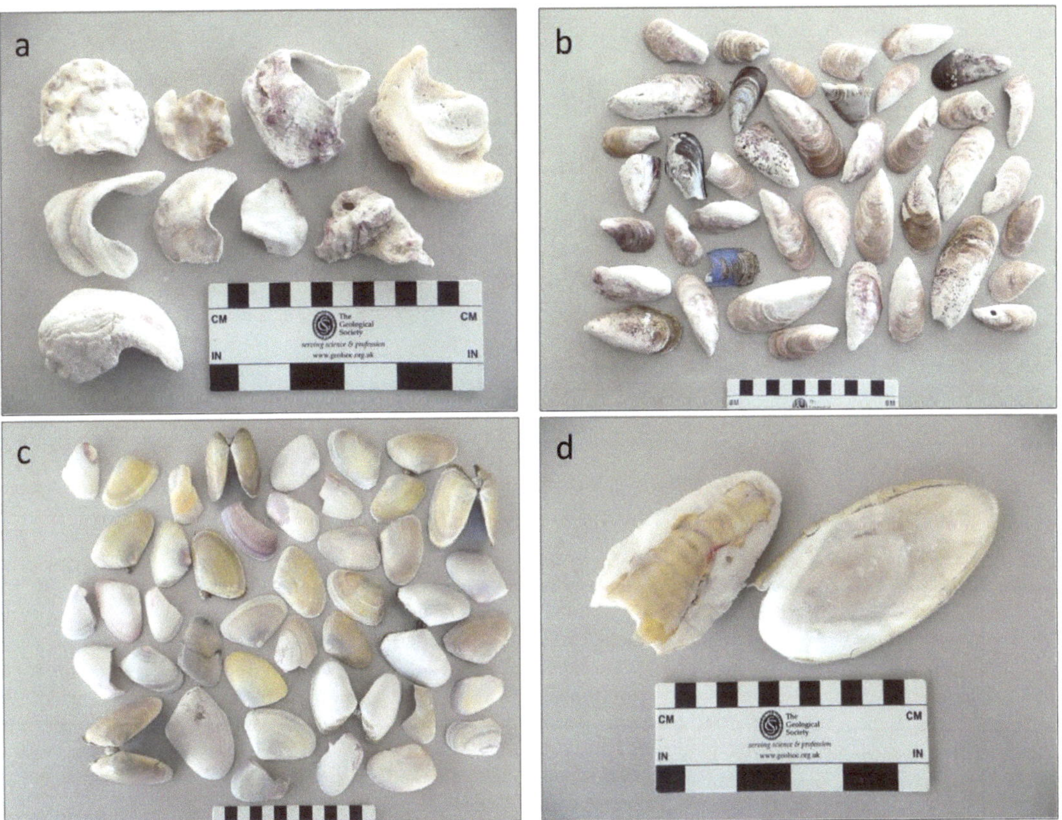

Figure 2. Shell samples collected from Oyster Bay. (**a**) Cape brooding oyster (*Ostrea atherstoni*, sample 1); (**b**) Brown mussel (*Perna perna*, sample 2); (**c**) Agulhas ridged nut clam (*Lembulus belcheri*, sample 3); (**d**) Southern cuttlefish (*Sepia australis*, sample 4).

The spectral signatures of sediment and shell samples were acquired under controlled environmental conditions in the laboratory using an Analytical Spectral Device (ASD)

FieldSpec®3spectrometer (ASD Inc., Boulder, CO, USA) and a light mug (Figure 3). The instrument measures wavelengths from 350 to 2500 nm and compares samples to a white reference panel. The 2 mm-sieved sediment and shell samples were placed in 100 mL glass bottles (Figure 3) and then placed on top of the light mug to measure the reflectance from each sample. Five spectral scans were captured for each sample to ensure spectral stability and an average reflectance was considered for further analysis. The spectrometer was calibrated using the white reference Spectralon® (Figure 3). The spectrometer was recalibrated after every 20 sample scans. The spectral measurement were stored in a notebook computer connected to the device. Figure 3 shows the laboratory setup used in this study.

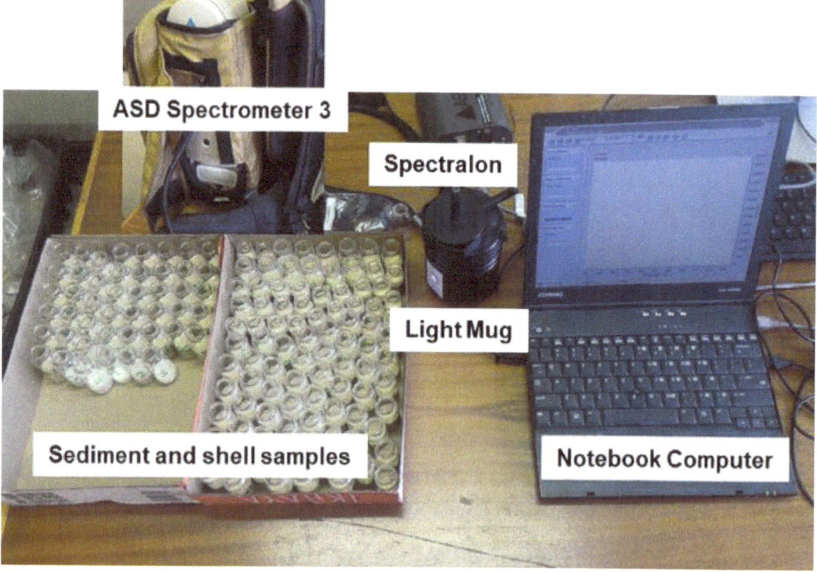

Figure 3. Laboratory spectral measurement system. The system involves an ASD spectroradiometer and a light mug device.

Following ASD data collection, the spectral data were first combined and converted from digital numbers to reflectance values using the ViewSpecPro® software. The spectral resolution was 1 nm which causes several data redundancy difficulties and affects the processing time. The spectral data also went through several pre-processing steps including removing noise-affected spectra located at the edge of the scans. First, the reflectance data less than 375 nm and greater than 2460 nm were removed from further analysis in order to disregard edge effects. Second, the moisture absorption spectral bands (at ~2500, 1950 and 1450 nm; [44]) were therefore eliminated from the final pre-processing stage. A correlation analysis was first performed on the data to identify the spectral wavelengths that show a significant association with different sediment properties. Based on the correlation analysis, specific wavelength regions were then selected. Linear regression analysis and partial least squares regression analysis were then performed on the selected spectral wavelengths based on the magnitude of the correlation coefficient. The data were partitioned into training and testing datasets. Approximately 70% (94 samples) of the dataset was used for the statistical model training, and 30% (40 samples) was used for model testing and validation.

3. Results

3.1. Site Geomorphology and Sediment Dynamics

Oyster Bay contains transgressive transverse dunes within the supratidal part of the beach (Figure 1b), and in the field these are observed to be asymmetric in profile and actively migrating towards the northeast, in the direction of the regional prevailing wind [45] and reflecting the relatively high sediment availability in Oyster Bay. The transgressive dunes show steep slipfaces (Figure 4b) and migrating free dunes over the beach surface (Figure 4c). Older vegetated dunes are left as residual eroded hummocks (Figure 4d).

Figure 4. Dune and beach morphology at Oyster Bay. (**a**) Dissipative beachface within the lower part of the beach system; (**b,c**) migrating transverse dune ridges within the supratidal zone; (**d**) erosional hummock of an older vegetated dune system, now isolated within the upper part of the beach.

3.2. Sediment Properties

Detailed laboratory analysis of sediment sample grain size data (Table 1) shows that the samples (n = 134) are remarkably uniform. In terms of texture, samples are dominantly (98%) medium sand with only one sample fine grained and two samples coarse grained. In terms of sorting, most samples (66%) are well sorted, 31% are moderately well sorted and 3% moderately sorted. For skewness, 98% are near symmetrical and 2% are coarse skewed. $CaCO_3$ values vary from 8.22% to 27.29%. For kurtosis, >99% are mesokurtic and only one sample is leptokurtic. The sediment samples were collected from backshore, beach, dune crest, ramp, slipface and interdune positions (Figure 1c). There are some statistically significant differences between grain size end-members (fine and coarse/very coarse sand) and $CaCO_3$ values between some of these sampling positions (Table 2), in particular in beach samples where wave action can contribute to sediment sorting and from supratidal dunes where wind transport leads to effective sediment sorting. As a result, there are some statistical differences between D_{10}, D_{90} and D_{50} values.

Table 1. Details of sediment samples examined in this study. Sample location codes are: BAB: Backshore, beach; BAC: Backshore, crest; BAI: Backshore, interdune; BAR: Backshore, ridge; BAS: Backshore, slipface; BAT: Backshore, trough; BEC: Beach, crest; BEI: Beach, interdune; BER: Beach, ridge; BES: Beach, slipface; BET: Beach, trough; CRI: Crest, interdune; CRR: Crest, ridge; CRS: Crest, slipface; CRT: Crest, trough; INT: Interdune, ridge; INS: Interdune, slipface; INT: Interdune, trough; RAS: Ramp, slipface; RAT: Ramp, trough; SLT: Slipface, trough. Sample numbers are indicated in Figure 1c.

Sample #	Location Code	Lat	Long	Elevation (m asl)	D_{10} (Microns)	D_{50} (Microns)	D_{90} (Microns)	Kurtosis	Skewness	$CaCO_3$ (%)	Fine sand (%)	Medium Sand (%)	Coarse Sand (%)	Very Coarse Sand (%)
006	BEC	−34.171938	24.642923	4.211	209	315	478	0.95	0.00	18.83	24.34	68.06	7.59	0.00
007	BET	−34.171984	24.642668	2.048	220	340	528	0.95	−0.01	13.78	18.57	68.07	13.35	0.00
008	SLT	−34.172052	24.642489	4.932	199	289	421	0.97	0.00	10.40	31.47	65.92	2.61	0.00
009	BES	−34.172016	24.642389	8.777	302	488	804	0.95	−0.02	21.40	3.16	49.36	44.73	2.75
010	BET	−34.172011	24.642128	7.816	226	350	547	0.96	−0.01	13.69	16.50	68.12	15.38	0.00
011	BET	−34.171679	24.641879	5.172	201	323	538	0.98	−0.04	16.20	25.10	61.41	11.89	0.88
012	BEC	−34.171517	24.642276	11.661	204	307	465	0.94	0.00	17.04	26.57	66.99	6.44	0.00
013	BEI	−34.171702	24.642363	11.661	198	288	420	0.97	0.00	9.89	31.79	65.61	2.60	0.00
014	CRS	−34.171690	24.642476	9.739	279	442	715	0.96	−0.02	21.98	5.30	57.36	36.33	1.01
015	SLT	−34.171666	24.642576	8.296	203	303	453	0.95	−0.01	12.30	27.39	67.18	5.43	0.00
016	BET	−34.171933	24.643384	6.134	220	340	529	0.95	−0.01	14.58	18.69	67.89	13.42	0.00
017	BEC	−34.172137	24.643902	9.739	254	392	609	0.96	−0.01	19.09	9.01	66.63	24.29	0.06
018	BEC	−34.171929	24.643805	7.816	207	311	470	0.95	−0.01	16.44	25.28	67.90	6.82	0.00
019	CRS	−34.171833	24.644029	12.382	248	382	589	0.95	−0.01	16.60	10.45	67.69	21.83	0.03
020	BET	−34.171374	24.643891	5.893	189	287	440	0.96	−0.01	12.66	34.09	61.50	4.41	0.00
021	BEC	−34.171120	24.643643	9.017	214	321	485	0.96	0.00	19.35	22.28	69.53	8.19	0.00
022	INR	−34.171363	24.643649	7.095	228	374	631	0.96	−0.03	18.76	15.15	60.88	23.40	0.50
023	INR	−34.171499	24.643289	9.498	217	351	583	0.97	−0.04	16.97	18.85	62.41	17.85	0.56
024	RAS	−34.171600	24.642989	9.979	218	343	543	0.96	−0.01	16.90	19.06	66.14	14.79	0.00
025	RAS	−34.171869	24.643010	6.134	204	318	503	0.95	−0.02	12.93	25.10	64.54	10.36	0.00
026	BAB	−34.172203	24.642861	8.537	199	302	459	0.95	−0.01	16.21	28.53	65.40	6.06	0.00

Table 1. *Cont.*

Sample #	Location Code	Lat	Long	Elevation (m asl)	D_{10} (Microns)	D_{50} (Microns)	D_{90} (Microns)	Kurtosis	Skewness	CaCO$_3$ (%)	Fine sand (%)	Medium Sand (%)	Coarse Sand (%)	Very Coarse Sand (%)
027	BAT	−34.171951	24.642091	1.087	219	339	528	0.95	−0.01	16.78	19.19	67.50	13.30	0.00
028	BAT	−34.171980	24.641696	6.134	226	372	631	0.97	−0.03	21.13	15.64	60.75	22.84	0.70
029	BAC	−34.171961	24.641441	7.095	211	314	472	0.95	0.00	14.73	23.86	69.23	6.91	0.00
030	BAT	−34.171942	24.641264	5.893	200	311	490	0.96	−0.02	13.61	26.92	64.19	8.89	0.00
031	BEC	−34.171955	24.641151	5.653	221	344	530	0.95	0.01	19.22	18.14	68.19	13.67	0.00
032	BES	−34.171565	24.641033	6.134	261	416	669	0.95	−0.01	20.08	7.88	60.64	30.94	0.54
033	BEC	−34.171277	24.640809	3.250	187	274	402	0.95	0.00	22.41	38.30	59.87	1.83	0.00
034	BAT	−34.171125	24.640865	8.296	202	307	466	0.95	0.00	17.58	27.05	66.48	6.47	0.00
035	RAS	−34.171038	24.640841	7.816	219	336	514	0.95	0.00	12.13	19.35	68.71	11.94	0.00
036	BAC	−34.171123	24.640999	5.653	210	316	481	0.96	0.00	20.16	23.92	68.25	7.83	0.00
037	BAT	−34.171193	24.641281	0.125	254	466	1020	1.06	−0.13	26.58	9.31	46.06	34.29	8.86
038	BAC	−34.171223	24.641788	4.692	191	284	425	0.96	0.00	19.23	34.60	62.35	3.04	0.00
039	BAR	−34.171376	24.641656	6.374	180	264	388	0.96	−0.01	12.38	42.81	55.96	1.22	0.00
040	BAT	−34.171621	24.641768	4.451	213	327	508	0.95	−0.01	14.72	21.97	67.04	11.00	0.00
041	BAC	−34.171713	24.641632	7.335	239	366	565	0.95	−0.01	18.44	13.12	68.71	18.16	0.00
042	BAB	−34.172004	24.639971	1.808	225	363	589	0.95	−0.01	14.92	16.26	63.45	20.20	0.09
043	BAC	−34.171775	24.639973	6.854	235	340	488	0.95	0.02	18.42	14.63	77.03	8.34	0.00
044	BAC	−34.171648	24.639890	9.258	219	328	494	0.96	−0.01	15.84	19.78	71.01	9.20	0.00
045	BAS	−34.171388	24.639973	8.296	253	393	611	0.96	−0.01	19.06	9.30	66.09	24.55	0.06
046	BAC	−34.171020	24.639773	10.700	213	322	490	0.96	0.00	19.00	22.25	68.96	8.79	0.00
047	BAS	−34.171247	24.640023	5.413	203	310	481	0.96	−0.02	10.27	26.40	65.66	7.94	0.00
048	RAT	−34.171330	24.640202	6.374	206	318	495	0.95	−0.01	15.96	24.55	66.09	9.36	0.00
049	RAS	−34.171756	24.640200	6.134	239	391	671	0.97	−0.05	17.20	12.77	59.31	26.55	1.37
050	BAC	−34.171887	24.640709	5.653	222	350	556	0.96	−0.01	17.06	17.60	66.07	16.32	0.01

Table 1. *Cont.*

Sample #	Location Code	Lat	Long	Elevation (m asl)	D_{10} (Microns)	D_{50} (Microns)	D_{90} (Microns)	Kurtosis	Skewness	CaCO$_3$ (%)	Fine sand (%)	Medium Sand (%)	Coarse Sand (%)	Very Coarse Sand (%)
051	BAC	−34.171602	24.640481	3.730	217	359	625	0.99	−0.06	20.63	18.40	59.74	20.10	1.31
052	RAT	−34.171232	24.640564	3.971	176	248	347	0.96	−0.01	15.27	51.21	48.62	0.17	0.00
053	RAT	−34.171263	24.640642	7.816	192	274	391	0.96	0.00	16.61	37.27	61.61	1.12	0.00
054	INT	−34.171117	24.640536	6.374	201	300	448	0.96	0.00	15.57	28.33	66.84	4.84	0.00
055	BAT	−34.171473	24.641231	5.893	207	319	498	0.95	−0.01	14.82	24.19	66.06	9.75	0.00
056	BAT	−34.171814	24.641265	7.335	205	326	530	0.95	−0.03	14.89	23.92	62.98	13.00	0.07
057	BAB	−34.172157	24.641878	3.730	197	289	427	0.97	0.00	13.48	31.65	65.29	3.06	0.00
058	BAB	−34.171930	24.639000	6.374	226	350	549	0.96	−0.01	15.02	16.61	67.71	15.67	0.00
059	CRI	−34.171679	24.639045	7.576	207	306	454	0.95	−0.01	17.67	26.14	68.32	5.54	0.00
060	BAC	−34.171248	24.639070	8.537	214	344	568	0.96	−0.03	20.12	20.34	62.49	17.06	0.11
061	BAC	−34.171124	24.639230	6.374	243	443	1240	1.09	−0.25	24.65	11.32	46.87	27.58	11.11
062	BAT	−34.171242	24.639314	0.606	195	291	438	0.96	−0.01	9.97	31.81	64.08	4.11	0.00
063	MAI	−34.171381	24.639556	5.893	219	345	551	0.96	−0.02	17.32	18.71	65.64	15.63	0.01
064	RAT	−34.171685	24.639376	4.451	217	354	620	1.00	−0.07	19.63	18.90	60.14	18.94	1.53
065	BAT	−34.171786	24.639684	7.816	216	329	504	0.96	0.00	16.76	20.70	68.83	10.47	0.00
066	INT	−34.171465	24.639731	3.730	214	351	616	0.99	−0.07	15.92	19.78	59.69	18.78	1.41
067	INR	−34.170996	24.639434	13.103	208	324	510	0.95	−0.01	18.77	23.52	65.21	11.27	0.00
068	INS	−34.171154	24.640406	8.777	219	327	488	0.96	0.00	18.75	19.95	71.52	8.53	0.00
069	CRI	−34.171594	24.629845	−5.641	238	406	722	0.98	−0.04	23.67	12.58	55.56	29.39	2.34
070	CRI	−34.171147	24.630392	12.863	246	396	639	0.96	0.00	18.05	11.02	61.88	26.90	0.21
071	CRI	−34.170616	24.630827	12.622	252	388	594	0.96	0.00	19.25	6.45	67.59	22.95	0.01
072	CRS	−34.170290	24.631101	10.700	262	392	583	0.95	0.00	19.47	7.53	70.20	22.27	0.01
073	BAR	−34.171394	24.631663	8.296	195	302	474	0.96	−0.02	18.67	29.59	6.97	7.43	0.00
074	BER	−34.172438	24.631051	−0.354	223	334	498	0.96	0.00	14.11	18.15	72.14	9.72	0.00

Table 1. *Cont.*

Sample #	Location Code	Lat	Long	Elevation (m asl)	D_{10} (Microns)	D_{50} (Microns)	D_{90} (Microns)	Kurtosis	Skewness	CaCO$_3$ (%)	Fine sand (%)	Medium Sand (%)	Coarse Sand (%)	Very Coarse Sand (%)
075	BAR	−34.170938	24.635953	6.134	220	358	581	0.95	0.00	22.10	17.65	62.85	19.47	0.03
076	BAT	−34.171616	24.635743	10.459	196	287	423	0.97	0.00	11.41	32.60	64.59	2.81	0.00
077	BAC	−34.171935	24.635784	4.692	215	331	512	0.95	−0.01	17.54	20.85	67.59	11.56	0.00
078	BER	−34.172500	24.636502	5.653	219	337	521	0.95	−0.01	14.71	19.23	68.12	12.65	0.00
079	BAT	−34.171834	24.636919	8.537	241	361	545	0.96	−0.01	13.34	12.68	71.64	15.68	0.00
080	BAR	−34.171110	24.637048	8.777	213	343	562	0.96	−0.02	20.47	20.49	62.92	16.54	0.05
081	CRS	−34.170847	24.638189	13.584	253	432	808	0.99	−0.08	27.25	9.38	53.07	33.00	4.49
082	BAB	−34.171690	24.638149	4.932	234	354	530	0.94	0.01	15.92	14.35	71.73	13.92	0.00
083	BAB	−34.172336	24.638428	7.335	239	429	865	0.99	−0.10	25.33	12.28	49.43	31.89	6.09
084	BAI	−34.171595	24.638524	2.048	197	294	442	0.96	−0.01	20.45	30.71	64.88	4.41	0.00
085	INR	−34.171029	24.638807	10.459	236	371	584	0.94	−0.01	20.34	13.49	65.94	20.54	0.04
086	CRI	−34.170733	24.638722	13.584	190	275	396	0.96	0.00	15.50	37.34	61.22	1.44	0.00
087	BET	−34.172485	24.640166	5.413	196	281	400	0.96	0.00	12.29	34.09	64.39	1.51	0.00
088	BET	−34.172541	24.641305	5.653	234	385	642	0.95	−0.01	19.48	13.60	60.30	25.74	0.36
089	BET	−34.172570	24.642399	1.567	275	436	699	0.95	−0.01	19.73	5.96	58.23	35.16	0.65
090	BET	−34.172851	24.643780	5.893	336	517	801	0.96	−0.01	24.00	1.02	45.16	51.74	2.09
091	BAB	−34.172520	24.644298	4.451	204	296	430	0.96	0.00	12.54	28.48	68.45	3.07	0.00
092	BAS	−34.172002	24.644281	6.134	193	272	385	0.96	0.00	8.26	37.81	61.32	0.87	0.00
093	BAS	−34.171981	24.644318	7.335	200	297	443	0.96	−0.01	8.86	29.30	66.25	4.45	0.00
094	BAC	−34.171375	24.644495	9.739	256	403	643	0.96	−0.01	17.99	8.82	63.06	27.87	0.25
095	BAR	−34.170761	24.644425	8.296	224	338	511	0.96	−0.01	17.33	17.64	70.80	11.57	0.00
096	BAT	−34.171188	24.644581	6.614	215	338	553	0.97	−0.04	13.67	20.53	64.14	14.33	0.54
097	BAI	−34.171828	24.644937	7.816	213	330	518	0.95	−0.01	13.86	21.65	66.14	12.21	0.00
098	BAT	−34.172261	24.644822	5.413	201	289	416	0.96	0.00	8.22	30.99	66.75	2.26	0.00

Table 1. *Cont.*

Sample #	Location Code	Lat	Long	Elevation (m asl)	D_{10} (Microns)	D_{50} (Microns)	D_{90} (Microns)	Kurtosis	Skewness	CaCO$_3$ (%)	Fine sand (%)	Medium Sand (%)	Coarse Sand (%)	Very Coarse Sand (%)
099	BAB	−34.172608	24.645214	6.134	217	329	504	0.96	−0.01	14.48	20.40	69.15	10.46	0.00
100	BER	−34.172899	24.645462	4.451	249	375	565	0.96	0.00	13.02	10.23	71.10	18.67	0.00
101	BAC	−34.172244	24.645432	8.056	235	353	535	0.95	−0.01	18.51	14.37	71.23	14.40	0.00
102	BAC	−34.171919	24.645569	9.258	225	349	541	0.96	0.00	21.15	16.78	68.35	14.88	0.00
103	BAS	−34.171447	24.645305	10.700	294	479	796	0.96	−0.02	22.23	3.75	50.59	42.96	2.71
104	BAT	−34.171306	24.645469	7.576	227	355	564	0.96	−0.02	14.56	16.22	66.43	17.32	0.02
105	RAS	-34.296997	24.645517	10.700	219	340	532	0.95	−0.01	18.10	18.95	67.40	13.65	0.00
106	CRS	−34.170927	24.645926	10.459	250	444	1030	1.11	−0.19	24.17	9.93	48.97	30.52	8.51
107	SLT	−34.170939	24.645957	9.498	214	314	460	0.95	0.00	10.60	22.94	71.26	5.80	0.00
108	CRR	−34.170890	24.646282	10.459	229	361	571	0.96	0.00	17.58	15.24	66.17	18.57	0.02
109	CRR	−34.171110	24.646571	11.901	218	351	573	0.96	−0.01	20.53	18.70	63.19	18.05	0.05
110	CRT	−34.170858	24.646711	7.335	195	293	442	0.96	0.00	12.43	31.15	64.41	4.41	0.00
111	CRS	−34.170820	24.646927	9.258	250	392	625	0.97	−0.02	20.51	10.04	64.54	25.21	0.21
112	CRT	−34.170819	24.647034	4.451	209	322	503	0.95	−0.01	11.89	23.57	66.05	10.37	0.00
113	CRR	−34.171350	24.647209	13.103	213	316	469	0.95	0.00	16.22	22.98	70.41	6.61	0.00
114	RAS	−34.171908	24.647452	19.832	179	265	391	0.95	0.00	12.67	42.77	55.89	1.34	0.00
115	CRR	−34.171724	24.647273	21.995	190	272	388	0.96	0.00	16.95	38.57	60.42	1.00	0.00
116	BAC	−34.171782	24.646935	20.313	204	308	467	0.95	−0.01	20.51	26.57	66.85	6.58	0.00
117	BAS	−34.171910	24.646599	12.622	200	295	437	0.96	0.00	12.42	29.66	66.46	3.88	0.00
118	BAC	−34.171930	24.646544	17.910	220	327	489	0.96	0.00	18.03	19.74	71.69	8.57	0.00
119	BAC	−34.172095	24.646437	12.622	244	372	571	0.96	−0.01	19.95	11.50	69.24	19.25	0.01
120	CRS	−34.171342	24.648271	4.451	197	280	393	0.96	0.00	8.99	34.25	64.66	1.09	0.00
121	CRS	−34.171682	24.648521	11.421	202	309	476	0.95	−0.01	13.09	26.68	65.85	7.47	0.00
122	CRS	−34.171691	24.648334	13.824	203	288	407	0.96	0.00	16.36	30.69	67.61	1.70	0.00

Table 1. *Cont.*

Sample #	Location Code	Lat	Long	Elevaticn (m asl)	D_{10} (Microns)	D_{50} (Microns)	D_{90} (Microns)	Kurtosis	Skewness	CaCO$_3$ (%)	Fine sand (%)	Medium Sand (%)	Coarse Sand (%)	Very Coarse Sand (%)
123	CRS	−34.171218	24.648048	13.103	266	422	678	0.94	−0.01	19.95	7.19	60.04	32.25	0.53
124	CRS	−34.171781	24.647969	22.716	189	264	372	0.96	0.01	19.57	41.90	57.63	0.47	0.00
125	CRS	−34.172032	24.647721	16.948	188	263	371	0.96	0.01	12.07	42.40	57.15	0.45	0.00
126	INR	−34.172320	24.647587	12.863	201	295	432	0.96	0.00	11.01	29.49	67.17	3.34	0.00
127	CRS	−34.172309	24.647358	16.468	301	466	733	0.96	−0.02	20.23	2.83	54.98	41.19	1.09
128	BAT	−34.172380	24.647111	9.979	214	330	512	0.95	−0.01	13.46	21.34	67.11	11.56	0.00
129	BAC	−34.172293	24.646948	12.382	213	313	458	0.95	0.00	17.74	23.50	70.80	5.70	0.00
130	BAT	−34.172286	24.646591	10.940	217	325	488	0.96	0.00	14.41	20.83	70.66	8.52	0.00
131	BAC	−34.172367	24.646384	10.459	212	317	478	0.96	0.00	14.05	23.17	69.29	7.54	0.00
132	BAB	−34.172595	24.646253	4.932	220	330	497	0.96	0.00	12.53	19.34	71.00	9.66	0.00
133	BAB	−34.172955	24.646132	6.854	285	500	941	0.96	−0.06	27.29	5.15	44.90	41.99	7.86
134	BAB	−34.172679	24.646923	4.692	208	312	471	0.95	−0.01	14.71	25.00	68.10	6.90	0.00
135	BER	−34.173066	24.647486	6.614	250	373	558	0.96	0.00	14.17	10.10	72.04	17.87	0.00
136	BAB	−34.172787	24.647920	5.172	216	316	463	0.96	0.00	13.70	22.17	71.79	6.04	0.00
137	BAT	−34.172667	24.648487	3.490	221	331	494	0.96	0.00	14.50	18.94	71.83	9.24	0.00
138	CRS	−34.172431	24.648988	6.614	229	333	485	0.96	0.00	16.86	16.69	75.25	8.06	0.00
139	CRT	−34.172160	24.649022	4.451	210	293	408	0.97	0.01	13.16	27.66	70.87	1.48	0.00

Table 2. Table of p-values obtained for different sediment properties from different sampling locations (Table 1), analysed using Fisher's partial least squares discriminant analysis. Significance levels are: ^ 0.1, * 0.05, ** 0.01 and *** 0.001.

Location	D_{10}	D_{50}	D_{90}	CaCO$_3$	Fine Sand	Medium Sand	Coarse Sand	Very Coarse Sand
Backshore, beach	0.0002 ***	0.0004 ***	0.0041 **	0.0455 *	0.0120 *	0.2313	0.0003 **	0.0148 *
Backshore, crest	0.0265 *	0.0318 *	0.0475 *	0.0005 **	0.1433	0.9129	0.0221 *	0.1460
Backshore, interdune	0.8668 ^	0.5823	0.5154	0.9509	0.7269	0.8188	0.3936	0.8481
Backshore, ramp	0.1903	0.3554	0.6308	0.9566	0.0472 *	0.5359	0.7108	0.9965
Backshore, slip face	0.4271	0.4077	0.5417	0.0813 ^	0.2634	0.7330	0.4244	0.9953
Backshore, trough	0.4534	0.8241	0.7296	0.0782 ^	0.3678	0.8872	0.9339	0.4140
Beach, crest	0.0113 *	0.0144 *	0.0817 ^	0.5547	0.0814 ^	0.1914	0.0192 *	0.0967 ^
Beach, interdune	0.0038 **	0.0141 *	0.0757 ^	0.1113	0.0803 ^	0.2343	0.0283 *	0.0674 ^
Beach, ramp	<0.0001 ***	<0.0001 ***	0.0022 *	0.0705 ^	<0.0001 ***	0.5811	0.0003 **	0.0241 *
Beach, slip face	0.0069 **	0.0076 **	0.0365 *	0.0067 **	0.0172 *	0.3404	0.0081 *	0.1989
Beach, trough	<0.0001 *	<0.0001 ***	0.0032 **	0.0002 ***	0.0004 **	0.1425	<0.0001 ***	0.0379 *
Crest, interdune	0.1626	0.3588	0.4935	0.0148 *	0.5158	0.7372	0.4968	0.4123
Crest, ramp	0.0007 ***	0.0036 **	0.0235 *	0.0031 **	0.0004 **	0.5245	0.0163 *	0.2081
Crest, slip face	0.0879 ^	0.0868 ^	0.1549	0.0015 **	0.0819 ^	0.6903	0.0820 ^	0.5586
Crest, trough	0.0001 ***	0.0019 **	0.0328 *	<0.0001 ***	0.0018 **	0.7333	0.0039 *	0.4294
Interdune, ramp	0.2232	0.2070	0.3184	0.9891	0.0532 ^	0.4740	0.2720	0.8547
Interdune, slip face	0.4023	0.2783	0.3484	0.0957 ^	0.2141	0.8496	0.2197	0.9095
Interdune, trough	0.4658	0.4445	0.6438	0.1765	0.2997	0.8822	0.3813	0.7004
Ramp, slip face	0.9154	0.7360	0.7257	0.0839 ^	0.9405	0.5169	0.5545	0.9972
Ramp, trough	0.4020	0.3912	0.3970	0.1098	0.1386	0.4083	0.6216	0.4729
Slip face, trough	0.6099	0.4454	0.4397	0.2822	0.4334	0.7691	0.3904	0.7306

Analysis of the covariation between different sediment properties shows that there are statistically significant relationships between several property types (Table 3). The properties that refer specifically to dimensional values of the grain size distribution (D_{10}, D_{50} and D_{90}) show evidence for very high correlation coefficients (<0.95) which is indicative of autocorrelation. Dimensionless parameters of skewness and kurtosis show more variable relationships but are also relatively strongly correlated (both positively and negatively) with grain size variables. The nondimensional parameter *span* broadly expresses the distributional range of particle sizes within the sample (Equation (1)) and thus has a high correlation coefficient with distributional parameters (Table 3). The independent parameter of CaCO$_3$ content shows a strong positive (negative) relationship with coarse (fine) sand because of the mechanical break up of marine shells over time, forming relatively large shell fragments mixed in with coarse mineral sand [46].

Averaged spectral characterisation of sediment samples from different geomorphic positions at Oyster Bay are presented in Figure 5. There are generally similar patterns seen at all positions, consistent with their generally similar sediment grain size compositions (Table 2), with some consistent variability in the water absorption bands. There is greatest variability in particular within the SWIR at ~1850–2400 nm. It is also notable that beach samples show somewhat more variability than samples from other positions, with higher reflectance values (compared to other positions) in the VNIR and lower values in the SWIR (Figure 5a). Based on the high correlation coefficients of beach samples with fine sand and CaCO$_3$ values (Table 2), we therefore speculate that this spectral variability of beach samples reflects the disproportionate influence of fine sand and CaCO$_3$ from shell fragments within these samples. The nature of these samples are now explored in more detail.

Table 3. Pearson Product Moment Correlation of different sediment properties across all samples (n = 134). F = fine, M = medium, C = coarse and VC = very coarse. Variable *span* is defined in the text. Significance levels are: ^ 0.1, * 0.05, ** 0.01 and *** 0.001.

	D_{10}	D_{50}	D_{90}	Kurtosis	Skewness	$CaCO_3$	F Sand	M Sand	C Sand	VC Sand
D_{10}	1.000									
D_{50}	0.951 ***	1.000								
D_{90}	0.754 ***	0.903 ***	1.000 ***							
Kurtosis	0.139	0.318 ***	0.634 ***	1.000						
Skewness	−0.252 **	−0.477 ***	−0.793 ***	−0.890 ***	1.000					
$CaCO_3$	0.571 ***	0.679 ***	0.698 ***	0.337 ***	−0.463 ***	1.000				
F sand	−0.914 ***	−0.918 ***	−0.761 ***	−0.156 ^	0.300 ***	−0.557 ***	1.000			
M sand	−0.201 *	−0.332 ***	−0.451 ***	−0.383 ***	0.478 ***	−0.411 ***	0.030	1.000		
C sand	0.928 ***	0.984 ***	0.872 ***	0.266 **	−0.443 ***	0.679 ***	−0.880 ***	−0.393 ***	1.000	
VC sand	0.414 ***	0.603 ***	0.853 ***	0.813 ***	−0.895 ***	0.566 ***	−0.367 ***	−0.529 ***	0.545 ***	1.000
Span	0.403 ***	0.642 ***	0.902 ***	0.782 ***	−0.941 ***	0.593 ***	−0.493 ***	−0.469 ***	0.616 ***	0.885 ***

3.3. Spectral Analysis of Sediment Samples

The spectral variation at the VNIR and SWIR bands can be examined in detail using the correlation matrices between selected sample particle size characteristics and $CaCO_3$ content. Here, we systematically calculate the correlation coefficient of D_{10}, $CaCO_3$, D_{90}, fine sand and kurtosis at 1 nm wavelength increments through the NVIR and SWIR wavebands (Figure 6). This shows that certain parts of the spectrum are associated with greater (positive or negative) correlation coefficient values and thus are more useful in terms of discriminating between different sediment properties at those wavelengths. For example, in the range ~700–1350 nm there is a clear discrimination between high positive correlations for D_{10} and D_{90} and high negative correlation for fine sand (Figure 6). Here, kurtosis and $CaCO_3$ shows no correlation. Likewise, in the range ~1850–2450 nm there is greater statistical discrimination between fine sand (highest positive correlation) and $CaCO_3$ (highest negative correlation) values. By contrast, the region ~1450–1700 nm is not useful for discriminating any sediment properties, because there are very low correlation coefficients throughout (i.e., all correlation coefficients are around zero).

The statistical relationships of $CaCO_3$ values, fine sand and grain size *span* to different wavelengths are described in Table 4, which shows the outputs of a linear regression model for each variable. The results highlight that certain wavelengths have a statistically significant relationship to some sediment properties. For example, $CaCO_3$ shows greatest significance in the wavelength range ~1052–1252 nm, which falls within the VNIR part of the spectrum. Fine sand has the greatest significance at shorter VNIR wavelengths (~852–952 nm), and *span* shows significance in isolated parts of the spectrum (2300, 2400 and 2447 nm) at the end of the SWIR range, which may be an artefact of sediment composition within the sample as a whole, e.g., [47]. Water absorption at the 1352 nm waveband has a strong signal and therefore this waveband is removed from the analysis (Table 4) in order to avoid erroneous overfitting.

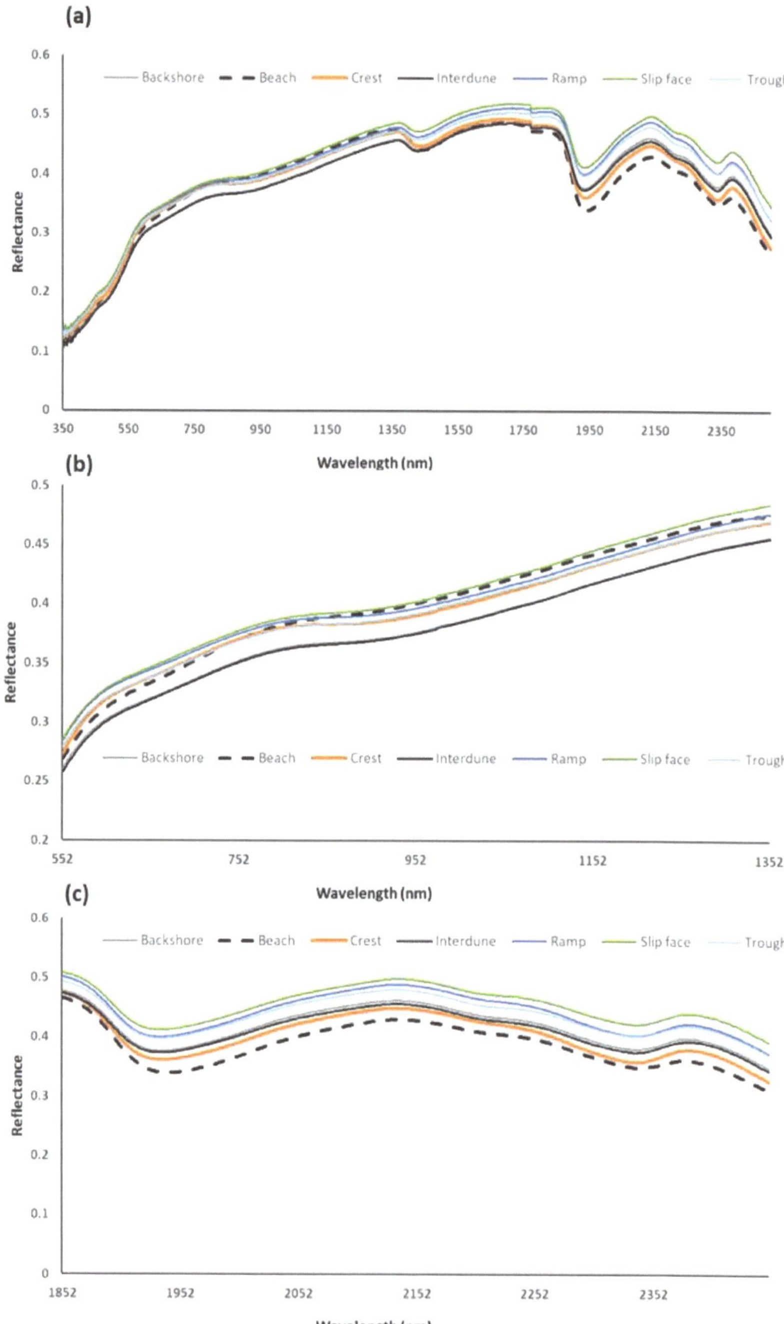

Figure 5. Results of spectral analysis of samples from different geomorphic positions at Oyster Bay. (**a**) Full average spectrum for samples from the different position; detailed results at (**b**) the VNIR (552–1352 nm) and (**c**) SWIR (1852–2450 nm) parts of the spectrum.

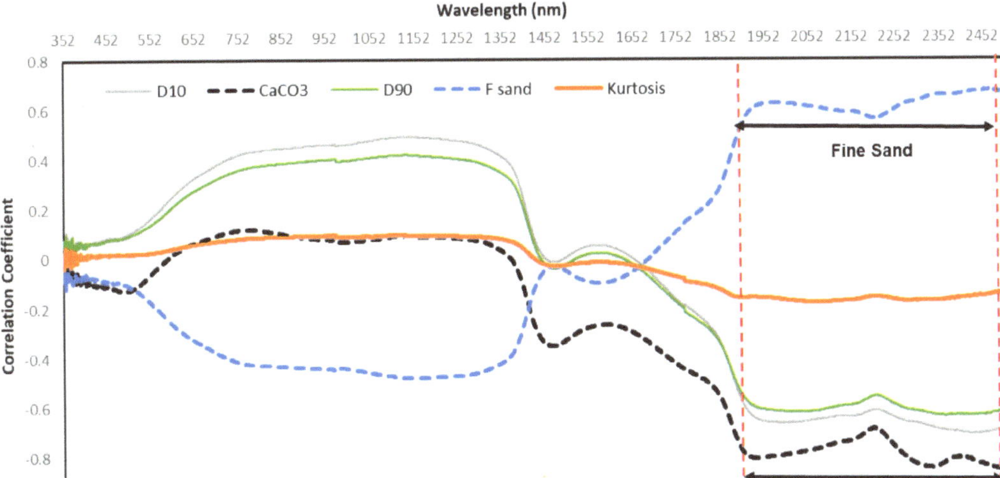

Figure 6. Correlation coefficient between reflectance values at different spectral wavelengths and sediment $CaCO_3$, and particle size characteristics.

Table 4. Analysis of *p*-values of linear regression of $CaCO_3$, fine sand and grain size *span* with selected wavelengths (see Figure 5). Significance levels are: ^ 0.1, * 0.05, ** 0.01 and *** 0.001.

Wavelength (nm)	$CaCO_3$	Fine Sand	Span
552	0.0928 ^	0.6264	0.9885
652	0.2507	0.9883	0.4891
752	0.7211	0.0928 ^	0.0734 ^
852	0.7343	0.0096 **	0.1257
952	0.9410	0.0026 **	0.2164
1052	0.0077 **	0.7944	0.3640
1152	0.0100 *	0.9415	0.2201
1252	0.0087 **	0.4923	0.1494
1462	0.0834 ^	0.2163	0.3150
1552	0.3162	0.8246	0.2073
1652	0.2605	0.9801	0.1364
1752	0.1525	0.8957	0.1300
1789	0.1064	0.0480 *	0.0158 *
1962	0.0264 *	0.7977	0.0528 ^
2028	0.4413	0.4095	0.2356
2082	0.9618	0.5296	0.6334
2152	0.4212	0.4293	0.7909
2200	0.5897	0.5675	0.6655
2252	0.3046	0.0789 ^	0.3075
2300	0.0098 **	0.4795	0.0085 **
2335	0.1032	0.8612	0.8394

Table 4. *Cont.*

Wavelength (nm)	CaCO$_3$	Fine Sand	Span
2338	0.1530	0.8069	0.7031
2350	0.7137	0.0696 ^	0.4356
2352	0.2443	0.0253 *	0.9282
2370	0.3290	0.2500	0.7780
2400	0.3596	0.3150	0.0041 **
2420	0.3237	0.1892	0.2231
2435	0.2855	0.0127 *	0.7258
2447	0.6500	0.0760 ^	0.0024 **
2450	0.2281	0.0470 *	0.5175
Adjusted R^2	0.8978	0.7510	0.5510
Overall *p*-value	2.2×10^{-16} ***	2×10^{-16} ***	4.88×10^{-16} ***

CaCO$_3$ values can be estimated using single wavelength relationships as:

$$CaCO_3 = 54.1 - 98.37 R_{2335} \left(R^2 = 0.718, \ n = 94 \right) \tag{2}$$

where R_{2335} is reflectance at the 2335 nm wavelength. A similar model performance can be achieved at the wavelength between 2038 and 2435 nm. A multilinear relationship can improve the model estimation as follows:

$$CaCO_3 = 38.99 + 460.23 R_{2200} - 1272.89 R_{2300} + 1158.99 R_{2335} - 409.99 R_{2370}$$
$$\left(R^2 = 0.872, \ n = 94 \right) \tag{3}$$

where R_{2200}, R_{2300}, R_{2335}, and R_{2335} are reflectances at the 2200, 2300, 2335 and 2370 nm wavelengths, respectively. Fine sand can be estimated using the following linear relationship:

$$F \ Sand = 179.06 R_{2450} - 41.01 \left(R^2 = 0.463, \ n = 94 \right) \tag{4}$$

where $F \ Sand$ is the fine sand percentage, and R_{2450} is reflectance at the 2450 nm wavelength. Fine sand values can also be estimated using the reflectance from the wavelengths ranging between 552 and 1789 nm. This can also be used to develop an improved model to estimate the fine sand percentage as follows:

$$F \ Sand = 21.55 - 364.86 R_{1462} + 369.07 R_{2082} \left(R^2 = 0.721, \ n = 94 \right) \tag{5}$$

where $F \ Sand$ is the fine sand percentage and R_{1462} and R_{2082} are reflectances at the 1462 and 2082 nm wavelengths, respectively. *Span* had very poor performing models when a single band was utilised. For example, the following model was the best performing single-wavelength model:

$$Span = 2.15 - 2.43 R_{2038} \left(R^2 = 0.183, \ n = 94 \right) \tag{6}$$

where R_{2038} is the reflectance at the 2038 nm wavelength. A multilinear model can be developed using highly correlated wavelengths ranging between 852 and 2450 nm, as shown below:

$$Span = 1.68 - 24.66 R_{852} + 29.62 R_{952} + 59.44 R_{2252} - 131.36 R_{2300} + 79.47 R_{2350} -$$
$$73.27 R_{2400} + 62.52 R_{2447} \left(R^2 = 0.516, \ n = 94 \right) \tag{7}$$

where R_{852}, R_{952}, R_{2252}, R_{2300}, R_{2350}, R_{2400}, and R_{2447} are reflectances at the 852, 952, 2252, 2300, 2350, 2400 and 2447 nm wavelengths, respectively.

Fine sand and *span* are not well predicted using single wavelength models (see Table 4), whereas $CaCO_3$ shows a much stronger relationship. The models were then validated using 40 independent samples that were not used in the model development (Figure 7). It is notable that a multilinear model leads to a better fit between measured and estimated values.

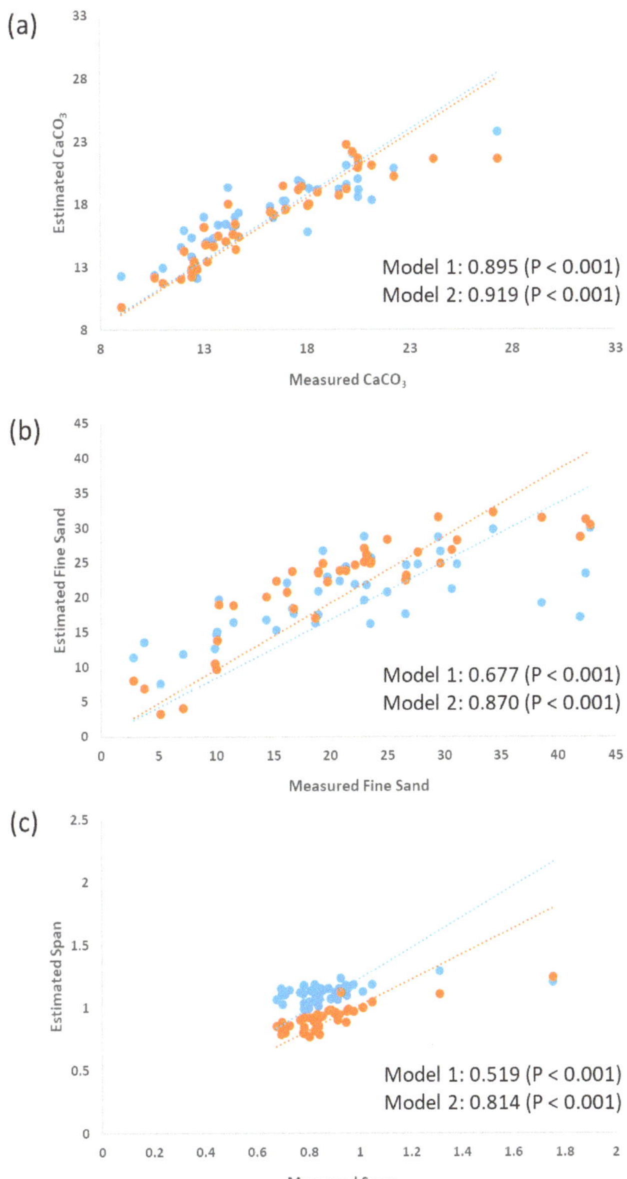

Figure 7. Relationship between estimated and measured (**a**) $CaCO_3$, (**b**) fine sand %, and (**c**) *span* using a linear single wavelength model (model 1) and multilinear model (model 2).

Cross-validation of the output of the linear regression model (Table 4) through comparison between predicted and measured samples is shown in Figure 8. R^2 values, adjusted for the number of variables considered in each model, are higher for $CaCO_3$ with decreasing values for fine sand and *span*. An increased degree of scatter reflects the inability of the model to describe all of the sample points, and this is particularly the case for *span* (see Figure 7). Thus, $CaCO_3$ and fine sand values show the most robust statistical relationships to the spectral measurement data.

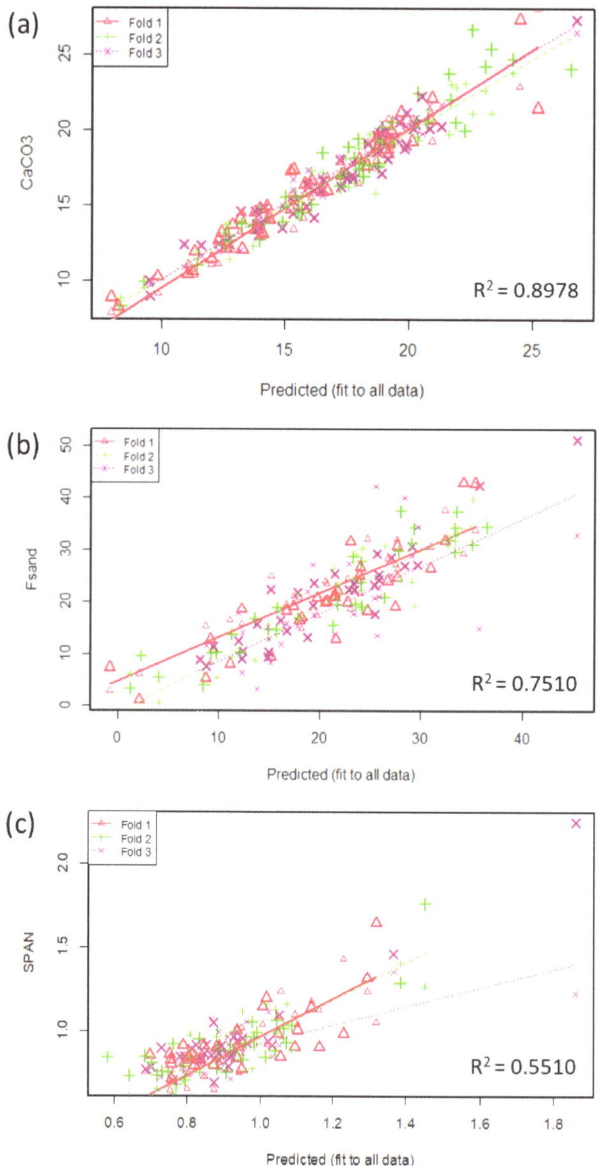

Figure 8. Cross validation results of (**a**) $CaCO_3$ values, (**b**) fine sand and (**c**) *span*. Vertical axes are measured values. Small symbols show cross-validation prediction values.

The spectral characteristics of shell samples 1–5 are presented in Figure 9. Throughout, this shows higher absorption values in the SWIR with consistent dips between all samples at the wavelengths ~1100, 1600 and 2000 nm. The latter may correspond to the water absorption wavelength at ~1950 nm. There is also a very slight jump in reflectance at the water absorption wavelength at ~1450 nm. In addition, the individual shell samples show some variability in the VNIR bands in particular, because of the different shell colours present (Figure 2).

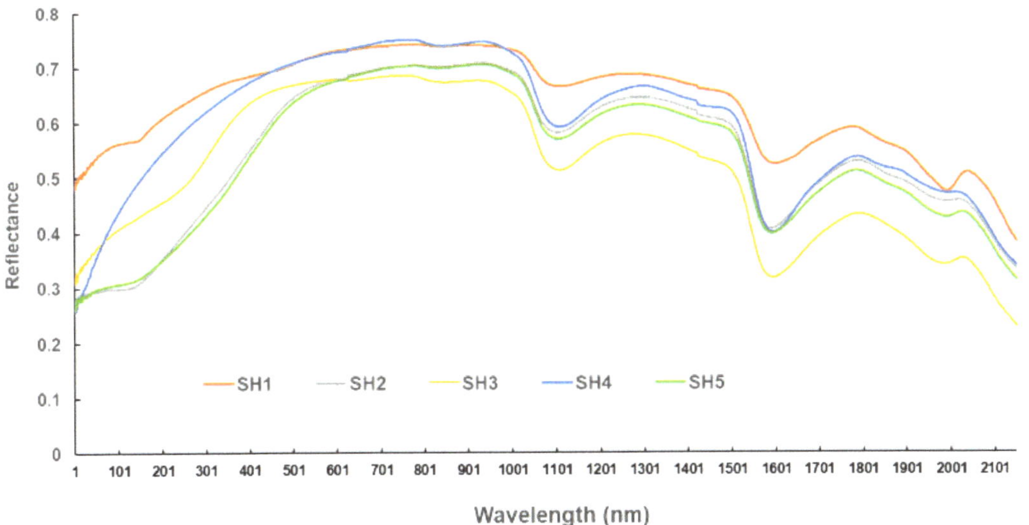

Figure 9. Spectral reflectance results for shell samples (SH) 1–5.

In order to consider whether there are any spectral differences between individual sediment samples with different values of $CaCO_3$, fine sand and *span*, the samples with the highest and lowest values of these parameters (Table 1) are compared to each other (Figure 10). All these samples reflect the aggregated patterns shown in Figure 5a, in which there are decreases in reflectance in the water absorption bands. The samples also show that, irrespective of individual values of $CaCO_3$, fine sand and *span*, there are similar reflectance values in the range 1400–1950 nm (see Figure 6). In more detail, comparison of the $CaCO_3$ values within individual samples shows that shorter wavelengths have a higher reflectance where higher $CaCO_3$ values are present, but that the sample with the lowest $CaCO_3$ values has a higher reflectance at longer wavelengths (Figure 10a). This is mirrored by the results for fine sand (Figure 10b), where the signature for the sample with the lowest amount of fine sand (i.e., the coarsest sample) is very similar to the sample with the highest amount of $CaCO_3$. The reason for this is that broken marine shells (as the source of $CaCO_3$ in the sample) give rise to coarse rather than fine sediment [46]. The parameter *span*, as a reflection of sediment sorting, tends to reflect the presence of coarser outliers in the sample (see the potential autocorrelation with coarse sediment in Table 3) and is therefore of less interpretive significance than either $CaCO_3$ or fine sand (see Figure 8).

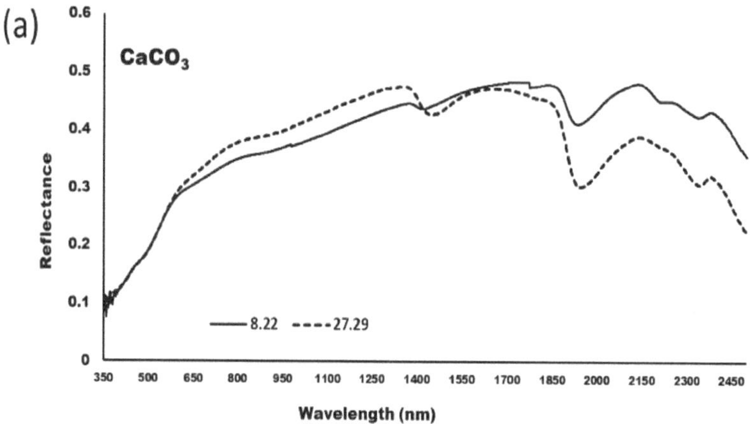

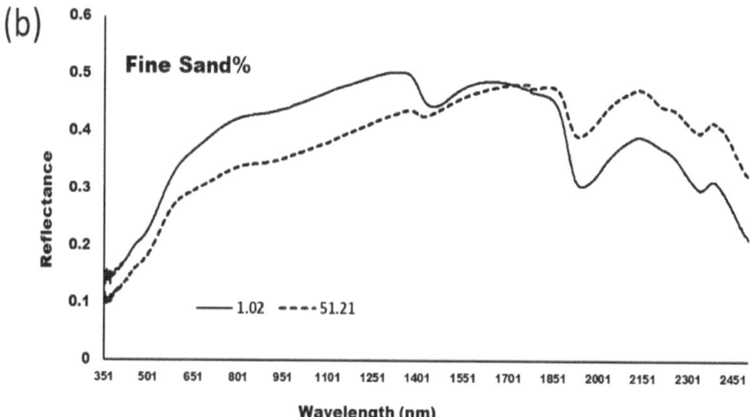

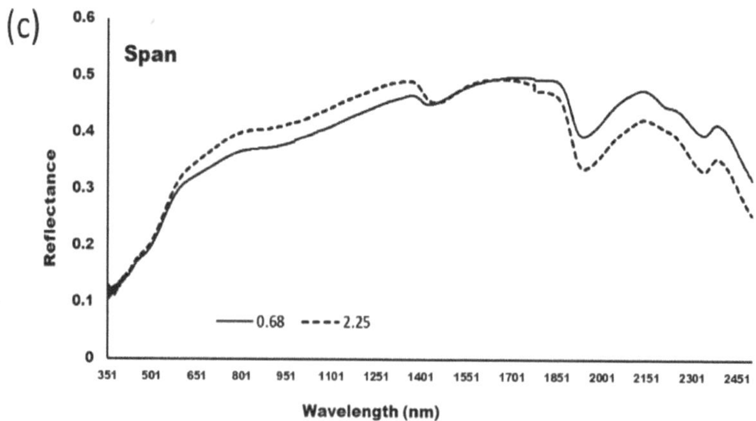

Figure 10. Examples of the spectral signature of individual sediment samples with the greatest range of (**a**) CaCO$_3$ values (highest: #133, lowest: #98); (**b**) *span* values (highest: #61, lowest #139); and (**c**) fine sand percentage (highest #52, lowest #90).

4. Discussion

The presence of a wave-dominated shoreface and wide supratidal zone with migrating transverse dunes (Figures 1 and 4) is typical of the south- and southeast-facing South African coast, in which the dunes have a net eastward migration rate of some 3–12 m yr^{-1} [39,48,49]. Sediment grain size analysis of field samples from Oyster Bay shows that overall they are fairly texturally uniform but with some significant differences in properties between different beach–dune sub-environments (Table 2) and a wide range of carbonate contents (Table 1). The values obtained for grain size properties are similar to other beach–dune systems in the region, e.g., [50,51]. The relatively limited textural and compositional differences mean that it is sometimes difficult to distinguish between such coastal samples, especially in wave-dominated shoreface environments and wind-affected supratidal environments, where sediments are relatively well sorted e.g., [2,4,11,52]. This is certainly the case with beach–dune sediments along the South African coast, e.g., [39,45,49,51,53]. However, detailed statistical analysis shows that different landforms and beach–dune settings at Oyster Bay have different sediment properties (Table 2). This is particularly the case with grain size endmembers (fine and coarse sand) and $CaCO_3$ values that reflect the presence of broken marine shells (not land snail shells) and thus a shoreface source. It is also notable that there are similar overall spectral signatures of all samples (Figure 5), irrespective of their depositional environment (Table 2), which means that the depositional environment of any one sediment sample cannot be resolved by their spectral signature alone. One potential reason for this is that all of the field sediment samples examined here are quartz-dominated (data not shown) and thus, variations in mineralogy cannot be considered as a significant control on their spectral signatures, unlike in previous studies, e.g., [29,47,54]. Although the spectral signature of interstitial water is a dominant feature in other previous studies of coastal sediments e.g., [25,26,32], we deliberately excluded this by drying the samples prior to analysis. This enabled the spectral data of the Oyster Bay samples to be a better representation of grain size and $CaCO_3$ properties (Figure 6), which is the primary aim of this study.

Spectral characteristics of sand systems (beaches, dunes and deserts) have been examined in several studies, e.g., [54,55], and these highlight the potential application of spectral analysis techniques to inform on, in particular, mineralogy and depositional environments [29,32,34–36,47]. Similar to this study, the VNIR part of the spectrum has been previously identified as the most useful in terms of sediment discrimination [55]. There are fewer studies, however, that have looked at grain size data and $CaCO_3$ content. In detail, the spectral reflectance of these samples, however, revealed some fundamental differences between $CaCO_3$ content, fine sand proportion and *span* (Figure 6). These properties also show statistically significant correlations with certain wavelengths (Table 4). However, despite the evidence for some differences in spectral reflectance at different wavelengths for samples with different values of $CaCO_3$ and fine sand (Figure 10), this does not mean that spectral reflectance can be used to predict values of these sediment properties in unknown samples. This is because measured reflectance values at any wavelength are the net result of all grains within the entire sample and not one single component such as shell fragments. In addition, detrital sediment samples of different provenance or found in different depositional environments could have a range of lithologies, water, organic content or other materials, such as microplastics, that may affect spectral reflectance, e.g., [56]. Previous field studies also show the spectral dominance of water absorption signals, e.g., [18,25], and these tend to drown out any signals related to sediment grain size or CaCO3, hence the methodology applied in this study.

These results and their caveats highlight that the potential for spatial mapping of sediment properties across beach–dune environments using hyperspectral imaging techniques may be challenging because of (1) the uncertainties associated with the interpretation of spectral signatures, even under laboratory conditions, and (2) the multiple environmental factors that may be present in a natural beach–dune environment and that may also affect spectral reflectance signatures, including microtopography, vegetation/algae, and salt and water content. Studies on tidal flats also highlight specific problems related to silt and mud particles and chlorophyll content [25,30,35], and these also have to be considered along coastlines that may contain many different types of depositional settings, as well as the spatial transitions between them.

5. Conclusions

This study, based on field samples from a South African beach–dune system, shows both the complexity and potential of hyperspectral techniques to analyse the properties of these samples (with respect to grain size and $CaCO_3$ content), and their limitations. The major conclusions from this study are:

Statistically, $CaCO_3$, fine sand and *span* are the most important sediment properties in terms of their ability to distinguish between coastal depositional environments (Table 3);

These properties in particular have distinctive spectral signatures in different parts of the VNIR and SWIR wavebands (Table 4);

Fine sand and $CaCO_3$ in particular are clearly distinguishable at ~1850–2450 nm in the SWIR waveband (Figure 6);

Shell content (giving rise to $CaCO_3$ values) and different shell types show somewhat different spectral signatures (Figure 9).

It is notable that previous studies have not described these sediment properties using such analytical techniques and in such a level of detail. The results from this study provide the basis for working towards the automated mapping of a beach–dune environment using hyperspectral satellite data, which must be seen as a long-term goal vital for ongoing monitoring of climate change-sensitive environments.

Author Contributions: Conceptualization, J.K.; Formal analysis, M.A.M.A.E.; Investigation, J.K.; Methodology, J.K.; Software, M.A.M.A.E.; Visualization, M.A.M.A.E.; Writing—original draft, J.K.; Writing—review & editing, M.A.M.A.E. All authors have read and agreed to the published version of the manuscript.

Funding: This research received no external funding.

Institutional Review Board Statement: Not applicable.

Informed Consent Statement: Not applicable.

Data Availability Statement: Data are available from the corresponding author.

Conflicts of Interest: The authors declare no conflict of interest.

References

1. De Lange, W.P.; Healy, T.R.; Darlan, Y. Reproducibility of sieve and settling tube textural determinations for sand-sized beach sediment. *J. Coast. Res.* **1997**, *13*, 73–80.
2. Kasper-Zubillaga, J.J.; Ortiz-Zamora, G.; Dickinson, W.W.; Urrutia-Fucugauchi, J.; Soler-Arechalde, A.M. Textural and compositional controls on modern beach and dune sands, New Zealand. *Earth Surf. Proc. Landf.* **2007**, *32*, 366–389. [CrossRef]
3. Kwarteng, A.Y.; Al-Hatrushi, S.M.; Illenberger, W.K.; McLachlan, A. Grain size and mineralogy of Al Batinah beach sediments, Sultanate of Oman. *Arab. J. Geosci.* **2016**, *9*, 557. [CrossRef]
4. Hallin, C.; Almström, B.; Larson, M.; Hanson, H. Longshore transport variability of beach face grain size: Implications for dune evolution. *J. Coastal Res.* **2019**, *35*, 751–764. [CrossRef]
5. Pyökäri, M. Beach Sediments of Crete: Texture, Composition, Roundness, Source and Transport. *J. Coast. Res.* **1999**, *15*, 537–553.
6. Smith, D.A.; Cheung, K.F. Empirical relationships for grain size parameters of calcareous sand on Oahu, Hawaii. *J. Coast. Res.* **2002**, *18*, 82–93.
7. Alsharhan, A.S.; El-Sammak, A.A. Grain size analysis and characterization of sedimentary environments of the United Arab Emirates coastal area. *J. Coast. Res.* **2004**, *20*, 464–477. [CrossRef]
8. Deidun, A.; Gauci, R.; Schembri, J.A.; Šegina, E.; Gauci, A.; Gianni, J.; Gutierrez, J.A.; Sciberras, A.; Sciberras, J. Comparative median grain size assessment through three different techniques for sandy beach deposits on the Maltese Islands (Central Mediterranean). *J. Coast. Res.* **2013**, *65*, 1757–1761. [CrossRef]
9. Poizot, E.; Anfuso, G.; Méar, Y.; Bellido, C. Confirmation of beach accretion by grain-size trend analysis: Camposoto beach, Cádiz, SW Spain. *Geo-Mar. Lett.* **2013**, *33*, 263–272. [CrossRef]

10. Choi, T.J.; Choi, J.Y.; Park, J.Y.; Um, H.Y.; Choi, J.H. The Effects of Nourishments Using the Grain-Size Trend Analysis on the Intertidal Zone at a Sandy Macrotidal Beach. *J. Coast. Res.* **2018**, *85*, 426–430. [CrossRef]
11. Gunaratna, T.; Suzuki, T.; Yanagishima, S. Cross-shore grain size and sorting patterns for the bed profile variation at a dissipative beach: Hasaki Coast, Japan. *Mar. Geol.* **2019**, *407*, 111–120. [CrossRef]
12. Pedreros, R.; Howa, H.L.; Michel, D. Application of grain size trend analysis for the determination of sediment transport pathways in intertidal areas. *Mar. Geol.* **1996**, *135*, 35–49. [CrossRef]
13. Edwards, A.C. Grain Size and Sorting in Modern Beach Sands. *J. Coast. Res.* **2001**, *17*, 38–52.
14. Gallagher, E.L.; MacMahan, J.; Reniers, A.J.H.M.; Brown, J.; Thornton, E.B. Grain size variability on a rip-channeled beach. *Mar. Geol.* **2011**, *287*, 43–53. [CrossRef]
15. Pradhan, U.K.; Sahoo, R.K.; Pradhan, S.; Mohany, P.K.; Mishra, P. Textural Analysis of Coastal Sediments along East Coast of India. *J. Geol. Soc. India* **2020**, *95*, 67–74. [CrossRef]
16. Deronde, B.; Kempeneers, P.; Forster, R.M. Imaging spectroscopy as a tool to study sediment characteristics on a tidal sandbank in the Westerschelde. *Estuar. Coast. Shelf Sci.* **2006**, *69*, 580–590. [CrossRef]
17. Castillo, E.; Pereda, R.; de Luis, J.M.; Medina, R.; Viguri, J. Sediment grain size estimation using airborne remote sensing, field sampling, and robust statistic. *Environ. Monit. Assess.* **2011**, *181*, 431–444. [CrossRef]
18. Manzo, C.; Valentini, E.; Taramelli, A.; Filipponi, F.; Disperati, L. Spectral characterization of coastal sediments using Field Spectral Libraries, Airborne Hyperspectral Images and Topographic LiDAR Data (FHyL). *Int. J. Appl. Earth Obs. Geoinf.* **2015**, *36*, 54–68. [CrossRef]
19. Ibrahim, E.; Kim, W.; Crawford, M.; Monbaliu, J. A regression approach to the mapping of bio-physical characteristics of surface sediment using in situ and airborne hyperspectral acquisitions. *Ocean. Dyn.* **2017**, *67*, 299–316. [CrossRef]
20. Smit, Y.; Ruessink, G.; Brakenhoff, L.B.; Donker, J.J.A. Measuring spatial and temporal variation in surface moisture on a coastal beach with a near-infrared terrestrial laser scanner. *Aeolian Res.* **2018**, *31*, 19–27. [CrossRef]
21. Kim, K.-L.; Kim, B.-J.; Lee, Y.-K.; Ryu, J.-H. Generation of a Large-Scale Surface Sediment Classification Map Using Unmanned Aerial Vehicle (UAV) Data: A Case Study at the Hwang-do Tidal Flat, Korea. *Remote Sens.* **2019**, *11*, 29. [CrossRef]
22. Jacq, K.; Giguet-Covex, C.; Sabatier, P.; Perrette, Y.; Fanget, B.; Coquin, D.; Debret, M.; Arnaud, F. High-resolution grain size distribution of sediment core with hyperspectral imaging. *Sediment. Geol.* **2019**, *393–394*, 105536. [CrossRef]
23. Ghanbari, H.; Jacques, O.; Adaïmé, M.É.; Gregory-Eaves, I.; Antoniades, D. Remote Sensing of Lake Sediment Core Particle Size Using Hyperspectral Image Analysis. *Remote Sens.* **2020**, *12*, 3850. [CrossRef]
24. Zander, P.D.; Wienhues, G.; Grosjean, M. Scanning Hyperspectral Imaging for In Situ Biogeochemical Analysis of Lake Sediment Cores: Review of Recent Developments. *J. Imaging* **2022**, *8*, 58. [CrossRef]
25. Ryu, J.-H.; Na, Y.-H.; Won, J.-S.; Doerffer, R. A critical grain size for Landsat ETM+ investigations into intertidal sediments: A case study of the Gomso tidal flats, Korea. *Estuar. Coast. Shelf Sci.* **2004**, *60*, 491–502. [CrossRef]
26. Small, C.; Steckler, M.; Seeber, L.; Akhter, S.H.; Goodbred, S., Jr.; Mia, B.; Imam, B. Spectroscopy of sediments in the Ganges–Brahmaputra delta: Spectral effects of moisture, grain size and lithology. *Remote Sens. Environ.* **2009**, *113*, 342–361. [CrossRef]
27. Legleiter, C.J.; Stegman, T.K.; Overstreet, B.T. Spectrally based mapping of riverbed composition. *Geomorphology* **2016**, *264*, 61–79. [CrossRef]
28. Ciampalini, A.; Consoloni, I.; Salvatici, T.; Di Traglia, F.; Fidolini, F.; Sarti, G.; Moretti, S. Characterization of coastal environment by means of hyper- and multispectral techniques. *Appl. Geogr.* **2015**, *57*, 120–132. [CrossRef]
29. Rejith, R.G.; Sundararajan, M.; Gnanappazham, L.; Loveson, V.J. Satellite-based spectral mapping (ASTER and Landsat data) of mineralogical signatures of beach sediments: A precursor insight. *Geocarto Int.* **2022**, 1–24, *in press*. [CrossRef]
30. Park, N.-W.; Jang, D.-H.; Chi, K.-H. Integration of IKONOS imagery for geostatistical mapping of sediment grain size at Baramarae beach, Korea. *Int. J. Remote Sens.* **2009**, *30*, 5703–5724. [CrossRef]
31. Williams, K.K.; Greeley, R. Laboratory and field measurements of the modification of radar backscatter by sand. *Remote Sens. Environ.* **2004**, *89*, 29–40. [CrossRef]
32. Van der Wal, D.; Herman, P.M.J. Regression-based synergy of optical, shortwave infrared and microwave remote sensing for monitoring the grain-size of intertidal sediments. *Remote Sens. Environ.* **2007**, *111*, 89–106. [CrossRef]
33. Rainey, M.P.; Tyler, A.N.; Gilvear, D.J.; Bryant, R.G.; McDonald, P. Mapping intertidal estuarine sediment grain size distributions through airborne remote sensing. *Remote Sens. Environ.* **2003**, *86*, 480–490. [CrossRef]
34. Choi, J.K.; Ryu, J.H.; Eom, J. Integration of spatial variables derived from remotely sensed data for the mapping of the tidal surface sediment distribution. *J. Coast. Res.* **2011**, *64*, 1653–1657.
35. Ibrahim, E.; Monbaliu, J. Suitability of spaceborne multispectral data for inter-tidal sediment characterization: A case study. *Estuar. Coast. Shelf Sci.* **2011**, *92*, 437–445. [CrossRef]
36. Park, N.-W. Geostatistical integration of field measurements and multi-sensor remote sensing images for spatial prediction of grain size of intertidal surface sediments. *J. Coast. Res.* **2019**, *SI90*, 190–196. [CrossRef]
37. Wepener, V.; Degger, N. South Africa. In *World Seas: An Environmental Evaluation: Volume II: The Indian Ocean to the Pacific*, 2nd ed.; Sheppard, C., Ed.; Elsevier: Amsterdam, The Netherlands, 2019; pp. 101–119.
38. Meeuwis, J.; van Rensburg, P.A.J. Logarithmic spiral coastlines: The northern Zululand coastline. *S. Afr. Geogr. J.* **1986**, *68*, 18–44. [CrossRef]

39. Knight, J.; Burningham, H. The morphodynamics of transverse dunes on the coast of South Africa. *Geo-Mar. Lett.* **2021**, *41*, 47. [CrossRef]

40. Roberts, D.L.; Botha, G.A.; Maud, R.R.; Pether, J. Coastal Cenozoic Deposits. In *The Geology of South Africa*; Johnson, M.R., Anhaeusser, C.R., Thomas, R.J., Eds.; Geological Society of South Africa/Council for Geoscience: Johannesburg/Pretoria, South Africa, 2006; pp. 605–628.

41. Claassen, D. Geographical controls on sediment accretion of the Cenozoic Algoa Group between Oyster Bay and St. Francis, Eastern Cape coastline, South Africa. *S. Afr. J. Geol.* **2014**, *117*, 109–128. [CrossRef]

42. Knight, J. The late Quaternary stratigraphy of coastal dunes and associated deposits in South Africa. *S. Afr. J. Geol.* **2021**, *124*, 995–1006. [CrossRef]

43. Butzer, K.W.; Helgren, D.M. Late Cenozoic evolution of the Cape coast between Knysna and Cape St. Francis, South Africa. *Quat. Res.* **1972**, *2*, 143–169. [CrossRef]

44. Rossel, R.A.V.; McBratney, A.B. Laboratory evaluation of a proximal sensing technique for simultaneous measurement of soil clay and water content. *Geoderma* **1998**, *85*, 19–39. [CrossRef]

45. Schumann, E.H.; Illenberger, W.K.; Goschen, W.S. Surface winds over Algoa Bay, South Africa. *S. Afr. J. Sci.* **1991**, *87*, 202–207.

46. Carter, R.W.G. Some problems associated with the analysis and interpretation of mixed carbonate and quartz beach sands, illustrated by examples from north-west Ireland. *Sediment. Geol.* **1982**, *33*, 35–56. [CrossRef]

47. Fang, Q.; Hong, H.; Zhao, L.; Kukolich, S.; Yin, K.; Wang, C. Visible and Near-Infrared Reflectance Spectroscopy for Investigating Soil Mineralogy: A Review. *J. Spectroscop.* **2018**, *2018*, 3168974. [CrossRef]

48. Lubke, R.A.; Sugden, J. Short-term change in mobile dunes at Port Alfred, South Africa. *Environ. Manag.* **1990**, *14*, 209–220. [CrossRef]

49. Knight, J.; Burningham, H. Sand dunes and ventifacts on the coast of South Africa. *Aeolian Res.* **2019**, *37*, 44–58. [CrossRef]

50. Harmse, J.T. Trend surface analysis of aeolian sand movement on the southwest African coast. *S. Afr. Geogr. J.* **1985**, *67*, 31–39. [CrossRef]

51. Illenberger, W.K. Variations of sediment dynamics in Algoa Bay during the Holocene. *S. Afr. J. Sci.* **1993**, *89*, 187–196.

52. Çelikoğlu, Y.; Yüksel, Y.; Kabdaşlı, M.S. Cross-shore sorting on a beach under wave action. *J. Coast. Res.* **2006**, *22*, 487–501. [CrossRef]

53. Olivier, M.J.; Garland, G.G. Short-term monitoring of foredune formation on the east coast of South Africa. *Earth Surf. Proc. Landf.* **2003**, *28*, 1143–1155. [CrossRef]

54. Bandfield, J.L.; Edgett, K.S.; Christensen, P.R. Spectroscopic study of the Moses Lake dune field, Washington: Determination of compositional distributions and source lithologies. *J. Geophys. Res.* **2002**, *107*, 5092. [CrossRef]

55. Sadiq, A.; Howari, F. Remote Sensing and Spectral Characteristics of Desert Sand from Qatar Peninsula, Arabian/Persian Gulf. *Remote Sens.* **2009**, *1*, 915–933. [CrossRef]

56. Moshtaghi, M.; Knaeps, E.; Sterckx, S.; Garaba, S.; Meire, D. Spectral reflectance of marine macroplastics in the VNIR and SWIR measured in a controlled environment. *Sci. Rep.* **2021**, *11*, 5436. [CrossRef] [PubMed]

Article

Multi-Classifier Pipeline for Olive Groves Detection

Priscilla Indira Osa [1] , **Anne-Laure Beck** [2,*] , **Louis Kleverman** [3] **and Antoine Mangin** [3]

1   MSc Environmental Hazards and Risks Management Program, Campus IMREDD Université Côte d'Azur, 9 Rue Julien Lauprêtre, 06200 Nice, France
2   ARGANS 260 Route du Pin Montard, 06410 Biot, France
3   ACRI-ST 260 Route du Pin Montard, 06410 Biot, France
*   Correspondence: albeck@argans.eu

Abstract: Pixel-based classification is a complex but well-known process widely used for satellite imagery classification. This paper presents a supervised multi-classifier pipeline that combined multiple Earth Observation (EO) data and different classification approaches to improve specific land cover type identification. The multi-classifier pipeline was tested and applied within the SCO-Live project that aims to use olive tree phenological evolution as a bio-indicator to monitor climate change. To detect and monitor olive trees, we classify satellite images to precisely locate the various olive groves. For that first step we designed a multi-classifier pipeline by the concatenation of a first classifier which uses a temporal Random-Forest model, providing an overall classification, and a second classifier which uses the result from the first classification. IOTA2 process was used in the first classifier, and we compared Multi-layer Perceptron (MLP) and One-class Support Vector Machine (OCSVM) for the second. The multi-classifier pipelines managed to reduce the false positive (FP) rate by approximately 40% using the combination RF/MLP while the RF/OCSVM combination lowered the FP rate by around 13%. Both approaches slightly raised the true positive rate reaching 83.5% and 87.1% for RF/MLP and RF/OCSVM, respectively. The overall results indicated that the combination of two classifiers pipeline improves the performance on detecting the olive groves compared to pipeline using only one classifier.

Keywords: pixel-based classification; Random-Forest; Multi-layer Perceptron; One-class Support Vector Machine

1. Introduction

Image classification, in remote sensing, refers to the assignation of thematic classes to image pixels [1] and classification approaches can be mainly divided into unsupervised classification and supervised classification [2]. Unsupervised classification has some benefits such as a minimum user involvement and less time consuming as there is no training process necessary [3]. However, unsupervised classification is usually less accurate compared to the supervised classifiers, especially when the radiometric distance between two classes is minor. On the other hand, while it has the drawback of having to prepare the training data, supervised classification is perceived as more suitable for accurate and complex classification tasks [4] and to identify spatial objects. Some examples of commonly used supervised pixel-based classification include Machine Learning algorithms such as Random Forest (RF), Support Vector Machine (SVM), and Artificial Neural Network (ANN) [5,6].

The algorithms mentioned above propose generally a binary or a multi-class classification approach, meaning that two or more classes are needed to train the model. In some cases, when the ground truth is limited or when users want to target only one land cover type, a one-class classification model could improve the final classification map [7]. For one-class classification, a training data set for the desired land cover type is needed and the model learns only from that positive data set and optimizes the spectral distances between the positive data and other objects [8]. Some examples of one-class classifiers include Support Vector Data Description (SVDD) and One-class Support Vector Machine (OCSVM) [7]. In order to classify satellite imagery, a method based on several classification models which is called Multiple-Classifier System (MCS) is perceived to be able to improve the accuracy of classification performance [9]. According to Brownlee [10], not only does MCS perform better than a single classifier, but it is also more robust especially for the overfitting issue. Du et al. [9] divided MCS into two main categories: concatenation system, and parallel system and some applications of the MCS concept are presented by Bühlmann [11] and Benediktsson [12]. The concatenation system uses the output of the first classifier as an input for the next, while for the parallel system, the results of several independent classifiers are combined based on different possible strategies to produce the final one [9].

In this study, we tested and applied our classification pipeline within SCO-Live, a CNES funded project. SCO-Live is a collaborative project between ACRI-ST, ARGANS and CAPG (Community of agglomeration of the country of Grasse) [13], which aims to use olive trees as a bio-indicator to monitor climate change. If some previous study monitored olive trees health and water status, none present an accurate approach to solely identify olive groves and to correlate their phenological change with climate change. SCO-Live project is the first to propose a complete processing chain and form identification to ecological analysis in the Southeast of France. To follow and analyze olive trees through time, we first need to identify olive groves' location for our area of interest. A field campaign to localize all olive groves is too expensive and time consuming without the insurance to locate them all. Satellite imagery allow us to visualize a large area (100 km × 100 km for Sentinel-2) and olive groves can be identify and extracted with a classification process. A temporal Random Forest algorithm is used within the IOTA2 processing chain (hereinafter referred to as the baseline pipeline).

Our objective is to improve the performance of olive groves detection from the baseline pipeline using the supervised concatenated multi-classifier pipeline. Furthermore, we also compared the performances between a binary (2-class) and a unary (1-class) approach.

2. Materials and Methods

2.1. Study Area

The area of interest is located in Southeast France (Figure 1) within the Alpes-Maritime over 272.1 km². The experiment is centred in Grasse and its surroundings including Saint-Cezaire-sur-Siagne, Cabris, and Saint-Vallier-de-Thiey. Olive trees can be found around these areas either in the wild, planted in the small-scale private garden, or as a grove in olive farms.

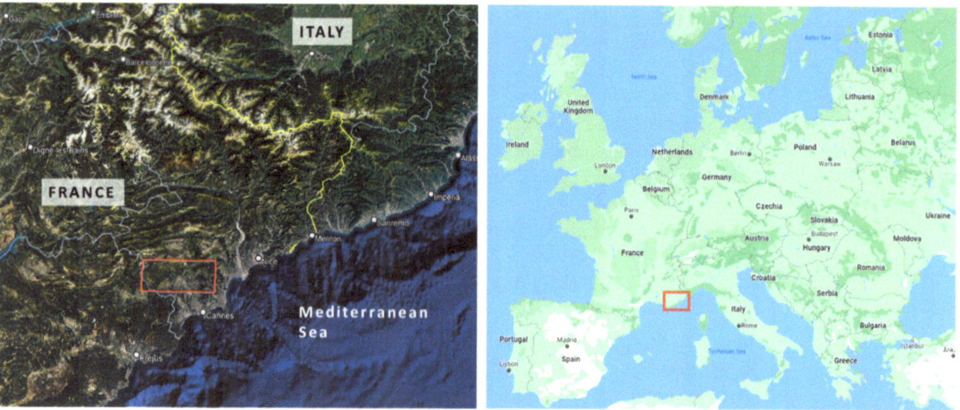

Figure 1. Area of interest identified by the red polygon, top left point: 43°43′15.86″ N 6°46′11.01″ E, bottom right point: 43°36′50.10″ N 7°4′45.83″ E in EPSG:4326/WGS84. (Google Earth).

2.2. Satellite Images

Our process is using Sentinel-2 Level 2A data downloaded from the Copernicus open access hub. Multiple images for the tile identified by the reference T32TLP were used to properly identify the olive grove's reflectance spectrum evolution according to the phenological stages throughout the year. In the Mediterranean area, olive phenological steps are described by Sanz–Cortés [14] and Torres [15] and EO data were selected according to the main stage of development as described in Table 1.

Table 1. Earth Observation data selection and their corresponding phenological stage.

Phenological Steps	File Name	Acquisition Date	Cloud Cover
Dormancy period after harvesting	S2B_MSIL2A_20190125T103319_N0211_R108_T32TLP_20190125T134253	25 January 2019	6.74%
Flowering period	S2A_MSIL2A_20190430T103031_N0211_R108_T32TLP_20190430T140106	30 April 2019	5.86%
Early-stage of fruit growth	S2B_MSIL2A_20190724T103029_N0213_R108_T32TLP_20190724T135539	24 July 2019	3.67%
Late-stage of fruit growth	S2B_MSIL2A_20190912T103019_N0213_R108_T32TLP_20190912T141218	12 September 2019	0.18%
Harvest season	S2A_MSIL2A_20191106T103231_N0213_R108_T32TLP_20191106T112019	6 November 2019	11.87%

The RPG (Registre Parcellaire Graphique) dataset and the SCO-Live project citizen field observations were used to create a reference and to evaluate the resulting classification map. The RPG data is provided by the French National Institute of Geographic and Forest Information (IGN). It is a vector file containing a detailed description of the various vegetation and crop types in France. The document is freely available on the IGN web site and updated every year. Olive groves are identified within the RPG 2019 shapefile by the code "OLI". The SCO-Live project relies on citizen science to collect field information. Citizen science is a process that involves individuals to perform some experiments requiring a low scientific level [16]. Another aspect of citizen science is to involve individuals in field data collection by providing all the tools required for intuitive data collection. Within the SCO-Live project, an application has been developed and distributed to allow citizens to contribute by identifying and locating olive trees and groves. The resulting data base gathers all observation made and stores the information in a georeferenced point vector file which contains for each location various information on olive trees exploitation such as crops type, stage of growth, possible damage, etc. All olive observations are located within a global 5 m spatial accuracy linked with the individual phone GPS accuracy.

All data set are georeferenced and projected to the coordinate reference system of EPSG:32632 (WGS84/UTM zone 32N).

2.3. Satellite Imagery Classification

Table 2 describes the different pipelines that are used and compared in this study.

Table 2. The list of pipelines used in this study.

Pipeline	Classifier(s)	Description
Baseline pipeline	Random Forest	The pipeline that used one classifier
Multi-classifier pipeline 1	Random Forest + Multi-layer Perceptron	Pipeline using a multi-classifier system random forest and a binary classification approach in its second classifier
Multi-classifier pipeline 2	Random Forest + One-class Support Vector Machine	Pipeline using a multi-classifier system, random forest and a unary classification approach in its second classifier

2.3.1. Baseline Pipeline

The baseline pipeline using Random-Forest (RF) [17] is implemented within the IOTA2 framework. IOTA2 (infrastructure for land use by automatic processing using Orfeo toolbox application) is an open-source land cover mapping framework developed by the French Center for Spatial Studies of Biosphere (CESBIO) [18]. The framework implements a classification pipeline using several libraries such as Orfeo Toolbox and Scikit-Learn to perform a temporal supervised pixel-based classification. IOTA2 process, as described in Figure 2, produces a land cover map where each pixel is assigned to a class by processing the surface reflectance taken at different times (multi-temporal images). Only the valid pixels (selected based on the various Sentinel-2 masks and the SCL—Scene CLassification—band) are processed and the different features listed in the configuration file are extracted. As part of the default features extracted, various spectral indices, such as NDVI (Normalized Difference Vegetation Index) and NDWI (Normalized Difference Water Index), are calculated to support land cover type identification.

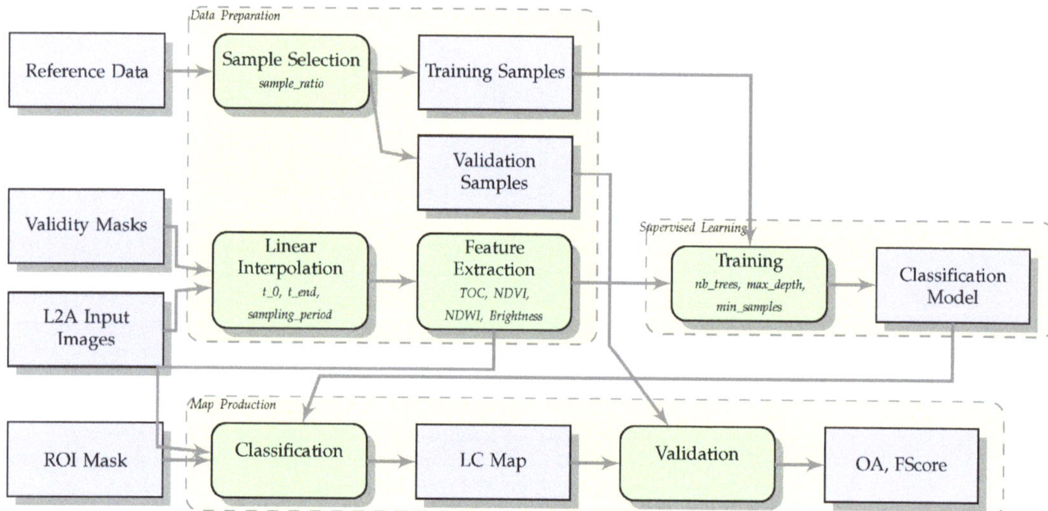

Figure 2. IOTA2 classification module flow chart [18].

The baseline pipeline classifies the input images into Urban, Forest, Nature, Grassland, and Olive classes. The reference data for the baseline pipeline was created by referencing the RPG dataset for Olive class and by photointerpretation and spectral indices analysis for other classes.

2.3.2. Multi-Classifier Pipeline

The multi-classifier pipeline presented in Figure 3 was built by combining the baseline pipeline with either a binary classification or a unary (one-class) classification. The first classification overclassifies all vegetation types to ensure that all possible vegetation types are captured in one vegetation class, and those pixels are re-classified by the second classifier to separate the vegetation into different classes. The final classification map is obtained by combining the result from the first and second classification processes.

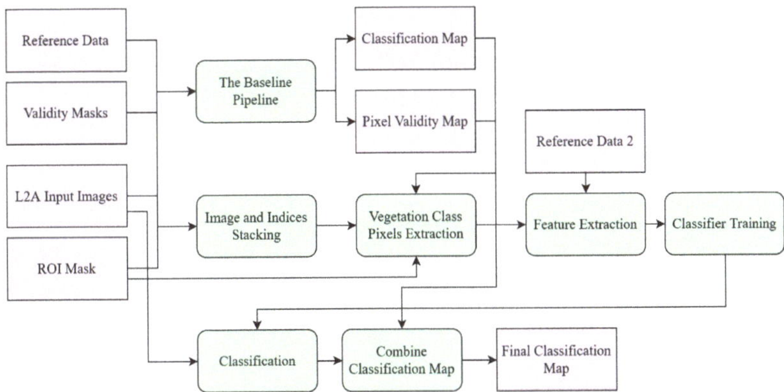

Figure 3. The multi-classifier pipeline.

In this study, two possibilities for the second classifier have been tested and are listed as follows: the Multi-layer Perceptron (MLP) [19] and the One-class Support Vector Machine (OCSVM) [20]. The MLP performs a binary classification using the architecture shown in Figure 4, and it consists of one input layer, two hidden layers and one output layer. The input layer has a number of nodes equal to the number of features corresponding to the five dates, ten bands and six spectral indices that were used. Hence, it made an input layer have 80 nodes. For the two hidden layers, the activation function used is Rectified Linear Unit (ReLU). After the activation function, a Batch Normalization layer (BatchNorm) is added to allow faster model training and allow the use of a larger learning rate. Before the output layer, a Softmax function is set to turn the output values into probabilities such that all nodes in the output class will amount to 1. The output layer consists of nodes equal to the number of classes, which is 2 because the MLP will perform binary classification between Olive and Non-olive classes. For the training phase, the Cross Entropy loss function was used, and the weights and biases of the model were adjusted by the Adam optimizer. OCSVM on the other hand, conducts a unary classification that focuses only on the identification of one class. The kernel used for the OCSVM model was the Radial Basis Function (RBF) kernel. The ν parameter, which is the noise tolerance when the model learns to set the boundary for the training data, was set to 0.5, and the γ parameter, which is the width of the Gaussian curve in the RBF kernel, was set to 0.1.

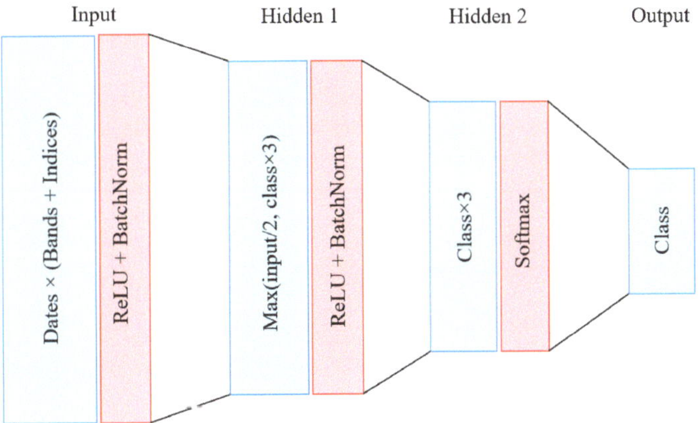

Figure 4. The architecture of Multi-Layer Perceptron (MLP) in this study.

The multi-classifier pipeline firstly classifies the pixels into Forest, Impervious, and Vegetation classes and then further classifies Vegetation pixels into Olive, and Non-olive classes. Impervious class includes the urban areas, bare land and other non-vegetation objects. Forest class comprises forest areas and dense dark vegetation areas. Vegetation class contains olive groves, trees and other vegetation that is not categorized as Forest. The reference data for the multi-classifier pipeline was also created by interpreting the satellite imagery, analyzing spectral indices, and referencing RPG dataset.

2.3.3. Evaluation Method

To evaluate the performance of the multi-classifier pipeline and the accuracy of the resulting maps, various methods are applied. First, we visually compared the three land cover maps produced by each pipeline and to highlight the differences between them, we performed raster subtraction focusing on Olive class. To realize this, the classes that are not olive were regrouped into one Non-olive class and we compared the olive class spatial distribution for the baseline classification and both MLP pipeline and OCSVM pipeline classification maps. The change in area was quantified from the subtraction raster to obtain the surface loss of each class.

To evaluate the performance of the new multi-classifier pipeline, we calculated the True Positive Rate (TPR) and the False Positive Rate (FPR) to check if there is an improvement in the olive trees detection with the multi-classifier pipeline compared to the baseline pipeline. Moreover, the comparison of MLP and OCSVM is also observed. TPR was calculated based on the SCO-Live project citizen field observations using the following equations:

$$True\ Positive\ Rate = \frac{Correctly\ predicted\ points}{Total\ points} \times 100\% \qquad (1)$$

A point is considered correct if there is at least one pixel of predicted Olive class in the point's location and its eight neighboring pixels as the spatial accuracy of GPS is considered.

On the other hand, the calculation of FPR was based on the non-olive vegetation polygons vector of the RPG dataset that are located inside the area of interest. We defined the incorrect pixels as the Olive class pixels inside those polygons, and computed FPR using following equations:

$$False\ Positive\ Rate = \frac{Incorrectly\ predicted\ pixels}{Total\ pixels} \times 100\% \qquad (2)$$

3. Results

3.1. Classification Pipeline Resulting Maps

From the three experimented pipelines, we have obtained various classification maps that are presented in Figure 5.

The baseline pipeline classified the input images into 5 classes: Urban, Forest, Nature, Grassland and Olive, while the two other ones assigned the pixels into Forest, Impervious, Olive and Non-olive classes. We observed significant visual changes between the map resulting from the one classification process and the ones resulting from the multi-classifier pipeline. We identified a significant decrease in the coverage of the olive class (represented in dark green on all land cover maps) in the multi-classifier pipeline's map. This can be confirmed particularly in the West, Northwest, and North of the area of interest.

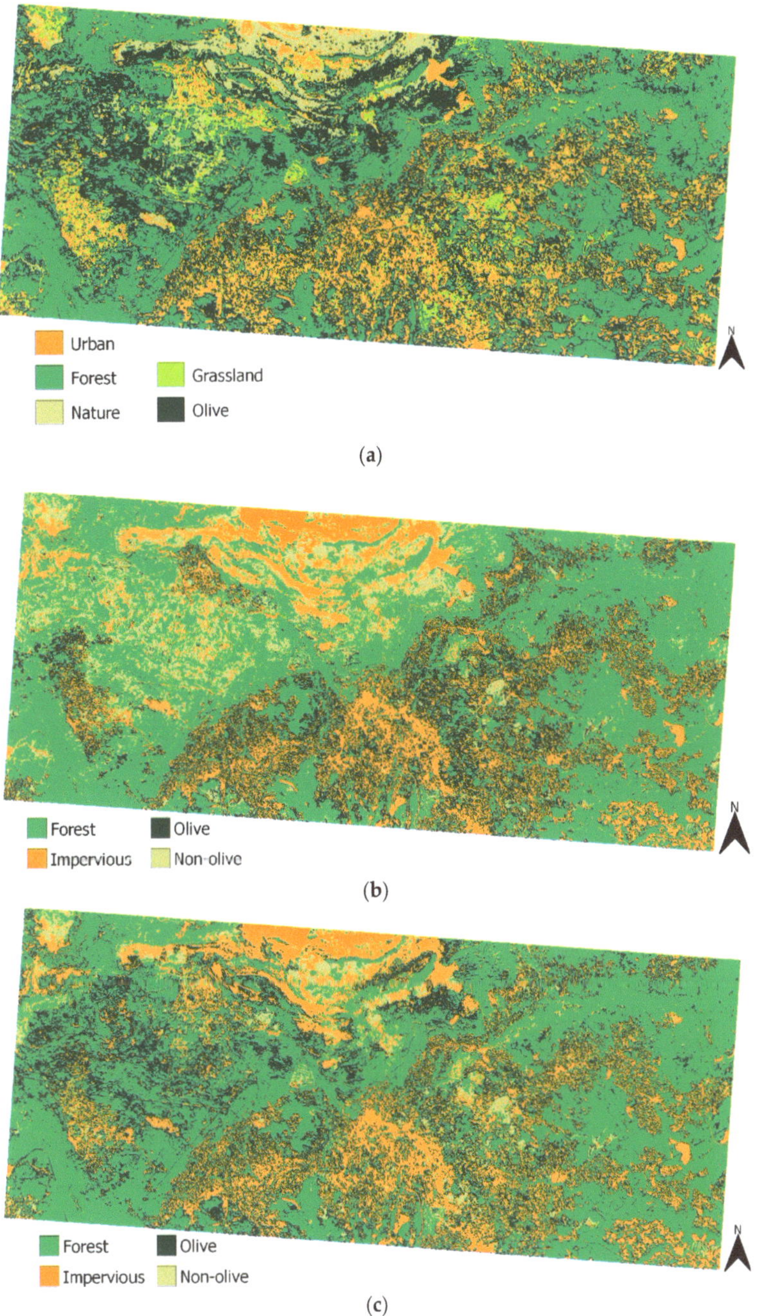

Figure 5. Classification maps from (**a**) the baseline pipeline; (**b**) the multi-classifier pipeline using Multi-Layer Perceptron (MLP); (**c**) the multi-classifier pipeline using One-class Support Vector Machine (OCSVM).

The differences in olive detection can be observed on the subtraction maps presented in Figure 6.

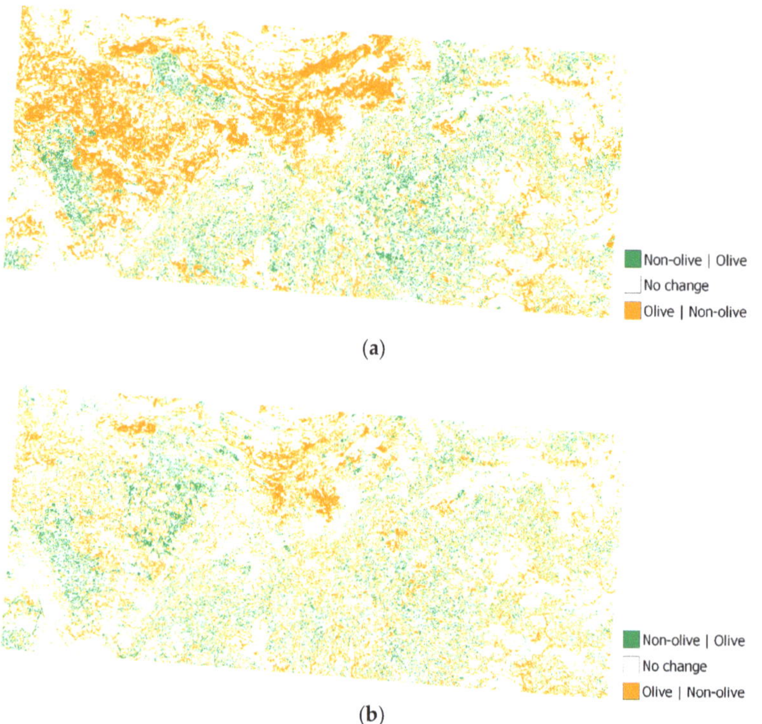

Figure 6. Raster subtraction of classification maps between: (**a**) the baseline pipeline and the multi-classifier pipeline using MLP; (**b**) the baseline pipeline and the multi-classifier pipeline using OCSVM. X | Y indicates class X in the baseline pipeline was classified as class Y in the multi-classifier pipeline.

We can confirm that the area of Olive pixels changing to Non-olive pixels, which are represented in orange color, is emphasized in the area where we observed the significant decrease of olive coverage. Furthermore, we also observed that there are more pixels that turned into Non-olive class than into Olive class as orange coverage is broader than the green one. The change of these areas was calculated and results are presented in Table 3. The quantification of area in orange reached approximately 62.5 km^2 and 42.9 km^2 for the baseline-MLP and the baseline-OCSVM respectively. On the other hand, the area that changed from Non-olive to Olive in both subtractions is about 16 km^2.

Table 3. The total area in km^2 based on each change category.

Raster Subtraction	Change Category	Area [km^2]
	Non-olive to Olive	16.8
Baseline–MLP	Olive to Non-olive	62.5
	No change	192.9
	Non-olive to Olive	16.3
Baseline–OCSVM	Olive to Non-olive	42.9
	No change	212.9

3.2. Performance Evaluation

The evaluation of performance intends to check if there is an improvement in the olive trees detection by the multi-classifier pipeline compared to the baseline pipeline. Table 4 shows the TPR of the baseline pipeline and the multi-classifier pipelines. The multi-classifier pipeline using MLP raised the TPR to 83.5% while the pipeline using OCSVM reached 87.1%. The improvement is also observed in FPR, which is described in Table 5. The multi-classifier pipeline decreased the FPR by approximately 40% and 13% using MLP and OCSVM respectively, compared to the FPR calculated in the baseline pipeline.

Table 4. True positive rate of the three pipelines.

Pipeline	True Positive Rate
IOTA2	81.7%
IOTA2 + MLP	83.5%
IOTA2 + OCSVM	87.1%

Table 5. False positive rate of the baseline pipeline and the multi-classifier pipelines.

Pipeline	False Positive Rate
IOTA2	42.5%
IOTA2 + MLP	1.4%
IOTA2 + OCSVM	28.9%

4. Discussion

The results have demonstrated that the significant difference between the baseline pipeline and the multi-classifier pipeline was highlighted in the change of pixels from Olive class to Non-olive class. As we simultaneously observed a decrease of FPR with both MLP and OCSVM, meaning that we reduced the number of wrongly classified pixels in Olive class, we can assume that most of pixels firstly classified as olive were misclassified. For confirmation, we conducted a field survey for some areas where the classification maps showed different predictions between baseline and multi-classifier pipeline. Most of our field observations validated that the pixels firstly classified as olive by the baseline pipeline but not by multi-classifier pipeline were indeed not olive trees in the field. In some areas, we could find green oak and white oak trees in the misclassified regions of the baseline pipeline. We hypothesize that these oak trees may possibly contribute to the confusion of the classification by the model in the baseline pipeline.

In addition to the decrease in misclassification of olive trees, we could also observe from the TPR calculation of the SCO-Live project field observation data that the correct detection of olive trees rose slightly despite the reduction of Olive class coverage in the multi-classifier pipeline. This suggests that the multi-classifier pipeline improves the overall performance of olive detection, which aligned with the theory of using more than one classifier in a system to target a better accuracy at the price of complexity [21]. Furthermore, the improvement of detection can also be linked to the division of classification tasks in the multi-classifier pipeline. By assigning the first classifier to predict all vegetations first and letting the second classifier to further classify the vegetations into olive trees or not, the pipeline performs hierarchical classification [22], where a specified land type is identified after a more a general classification.

With SVM as a one-class approach and MLP as a two-class approach, the results proved that both have raised the performance of olive detection. In our use-case, because we need to minimize the noise around of the classified olive trees pixels, we perceived that the two-class MLP is more suitable for our project as it lowered the FPR significantly while maintaining a good TPR. However, we found that using one-class SVM in our data can achieve even better results in detecting the olive trees (positive data). Therefore, the use of one-class approach can be an interesting and promising solution for other use cases where we only have data for one category of land cover.

5. Conclusions

Pixel-based classification is widely used for satellite imagery analysis. In this paper, we presented the use of multi-classifiers to compare its results and performances with the scenario using only one classifier. Within the frame of the SCO-Live project that aims to identify olive groves to monitor climate change impact, we evaluated our results based on the SCO-Live database and field data. The results indicated that multi-classifier pipelines (MLP and OCSVM) showed better performances in both detecting olive trees and lowering the misclassification compared to single classifier pipeline. In addition, we observed that using both one-class and two-class approaches as the second classifier in multi-classifier pipeline improves the quality of detection. However, from our field campaign, we found that a confusion between oak trees and olive trees often occurs during the classification process. A further and more complete spectral analysis of both species might highlight some spectral differences at specific season allowing a better discrimination of both types. Furthermore, with the newest version of the SCO-Live database, we can access more detailed information on the various farming conditions (irrigation, health) that affect olive trees' spectral response. This new information needs to be considered to further improve the resulting classification. Moreover, quantifying the computational cost and comparing other pairs of classifiers in the pipeline will be an interesting experiment to give more insight on the multi-classifier pipeline.

Author Contributions: Conceptualization, P.I.O. and A.-L.B.; methodology, P.I.O. and A.-L.B.; software, P.I.O. and L.K.; validation, P.I.O. and A.-L.B.; formal analysis, P.I.O.; resources, A.M.; data curation, P.I.O.; writing—original draft preparation, P.I.O.; writing—review and editing, P.I.O. and A.-L.B.; supervision, A.M.; project administration, A.M. All authors have read and agreed to the published version of the manuscript.

Funding: This research received no external funding.

Institutional Review Board Statement: Not applicable.

Data Availability Statement: Publicly available satellite imagery was analyzed in this study. This data can be found here: [https://scihub.copernicus.eu/dhus/#/home].

Conflicts of Interest: The authors declare no conflict of interest.

References

1. Knudby, A. [En ligne]. Remote Sensing. 2021. Available online: https://ecampusontario.pressbooks.pub/remotesensing/ (accessed on 8 September 2022).
2. Alloghani, M.; Al-Jumeily, D.; Mustafina, J.; Hussain, A.; Aljaaf, A.J. A Systematic Review on Supervised and Unsupervised Machine Learning Algorithms for Data Science. In *Supervised and Unsupervised Learning for Data Science*; Berry, M.W., Mohamed, A., Yap, B.W., Eds.; Springer International Publishing: Cham, Switzerland, 2020; pp. 3–21. [CrossRef]
3. Mohammady, M.; Moradi, H.R.; Zeinivand, H.; Temme, A. A comparison of supervised, unsupervised and synthetic land use classification methods in the north of Iran. *Int. J. Environ. Sci. Technol.* **2014**, *12*, 1515–1526. [CrossRef]
4. Singh, A.; Thakur, N.; Sharma, A. A review of supervised machine learning algorithms. In Proceedings of the 2016 3rd International Conference on Computing for Sustainable Global Development (INDIACom), New Delhi, India, 16–18 March 2016; pp. 1310–1315.
5. Zhang, X.; Han, L.; Han, L.; Zhu, L. How Well Do Deep Learning-Based Methods for Land Cover Classification and Object Detection Perform on High Resolution Remote Sensing Imagery? *Remote Sens.* **2020**, *12*, 417. [CrossRef]
6. Richards, J.A. Supervised Classification Techniques. In *Remote Sensing Digital Image Analysis: An Introduction*; Springer: Berlin/Heidelberg, Germany, 2013; pp. 247–318. [CrossRef]
7. Song, B.; Li, P.; Li, J.; Plaza, A. One-Class Classification of Remote Sensing Images Using Kernel Sparse Representation. *IEEE J. Sel. Top. Appl. Earth Obs. Remote Sens.* **2016**, *9*, 1613–1623. [CrossRef]
8. Perera, P.; Oza, P.; Patel, V.M. One-Class Classification: A Survey. *arXiv* **2021**. [En ligne]. Available online: http://arxiv.org/abs/2101.03064 (accessed on 8 September 2022).
9. Du, P.; Xia, J.; Zhang, W.; Tan, K.; Liu, Y.; Liu, S. Multiple Classifier System for Remote Sensing Image Classification: A Review. *Sensors* **2012**, *12*, 4764–4792. [CrossRef] [PubMed]

10. Brownlee, J. Ensemble Learning Methods for Deep Learning Neural Networks, Machine Learning Mastery. 18 December 2018. Available online: https://machinelearningmastery.com/ensemble-methods-for-deep-learning-neural-networks/ (accessed on 8 September 2022).
11. Bühlmann, P. Bagging, Boosting and Ensemble Methods. In *Handbook of Computational Statistics: Concepts and Methods*; Gentle, J.E., Härdle, W.K., Mori, Y., Eds.; Springer: Berlin/Heidelberg, Germany, 2012; pp. 985–1022. [CrossRef]
12. Benediktsson, J.A.; Chanussot, J.; Fauvel, M. Multiple Classifier Systems in Remote Sensing: From Basics to Recent Developments. In *Multiple Classifier Systems*; Springer: Berlin/Heidelberg, Germany, 2007; pp. 501–512. [CrossRef]
13. SCOlive. Available online: https://www.spaceclimateobservatory.org/scolive (accessed on 8 September 2022).
14. Sanz-Cortes, F.; Martinez-Calvo, J.; Badenes, M.L.; Bleiholder, H.; Hack, H.; Llacer, G.; Meier, U. Phenological growth stages of olive trees (Olea europaea). *Ann. Appl. Biol.* **2002**, *140*, 151–157. [CrossRef]
15. Torres, M.; Pierantozzi, P.; Searles, P.; Rousseaux, M.C.; García-Inza, G.; Miserere, A.; Bodoira, R.; Contreras, C.; Maestri, D. Olive Cultivation in the Southern Hemisphere: Flowering, Water Requirements and Oil Quality Responses to New Crop Environments. *Front. Plant Sci.* **2017**, *8*, 1830. [CrossRef]
16. Mahajan, S.; Kumar, P.; Pinto, J.A.; Riccetti, A.; Schaaf, K.; Camprodon, G.; Smári, V.; Passani, A.; Forino, G. A citizen science approach for enhancing public understanding of air pollution. *Sustain. Cities Soc.* **2019**, *52*, 101800. [CrossRef]
17. Breiman, L. Random forests. *Mach. Learn.* **2001**, *45*, 5–32. [CrossRef]
18. Inglada, J.; Vincent, A.; Arias, M.; Tardy, B.; Morin, D.; Rodes, I. Operational High Resolution Land Cover Map Production at the Country Scale Using Satellite Image Time Series. *Remote Sens.* **2017**, *9*, 95. [CrossRef]
19. Popescu, M.-C.; Balas, V.E.; Perescu-Popescu, L.; Mastorakis, N. Multilayer perceptron and neural networks. *WSEAS Trans. Circuits Syst.* **2009**, *8*, 579–588.
20. Schölkopf, B.; Williamson, R.C.; Samola, A.J.; Shawe-Taylor, J.; Platt, J. Support vector method for novelty detection. In *Advances in Neural Information Processing Systems*; Max Planck Institute for Intelligent Systems: Denver, CO, USA, 2000; pp. 582–588. Available online: https://is.mpg.de (accessed on 12 September 2022).
21. Kuncheva, L.I. *Combining Pattern Classifiers: Methods and Algorithms*; John Wiley & Sons: Hoboken, NJ, USA, 2014.
22. Silla, C.N.; Freitas, A.A. A survey of hierarchical classification across different application domains. *Data Min. Knowl. Discov.* **2010**, *22*, 31–72. [CrossRef]

Article

A Classification Feature Optimization Method for Remote Sensing Imagery Based on Fisher Score and mRMR

Chengzhe Lv [1], Yuefeng Lu [1,2,3,*], Miao Lu [4], Xinyi Feng [1], Huadan Fan [1], Changqing Xu [1] and Lei Xu [5]

[1] School of Civil and Architectural Engineering, Shandong University of Technology, Zibo 255049, China
[2] State Key Laboratory of Resources and Environmental Information System, Institute of Geographical Sciences and Natural Resources Research, Chinese Academy of Sciences, Beijing 100101, China
[3] Hunan Provincial Key Laboratory of Geo-Information Engineering in Surveying, Mapping and Remote Sensing, Hunan University of Science and Technology, Xiangtan 411201, China
[4] Key Laboratory of Agricultural Remote Sensing, Ministry of Agriculture and Rural Affairs/Institute of Agricultural Resources and Regional Planning, Chinese Academy of Agricultural Sciences, Beijing 100081, China
[5] China Railway Design Corporation, Tianjin 300308, China
* Correspondence: yflu@sdut.edu.cn; Tel.: +86-0533-278-0964

Citation: Lv, C.; Lu, Y.; Lu, M.; Feng, X.; Fan, H.; Xu, C.; Xu, L. A Classification Feature Optimization Method for Remote Sensing Imagery Based on Fisher Score and mRMR. *Appl. Sci.* **2022**, *12*, 8845. https://doi.org/10.3390/app12178845

Academic Editors: Giovanni Randazzo, Anselme Muzirafuti and Stefania Lanza

Received: 10 August 2022
Accepted: 31 August 2022
Published: 2 September 2022

Publisher's Note: MDPI stays neutral with regard to jurisdictional claims in published maps and institutional affiliations.

Abstract: In object-oriented remote sensing image classification experiments, the dimension of the feature space is often high, leading to the "dimension disaster". If a reasonable feature selection method is adopted, the classification efficiency and accuracy of the classifier can be improved. In this study, we took GF-2 remote sensing imagery as the research object and proposed a feature dimension reduction algorithm combining the Fisher Score and the minimum redundancy maximum relevance (mRMR) feature selection method. First, the Fisher Score was used to construct a feature index importance ranking, following which the mRMR algorithm was used to select the features with the maximum correlation and minimum redundancy between categories. The feature set was optimized using this method, and remote sensing images were automatically classified based on the optimized feature subset. Experimental analysis demonstrates that, compared with the traditional mRMR, Fisher Score, and ReliefF methods, the proposed Fisher Score–mRMR (Fm) method provides higher accuracy in remote sensing image classification. In terms of classification accuracy, the accuracy of the Fm feature selection method with RT and KNN classifiers is improved compared with that of single feature selection method, reaching 95.18% and 96.14%, respectively, and the kappa coefficient reaches 0.939 and 0.951, respectively.

Keywords: object-oriented; feature selection; Fisher Score; mRMR

1. Introduction

The spectra, textures, and geometry of high-resolution remote sensing images are very rich, and different features describe ground objects from different angles [1,2]. To give full play to the advantages of the spectral, texture, and geometric features of high-resolution remote sensing images, object-oriented classification usually allows more features to participate in classification. If all features participate in classification, the processing speed is greatly reduced, while the classification accuracy is reduced in the case of limited training samples [3,4]. Therefore, how to select the optimal features from the feature space to participate in classification is the primary problem to be solved in the field of high-resolution image object-oriented classification [5,6]. Feature selection is an important task in data mining and machine learning and can effectively reduce the dimension of data and improve the performance of algorithms [7,8]. With the increase in data, feature selection has become an indispensable part of data processing [9]. The purpose of feature selection is to remove irrelevant or redundant features, retain useful features, and obtain appropriate feature subsets [10]. Feature selection methods can be divided into three series: filter, wrapper, and embedded [11]. Among them, filter methods directly evaluate the statistical performance of all the training data, as this is independent of the subsequent learning algorithm. Although it has the advantage of fast speed, it has a large performance deviation from the subsequent learning algorithm and is not effective when considering big data features [12]. Wrapper methods evaluate a subset of features with respect to the training accuracy of the subsequent learning algorithm and have the advantage of small deviation, but this type of method is large in size and involves significant computational burden [13]. Embedded methods combine the advantages of the above methods to some extent, but the difficulty with this type of method is the need to construct a suitable function optimization model [14].

From the above analysis, it is clear that the various types of methods have limitations in feature selection. In order to address these limitations, we selected the Fisher Score [15,16] and mRMR [17] as filter methods for comparison with decision tree [18,19] and random forest methods, respectively. The RF [20,21], *k*-nearest neighbors (*k*NN) and support vector machine (SVM) approaches were combined for image classification. Filter methods can be divided into unsupervised, semi-supervised, and supervised feature selection methods [22,23]. At present, supervised feature selection methods include Relief-F [24], mRMR, and Fisher Score. The Relief-F algorithm is a typical filtered feature optimization algorithm, which calculates the weights of feature variables, ranks them, and then extracts the optimal set of features. The Relief-F algorithm is highly efficient and suitable for most data. The mRMR algorithm is a feature optimization method based on mutual information theory, which is used to maximize the correlation between a selected feature subset and the category, while ensuring that the redundancy between the selected features is as small as possible [25,26]. The Fisher Score is an effective criterion for judging the sample features, derived from Fisher's linear discriminant, which finds feature subsets in the feature set space that maximize the distance between different categories of data points while minimizing the distance between those in the same category. Based on the above, we chose to combine the Fisher Score and mRMR algorithm to downscale the feature space of remote sensing images, where the Fisher Score is used to calculate the ratio of the variance within each feature class and the variance between each feature class, while the mRMR algorithm is used to filter out those features with the greatest relevance to the target category and the least redundancy between them. Finally, the filtered features are used as feature subsets. In this study, the feature dimension of the remote sensing image was reduced by combining two feature selection methods, and the optimal feature subset was obtained through feature dimension reduction, which can reduce the classification time of classifier and improve the classification accuracy of the image. We also selected different types of feature selection methods to verify the ability of the Fm feature dimensionality reduction. In addition, we utilized a variety of classifiers and selected the one suitable for Fm by comparing their overall classification accuracy.

2. The Study Area and the Data Source

2.1. Study Area

The study area is located in Guang'an area, Sichuan Province, China, between $106°38'$–$106°41'$ E and $30°27'$–$30°29'$ N. According to a ground cover map of the study area, the ground objects in this area can be classified as water, vegetation, bare ground, buildings, and roads. The location of the study area is shown in Figure 1.

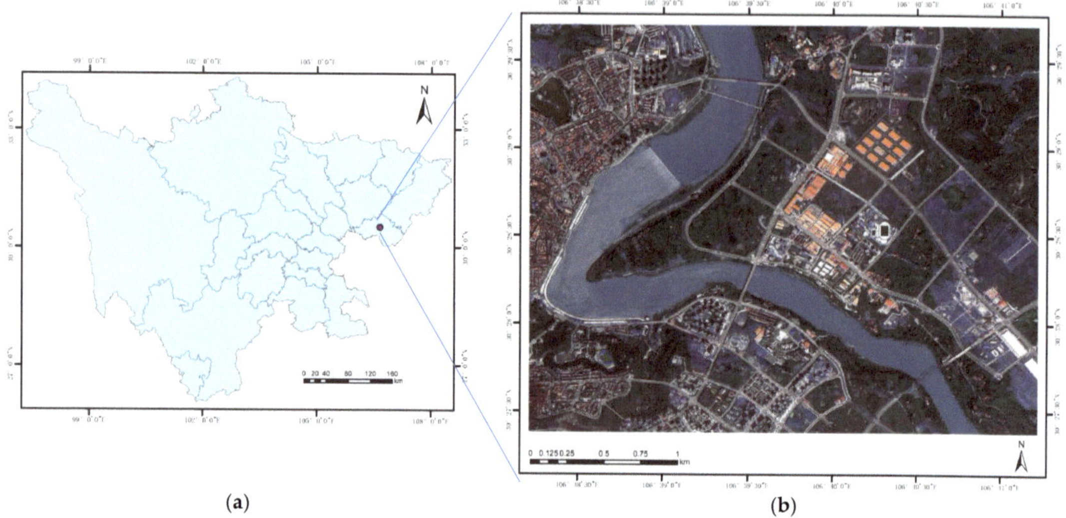

Figure 1. Location and imaging of the study area: (**a**) administrative boundary map of Sichuan Province; (**b**) pre-processed image.

2.2. Data Source and Preprocessing

The data came from the China Centre for Resources Satellite Data and Application (https://data.cresda.cn/#/home, accessed on 10 January 2022). The data used were multi-spectral and panchromatic ortho-corrected images obtained by the GF-2 satellite in August 2020, including multi-spectral data at 4 m resolution (four bands of red, green, blue and near-red) and panchromatic data at 1 m resolution [27]. Radiation calibration, atmospheric correction, geometric rectification, and alignment were performed on the CF-2 images using the ENVI software, while the NNDiffuse Pan-Sharpening fusion algorithm was used to generate multi-spectral remote sensing data with 1 m resolution.

3. Research Methods

Object-oriented classification methods based on feature selection mainly include the steps of image pre-processing, multi-scale segmentation, construction of initial feature space, and image classification. The technical process is depicted in Figure 2. Firstly, the image data were preprocessed based on ENVI5.3. The detailed preprocessing process is shown in Section 2.2. Secondly, eCognition9.0 was used to segment the image, and then some objects were selected as training samples to calculate the eigenvalues of spectral, texture and geometric features of each sample. Third, based on PyCharm software, five feature selection methods were used to screen out feature subsets. Finally, four machine learning classifiers were used to train the training samples and classify the images. The accuracy of the classified remote sensing images was evaluated using validation samples.

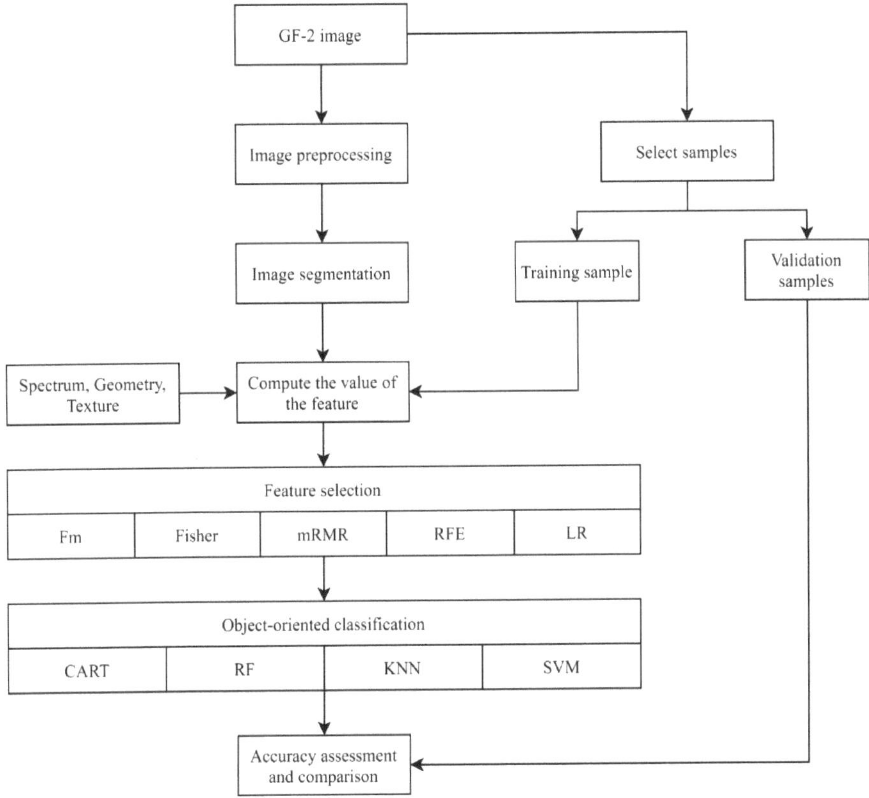

Figure 2. Technical flow chart.

3.1. Build the Feature Space

Based on the ground object types in the study area as well as empirical knowledge, the initial feature space constructed in this study contained 32 features. Spectral features included the mean and standard deviation in bands 1–4 of the GF-2 images; geometric features included the area, length, and width of objects; and texture features included homogeneity, contrast, heterogeneity, angular second moment, entropy, and the correlation between the gray-level co-occurrence matrix (GLCM) and gray-level difference vector (GLDV). The feature information is shown in Table 1.

Table 1. Feature information.

Feature Type	Feature Name	Number of Features
Spectrum	Mean value of bands 1–4, Standard deviation of bands 1–4, Brightness, Max. diff, NDVI, NDWI	12
Geometry	Area, Length, Width, Length/Width, Density, Compactness, Border length, Number of pixels	8
Texture	Homogeneity, Contrast, Dissimilarity, Ang. 2nd moment, Entropy, Correlation, StdDev, Mean	12

3.2. Feature Selection

There are many kinds of image features. Choosing appropriate features can improve the accuracy and efficiency of object-oriented automatic classification. The principle of feature selection is to reduce the total quantity of data while not reducing the classification-related information by obtaining a small subset of features to achieve the purpose of feature optimization.

(1) Fisher Score feature weight calculation. The Fisher Score provides an effective method for feature selection, which mainly identifies features with strong performance. When it is as small as possible within a class and as large as possible between classes, the optimal feature subset can be selected [28–30]. Let the inter-class variance of the kth feature in the data set be expressed by $S_B^{(k)}$. Then, the calculation formula is shown in Equation (1) [28,29].

$$S_B^{(k)} = \sum_{i=1}^{c} \frac{n_i}{n} \left(m_i^{(k)} - m^{(k)} \right)^2 \tag{1}$$

where c denotes the number of sample classes, n denotes the total number of samples, n_i denotes the number of samples in the ith class of the sample, $m_i^{(k)}$ denotes the mean of the values taken by the samples in the ith class on the kth feature, and $m^{(k)}$ denotes the mean of the values taken by the samples in all classes on the kth feature. Let the intra-class variance of the kth feature on the data set be denoted by n_i. Then, the formula is shown in Equation (2) [28,29]:

$$S_w^{(k)} = \frac{1}{n} \sum_{i=1}^{C} \sum_{x \in w_i} \left(x^{(k)} - m_i^{(k)} \right)^2 \tag{2}$$

where $x^{(k)}$ denotes the value of sample x on the kth feature and w_i denotes the ith class sample. The weight coefficient of the kth feature on the data set is denoted by $J_{fisher}(k)$. The calculation formula is shown in Equation (3) [28,30]:

$$J_{fisher}(k) = \frac{S_B^{(k)}}{S_w^{(k)}} \tag{3}$$

(2) mRMR filtering feature subset. The mRMR algorithm is a heuristic feature selection algorithm which calculates the correlation between features and attributes based on an evaluation function, ranks the original features, and obtains a feature set with high correlation and few redundant features [31–33].

The mutual information [34] is first calculated in order to determine the correlations between features and between features and categories. The mutual information formula for variables M and N is [32]:

$$I(M; N) = \sum_{m \in M} \sum_{n \in N} p(m, n) log \frac{p(m, n)}{p(m)p(n)} \tag{4}$$

where $p(m)$ and $p(n)$ denote the probability density functions of the random variables m and n, and $p(m, n)$ denotes the joint probability density function of the random variables m and n. The greater the mutual information, the greater the correlation between M and N. A feature subset S containing K features is searched to maximize the correlation between the K features and a category c. The maximum correlation is calculated as shown in Equation (5) [31,32]:

$$maxD(S, c), D = \frac{1}{|S|} \sum_{x_i \in S} I(x_i; c) \tag{5}$$

The correlation between feature set S and class c is determined by the average of all mutual information values between each feature x_i and class c, and k sets with maximum average mutual information are selected. Subsequently, the redundancy between the k features is eliminated, where the minimum redundancy is calculated as shown in Equation (6) [31,32]:

$$minR(S), R = \frac{1}{|S|^2} \sum_{x_i, x_j \in S} I(x_i, x_j) \tag{6}$$

The maximum correlation and minimum redundancy are combined to form the mRMR algorithm, and the formula for calculating D and R using the operator $\Phi(D, R)$ is shown in Equation (7) [17,31]:

$$max\Phi(D, R), \Phi = D - R \tag{7}$$

Using this feature selection criterion, the features are selected by maximizing the defined operator $\Phi()$, using an incremental search method. Based on the feature subset S_{k-1}, the kth feature is calculated from the remaining feature space $X - S_{k-1}$, which is made to maximize $\Phi()$ using the following equation, that is, the incremental feature selection optimization formula [17,31]:

$$\max_{x_j \in X - S_{k-1}} \left[I\left(x_j; c\right) - \frac{1}{k-1} \sum_{x_i \in S_{k-1}} I\left(x_j; x_i\right) \right] \tag{8}$$

The weight of each feature is calculated according to the Fisher Score, and features with higher weight have better classification ability. As the correlation between features is not calculated, redundant features cannot be removed. However, the mRMR algorithm can obtain the feature subset that has the maximum correlation with the target category and the least redundancy, but it cannot obtain the weight coefficient of each feature, and the extracted feature subset cannot reflect the difference of the effect of different features on the classification.

Firstly, the Fisher Score calculation method was used to build the ranking rules of feature index importance, and the features with larger weight were selected by calculating the weight value of each feature. The feature vector with a high weight can be used as the dominant vector of the classification set, and the feature vector with a low weight has less influence on the classification result. Then, the mRMR algorithm was used to calculate the selected features, and the features with the maximum correlation and the minimum redundancy between the categories were selected. Therefore, by combining the Fisher Score and mRMR algorithms for feature dimension reduction, an optimal feature subset can be obtained.

In addition to the above two methods, in order to verify the reliability in the experiment, the commonly used recursive feature elimination (RFE) algorithm (a wrapped feature selection method) and logistic regression (LR) algorithm (an embedded feature selection method) were selected.

(3) RFE is a greedy algorithm [35]. It takes the whole data set as the starting point of the search and uses a feature ordering approach to select backward sequences from the whole set, eliminating one feature with the lowest ranking each time, until the feature subset that is most important for the classification results is selected. In the iterative process of the above steps, the order in which features are eliminated depends on their importance. The RFE algorithm requires a suitable classifier for modeling and prediction, for which the linear regression model was used in our experiment.

(4) LR is a machine learning model with simple form and good interpretability [36]. The LR model studies the multiple regression relationship formed between one dependent variable and multiple independent variables. Assuming a vector $x = (x_1, x_2, \ldots, x_n)$ of n independent variables, representing n characteristics of each sample, and letting the conditional probability $p(y = 1|x) = p$ be the probability of occurrence of an event x relative to an observed quantity, the LR model [36] can be expressed as

$$p(y = 1|x) = \frac{1}{1 + e^{-g(x)}} \tag{9}$$

where $g(x) = w_0 + w_1 x_1 + \ldots + w_n x_n$, $w_0, w_1, \ldots, w_n$ are the weights estimated with maximum likelihood.

3.3. Image Classification

In the classification process, the choice of classifier is an important factor determining the classification results. The CART decision tree, RF, k-nearest neighbors (kNN), and support vector machine (SVM) methods comprise four different classification algorithms.

The basic principle of CART is to form the test variables and the target variables into a data set, select the optimal segmentation features by calculating the Gini coefficient, then build a binary tree according to the feature values. These steps are cycled until the sample set to be classified reaches a stopping condition. There are two conditions for stopping: One is that there are no more feature variables for the target classification, and the other is that all samples of a given node belong to the same class. If the sample points of the stop classification node are of multiple classes, the node is specified as the class with the highest number of subclasses, and a new leaf is created within that class [37]. The binomial tree structure of the CART decision tree greatly improves the operational efficiency compared to the multinomial tree structure of the traditional decision tree [38].

The RF algorithm is an integrated classifier based on multiple decision trees. Through a bootstrap sampling method, a subset of samples is randomly selected from the original data as training samples, and decision trees are constructed for each training sample separately. A randomly selected feature is used as a node ($m < N$) of the decision tree, which is split and grown based on the amount of feature information. The training process is iterated until the maximum tree depth set by the user is reached or the splitting cannot continue [39]. An RF consists of N decision trees, and voting is used for each decision tree to obtain the final classification result [40]. RF has the advantage of high prediction accuracy, coupled with the fact that it is less prone to overfitting. Therefore, it has been widely used for image classification in high-resolution remote sensing data sets.

The kNN classification algorithm is a relatively simple machine learning algorithm [41]. In remote sensing image classification, this method determines the nearest k neighbors by calculating the distance between the samples to be classified and the training samples, then judges according to the categories of these k neighbors selected. The category to which the k neighbors belong the most is selected, and the samples to be classified are considered to belong to this category.

The SVM is a new machine learning method developed on the basis of statistical learning theory [42,43]. It is a non-parametric classifier. Based on the structural risk minimization criterion, the SVM solves image classification and regression problems by finding the optimal classification hyperplane in the high-dimensional feature space. According to the limited sample information, the best compromise between learning accuracy and learning effect can be obtained. The support vector machine has the advantages of simple implementation and high operational efficiency.

The above four classification algorithms each have their own advantages. In this study, all four algorithms are used to classify the optimized feature combination and the unoptimized full feature combination.

4. Object-Oriented Classification Process

The image segmentation in this study used the multi-scale segmentation [44] algorithm, where the segmentation parameters were determined by control variates. The basic principle of the control variates method is that all other parameters are unchanged, while only one of them is adjusted, and the best segmentation parameter combination is determined by adjusting the parameter values until each segmentation parameter is determined. Firstly, the shape factor and compactness factor are set as a fixed value, and then different segmentation scale parameters are set. The smaller the segmentation scale parameter, the larger the segmentation degree, and the more objects after segmentation. When the segmentation parameter is large, the image is undersegmented, and several ground objects are segmented into one object. After comparison, we found that when the segmentation scale is 80, the segmentation result is the best, and all different ground objects are divided. The size of shape factor and compactness factor also affect the segmentation result. A small form factor leads to poor segmentation of results, while a large form factor leads to excessively fragmented results. The compactness factor uses the shape criterion to optimize the results of the affected objects considering the overall compactness. When the shape factor is set to 0.1 and the compactness factor is set to 0.5, the segmentation effect of the experimental study area is better. After experimental analysis, when the segmentation scale, shape factor, and compactness factor were 80, 0.1, and 0.5, respectively, a relatively good segmentation effect was obtained, as depicted in Figure 3.

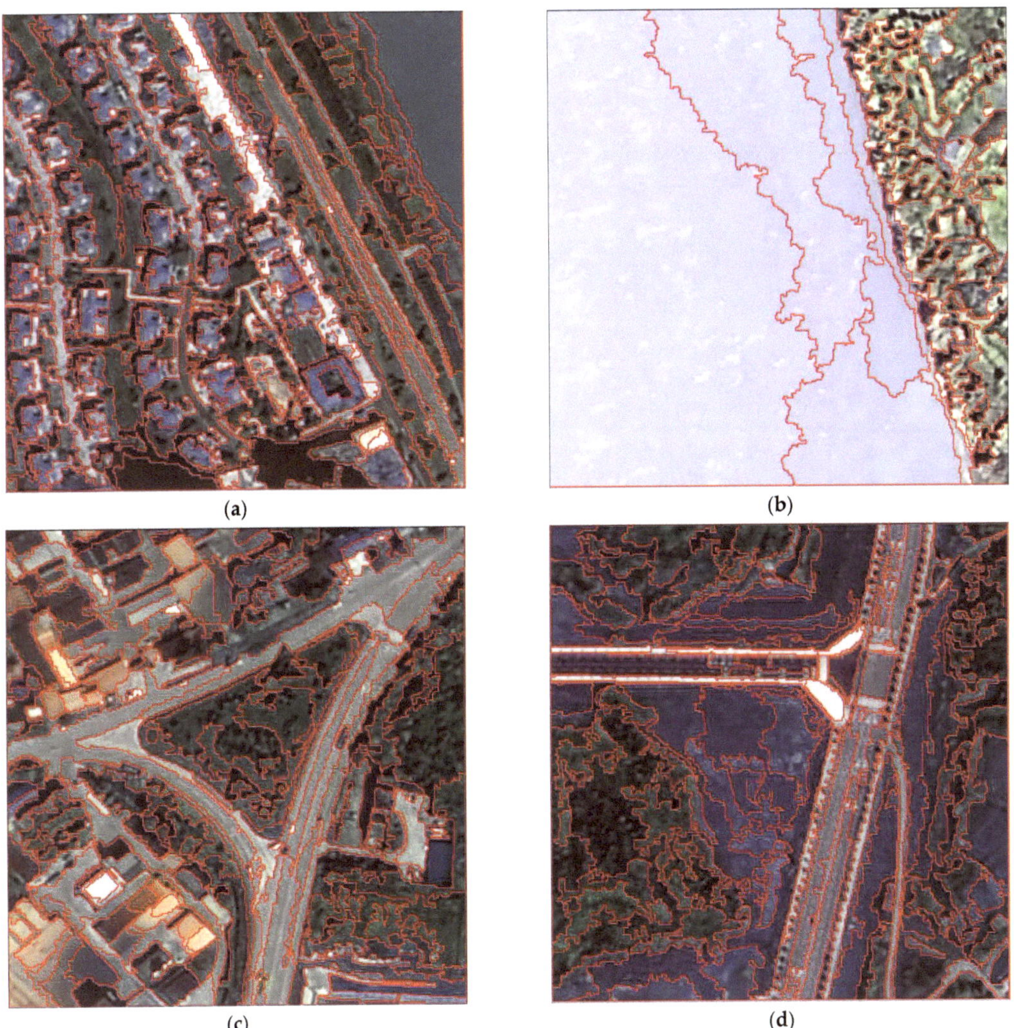

Figure 3. Typical ground object segmentation plot: (**a**) buildings; (**b**) water; (**c**) roads; (**d**) bare land.

4.1. Feature Selection Results

As the classification results of image classification are influenced by the number of samples and spatial location, stratified random sampling was adopted for each category of features, such that the number of samples in each category was proportional to the total area of the category. We selected 2/3 of the segmented objects to extract the features, including texture, geometric, and spectral feature values, while the remaining samples were used for accuracy testing.

To explore the importance ranking of the relevant features, the top 15 features obtained with the five feature selection methods are listed in Table 2, while the proportions of different types of features in different subsets are shown in Table 3. The results show that the features screened by different feature selection methods presented significant differences. In general, spectral and texture features accounted for a large proportion of the top 15 features. Figure 4 shows the correlation coefficient matrix of the top 15 features of different feature selection methods. The darker the grid color, the smaller the correlation coefficient between the features, and the more negative the correlation between the two features. On the contrary, the larger the correlation coefficient between features, the more positive the correlation between the two features.

Table 2. Top 15 features using various FS methods.

Fm	Fisher	mRMR	REF	LR
Standard_G	NDVI	Standard_R	GLCM_Entropy	GLCM_Ang_2nd moment
Density	NDWI	NDVI	Compactness	GLCM_Correlation
Standard_R	Mean_NIR	Length/Width	Standard_B	NDWI
Mean_B	Mean_B	GLCM_StdDev	Standard_R	Width
Width_Pxl	Area_Pxl	Standard_G	LengthWidt	GLCM_Mean
Border_length	Standard_B	Density	GLCM_StdDev	length
NDVI	Standard_NIR	Compactness	GLCM_Dissimilarity	GLDV_Entropy
Max_diff	Max_diff	GLCM_Correlation	GLCM_Mean_	GLCM_Homogeneity
GLDV_Entropy	Mean_R	GLCM_Dissimilarity	Mean_B	GLCM_StdDev
Standard_B	Width	Mean_B	Mean_G	Density
NDWI	Mean_G	Width_Pxl	Mean_NIR	max_diff
Mean_NIR	GLCM_Homogeneity	NDWI	Mean_R	Standard_NIR
GLDV_Ang_2nd moment	Standard_G	Number_of_	Standard_G	Length/Width
Mean_R	GLCM_Ang_2nd moment	Standard_B	GLDV_Entropy	NDVI
GLCM_Homogeneity	Brightness	GLCM_Mean	Standard_NIR	Standard_G

Table 3. Summary of the characteristics in the different categories of the top 15 characteristics according to Table 2.

Feature Selection Method	Feature Description	Spectral	Geometric	Texture
Fm	Number of features	9	3	3
	Top 15 feature ratios	60.00%	20.00%	20.00%
Fisher	Number of features	11	2	2
	Top 16 feature ratios	73.33%	13.33%	13.33%
mRMR	Number of features	7	5	4
	Top 17 feature ratios	46.67%	33.33%	26.67%
REF	Number of features	8	2	5
	Top 18 feature ratios	53.33%	13.33%	33.33%
LR	Number of features	5	4	6
	Top 19 feature ratios	33.33%	26.67%	40.00%

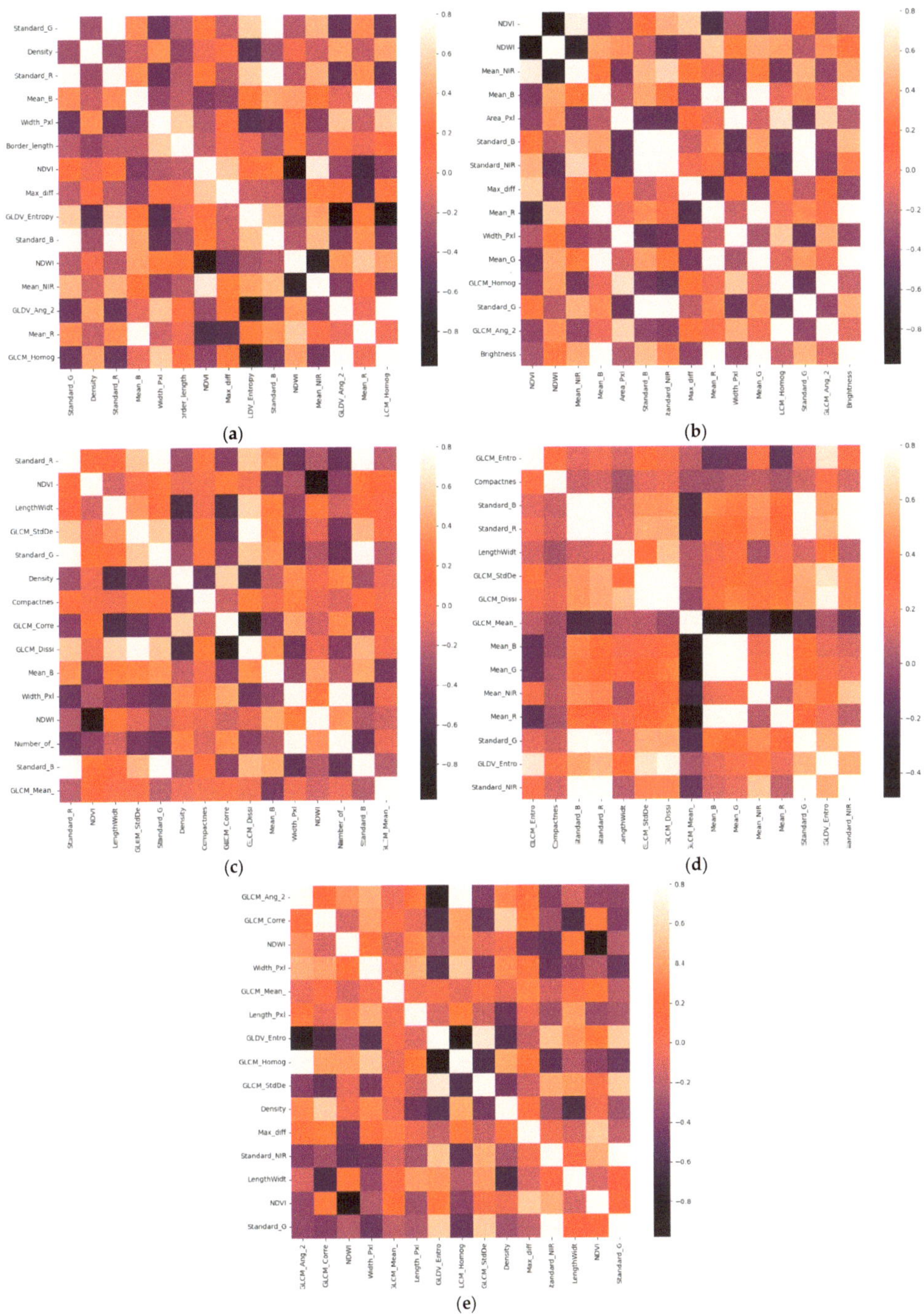

Figure 4. The correlation matrix of the top 15 features of the five feature selection methods: (**a**) Fm; (**b**) Fisher; (**c**) mRMR; (**d**) RFE; (**e**) LR.

Overall, the spectral features appeared significantly more frequently than the geometric and texture features. In the filtered feature selection method, NDVI, NDWI, Mean_B, Standard_B, Standard_G, and Width features were all present, while the number of texture features was more than 1/3. The mRMR and RFE algorithms selected seven features among the first 15 features which were the same, showing strong consistency. In addition, the LR algorithm appeared to choose more geometric and texture features, reaching 50% of the features in each category.

From the feature extraction results, it can be seen that spectral features were the most numerous. From the remote sensing images of the research area, the vegetation and water areas were larger than other land features, and NDVI and NDWI can effectively extract vegetation and water bodies. Secondly, there were many geometric features. Ground objects are usually characterized by large area and complex spectral features. It is difficult to distinguish objects such as roads and construction land from other ground objects only using spectral features, but they can be effectively classified using geometric features.

4.2. Comparison of Classification Results

As shown in Figure 5 below, based on the four classification methods, the trend of overall accuracy was obtained by continuously increasing the number of feature fields, and the feature selection methods were compared. As the number of features increased, the overall accuracy gradually improved. When the number of features reached about 15, the classification accuracy decreased slightly with an increase in the number of features, then remained stable. Therefore, in the process of object-oriented classification experiments, when too many features are involved in the classification, it may not be possible to achieve the optimal classification results, and instead the classification accuracy and classification efficiency are reduced. Overall, filtered feature selection methods presented better results than wrapped methods, while embedded feature selection methods presented the worst results. Furthermore, the SVM classification results were relatively stable, and the impact of different feature selection methods on the classification accuracy was smaller than the use of other classification methods.

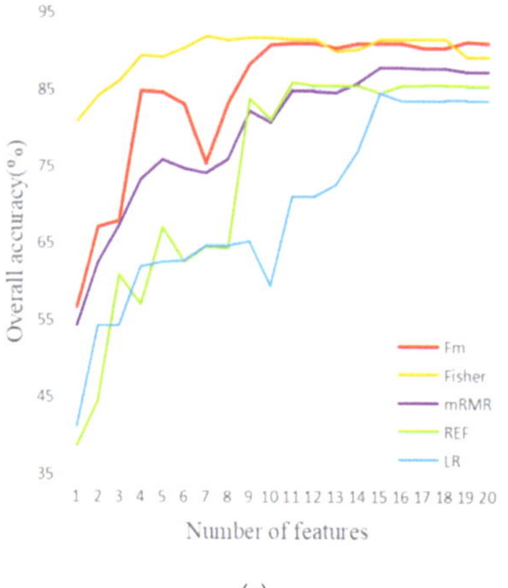

(a)

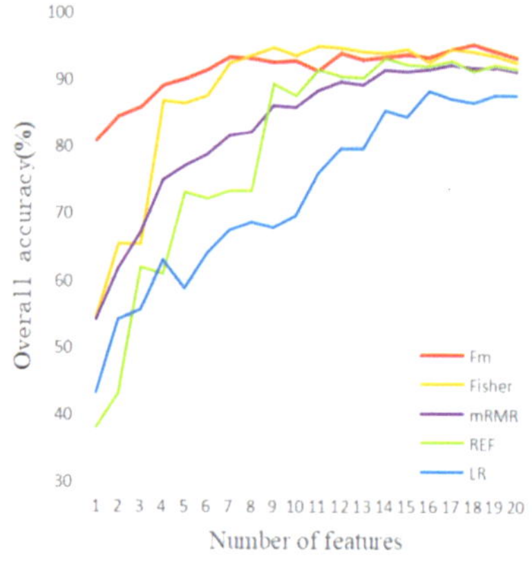

(b)

Figure 5. *Cont.*

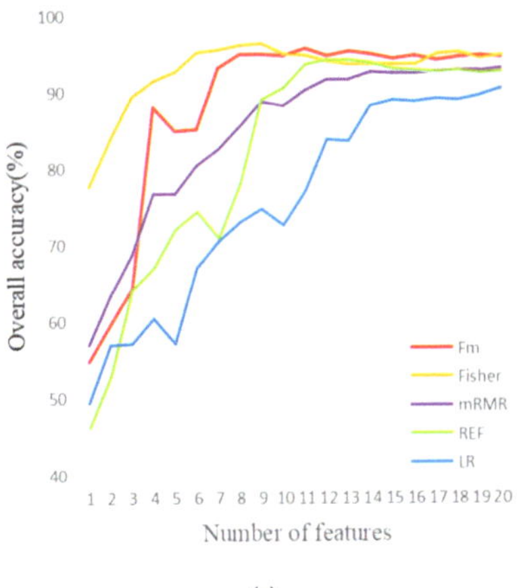
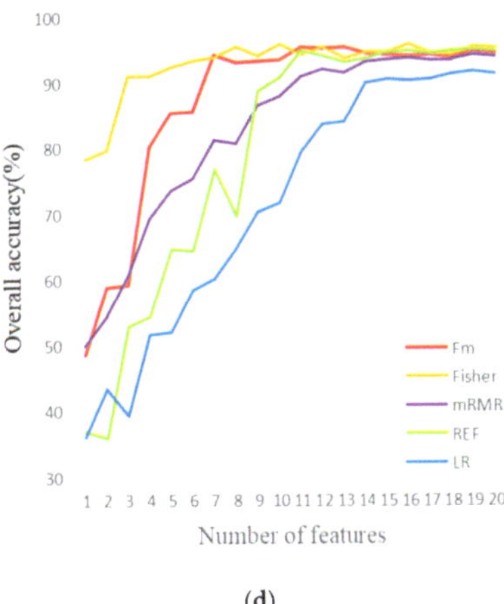

(c) (d)

Figure 5. The variation trend of number of features and overall accuracy of five feature selection methods with different classifiers: (**a**) CART; (**b**) RF; (**c**) KNN; (**d**) SVM.

According to Figure 6, the proposed Fm was found to have higher overall accuracy than the other feature selection methods with both RF and *k*NN classifiers, with accuracies of 95.18% and 96.14%, respectively. Although it was not optimal with the CART and SVM classifiers, the overall accuracy still achieved good results. This indicates that the combined scheme of Fisher Score and mRMR algorithm can obtain high-accuracy classification results with specific classifiers and can outperform both wrapped and filter feature selection methods.

	Fm	Fisher	mRMR	RFE	LR
CART	91.01	91.52	87.73	85.74	84.4
RT	95.18	94.8	92.15	93.06	88.24
KNN	96.14	95.56	92.96	94.02	90.36
SVM	95.37	94.22	95.76	95	91.71

Figure 6. The optimal accuracy of four features with different classifiers was selected.

As shown in the bar chart, the overall accuracy of Fm is better than the other four feature selection methods. In the CART classifier, the overall accuracy of Fm is 3.28%, 5.27%, and 6.61% higher than mRMR, RFE, and LR, respectively, and 0.41% lower than Fisher Score. In the SVM classifier experiment, the overall accuracy of Fm is higher than Fisher Score, RFE, and LR, respectively, and the overall accuracy of Fm is 0.39% lower than mRMR. Among RT and KNN classifiers, Fm achieves the highest overall accuracy, which is 0.38%, 3.03%, 2.12%, and 6.94% higher than Fisher Score, mRMR, REF, and LR and 0.58%, 3.18%, 2.12%, and 5.78% higher than Fisher Score, mRMR, and LR, respectively.

In the case of limited samples, excessive features do not improve the classification accuracy of the image. However, the classification accuracy can be improved to a certain extent by taking into account the correlation between features through the mRMR algorithm or only considering the separability of a single feature through the Fisher Score algorithm. Our experiments showed that the proposed Fm method can effectively improve the classification accuracy of high-resolution remote sensing images with the RF and kNN classifiers.

4.3. Classification Results

For further analysis, the optimal combinations of the four classification methods and five feature selection methods were selected—Fisher–CART, Fm–RF, Fm–kNN, and mRMR–SVM—and the classification graphs of these four combinations are shown in Figure 7. Meanwhile, in order to analyze the accuracy of the classified ground objects, the overall accuracy, as well as kappa coefficients, were calculated to evaluate the classification results, based on the decoded flags and visually decoded sample points, as shown in Table 4, as well as the specific land-cover classification accuracies.

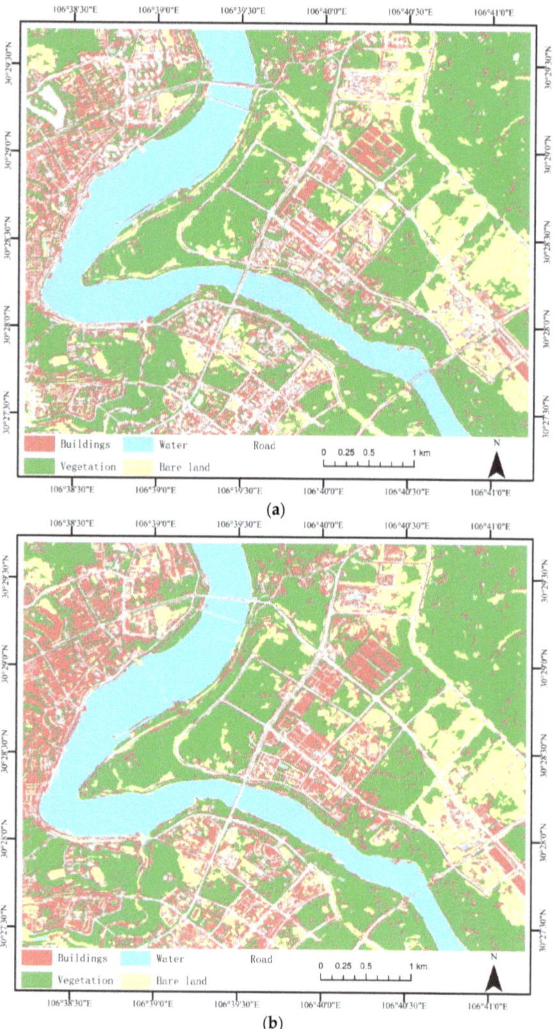

Figure 7. *Cont.*

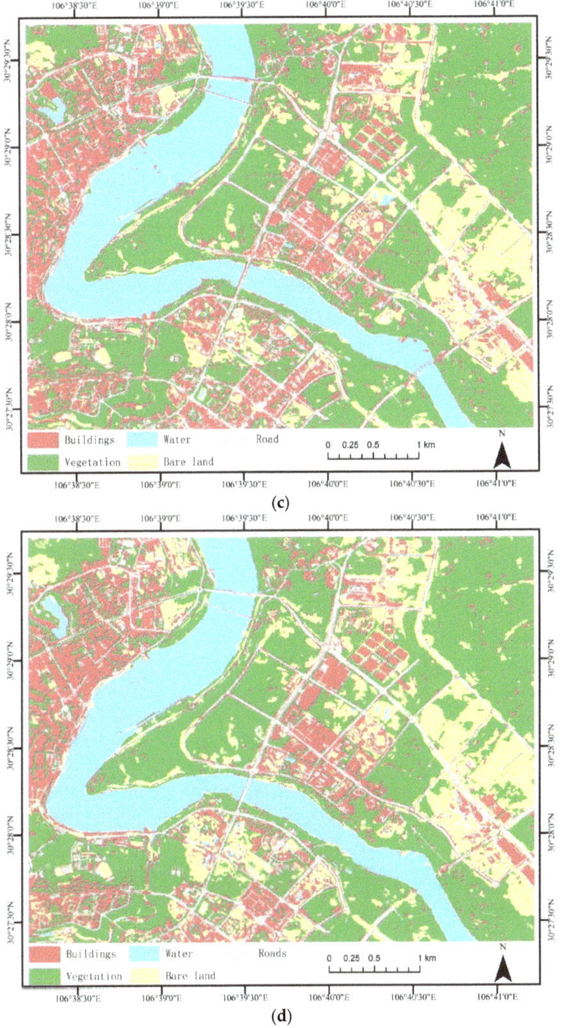

Figure 7. Graph of optimal classification results for different classifiers (**a**) Fisher-CART; (**b**) Fm-RF; (**c**) Fm-KNN; (**d**) mRMR-SVM.

Table 4. Accuracy of ground cover classification.

	Fisher-CART		Fm-RF		Fm-KNN		mRMR-SVM	
	Producer's accuracy	User's accuracy	Producer's accuracy	User's accuracy	Producer's accuracy	User's accuracy	Producer's accuracy	User's accuracy
Roads	95.00%	76.61%	97.00%	93.26%	93.00%	95.87%	97.00%	89.81%
Buildings	74.00%	91.02%	85.00%	91.40%	92.00%	91.08%	85.00%	94.44%
Water	92.93%	97.87%	93.94%	97.90%	96.96%	98.96%	97.98%	98.97%
Bare land	97.14%	94.44%	100%	90.90%	98.57%	94.52%	98.57%	97.18%
Vegetation	99.33%	98.68%	99.33%	99.33%	99.33%	98.67%	99.33%	98.03%
Overall accuracy	91.52%		95.18%		96.14%		95.76%	
Kappa	0.8923		0.939		0.951		0.9461	

From the analysis in Table 4, the producer's and user's accuracies for water, bare land, and vegetation in all four scenarios were greater than 94%. The water and vegetation extraction was improved, followed by that of bare land. The reason for this is that sparse grass can be classified as vegetation and bare land, and it is difficult to accurately determine which type the associated features belong to. The extraction effect of buildings and roads was relatively poor as the resolution of the images was high, and some narrow roads were interspersed among the buildings, making it easy to divide the roads and buildings together during segmentation, causing confusion between the two types of features. From the overall accuracy of the four schemes, the overall accuracy of Fm–RF, Fm–*k*NN, and mRMR–SVM were all greater than 95%, which indicates that Fm can better combine and optimize feature subsets and improve the classification ability of the used feature sets. All of the feature selection methods based on Fm could achieve effective surface feature information extraction. Meanwhile, the RFE and LR classification methods did not present high classification accuracy. Wrapper feature selection methods rely on feature models and specific machine learning algorithms, and the optimal feature combinations change as the learners change, which, in some cases, can have detrimental effects. In our experiment, there were negative values in the NDVI and NDWI feature values, and there was a data imbalance; for the embedded LR feature selection method, it is difficult to solve the data imbalance problem. In conclusion, both the single filtering feature selection methods and the combination of the proposed two filtering feature selection methods presented good performance, and filtering feature selection methods can more easily obtain better classification results when performing object-oriented classification.

4.4. Validation of the Fm Method

In order to test the effectiveness of the Fm method, we selected different research areas and GF-2 images and carried out experiments. The classification results and overall accuracy with different classifiers are shown in the Figure 8.

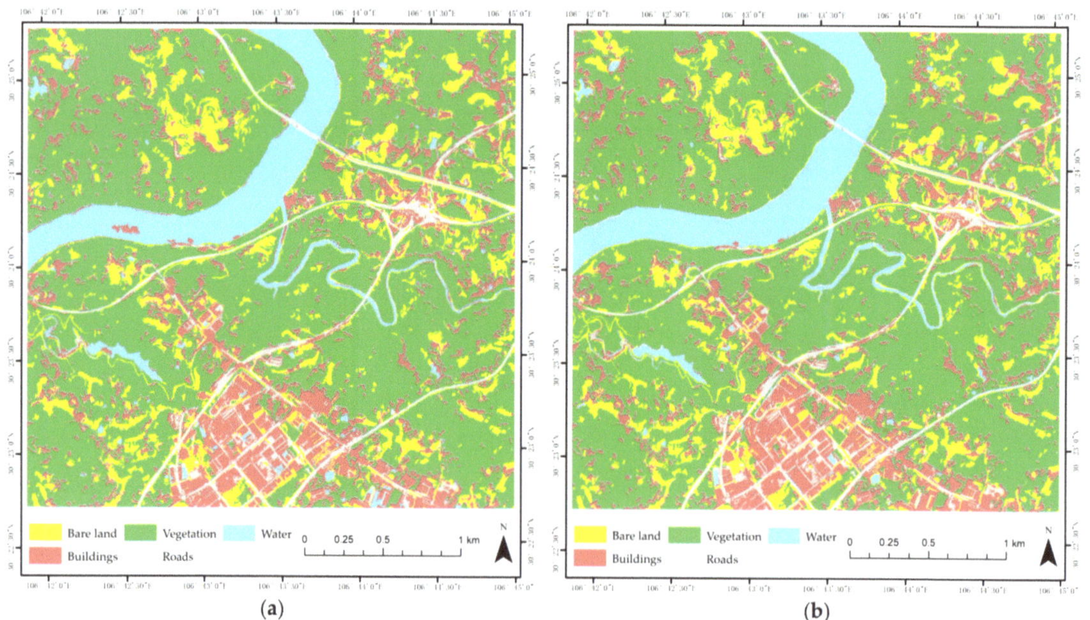

Figure 8. *Cont.*

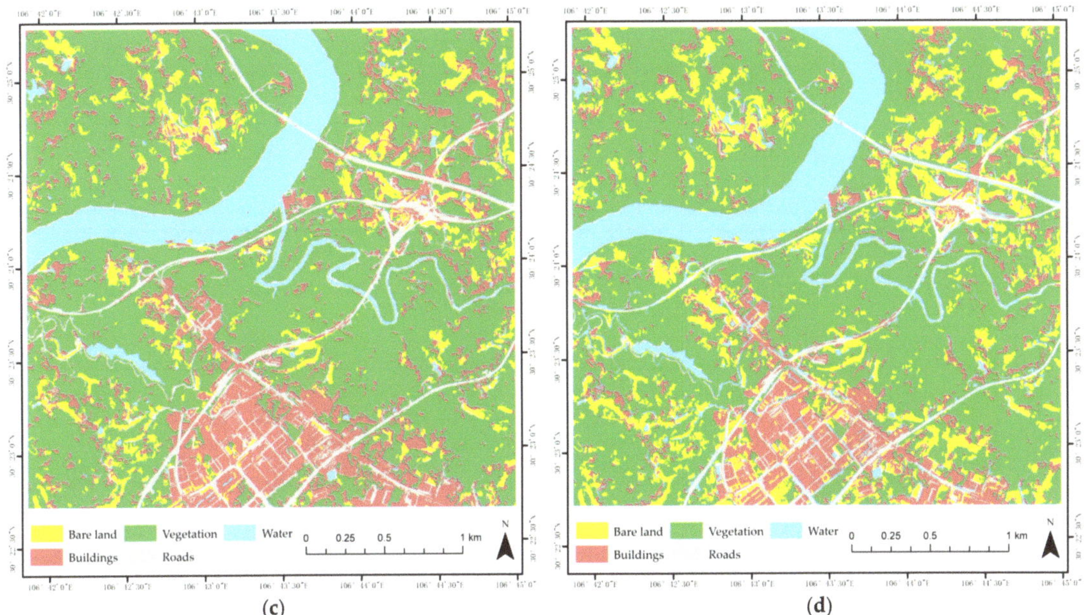

Figure 8. Fm classification image with different classifiers: (**a**) CART; (**b**) RF; (**c**) KNN; (**d**) SVM.

As shown in Table 5, Fm achieves better classification results in the different study area. The overall accuracy of Fm with CART, RF, KNN, and SVM classifiers were 88.67%, 92.04%, 91.08%, and 88.68%, respectively. The kappa coefficient also reached 0.8545, 0.8979, 0.8852, and 0.8546, respectively. The overall accuracy of RF and KNN is better than that of the CART and SVM, which is consistent with the experimental results above. Verified experiments show that Fm can effectively reduce the dimensionality of high-dimensional data and obtain the optimal feature subset. RF and KNN classifiers are more suitable for image classification combined with Fm.

Table 5. Overall accuracy and kappa coefficient of Fm with different classifiers.

	CART	RF	KNN	SVM
Overall accuracy	88.67%	92.04%	91.08%	88.68%
Kappa	0.8545	0.8979	0.8852	0.8546

5. Conclusions

In this paper, an algorithm combining the Fisher Score and mRMR algorithms is proposed to address the problem of high dimensionality of the feature space in object-oriented classification. Although Fisher Score and mRMR feature selection methods have good applicability in feature screening, a single method cannot take into account the redundancy between features and the correlation between features and categories at the same time. The Fisher Score algorithm does not take into account the redundancy between features, and the mRMR algorithm cannot reflect the differences in the role of different features in classification. The combination of the Relief and mRMR algorithms can effectively make up for their shortcomings. After the experiments involving four different machine learning classification methods, the overall accuracy of Fm combined with RF and KNN is better than that of RF and CART.

Through a comparative test of four kinds of classifiers, we determined that: (1) Feature selection can allow for the elimination of redundant features, and high classification accuracy can still be achieved when using a small number of features. Accordingly, filtered feature selection methods were found to perform better than wrapped and embedded feature selection methods. (2) Two classifiers—RF and SVM—exhibited better stability than the other two classifiers as the number of features increased during the experiment. (3) In this study, the proposed Fm feature selection method was used in the classification experiment and showed the best performance when used with the RF and *k*NN classifiers, allowing for better optimization of the feature set. The final classification accuracy and efficiency were improved obviously by using the Fm feature subset. The overall accuracy of the Fm–RF and Fm–*k*NN approaches reached 95.18% and 96.14%, respectively. The kappa coefficient reached 0.939 and 0.951, respectively. Except for the mapping accuracy of Fm-RF construction land, the mapping accuracy and user accuracy of Fm-RF and Fm-KNN both reached more than 91%.

Author Contributions: Conceptualization, Y.L. and C.L.; methodology, C.L.; software, H.F.; validation, X.F. and C.X.; formal analysis, L.X.; investigation, M.L.; data curation, X.F.; writing—original draft preparation, C.L.; writing—review and editing, Y.L.; supervision, L.X.; project administration, M.L. All authors have read and agreed to the published version of the manuscript.

Funding: This research was funded by the Major Project of High Resolution Earth Observation System of China (no. GFZX0404130304); the Open Fund of Hunan Provincial Key Laboratory of Geo-Information Engineering in Surveying, Mapping and Remote Sensing, Hunan University of Science and Technology (no. E22201); the Agricultural Science and Technology Innovation Program (ASTIP no. CAAS-ZDRW202201); a grant from the State Key Laboratory of Resources and Environmental Information System; and the Innovation Capability Improvement Project of Scientific and Technological Small and Medium-Sized Enterprises in Shandong Province of China (no. 2021TSGC1056).

Institutional Review Board Statement: Not applicable.

Informed Consent Statement: Not applicable.

Data Availability Statement: Data and code from this research will be available from the authors upon request.

Acknowledgments: The authors sincerely thank the comments from anonymous reviewers and members of the editorial team.

Conflicts of Interest: The authors declare no conflict of interest.

References

1. Muzirafuti, A.; Cascio, M.; Lanza, S.; Randazzo, G. UAV Photogrammetry-based Mapping of the Pocket Beaches of Isola Bella Bay, Taormina (Eastern Sicily). In Proceedings of the 2021 International Workshop on Metrology for the Sea; Learning to Measure Sea Health Parameters (MetroSea), Reggio Calabria, Italy, 4–6 October 2021; pp. 418–422.
2. Randazzo, G.; Italiano, F.; Micallef, A.; Tomasello, A.; Cassetti, F.P.; Zammit, A.; D'Amico, S.; Saliba, O.; Cascio, M.; Cavallaro, F.; et al. WebGIS Implementation for Dynamic Mapping and Visualization of Coastal Geospatial Data: A Case Study of BESS Project. *Appl. Sci.* **2021**, *11*, 8233. [CrossRef]
3. Hong, L.; Feng, Y.F.; Peng, H.Y.; Chu, S.S. Classification of high spatial resolution remote sensing imagery based on object-oriented multi-scale weighted sparse representation. *Acta Geod. Cartogr. Sin.* **2022**, *51*, 224–237. [CrossRef]
4. Wang, H.; Wang, C.B.; Wu, H.G. Using GF-2 Imagery and the Conditional Random Field Model for Urban Forest Cover Mapping. *Remote Sens. Lett.* **2016**, *7*, 378–387. [CrossRef]
5. Zhang, X.Y.; Feng, X.Z.; Jiang, H. Feature set optimization in object-oriented methodology. *J. Remote Sens.* **2009**, *13*, 664–669. [CrossRef]
6. Stefanos, G.; Tais, G.; Sabine, V.; Morutz, S.; Stamatis, K.; Eleonore, W. Less is more: Optimizing classification performance through feature selection in a very-high-resolution remote sensing object-based urban application. *GISci. Remote Sens.* **2018**, *55*, 221–242.
7. Xue, B.; Zhang, M.; Browne, W.N.; Yao, X. A Survey on Evolutionary Computation Approaches to Feature Selection. *IEEE Trans. Evol. Comput.* **2016**, *20*, 606–626. [CrossRef]
8. Li, J.; Cheng, K.; Wang, S.; Morstatter, F.; Trevino, R.P.; Tang, J.; Liu, H. Feature Selection: A Data Perspective. *ACM Comput. Surv. (CSUR)* **2017**, *50*, 1–45. [CrossRef]
9. Dokeroglu, T.; Deniz, A.; Kiziloz, H.E. A Comprehensive Survey on Recent Metaheuristics for Feature Selection. *Neurocomputing* **2022**, *494*, 269–296. [CrossRef]
10. Saúl, S.F.; Carrasco-Ochoa, J.A.; José, M.T. A new hybrid filter-wrapper feature selection method for clustering based on ranking. *Neurocomputing* **2016**, *214*, 866–880.

11. Zhao, L.; Gong, J.X.; Huang, D.R.; Hu, C. Fault feature selection method of gearbox based on Fisher Score and maximum information coefficient. *Control. Decis.* **2021**, *36*, 2234–2240. [CrossRef]

12. Zhou, Y.; Zhang, R.; Wang, S.X.; Wang, F.T. Feature Selection Method Based on High-Resolution Remote Sensing Images and the Effect of Sensitive Features on Classification Accuracy. *Sensors* **2018**, *18*, 2013. [CrossRef]

13. Liu, W.; Wang, J.Y. Recursive elimination–election algorithms for wrapper feature selection. *Appl. Soft Comput. J.* **2021**, *113*, 107956. [CrossRef]

14. Li, M.; Kamili, M. Research on Feature Selection Methods and Algorithms. *Comput. Technol. Dev.* **2013**, *23*, 16–21. [CrossRef]

15. Wu, D.; Guo, S.C. An improved Fisher Score feature selection method and its application. *J. Liaoning Tech. Univ. (Nat. Sci.)* **2019**, *38*, 472–479.

16. Gu, Q.Q.; Li, Z.H.; Han, J.W. Generalized Fisher Score for Feature Selection. *arXiv* **2012**, arXiv:1202.3725.

17. Cheng, X.M.; Shen, Z.F.; Xing, T.Y.; Xia, L.G.; Wu, T.J. Efficiency and accuracy analysis of multi-spectral remote sensing image classification based on mRMR feature optimization algorithm. *J. Geo-Inf. Sci.* **2016**, *18*, 815–823. [CrossRef]

18. Chen, S.L.; Gao, X.X.; Liao, Y.F.; Deng, J.B.; Zhou, B. Wetland classification method of Dongting Lake district based on CART using GF-2 image. *Bull. Surv. Map.* **2021**, *6*, 12–15. [CrossRef]

19. Gómez, C.; Wulder, M.A.; Montes, F.; Delgado, J.A. Modeling Forest Structural Parameters in the Mediterranean Pines of Central Spain using QuickBird-2 Imagery and Classification and Regression Tree Analysis (CART). *Remote Sens.* **2012**, *4*, 135–159. [CrossRef]

20. Gu, H.Y.; Yan, L.; Li, H.T.; Jia, Y. An Object-based Automatic Interpretation Method for Geographic Features Based on Random Forest Machine Learning. *Geomat. Inf. Sci. Wuhan Univ.* **2016**, *41*, 228–234. [CrossRef]

21. Dennis, C.D.; Steven, E.F.; Monique, G.D. A comparison of pixel-based and object-based image analysis with selected machine learning algorithms for the classification of agricultural landscapes using SPOT-5 HRG imagery. *Remote Sens. Environ.* **2012**, *118*, 259–272.

22. Voisin, A.; Krylov, V.A.; Moser, G.; Serpico, S.B.; Zerubia, J. Supervised Classification of Multisensor and Multiresolution Remote Sensing Images with a Hierarchical Copula-Based Approach. *IEEE Trans. Geosci. Remote Sens.* **2014**, *52*, 3346–3358. [CrossRef]

23. Paradis, E. Probabilistic unsupervised classification for large-scale analysis of spectral imaging data. *Int. J. Appl. Earth Obs. Geoinf.* **2022**, *107*, 102675. [CrossRef]

24. Liu, S.; Jiang, Q.G.; Ma, Y.; Xiao, Y.; Li, Y.H.; Cui, C. Object-oriented Wetland Classification Based on Hybrid Feature Selection Method Combining with Relief F/Mult-objective Genetic. *Trans. Chin. Soc. Agric. Mach.* **2017**, *48*, 119–127. [CrossRef]

25. Zhang, W.Q.; Li, X.R.; Zhao, L.Y. Discovering the Representative Subset with Low Redundancy for Hyperspectral Feature Selection. *Remote Sens.* **2019**, *11*, 1341. [CrossRef]

26. Wang, L.; Gong, G.H. Multiple features remote sensing image classification based on combining ReliefF and mRMR. *Chin. J. Stereol. Image* **2014**, *19*, 250–257. [CrossRef]

27. Wu, Q.; Zhong, R.F.; Zhao, W.J.; Song, K.; Du, L.M. Land-cover classification using GF-2 images and airborne lidar data based on Random Forest. *Int. J. Remote Sens.* **2019**, *40*, 2410–2426. [CrossRef]

28. Shao, L.Y.; Zhou, Y. Application of improved oversampling algorithm in class-imbalance credit scoring. *Appl. Res. Comput.* **2019**, *36*, 1683–1687. [CrossRef]

29. Zhu, J.F.; Li, F.; Lu, B.X. Comparative Study of Fisher and KNN Discriminant Classification Algorithms Based on Clustering Improvement. *J. Anhui Agric. Sci.* **2019**, *47*, 250–252, 257.

30. Xu, X.Y.; Zhao, L.Z.; Chen, X.Y.; He, Z.C. Design of Convolutional Neural Network Based on Improved Fisher Discriminant Criterion. *Comput. Eng.* **2020**, *46*, 255–260, 266. [CrossRef]

31. Huang, L.S.; Ruan, C.; Huang, W.J.; Shi, Y.; Peng, D.L.; Ding, W.J. Wheat Powdery mildew monitoring based on GF-1 remote sensing image and relief-mRMR-GASVM model. *Trans. Chin. Soc. Agric. Eng.* **2018**, *34*, 167–175, 314. [CrossRef]

32. Özyurt, F. A fused CNN model for WBC detection with MRMR feature selection and extreme learning machine. *Soft Comput.* **2020**, *24*, 163–8172. [CrossRef]

33. Huang, L.; Xiang, Z.J.; Chu, H. Remote sensing image classification algorithm based on mRMR selection and IFCM clustering. *Bull. Surv. Map.* **2019**, *4*, 32–37. [CrossRef]

34. Zhang, X.L.; Zhang, F.; Zhou, N.; Zhang, J.J.; Liu, W.F.; Zhang, S.; Yang, X.J. Near-Infrared Spectral Feature Selection of Water-Bearing Rocks Based on Mutual Information. *Spectrosc. Spectr. Anal.* **2021**, *41*, 2028–2035. [CrossRef]

35. Wu, C.W.; Liang, J.H.; Wang, W.; Li, C.S. Random Forest Algorithm Based on Recursive Feature Elimination. *Stat. Decis.* **2017**, *21*, 60–63. [CrossRef]

36. Fan, T.C.; Jia, Y.F.; Li, Y.F.; Zhao, J.L. Prediction of Gully Distribution Probability in Yanhe Basin Based on Remote Sensing Image and Logistic Regression Model. *Res. Soil Water Conserv.* **2022**, *29*, 316–321. [CrossRef]

37. Luo, H.X.; Li, M.F.; Dai, S.P.; Li, H.L.; Li, Y.P.; Hu, Y.Y.; Zheng, Q.; Yu, X.; Fang, J.H. Combinations of Feature Selection and Machine Learning Algorithms for Object-Oriented Betel Palms and Mango Plantations Classification Based on Gaofen-2 Imagery. *Remote Sens.* **2022**, *14*, 1757. [CrossRef]

38. Lu, L.Z.; Tao, Y.; Di, L.P. Object-Based Plastic-Mulched Landcover Extraction Using Integrated Sentinel-1 and Sentinel-2 Data. *Remote Sens.* **2018**, *10*, 1820. [CrossRef]

39. Yang, H.B.; Li, F.; Wang, W.; Yu, K. Estimating Above-Ground Biomass of Potato Using Random Forest and Optimized Hyperspectral Indices. *Remote Sens.* **2021**, *13*, 2339. [CrossRef]

40. Wang, G.Z.; Jin, H.L.; Gu, X.H.; Yang, G.J.; Feng, H.K.; Sun, Q. Remote Sensing Classification of Autumn Crops Based on Hybrid Feature Selection Model Combining with Relief F and Improved Separability and Thresholds. *Trans. Chin. Soc. Agric. Mach.* **2021**, *52*, 199–210. [CrossRef]
41. Garg, R.; Kumar, A.; Prateek, M.; Pandey, K.; Kumar, S. Land Cover Classification of Spaceborne Multifrequency SAR and Optical Multispectral Data using Machine Learning. *Adv. Space Res.* **2021**, *69*, 1726–1742. [CrossRef]
42. Zhang, S.; Huang, H.; Huang, Y.; Cheng, D.; Huang, J. A GA and SVM Classification Model for Pine Wilt Disease Detection Using UAV-Based Hyperspectral Imagery. *Appl. Sci.* **2022**, *12*, 6676. [CrossRef]
43. Hu, J.M.; Dong, Z.Y.; Yang, X.Z. Object-oriented High-resolution Remote Sensing Image Information Extraction Method. *Geospat. Inf.* **2021**, *19*, 10–13, 18, 157. [CrossRef]
44. Hao, S.; Cui, Y.; Wang, J. Segmentation Scale Effect Analysis in the Object-Oriented Method of High-Spatial-Resolution Image Classification. *Sensors* **2021**, *21*, 7935. [CrossRef] [PubMed]

Article

Determination of Forest Structure from Remote Sensing Data for Modeling the Navigation of Rescue Vehicles

Marian Rybansky 

Faculty of Military Technology, University of Defence, Kounicova 65, 662 10 Brno, Czech Republic; marian.rybansky@unob.cz; Tel.: +420-973-445-298

Abstract: One of the primary purposes of forest fire research is to predict crisis situations and, also, to optimize rescue operations during forest fires. The research results presented in this paper provide a model of Cross-Country Mobility (CCM) of fire brigades in forest areas before or during a fire. In order to develop a methodology of rescue vehicle mobility in a wooded area, the structure of a forest must first be determined. We used a Digital Surface Model (DSM) and Digital Elevation Model (DEM) to determine the Canopy Height Model (CHM). DSM and DEM data were scanned by LiDAR. CHM data and field measurements were used for determining the approximate forest structure (tree height, stem diameters, and stem spacing between trees). Due to updating the CHM and determining the above-mentioned forest structure parameters, tree growth equations and vegetation growth curves were used. The approximate forest structure with calculated tree density (stem spacing) was used for modeling vehicle maneuvers between the trees. Stem diameter data were used in cases where it was easier for the vehicle to override the trees rather than maneuver between them. Although the results of this research are dependent on the density and quality of the input LiDAR data, the designed methodology can be used for modeling the optimal paths of rescue vehicles across a wooded area during forest fires.

Keywords: forest fire; rescue vehicle; vegetation structure; optimal pathfinding; canopy height model (CHM)

Citation: Rybansky, M. Determination of Forest Structure from Remote Sensing Data for Modeling the Navigation of Rescue Vehicles. *Appl. Sci.* **2022**, *12*, 3939. https://doi.org/10.3390/app12083939

Academic Editors: Giovanni Randazzo, Stefania Lanza and Anselme Muzirafuti

Received: 22 March 2022
Accepted: 11 April 2022
Published: 13 April 2022

Publisher's Note: MDPI stays neutral with regard to jurisdictional claims in published maps and institutional affiliations.

1. Introduction

Forest fires are very frequent crisis situations, especially in dry or arid landscapes [1]. The prediction of forest fire occurrence depends on knowledge of the factors that affect the fires and on the technologies that facilitate the monitoring and modeling the spread of the fire. Ganteaume et al. [2] analyzed the most common human and environmental factors driving forest fire ignition. The primary factors that directly cause forest fires are natural (lightning strikes, seismic and volcanic activity, etc.) or human (carelessness and activities such as arson, slash-and-burn agriculture, fire-fallow cultivation, machinery sparks, discarded glass bottles or cigarette butts, military activity, etc.). The factors that determine fire spread are as follows: forest type and structure (distances between trees, *DBH*, canopy height, tree crown density, etc.); meteorological conditions (precipitations, temperature, wind speed, air humidity, cloudiness, soil moisture, etc.); topographic (morphological shapes of terrain, orientation of relief slopes, etc.); geological and pedological (underground structure, soil structure, and terrain surface color); and season and time of day, which determine the amount of available sunlight and temperature, etc. The technologies that facilitate the monitoring and modeling of a forest structure and the spread of forest fires include the sensor types for vegetation data collection and forest structure determination and technologies for monitoring and modeling the spread of fire. Blair, Rabine, and Hofton [3] described the Laser Vegetation Imaging Sensor (LVIS), which operates at altitudes of up to 10 km aboveground and is capable of producing data for topographic mapping with dm accuracy and vertical height and structure measurements of vegetation. The LVIS instrument is also suitable for subcanopy ground elevation mapping. Lim et al. [4] described many of the initial studies of the application of LiDAR for forestry focused on verifying through statistical analysis that LiDAR could be used to accurately measure forest attributes. The focus has been on canopy tree heights given the nature of this attribute as a predictor variable for other forest attributes, such canopy density. Hyyppä et al. [5] analyzed existing algorithms and methods of airborne laser scanning that are used for extraction of the canopy height and individual tree information. Aschoff and Spiecker [6] described an algorithm for detecting trees in a semiautomatic way. Gobakken and Næsset [7] analyzed the effects of forest growth on laser-derived canopy metrics. Carson et al. 2004 [8], Ahlberg et al. [9], and Su et al. [10] provided the overview of LiDAR applications in forestry. By comparing these methods based on laser scanning, it can be stated that, at present, an approximate forest structure for modeling the movement of rescue vehicles can be determined. At the metric density of the DSM (CHM), the error in tree positioning can reach values in decimeters, sometimes up to meters, depending on the structure and type of the canopy. At the decimeter density of the DSM points out, it is possible to calculate the tree position errors in centimeters to decimeters. When comparing the possibilities of using LiDAR and aerial optical images, both methods have advantages and disadvantages. Aerial images provide both spatial and image information, but they do not allow, unlike LiDAR, full automation to determine the forest structure.

However, the use of LiDAR and aerial optical images may be problematic in the area of fire because of the clouds or smoke generated by the fire. In these situations, radar methods can be used to measure forest parameters. The mapping of forest units by radar is described, for example, by Martoni et al. in [11]. Kugler et al. [12] compared the LiDAR and radar methods for determining the heights of the forest in three areas: boreal, temperate, and tropical. The correlations achieved confirm the possibility of combining the use of both forest mapping methods. Additionally, Cazcarra-Bes et al. [13] described the possibilities of the horizontal and vertical forest structure mapping from radar using data obtained by synthetic aperture radar tomography. The use of radar methods is, however, limited in terms of the accuracy of the determination of the characteristics of individual trees. Furthermore, Landsat or Sentinel 2 global satellite data can be used to monitor forests before and during a fire. Sentinel 2 with a multispectral instrument (MSI) with 13 spectral channels in the visible/near-infrared (VNIR) and shortwave infrared spectral range (SWIR) and three bands for vegetation mapping can provide the crisis management with actual data in the shortest possible time, especially during a forest fire. The data accuracy (about 20 m) does not allow a more accurate mapping of the internal forest structure—see, e.g., Puletti et al. [14].

Technologies for monitoring and modeling the spread of a fire are divided into stages: prediction, during the fire, and post-fire [15]. Milz and Rymdteknik [16] described the technologies of detection and the spread of the forest fires by using satellite-borne remote sensing techniques. However, the technologies of fire monitoring and distribution are limited by the availability of up-to-date data from satellites, planes, UAVs, or terrestrial observations. We also need to know the prediction of the spread of a fire to deploy rescue vehicles. Koo et al. [17] described possible solutions using a physical model for the forest fire spread rate. This model successfully evaluated wind and slope effects of a fire on forest vegetation.

The above-mentioned factors and technologies are very important for the teams (fire brigades, military units, health services, and police) that are deployed to rescue people and reduce the damage during forest fires. Remote sensing support is very important for rescue units when they are moving across vegetation before and during a fire and, also, for the decision to deploy aircraft. We can use LiDAR and aerial image data to create a navigation analysis for rescue (fire brigade or military) vehicles—see also [18–20]. These data can be supplemented by active fire scenes using infrared sensors or aerial or UAV images. Among the most effective data sources for Cross-Country Movement (CCM) navigation and optimal pathfinding across a forest are LiDAR data and the products of its analysis.

A prerequisite for the success of this analysis is an up-to-date picture of vegetation data obtained by laser scanning. This precondition is especially crucial for forest stands, where data become quickly outdated due to vegetation growth.

The primary focus of this article was the LiDAR data update for the forest stand structure, a simulation of the creation of a forest structure with the subsequent creation of a model for navigating the movement of a rescue vehicle between trees as obstacles in the terrain. The reason for designing the method of detecting the current forest structure was that LiDAR data in the Czech Republic is gradually becoming obsolete as a result of tree growth. The following procedure was chosen: (1) Selection of the most common type of forest stands in the territory of the Czech Republic with the predominant spruce tree (Picea abies). Obtaining LiDAR data characterizing DSM with a density of 1 × 1 m. (2) Obtaining inventory data on the growth of spruce trees from MENDEL University, Brno. (3) Detection of DSM accuracy by geodetic and photogrammetric method. (4) Corrections of tree heights due to DSM density and tree growth. (5) Creation of forest structure by random distribution. (6) Selecting a simulated area where a fire could occur (older, drier forest). (7) Calculating the simulated shortest route for a particular vehicle (outside the area of the fire). The research results presented in this paper represent a new methodology of updating a digital surface model (DSM) or canopy height model (CHM) using the equations of tree growth and vegetation growth curves. DSM and Digital Elevation Model (DEM) data evaluated for forestry passability were scanned by LiDAR in 2013. CHM data and field measurements were used for determining the approximate forest structure (tree height, *DBH*, and stem spacing between trees). The described methods were tested on a spruce forest stand composed only from one type of tree—Sitka spruce (Picea abies), situated approximately 300 m south of the village of Brno-Utechov (see Figure 1), where the heights of trees in the Krtiny Training Forest Enterprise (TFE) area were detected and measured. This spruce forest was chosen because of the availability of a series of aerial photos and LiDAR data. Additionally, the forest is highly representative, as it contains the tree species most commonly found in many Central European countries. The current age of this forest is about 30 years.

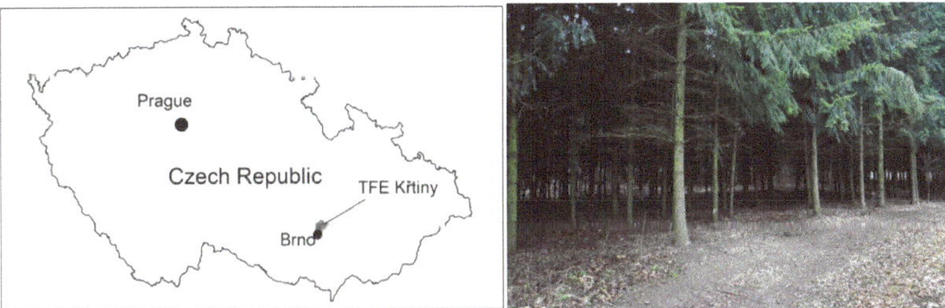

Figure 1. Location and character of TFE Krtiny.

The approximate forest structure with the calculated tree density (stem spacing) was used for modeling fire brigade vehicle maneuvers between the trees. *DBH* data were used in cases where it was easier for the vehicle to override the trees rather than maneuver between them. *DBH* data were also used for the calculation of the distances between trees. Due to the availability of a DSM with a density of 1 × 1 m, it was impossible to precisely determine the locations of individual trees, so a random simulated forest structure was created based on the number of trees per hectare.

The article describes the methodology of calculation of the rescue vehicle movement using simulated areas of a burning forest. In the case of a real fire, we can use the above-mentioned LiDAR data (DSM data) or current data from different sensors. The type of sensors and the accuracy of the data obtained will have a significant impact on the terrain analysis and search algorithm for optimal rescue vehicle routes. Only some scattered and low-resolution data of the fire can help. If the optimal route for a special vehicle (fire-resistant rescue vehicle, tank, etc.) that will move through a burning forest is to be calculated, we will need detailed vegetation and elevation data with a meter or decimeter resolution for reconnaissance of fallen trees, boulders, etc.

2. Materials and Methods

Interventionary studies involving animals or humans, and other studies that require ethical approval, must list the authority that provided approval and the corresponding ethical approval code. For modeling forest passability for fire brigade vehicles, it was first necessary to specify a forest structure using the DSM of a forest created from a LiDAR data source using DEM, photogrammetric method, and tachymetry to correct the tree heights. In order to update the CHM and determine the above-mentioned forest structure parameters, tree growth equations and vegetation growth curves were used. However, DSM forest data change quite rapidly, so it was necessary to adjust the CHM and forest structure model. Having an updated forest structure, we finally created a model of its passability by a chosen fire brigade vehicle Tatra 815.

2.1. Forest Structure Determination

The most important elements of forest structure determination for modeling vehicle passability across vegetation are the average distances between trees and *DBH*. When determining the forest structure while not using the most recently acquired LiDAR data, it can be assumed that the older trees are taller (see [21–24] and Figure 2), the diameter of the trunks grow, the distances and between trees also grow over time, but the number of trees per unit area decreases. In order to derive the age, distances, and *DBH* of trees from their height, a homogeneous forest, composed only from one type of spruce, was chosen. All following equations and vegetation growth curves were provided by the Mendel University in Brno and obtained from inventory data.

The number of trees per square unit N·ha-1 depends on the age of vegetation, slope and other parameters (see Fatehi et al. [25]). We can express it using Formula (1) [26,27]:

$$N = B \cdot t^{-m},$$

(1)

where *B* and *m* are the constants of vegetation stand quality, and *t* is an age.

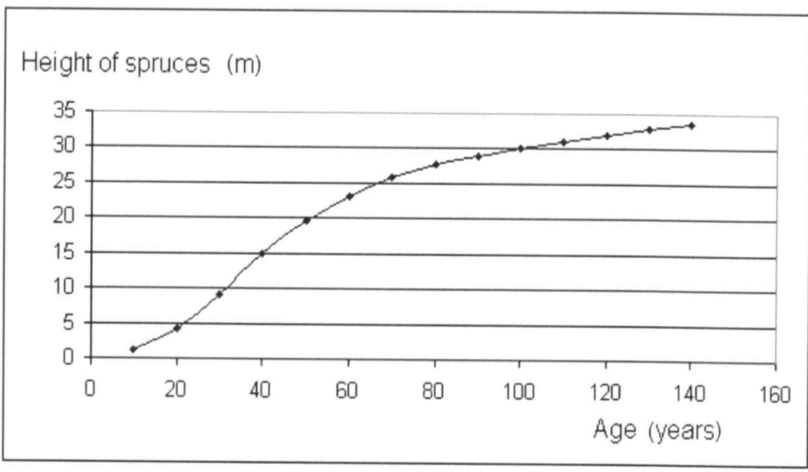

Figure 2. Process of height growth of a spruce forest [28].

When we know the age of the vegetation (which can be derived from the height) we can determine the number of trees per hectare (*N*)—see [21]. The number of trees per unit area is also highly important for determining the average distances between trees within a given forest unit—Mean Tree Spacing (*MTS*). We can express the *MTS* using Formula (2) [21,28]:

$$MTS \cong \sqrt{\frac{40,000}{\pi \cdot N}} - MDBH \tag{2}$$

where *N*—number of trees per 1 ha, and *MDBH*—mean *DBH*.

The last important element of forest structure is *DBH*. Tree trunks are measured at the height of an adult's breast. However, this is defined differently in different countries and situations. The convention is now 1.3 m above ground level. The *DBH* of trees is a function of the *N*—number of trees per 1 ha, the age of vegetation, slope, and other parameters. We can express it using Formula (3):

$$N = B \cdot DBH^{-k}, \ DBH = (N/B)^{k} \tag{3}$$

where *B* and *k* are the constants of the vegetation stand quality. Each type of tree has its own constant *B* and *k*—see also [26].

All of the above-described methodology defining the relationships between forest structure parameters were applied in the context of obsolete LiDAR data acquired in 2013. The procedure for determining the individual parameters of the forest structure is shown in Figure 3.

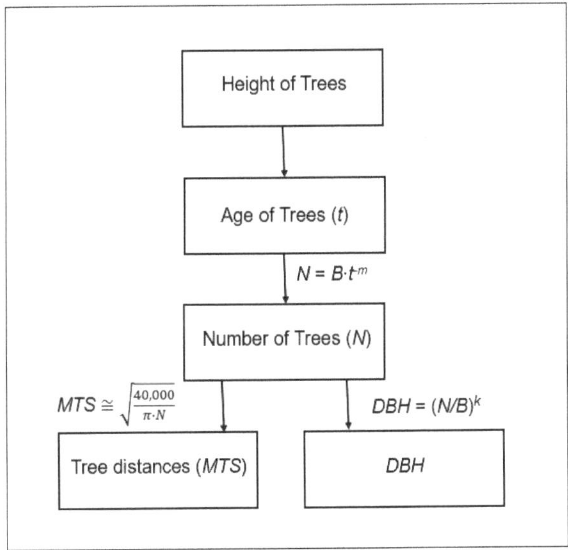

Figure 3. Dependence of forest structure determination.

The heights of the trees are extracted from LiDAR or DSM data. The ages of trees are possible to determine from the inventory data or from the growth equations and parameters. Additionally, the number of trees and *DBH* can be directly determined from the inventory data or calculated from *N*.

2.2. Forest Structure Updating

For determining the CHM while not using the most recently acquired LiDAR data (from 2013), it was necessary to recalculate the tree height according to spruce age—see Figure 2 and Formula (1). For determining the forest structure parameters, LiDAR data and the derived DSM (CHM) were used. Since the default DSM data density was 1 m × 1 m, it was necessary to verify how the actual spruce heights differ from the heights determined by the DSM. Verification was done using a photogrammetric evaluation of the aerial photographs and geodetic method (tachymetry) for accuracy verification—see also [21]. Geodetic measurements of the positions of 116 trees and their heights were carried out by a total station Leica TC 1500. Geodetic survey of the tree heights was taken as the most accurate measurement method. Those tree height estimations were determined photogrammetrically in different time periods with the aid of color aerial photographs. The Military Geographical and Hydrometeorological Office (VGHMÚř) Dobruška took them at regular photographing periods in the Czech Republic (2003, 2006, 2009, 2012, and 2014). A detailed description of the photogrammetric evaluation was given in [21]. Tree height LiDAR data do not match those more accurate from aerial photographs due to the fact that the density of the reflected laser beams (1 m × 1 m) is not sufficient enough to catch the peaks of trees—see the red dots in Figure 4. The LiDAR average tree height is 6 m less than the average height determined by the photogrammetric method [22]. Photogrammetric evaluation and tachymetric verification revealed that, because of the lower density of the LiDAR data (1 m × 1 m), the treetops were not captured, and DSM needed to be corrected (increase in height)—see the red lines in Figure 4.

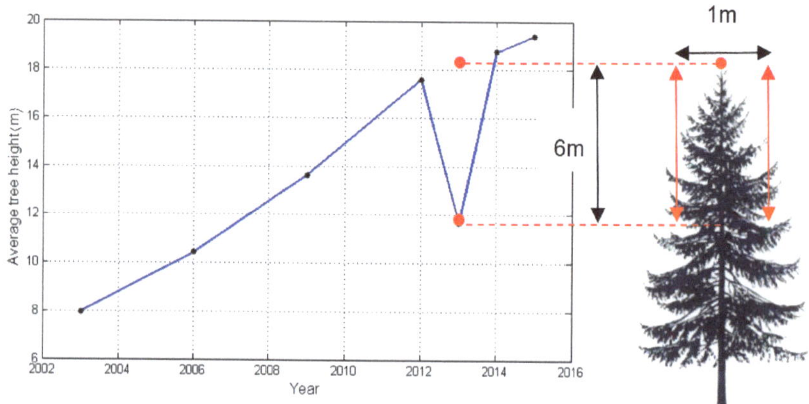

Figure 4. Dependence of a forest structure determination.

Figure 4 shows a series of average spruce heights from 2003, 2006, 2009, 2012, and 2014 determined for the photogrammetry (blue points) and the average height of the same forest determined from the DSM in 2013 from the LiDAR data (red point). The height difference for 2013 was about 6 m; that is, we needed to adjust the values of the DSM heights (CHM) for this constant. Due to the corrected elevations of the DSM, it was possible to define a new forest structure (see the methodology above) and calculate the tree distances and stem diameters. For the forest structure simulation, the normal Gauss distribution of distances and *DBH* were used. The mean values of height, distance, and *DBH* were used at 18 m, 4 m, and 0.25 m.

2.3. Passability Model

To find the optimal route through the forest, it is necessary to know: parameters of the vehicle, parameters of the trees, start and end points of a route, and impassable areas. The most important parameters of a vehicle are vehicle width (*VW*), length, turning radius, and tolerance (*T*)—see Figure 5.

(a)   (b)

Figure 5. Fire and tested vehicles. (**a**) Fire vehicle Tatra 815 4 × 4, Length/Width/Height: 7950/2550/3000 mm. (**b**) Tested terrain vehicle Tatra 815 8 × 8, Length/Width/Height: 8950/2550/3300 mm.

T determines the minimum distance of the vehicle from a trunk to pass between two trees safely. To simplify the passability model, T also replaces the effect of other vehicle parameters (length, turning radius, etc.). In turn, tree parameters refer to those characteristics that are key to finding the optimal route in the forest. In our case, those parameters are stem simulated coordinates, mean tree spacing (*MTS*), mean *DBH*, mean riding corridor (*MRC*), and a *VW*. We can express the relationship between *MTS*, *MDBH*, and *MRC* (see Figure 6) using Formula (4):

$$MRC = MTS - MDBH. \tag{4}$$

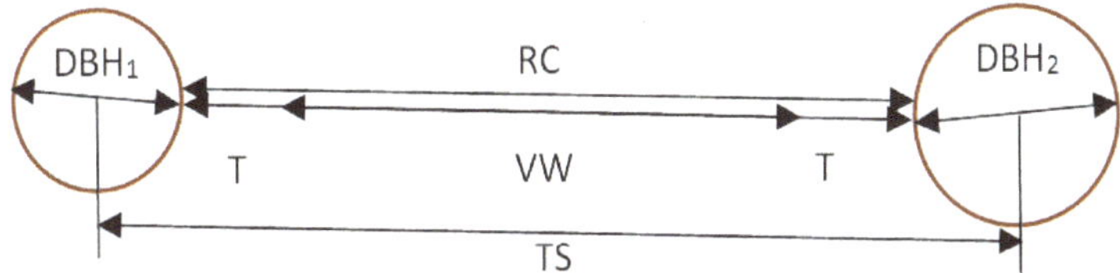

Figure 6. The relationship between particular tree spacing, *DBH*, and riding corridor.

If we do not know the exact coordinates of each tree (which is the usual situation), we can generate the simulated forest structure from average values—see the procedure above. We use the process of random tree deployment to determine the probability of crossing the forest.

The start and end points of a route can be substituted by an initial and final area in a forest region. It is usually not possible to go between these points or areas directly, and we must maneuver between tree stems. Impassable objects (steep slopes, rocks, lakes, rivers, burning forest, etc.) can be obtained from GIS databases or using aerial or satellite images. For the purposes of modeling vehicle mobility in the above-described forest, the impassable areas of the simulated burning polygons were chosen, though other objects (obstacles) were not considered. To search the optimal vehicle route, the following algorithms were used (see Figure 7): Voronoi graph and Delaunay triangulation, Dijkstra algorithm, and optimization of the fractional line.

Figure 7 shows the positions of individual trees (Vertex M)—blue points. The closest two trees to the given tree create the Delaunay triangle—Figure 7b, and the most secure route sections (Voronoi edges) are intersected in the Voronoi nods—blue lines in Figure 7c. The Dijkstra algorithm was used to find the shortest routes from the nod of the graph given to all other nods—see also [23,24]. Using Figure 7c, we can simulate a forest path (see Figure 7d). Trees that were obstacles are marked in red, and Voronoi nods (pale blue points) are connected with Voronoi edges (dark blue lines). All Voronoi edges are rated by weights. These weights may represent the distances or time for which a vehicle passes through the Voronoi edges. In the event that we search for the shortest route from point 1 to point 15, the condition of the minimum sum of the Voronoi edges (weighing) is to be compliant with the shortest route—red line in Figure 7d passing through Voronoi nods 1-3-7-6-11-12-14-15, since the sum of route segment values (3.0 m + 3.3 m + 0.8 m + 3.8 m + 1.2 m + 2.1 m + 1.6 m) = 15.8 m is the smallest (shortest) compared to all possible routes connecting points 1 and 15.

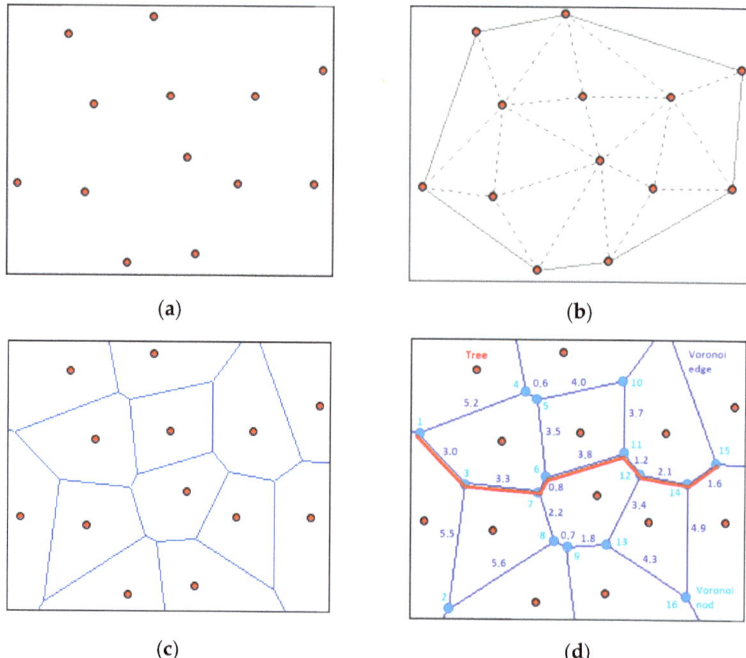

(a)

(b)

(c)

(d)

Figure 7. The relationship between the Delaunay triangulation and Voronoi graph (scale 1:200). (**a**) Set of trees generated from average *MTS* values. (**b**) The area is divided into elementary triangles using Delaunay triangulation (each reference tree has its two closest neighbors). (**c**) Axes of triangle sides create so-called Voronoy edges and Voronoy cells. (**d**) Voronoy edges represent the safest.

3. Results

The result of the creation of a forest structure from data obtained from the original forest (see Figure 8) by generating random tree positions is shown in Figure 9. The size of the analyzed forest area was 140 × 80 m (11,200 m^2). The length of the vehicle's passage was 212 m. The direct path between the starting point and the target is shown by a black line. This path is generally unrealistic due to the tree stem obstacles (displayed as green points). All possible paths (blue closed Voronoi polygons) that match the tree distances and vehicle parameters have been computerized using Dijkstra algorithm and displayed in Figure 8 using our own software tools. Unfinished Voronoi polygons (ending between trees) are nonbinding paths where the width of the vehicle does not allow passage between trees. We can choose any of these blue passable routes, but only one will be the shortest (fastest)—the red highlighted route. This route traverses around (between) the burning forest polygons. The simulation of the polygons displaying the fire areas was done completely at random by adding the points around the impassable zones (orange areas). These areas can be complemented e.g., by satellite images or aerial photos. If we wanted to avoid these risky places, we would have to create a security zone around the burning polygons—so-called buffers—using GIS tools.

Figure 8. The Sitka spruce forest near Utechov, where the heights of trees and *DBH* data were measured (scale 1:1000, center coordinates: φ = 49.282744, λ = 16.632918)—(Ortophoto Mapy.cz).

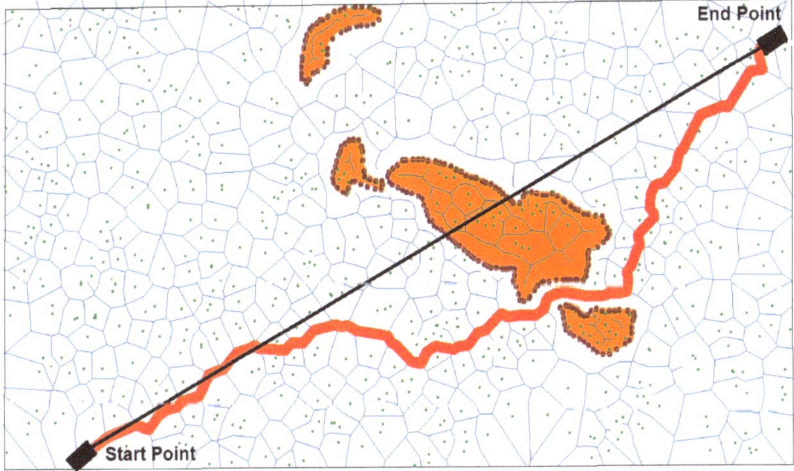

Figure 9. Navigation on the shortest vehicle route (red track) avoiding the burning forest (orange areas). Randomly spaced trees, approx. scale 1:1000.

There are also displayed the routes inside the areas of fires (blue lines inside the orange polygons)—see Figure 9. These routes can be used later when the fires end, but they are primarily not included into the calculation of the shortest route. For some types of vehicles, such as tanks, we can also choose the route through the burning area and calculate the route segments inside the orange polygons. The influence of other elements of the terrain (slope gradient, soil properties, terrain surface roughness, forest paths, etc.) are not calculated.

The above-mentioned result of seeking an optimal forest path partially affected by a fire may be modified in case when the forest structure is regular triangular or rectangular. There are displayed all the possible routes and the shortest route—red line in the triangular forest structure in Figure 10.

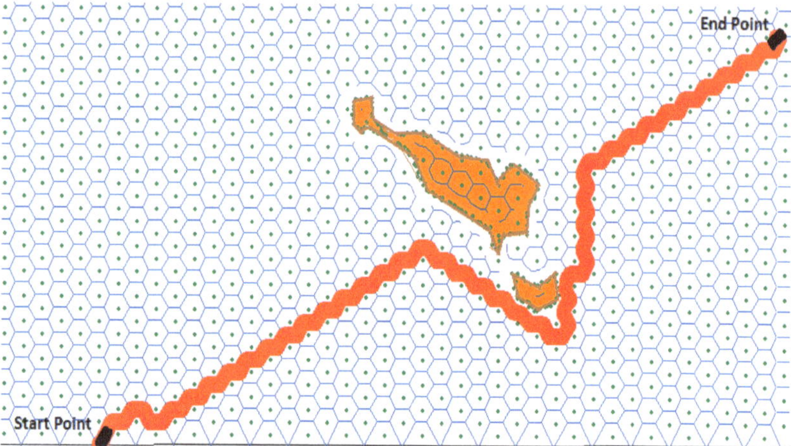

Figure 10. Navigation on the shortest vehicle route—regular triangular forest structure—scale 1:1500.

The methodology for finding the optimal rescue vehicle path was based on verifying the input data. Determining a forest structure from DSM data can be very unreliable, especially when LiDAR data are obsolete. Therefore, the use of growth curves of trees and derived vegetation parameters were used. These parameters were determined on a relatively small area. Verification of the vegetation parameters by photogrammetric and geodetic pathways lasted several months. The results of the presented model are valid for the designated coniferous forest. The general application of the optimal vehicle route determination will depend on the type of trees (coniferous, deciduous, and mixed). The author assumes that the presented model for finding the optimal route of a vehicle will be better utilized with the development of mapping methods aimed at determining the exact coordinates of the trees.

4. Discussion

The described methodology for determining the possibility of moving the fire brigade vehicles in forest vegetation can be used if tree position data or forest structure (generated from photogrammetric data or from LiDAR data) are available. In both cases, the same algorithm can be used to find the optimum forest path. In case we have more precise tree coordinates (from terrestrial or remote sensing sensors), the calculated route of the vehicle will be more reliable. Although the methods of directly determining the exact tree position by remote sensing data are constantly developing, the forest structure is often determined using DSM (CHM) methods. This is due to the financial cost of the high density of LiDAR data, as well as the personnel demand for data acquisition using photogrammetric methods. The quality of the photogrammetric evaluation depends on the scale of the images and the evaluator's experience. The main problem is to target the marker at the tree's top point, which is above the tree trunk. The accuracy of tree position evaluation is higher for coniferous trees than for leafy vegetation. The disadvantage of the photogrammetric method is the lower performance of manual evaluation compared to the possibility of automated evaluation of LiDAR data. LiDAR methods are faster than photogrammetric methods, and they allow a more efficient assessment of the forest structure and determination of the possibilities of vegetation passability without a manual evaluation. LiDAR methods can also be better combined with other remote sensing data sources (infrared, multispectral, radar, etc.). For example, an infrared spectrum can be used to map environmental and fire temperature characteristics, and at night, multispectral imagery can be used to classify species, and radar data is appropriate for mapping a burning forest covered with smoke or clouds. However, these methods have a disadvantage when scanning vegetation with a low density of DSM elevation points (smaller than 1 m × 1 m) and in the case of DTM data absence. On the other hand, the repeated photogrammetric evaluation of the representative forest stands and the data from DSM could bring about a new approach for forest growth analysis and could be a sufficient method for DSM updating in different growth conditions of forest stands.

The results of this experiment showed that this method is fully applicable for the DSM generated from LiDAR data. The method can be appropriately implemented as a relatively inexpensive updating tool for GIS technology between two laser scanning campaigns of a territory. This method can also be refined using the growth curves of individual types of trees. Forest growth characteristics are very important due to the age of LiDAR DSM data. The results of photogrammetric measurements from aerial images taken at consecutive time intervals and statistical calculations show that the growth curves of the trees are initially steeper, but vegetation growth later slows. It is also necessary to investigate the relationships between the natural environment factors and specific canopy growth. The above-mentioned DSM updating method could be used for many applications, e.g., in forestry, military, etc.—see, e.g., [29–33]. It should be noted that the tree height correction values decrease with the increasing density of the LiDAR data. At the DSM density 1 m × 1 m, the average correction is approx. +6 m. At the DSM points density of 1 dm × 1 dm, it is possible to estimate the average height corrections of spruce trees in decimeters, depending on the age of the vegetation. Height corrections of the DSM can significantly affect the computationally generated forest structure and, hence, the vehicle motion models. The resulting model of forest crossing by a vehicle will depend, to a large extent, on the quality of the forest structure data. This study focused on a spruce forest—the predominant tree species in Central Europe. In general, it can be said that the species of vegetation may be variable in different forest groups. From this point of view, the study presented in this article can be considered as partially applicable. Using LiDAR/DSM data, the determination of the deciduous forest structure and positioning of the tree trunks will be more difficult, especially due to the crown surface diversity. From this point of view, it can be assumed that the model of finding the optimal vehicle path through the deciduous forest will be less reliable. The success of these models will largely depend on the resolution, coverage, and actuality of LiDAR data, as well as on the accuracy of the forest fire localization data. It should be noted that the use of this methodology in practice has a number of limitations resulting from data that cannot include all objects in the forest, such as lying trees, stones, low tree branches, etc. [34].

Tree branches can be an important obstacle to the movement of rescue vehicles. It mainly concerns young forests or deciduous forests, where branches are thicker and located below the ground. In coniferous stands, the lower branches of older trees are dry and thin and do not represent a major obstacle for heavy wheeled or tracked vehicles. Below is Table 1, containing measured data of tree branching; the lowest branches were about 1–2 cm thick. The measurement of tree canopy branching was performed only on trees for which resistance tensile forces were measured, not on all the trees in the area.

The problem is how to get the lower branch data. For this purpose, we plan to use LiDAR data with a resolution in cm [35–37] and use the last but one reflection for this measurement. Additionally, terrestrial LiDAR could help to solve this problem—we tested it on a small area in March 2022—see Figure 11 below.

The author assumes that, in the near future, it will be possible to solve the coordinates of trees, their *DBH*, and the characteristics of tree crown branches.

The spread of fire, depending on a number of factors, can be very variable, and actual data from burning areas will not always be available. Additionally, visibility can be significantly affected by smoke and the daytime. It should be noted that the calculation of the optimal vehicle route was based on the width of the vehicle. The reliability of route determination also depends on other vehicle parameters, such as vehicle length and height and minimum turning radius. This model did not even include a case where the vehicle would go back (e.g., in case of a spreading fire). For the more accurate calculations of a vehicle route, the impact of the side slope should be also considered. Notable was also the driver's ability to overcome difficult terrain and to maneuver between trees in crisis situations.

Table 1. Branch position of selected trees.

Tree Num.	Tree Stem Diameter *DBH* (cm)	Height of Dry Branches (cm) (ɸ 1–2 cm)	Height of Semi-Dry Branches (cm)	Height of Green Branches (cm)	Tree Height (cm)
1	21.0	310	820	1080	1750
2	19.5	350	800	1040	1600
3	12.8	460	690	950	1460
4	19.0	490	870	1150	1730
5	13.9	none	720	930	1590
6	17.0	230	450	990	1710
7	17.8	380	510	650	1710
8	12.4	550	850	920	1470
9	22.1	170	500	1030	1730
10	18.3	330	770	900	1640
11	19.0	400	570	840	1640
12	16.7	560	830	970	1620
13	25.3	360	830	960	2050
14	23.7	none	340	1040	1870
15	22.9	350	550	930	1730
16	15.3	none	370	780	1525
17	14.3	260	470	750	1490
18	22.0	280	580	770	1550
19	10.8	none	380	650	1100
20	10.5	none	320	515	1030
21	14.5	770	990	1180	1770

Figure 11. Deciduous and coniferous trees—segmentation of trunks and branches (crowns).

5. Conclusions

The primary aim of this article was to introduce the theoretical aspects, methods, and results of modeling the possibilities of firefighting rescue vehicle mobility in forest areas during fires using remote sensing data. The main result of the presented research is the methodology of the forest structure creation from DSM data, updating due to the growing vegetation parameters, as well as the proposal of the methodology of finding the optimum path of the vehicle to cross the forest, which is considerably more difficult than navigation on the roads.

Although the presented methods are approximate and their applications depend on a number of other factors, the author of the article believes that the presented methods and research results will be applicable in relation to the severity of damage caused by fire. The author also expects further developments of the vehicle navigation methodology in forest regions and the calculations of other factors influencing the search for optimal rescue vehicle routes (low vegetation, lower branches of trees, inclination of slopes, soil influence, terrain surface, etc.). It will also be important to develop the theory and modeling of fire spread in forest areas and to link these models to rescue vehicle navigation.

Funding: This research received no external funding.

Institutional Review Board Statement: Not applicable.

Informed Consent Statement: Not applicable.

Data Availability Statement: Experimental data for Figure 2 and Table 1 were provided by Mendel University in Brno, for Figure 11 by the Czech Technical University in Prague and the Czech University of Life Sciences Prague.

Acknowledgments: This paper is the particular result of the defense research project DZRO VAROPS managed by the University of Defence in Brno, NATO-STO Support Project (CZE-AVT-2019), and specific research project 2021-23 at Department K-210 managed by the University of Defence, Brno.

Conflicts of Interest: The authors declare no conflict of interest. The funders had no role in the design of the study; in the collection, analyses, or interpretation of the data; in the writing of the manuscript; or in the decision to publish the results.

References

1. Vacchiano, G.; Foderi, C.; Berretti, R.; Marchi, E.; Motta, R. Modeling anthropogenic and natural fire ignitions in an inner-alpine valley. *Nat. Hazards Earth Syst. Sci.* **2018**, *18*, 935–948. [CrossRef]
2. Ganteaume, A.; Camia, A.; Jappiot, M.; San-Miguel-Ayanz, J.; Long-Fournel, M.; Lampin, C. A Review of the Main Driving Factors of Forest Fire Ignition Over Europe. *Environ. Manag.* **2012**, *51*, 651–662. [CrossRef]
3. Blair, J.; Rabine, D.L.; Hofton, M.A. The Laser Vegetation Imaging Sensor: A medium-altitude, digitisation-only, airborne laser altimeter for mapping vegetation and topography. *ISPRS J. Photogramm. Remote Sens.* **1999**, *54*, 115–122. [CrossRef]
4. Lim, K.; Treitz, P.; Wulder, M.; St-Onge, B.; Flood, M. LiDAR remote sensing of forest structure. *Prog. Phys. Geogr. Earth Environ.* **2003**, *27*, 88–106. [CrossRef]
5. Hyyppä, J.; Hyyppä, H.; Litkey, P.; Yu, X.; Haggrén, H.; Rönnholm, P.; Pyysalo, U.; Pitkänen, J.; Maltamo, M. Algorithms and methods of airborne laser scanning for forest measurement. *Int. Arch. Photogramm. Remote Sens.* **2004**, *36*, 82–89.
6. Aschoff, T.; Spiecker, H. Algorithms for the automatic detection of trees in laser scanner data. *Int. Arch. Photogramm. Remote Sens.* **2004**, *36*, 71–75.
7. Gobakken, T.; Næsset, E. Effects of forest growth on laser derived canopy metrics. *Int. Arch. Photogramm. Remote Sens.* **2004**, *36*, 224–227.
8. Carson, W.W.; Andersen, H.E.; Reutebuch, S.E.; McGaughey, R.J. Lidar Applications in Forestry—An Overview. In Proceedings of the ASPRS Annual Conference, Denver, Colorado, 24–28 May 2004.
9. Ahlberg, S.; Söderman, U.; Tolt, G. *High Resolution Environment Models from Sensor Data, In Defence Imagery Exploitation*; The United Kingdom's Ministry of Defence (MOD): London, UK, 2006.
10. Su, Y.; Guo, Q.; Fry, D.L.; Collins, B.M.; Jacob, M.K.; Flanagan, P.; Battles, J.J. A Vegetation Mapping Strategy for Conifer Forests by Combining Airborne Lidar Data and Aerial Imagery 2015. *Can. J. Remote Sens.* **2015**, *12*, 53. [CrossRef]
11. Martone, M.; Rizzoli, P.; Wecklich, C.; González, C.; Bueso-Bello, J.-L.; Valdo, P.; Schulze, D.; Zink, M.; Krieger, G.; Moreira, A. The global forest/non-forest map from TanDEM-X interferometric SAR data. *Remote Sens. Environ.* **2018**, *205*, 352–373. [CrossRef]
12. Kugler, F.; Schulze, D.; Hajnsek, I.; Pretzsch, H.; Papathanassiou, K.P. TanDEM-X Pol-InSAR Performance for Forest Height Estimation. *IEEE Trans. Geosci. Remote Sens.* **2014**, *52*, 6404–6422. [CrossRef]
13. Cazcarra-Bes, V.; Tello-Alonso, M.; Fischer, R.; Heym, M.; Papathanassiou, K. Monitoring of Forest Structure Dynamics by Means of L-Band SAR Tomography. *Remote Sens.* **2017**, *9*, 1229. [CrossRef]
14. Puletti, N.; Chianucci, F.; Castaldi, C. Use of Sentinel-2 for forest classification in Mediterranean environments. *Ann. Silvic. Res.* **2018**, *42*, 32–38. [CrossRef]
15. Gitas, I.Z.; Polychronaki, A.; Katagis, T.; Mallinis, G. Contribution of remote sensing to disaster management activities: A case study of the large fires in the Peloponnese, Greece. *Int. J. Remote Sens.* **2008**, *29*, 1847–1853. [CrossRef]
16. Milz, M.; Rymdteknik, A. *Study on Forest Fire Detection with Satellite Data*; Lulea Teknicka Universitet: Kiruna, Sweden, 2013; Available online: https://rib.msb.se/Filer/pdf/26593.pdf (accessed on 10 January 2022).
17. Koo, E.; Pagni, P.; Woycheese, J.; Stephens, S.; Weise, D.; Huff, J. A Simple Phisical Model for Forest Fire Spread Rate. *Fire Saf. Sci.* **2005**, *8*, 851–862. [CrossRef]
18. Ahlvin, R.B.; Haley, P.V. *NRMM II Users Guide*, 2nd ed.; Army Corps of Engineers; Procedural Guide for Preparation of DMA Cross-Country Movement (CCM) Overlays; Student handbook; DMA: Fort Belvoir, VA, USA, 1993; Volume 1.
19. Rybansky, M.; Vala, M. Analysis of relief impact on transport during crisis situations. *Morav. Geogr. Rep.* **2009**, *17*, 19–26.

20. Rybansky, M.; Hofmann, A.; Hubacek, M.; Kovarik, V.; Talhofer, V. Modelling of cross-country transport in raster format. *Environ. Earth Sci.* **2015**, *74*, 7049–7058. [CrossRef]
21. Rybansky, M.; Zerzán, P.; Břeňová, M.; Simon, J.; Mikita, T. Methods for the update and verification of forest surface model. In *International Archives of the Photogrammetry, Remote Sensing and Spatial Information Sciences—ISPRS Archives*; International Society for Photogrammetry and Remote Sensing: Praha, Czech Republic, 2016; pp. 51–54.
22. Rybansky, M.; Brenova, M.; Cermak, J.; Van Genderen, J.; Sivertun, Å. Vegetation structure determination using LIDAR data and the forest growth parameters. In Proceedings of the IOP Conference Series: Earth and Environmental Science, Kuala Lumpur, Malaysia, 13–14 April 2016; Volume 37, p. 12031.
23. Rybansky, M. Modelling of the optimal vehicle route in terrain in emergency situations using GIS data. In Proceedings of the IOP Conference Series: Earth and Environmental Science, Kuching, Malaysia, 26–29 August 2013; Volume 18, p. 12131.
24. Parsakhoo, A.; Jajouzadeh, M. Determining an optimal path for forest road construction using Dijkstra's algorithm. *J. For. Sci.* **2016**, *62*, 264–268. [CrossRef]
25. Fatehi, P.; Damm, A.; Leiterer, R.; Bavaghar, M.P.; Schaepman, M.E.; Kneubühler, M. Tree Density and Forest Productivity in a Heterogeneous Alpine Environment: Insights from Airborne Laser Scanning and Imaging Spectroscopy. *Forests* **2017**, *8*, 212. [CrossRef]
26. Matthews, R.W.; Jenkins, T.A.R.; Mackie, E.D.; Dick, E.C. *Forest Yield: A Handbook on Forest Growth and Yield Tables for British Forestry*; Forestry Commission: Edinburgh, Scotland, 2016; pp. 1–92, ISBN 978-0-85538-942-0.
27. Rybansky, M. *The Cross–Country Movement—The Impact and Evaluation of Geographic Factors*; CERM: Brno, Czech Republic, 2009; p. 113, ISBN 978-80-7204-661-4.
28. Simon, J.; Kadavý, J.; Macků, J. *Forest Economic Adjusting*; MZLÚ: Brno, Czech Republic, 1998. (In Czech)
29. Hubacek, M.; Kovarik, V.; Kratochvil, V. Analysis of Influence of Terrain Relief Roughness on Dem Accuracy Generated from Lidar in the Czech Republic Territory. *ISPRS Int. Arch. Photogramm. Remote Sens. Spat. Inf. Sci.* **2016**, *XLI-B4*, 25–30. [CrossRef]
30. Cibulová, K.; Sobotková, Š. Different Ways of Judging Trafficability. *Adv. Mil. Technol.* **2006**, *1*, 77–87.
31. Hošková-Mayerová, S.; Talhofer, V.; Hofmann, A. Mathematical modell used in decision-making process with respect to the reliability of geo database. *Procedia Soc. Behav. Sci.* **2010**, *9*, 1652–1657. [CrossRef]
32. Stodola, P.; Mazal, J. Optimal Location and Motion of Autonomous Unmanned Ground Vehicles. *WSEAS Trans. Signal Processing* **2010**, *6*, 68–77.
33. Pokonieczny, K. Automatic military passability map generation system. In Proceedings of the International Conference on Military Technologies (ICMT), Brno, Czech Republic, 31 May–2 June 2017; pp. 285–292.
34. Abdullahi, S.; Kugler, F.; Pretzch, H. Prediction of stem volume in complex temperate forest stands using TanDEM-X SAR data. *Remote Sens. Environ.* **2016**, *174*, 197–211. [CrossRef]
35. Štroner, M.; Urban, R.; Línková, L. A New Method for UAV Lidar Precision Testing Used for the Evaluation of an Affordable DJI ZENMUSE L1 Scanner. *Remote Sens.* **2021**, *13*, 4811. [CrossRef]
36. Surový, P.; Kuželka, K. Acquisition of Forest Attributes for Decision Support at the Forest Enterprise Level Using Remote-Sensing Techniques—A Review. *Forests* **2019**, *10*, 273. [CrossRef]
37. Krůček, M.; Král, K.; Cushman, K.; Missarov, A.; Kellner, J.R. Supervised Segmentation of Ultra-High-Density Drone Lidar for Large-Area Mapping of Individual Trees. *Remote Sens.* **2020**, *12*, 3260. [CrossRef]

applied sciences

Article

Gradiometry Processing Techniques for Large-Scale of Aeromagnetic Data for Structural and Mining Implications: The Case Study of Bou Azzer Inlier, Central Anti-Atlas, Morocco

Ayoub Soulaimani [1,*], Saïd Chakiri [1], Saâd Soulaimani [2,*], Zohra Bejjaji [1], Abdelhalim Miftah [3,4,*] and Ahmed Manar [5]

[1] Geosciences Laboratory, Department of Geology, Faculty of Sciences, Ibn Tofaïl University, BP 133, Kénitra 14000, Morocco; chakiri@uit.ac.ma (S.C.); zohra.bejjaji@uit.ac.ma (Z.B.)

[2] Resources Valorization, Environment and Sustainable Development Research Team (RVESD), Department of Mines, Mines School of Rabat, Av Hadj Ahmed Cherkaoui, BP 753, Rabat 10090, Morocco

[3] Research Team of Geology of the Mining and Energetics Resources, Faculty of Sciences and Technology, Hassan First University of Settat, BP 577, Settat 26000, Morocco

[4] Laboratory of Intelligent Systems Georesources and Renewable Energies, Faculty of Science and Technology, Sidi Mohamed Ben Abdellah University, BP 2202, Fes 30000, Morocco

[5] Ministry of Energy Transition and Sustainable Development, Directorate of Geology, BP 6208, Rabat 10090, Morocco; a2manar@yahoo.fr

* Correspondence: ayoub.soulaimani@uit.ac.ma (A.S.); soulaimani@enim.ac.ma (S.S.); a.miftah@uhp.ac.ma (A.M.)

Abstract: Due to its unique geographic and tectonic location, the Bou Azzer inlier has drawn increased interest in mining studies. The inlier's basement structure remains subject to investigation meanwhile faults and igneous rocks affect the local geology. In order to comprehend the Bou Azzer inlier's structure, we use aeromagnetic data. The edge enhancement method described in this work is based on the gradiometry tensor analysis (GTA) of aeromagnetic data, which yields estimated magnetic tensors, rotational invariants, horizontal invariants, computed strike lines, and Eigensystems. This study's primary objective is to use GTA to define structural boundaries in complicated geological and tectonic environments. The vertical and horizontal positions of the geological border's limits have been determined via analysis of the acquired answers. The borders of the anomalous sources are marked in space by the lowest eigenvalue. According to the research, the inlier demonstrates potential for further mineralization with regard to its complicated structure, which is mostly dominated by WNW-ESE, ENE-WSW, NE-SW and E-W trending lineaments with varying depths between 3.45 and 9.06 km. Certainly, the derived structural scheme has enabled the identification of various formations that may be favorable for the circulation of mineralizing fluids, facilitating the concentration of economically valuable mineral deposits, similar to existing metal reserves in the examined area.

Keywords: geophysics; gradient tensor analysis; rotational invariants; horizontal invariants; eigensystems

Citation: Soulaimani, A.; Chakiri, S.; Soulaimani, S.; Bejjaji, Z.; Miftah, A.; Manar, A. Gradiometry Processing Techniques for Large-Scale of Aeromagnetic Data for Structural and Mining Implications: The Case Study of Bou Azzer Inlier, Central Anti-Atlas, Morocco. *Appl. Sci.* **2023**, *13*, 9962. https://doi.org/10.3390/app13179962

Academic Editor: Giuseppe Lacidogna

Received: 1 August 2023
Revised: 27 August 2023
Accepted: 27 August 2023
Published: 4 September 2023

1. Introduction

Enhancing the data to highlight key aspects is one of the main goals in the analysis of aeromagnetic data. In addition to providing information on lithological changes and structural trends, aeromagnetic prospecting also defines lateral changes in susceptibility contrast. Gradients themselves can be used to infer the properties of subsurface structures.

From a mining point of view, the Bou Azzer inlier hosts two important mines: the cobalt mine of Bou Azzer associated with the ophiolitic complex whose mineralization is in the form of lenses and veins, and the mine of Bleïda marked by the presence of significant copper indices in the form of lenses.

Aeromagnetic data are analysed using gradient tensor and other operators to identify many positions of the geological borders. The greatest amplitudes like horizontal gradients which can be replaced by horizontal components of gradient tensor were employed by [1] with the intention of identifying geologic near-vertical borders from aeromagnetic anomalies. The boundaries displayed the highest values signifying sharp disparities in lateral susceptibility. Gradiometers have enabled the recent ability to conduct measurements of all the tensor elements. For the purpose of distinguishing structural characteristics from potential field gradient tensor data, many mapping approaches have been devised. Researchers in [2] calculated the gravity–magnetic tensor invariants across a sophisticated model simulating a basement surface with both horizontal and vertical faults.

In the interest of estimating source depth, ref. [3] demonstrated that the greater negative eigenvalue is analogous to the greatest negative bend acquired from typical potential field anomalies. Geologic contacts were identified [4] by means of the horizontal gravity–magnetic gradient tensor. Horizontal gravity gradients that have been seen or calculated from magnetic data are used to define the eigenvalues of the Hessian matrix in [5–7] as described by a linear feature analysis. Refs. [8,9] imaged subsurface geology using invariants derived from horizontal gravity gradient components.

A method for using the gravity–magnetic gradient tensor was developed by [8]. According to [10], the eigenvector of the entire tensor gravity–magnetic gradient matrix that corresponds to the least eigenvalue may be used to predict the striking direction of quasi 2D entities. According to [11], the sub-vertical plugs, dikes, or diatremes connected to alkaline encroachments are imaged by the rotational and horizontal invariants of the gravity–magnetic gradient tensor. The different tensor components were merged into invariants by [12], who also specified information on geologic contact and body shape.

To identify the margins of anomalous sources, we compute and map strike lineaments (θ) using processed rotational invariants (I_0, I_1, I_2) and horizontal invariants (I_{h1}, I_{h2}). This technique was used to construct the gradient tensor of the aeromagnetic data from Bou Azzer.

2. Geological Background

The Bou Azzer-El Graara Inlier (BAEI) (Figure 1A–C) is linked with two nearby depressions, Bou Azzer then El Graara, situated 45 km southwest of Ouarzazate city. It spreads approximately 20 km from Taznakht to Bleïda in a WNW-ESE direction. The Tissoukine and Tasla synclines surround it from the north and south, respectively. The BAEI was formed during the excavation of the Anti-Atlas Major Accident (AAMA), representing an amount of the earliest Pan-African seams.

In the southern inlier part, the Tachdamt-Bleïda Set, it is thought that remains of the Tonian platform came into existence on the northern border of the West African Craton (WAC). Based on recent tectonic and lithological classifications along with newly acquired geochronological information, it encompasses the most ancient rock formations within the BAEI. During this period of rifting, there is a distinctive emission of tholeiitic to alkaline basalts, initially approximated at 788 ± 10 Ma and more lately, around 883 Ma [13].

The Assif n'Bougmmane-Takroumt complex's metamorphic rock formations stretch from west-northwest to east-southeast. These formations create separate elevated areas in Bou Azzer, Tazigzaout, Oumlil, and the primary elevated region of the Bougmmane massif situated in the Takroumt region, located north of the Tachdamt-Bleïda group. These metamorphic rocks were previously thought to belong to the Paleoproterozoic (PI) basement. However, recent geochronologic data suggest that they are younger and are now attributed to Cryogenian dates [14,15].

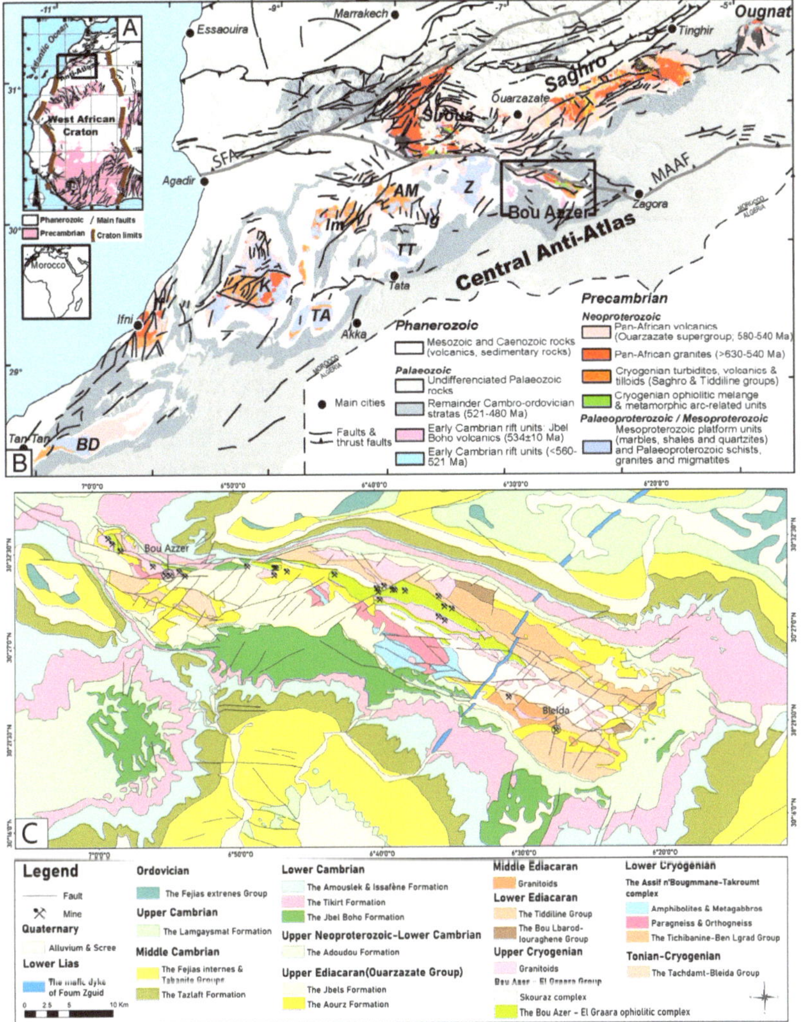

Figure 1. The location of the studied area (Bou Azzer inlier) at (**A**) The West African Craton level modified after [16] and (**B**) at the Anti-Atlas level modified after [16]. (**C**) Geological map of the study area modified after [15].

They consist of amphibolites, metagabbros, augen gneiss, and micaschists in their lithology [15]. The protoliths of these rocks have geochronological dates ranging between 730 million years and 752 million years ago within the Tazigzaout Complex [17], and from 745 million years to 755 million years ago in both the Bou Azzer massif and the Oumlil massif [15]. Furthermore, there are extra geological characteristics, including a mafic intrusion within the Bougmmane Complex established 700 million years ago, along with two veins of micas leucogranodiorite in the Tazigzaout massif, dated at 705 million years ago and 701 million years ago [18], and another at 695 Ma in the Oumlil massif, which all cross-cut these rocks [15]. These findings suggest that an orthogneissification event occurred earlier than 700 Ma during the Lower Cryogenian period.

Similarly, to the northeastern side of the BAEI, within the Tichibanine-Ben Lgrad area, one can find rocks formed from volcanic arcs, including rhyolites, microgabbros and basalts, alongside volcano-sedimentary formations like cinerites, sandy siltstones, dacite and felsic tuffs. The dating of the dacite and felsic tuffs, using U/P on zircon, reveals ages of 767 million years and 761 million years, respectively [19].

Ref. [20] was the first to characterize the famed Neoproterozoic supracrustal ophiolite found at BAEI. It is composed of ophiolite slices from ultramafic to mafic, connected to mélanges of volcanic sediments that have been invaded by intrusions related to subduction [21]. Ultramafic cumulates of wehrlites, harzburgites and dunite lenses, and clinopyroxenolites, a crustal gabbroic unit's section, basaltic pillow lavas a complex of sheeted dykes are all parts of the BAO series. The BAO and the Khzama orb in the Siroua window are examples of the pan-African oceanic lithosphere, which was thrust onto the northern boundary of the WAC (West African Craton) as stated in reference [22]. Its precise age remains a topic of debate despite various speculative indirect estimates [16].

However, the Khzama plagiogranite within the Siroua ophiolite provides the most accurate age estimation for the Central Anti-Atlas ophiolite, pinpointing it at 762 million years old [23]. Around 660 Ma to 640 Ma, a series of quartz diorite intrusions (located in the Bou Azzer massifs, Ousdrat, Taghouni, Ait Ahmane and Bou Frokh) intruded Cryogenian rocks in the Bouazzer region. These intrusions occurred during the Pan-African paroxysm, characterized by sinistral deformation [24].

The Bou Azzer-El Graara Cryogenian foundational layer is covered by non-metamorphic clastic and volcanic deposits from the Ediacaran period in an unconformable manner. The lower units encompass the Tiddiline and Bou Lbarod-Iouraghene Groups, formed approximately 625 million years ago, primarily consisting of andesite and ignimbrite within an active margin setting. Additionally, the age of Tiddiline deposits, defined by fault boundaries and interspersed with rhyolitic layers, date back 606 million years. Before the deposition of the 580 million year old Bleïda granodiorite in a later Pan-African transpressional event, the Tiddiline deposits experienced uplift and folding [25].

The Ouarzazate Group, a large volcano-sedimentary post-collisional complex, corresponds to the Upper Ediacaran period [15]. Within this time frame (580–570 million years ago), the Aourz geological background dominate with volcanic ignimbritic facies having a dacitic-rhyolitic formation, and the Jbels composition is branded by alternating andesitic rhyolites-ignimbrites and flows.

Above the Late Ediacaran Ouarzazate Group, there is a paraconformable overlay of rich-carbonate Adoudou composition. Taroudant Set, representing a transgressive unit from the Early Paleozoic to Late Proterozoic, comprises several marine deposits. Concurrently, the volcanic complex at Jbel Boho, known as Alougoum, was assigned an age ranging between 529 million years and 531 million years [26]. The Paleozoic succession persists with the Tata Group during the Cambrian period, succeeded by the internal Feijas, Tabanit, external Feijas, first and second Bani, as well as Ktaoua transgression groups spanning from the Cambrian to the Ordovician era.

Understanding the Mesozoic and Cenozoic evolution of the Anti-Atlas region is limited by the absence of relevant sedimentary successions. The Anti-Atlas has experienced the influence of various Lower Liassic mafic dykes aligned in a northeast-southwest direction, which are associated with the CAMP magmatic event, signifying the breakup of Pangea during the Triassic-Jurassic transition. Notably, the Foum Zguid dyke, which crosses the BAEI, stands out as the most prominent of these dykes.

The present Anti-Atlas topography emerged during the Neogene period, coinciding with the High Atlas uplift, following several burial and exhumation phases [27].

3. Material and Methods

3.1. Description of the Aeromagnetic Data

This study made use of geophysical data acquired during a comprehensive survey of Morocco, conducted by Canadian Fugro Airborne Surveys Corporation for MMETSD between October 1998 and May 1999 [28]. The survey was an integral component of the National Geological Mapping Project, encompassing the entire Anti-Atlas region.

The latter refers to a comprehensive initiative with two primary objectives: first, to enhance searching for minerals in the Anti-Atlas area, which stands out as one of the most important metallogenic regions in Morocco, and second, to consolidate geoscientific databases. The study made use of a high-resolution data collection generated from an aerial geophysical survey on the part of providing accurate structural and geology cartography of the research region. Specifically, this research focuses on aeromagnetic data obtained from a rectangular region measuring 40 km by 69 km surrounding the BAEI (Figure 1). The data were captured using an optically pumped cesium vapor magnetometer that operates using pumping, featuring a level of sensitivity of 0.01 nT, and a data collection frequency of 10 samples per second was applied.

For the airborne magnetic survey, the N30° E flying path was utilized, set apart at intervals of 500 m. The magnetic sensor's height above the terrain was set to approximately 30 m, and the tie lines were perpendicular with N120° E direction, a spacing of 4000 m. These parameters allowed for the collection of a substantial amount of magnetic field data, covering a total distance of 5700 km. As an advantage of these survey settings, a high-resolution aeromagnetic survey of the research region was effectively conducted. The preliminary handling of the unprocessed data involved standard levelling, noise reduction, diurnal adjustment, and other essential steps.

The contractor performed the gridding of the magnetic data using the minimal curvature technique, yielding a standardized square grid with a grid spacing of 250 m (half of the line interval, regardless of the data sampling rate) [29]. Aeromagnetic maps of Bou Azzer Inlier Reduced to the north pole and the Residual Aeromagnetic map of Bou Azzer Inlier (Figure 2), computed by [28], were referenced using the southern Moroccan Lambert metric coordinate projection. The residual magnetic field performed by [28] shows a very agitated magnetic relief and a certain heterogeneity with the presence of several anomalies (Figure 2A), the observed class ranging from −900 nT to 1800 nT.

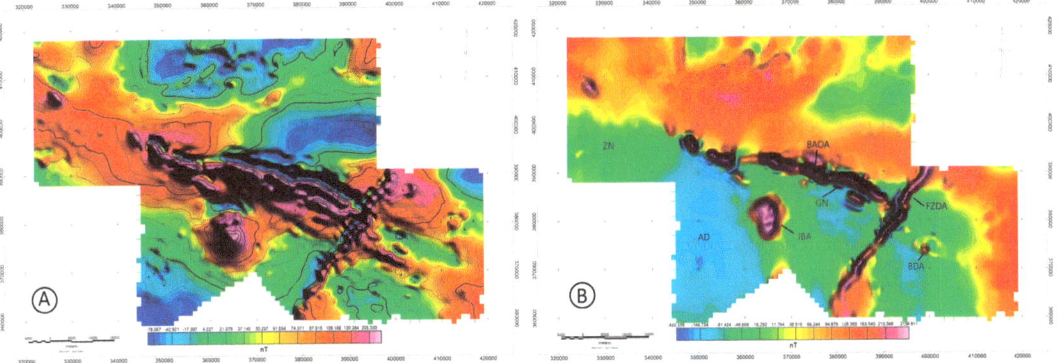

Figure 2. (**A**) Residual Aeromagnetic map of Bou Azzer Inlier. (**B**) Aeromagnetic map of Bou Azzer Inlier Reduced to the north pole [28].

3.2. Gradiometric Tensor

Aeromagnetic assessments are frequently employed in geophysical local surveys aimed at unexploded ordinance surveillance, geology mapping, and mineral prospecting (UXO). Using a single total field magnetometer, aeromagnetic traditional surveys only yield assessments of the total magnetic intensity (TMI).

Different sensors provide precise measurements of the entire field gradient. These yields detailed data about shallow geological characteristics, which remain unaltered by diurnal fluctuations and the magnetic field background of the region. As an illustration of the gradient tensor measurement's power, the Canadian Geological Survey initiated an aeromagnetic gradiometer program in Canada, developing short baseline aeromagnetic gradiometers back in 1975 [30–32].

In recent years, there has been the development of gradiometers that use supercon-ducting quantum interference devices (SQUIDs). These devices have been demonstrated to be successful for observing geomagnetic phenomena. IPHT Jena has successfully created and tested full tensor magnetic gradiometer systems based on LTSQUIDs (low-temperature superconducting quantum interference devices) for various targets [33,34].

Vector surveys offer significant advantages, and the interpretability of magnetic sur-veys is greatly improved through the use of the complete magnetic gradient tensor (MGT), This describes how the three field components change in space along three perpendicu-lar axes. This comprehensive MGT opens up possibilities for applications like magnetic navigation, air anti-submarine warfare, UXO detection and more. As a result, multiple geophysical firms are in the process of creating complete-tensor magnetic gradiometer systems. However, there are additional engineering obstacles that need to be addressed to create a functional airborne SQUID gradiometer system [34].

The Magnetic Gradient Tensor (MGT) can be derived by transforming the potential field. Based on potential field theory, it is feasible to calculate the various potential fields and gradient elements overhead magnetic sources [35]. Consequently, gridded data for Total Magnetic Intensity (TMI) could be employed to generate the components of the gradient tensor [36].

The gradient tensor can alternatively be obtained by measuring the total-field gradients in either the vertical or horizontal directions [37]. There is no requirement to perform derivative operations, which could be unstable on Fourier domain, toward computing the gradient tensor from the total field gradient. Utilizing data on triaxial magnetic gradients allows for a more effective determination of gradient tensor components. However, in [37], triaxial aeromagnetic gradients were not taken into consideration, and instead, the observed vertical gradient was derived by means of the horizontal gradient.

While gradiometer (or tensor) data are becoming more prevalent, especially in aerial gravity surveys, magnetic data are still often collected as a single value [38]. Gradiometer data involve the collection of nine gradient magnetic or gravity values for each point, instead of just one overall field magnitude [39]. This implies that more data sets are available for modeling, which can aid in resolving ambiguity issues during the modeling process.

3.3. The Art of Gradient Tensor Analysis

Tensors are an extension to the concepts of scalars, vectors and matrices. A tensor is represented as an organized multidimensional array of numerical values. A practical example of this is the magnetic gradient tensor (Figure 3) [37]:

$$B = \begin{bmatrix} \frac{dB_x}{dx} & \frac{dB_x}{dy} & \frac{dB_x}{dz} \\ \frac{dB_y}{dx} & \frac{dB_y}{dy} & \frac{dB_y}{dz} \\ \frac{dB_z}{dx} & \frac{dB_z}{dy} & \frac{dB_z}{dz} \end{bmatrix} = \begin{bmatrix} B_{xx} & B_{xy} & B_{xz} \\ B_{yx} & B_{yy} & B_{yz} \\ B_{zx} & B_{zy} & B_{zz} \end{bmatrix} \tag{1}$$

where B_x, B_y and B_z are: x, y and z magnetic field components. The sheer volume increase in measured data presents many opportunities for new input into source detection and forward modelling algorithms. Tensors can be categorized based on their rank or order [40]. This categorization is evident in the count of components that a tensor has within an N-dimensional space. Hence, a tensor with order p comprises N^p components.

For instance, in a three-dimensional Euclidean space, the number of components in a tensor is 3^p. From this, for example:

- A zero-order tensor ($p = 0$) contains a single element and is referred to as a scalar. Physical quantities that only possess magnitude are depicted using scalars;
- A tensor of order one ($p = 1$) consists of three elements and is termed a vector. Quanti-ties that encompass both magnitude and direction are depicted using vectors. B_x is an example of a first rank tensor component;

- A tensor of order two ($p = 2$) comprises nine elements and is commonly illustrated using a matrix. $B_{xx} = \frac{dB_x}{dx}$ is an example of a second rank tensor component;
- A tensor of order three ($p = 3$) has twenty-seven components. $B_{xxy} = \frac{d^2 B_x}{dx\,dy}$ is an example of a third rank tensor component.

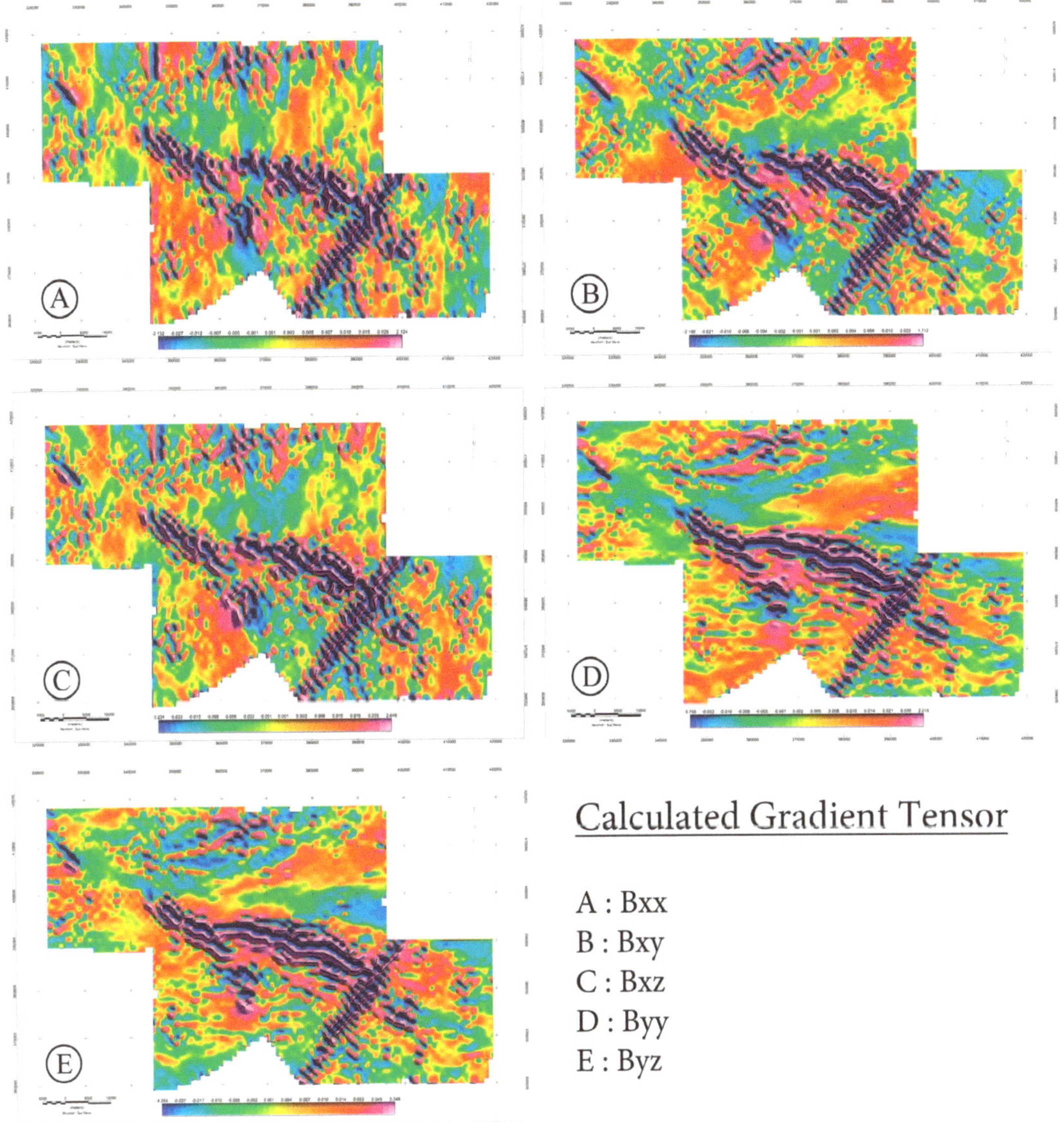

Figure 3. Calculated gradient tensor from Bou Azzer aeromagnetic datasets.

A structure tensor is a matrix derived from the gradient of a function. It is a second order tensor (Figure 3) (has components such as B_{xy}), and has 2D and 3D forms (can be represented by either two or three variables). The 3D form is used in gradiometer surveys. Assume that B is a function of three variables (x, y, z). We can recognize that since

$H = -\nabla \varphi$, this implies that (in SI units) $B_x = -\mu_0 \frac{d\varphi}{dx}$, $B_y = -\mu_0 \frac{d\varphi}{dy}$, $B_z = -\mu_0 \frac{d\varphi}{dz}$ Therefore, the structure tensor would be [41]:

$$B = \nabla \otimes \mu_0 \nabla \varphi = \begin{bmatrix} \frac{d}{dx} \\ \frac{d}{dy} \\ \frac{d}{dz} \end{bmatrix} \begin{bmatrix} -\mu_0 \frac{d\varphi}{dx} & -\mu_0 \frac{d\varphi}{dy} & -\mu_0 \frac{d\varphi}{dz} \end{bmatrix} = -\mu_0 \begin{bmatrix} \frac{d^2\varphi}{dx^2} & \frac{d^2\varphi}{dxdy} & \frac{d^2\varphi}{dxdz} \\ \frac{d^2\varphi}{dydx} & \frac{d^2\varphi}{dy^2} & \frac{d^2\varphi}{dydz} \\ \frac{d^2\varphi}{dzdx} & \frac{d^2\varphi}{dzdy} & \frac{d^2\varphi}{dz^2} \end{bmatrix} \tag{2}$$

$$\implies B = \begin{bmatrix} B_{xx} & B_{xy} & B_{xz} \\ B_{yx} & B_{yy} & B_{yz} \\ B_{zx} & B_{zy} & B_{zz} \end{bmatrix} \tag{3}$$

where $\otimes$ is the dyadic product. Since the magnetic field represents a gradient of potential, where the tensor elements correspond to second-order derivatives of the scalar potential (φ). As a result, the tensor components exhibit symmetry. Therefore:

$$B_{xy} = B_{yx}, \; B_{yz} = B_{zy}, \; B_{xz} = B_{zx} \tag{4}$$

According to Laplace's equation:

$$\nabla^2 \varphi = 0 \tag{5}$$

$$\nabla \times B = 0 \tag{6}$$

From this it can be seen that:

$$B_{xx} + B_{yy} + B_{zz} = 0 \tag{7}$$

Based on previous equations we have:

$$B = \begin{bmatrix} B_{xx} & B_{xy} & B_{xz} \\ B_{yx} & B_{yy} & B_{yz} \\ B_{zx} & B_{zy} & B_{zz} \end{bmatrix} = \begin{bmatrix} B_{xx} & B_{xy} & B_{xz} \\ B_{xy} & B_{yy} & B_{yz} \\ B_{xz} & B_{yz} & -B_{xx} - B_{yy} \end{bmatrix} \tag{8}$$

This means that there exist only five distinct and unconnected tensor components with magnetic and gravity data. For magnetic data these are $B_{xx}, B_{xy}, B_{xz}, B_{yy}, B_{yz}$.

Refs. [42,43] explore a different type of tensor representation that is based on amplitudes and phases.

A tensor measurement can be transformed using eigenvalues and eigenvectors. Each reading is divided into its orthogonal rotation matrix, invariant eigenvalue amplitudes, and local to the survey reference frame eigenvectors. The amplitude and phase of the tensor are represented by the eigenvalue amplitudes and eigenvector rotations. The amplitude-phase form respects the inherent physical features of tensors while enabling extremely quick and reliable processing of tensor data.

A thorough summary of eigenvector analysis of the tensor is provided by [36]. The link between these values is as follows if we define a tensor measurement as the matrix B with a scalar eigenvalue and eigenvector v:

$$Bv = \lambda v \tag{9}$$

The eigenvalues are determined through the solution of the characteristic equation. $\text{Det}(B - \lambda I)$. Expanding this, we get:

$$\lambda^3 + I_1 \lambda - I_2 = 0 \tag{10}$$

where:

$$I_1 = B_{yy}B_{zz} + B_{xx}B_{yy} + B_{zz}B_{xx} - B_{xy}^2 B_{yy}^2 = \lambda_1\lambda_2 + \lambda_1\lambda_3 + \lambda_2\lambda_3 = \frac{-\left(\lambda_1^2 + \lambda_2^2 + \lambda_3^2\right)}{2} \tag{11}$$

$$I_2 = \det(B) = B_{xx}B_{yy}B_{zz} + B_{xx}B_{yz}^2 + B_{zz}B_{xy}^2 - 2B_{xy}B_{xz}B_{yz}B_{xy}^2 = \lambda_1\lambda_2\lambda_3 \tag{12}$$

where $\lambda_1 \geq \lambda_2 \geq \lambda_3$. Applying eigenvector analysis to our tensor equation, as shown in the previous equation, we obtain 3 eigenvalues and 3 eigenvectors. The rotation matrix R which has as its columns the eigenvectors $[\hat{v}_1, \hat{v}_2, \hat{v}_3]$, diagonalises B when applied to it. It is straightforward to verify that the following holds:

$$R^T B R = R^T \begin{bmatrix} B_{xx} & B_{xy} & B_{xz} \\ B_{yx} & B_{yy} & B_{yz} \\ B_{zx} & B_{zy} & B_{zz} \end{bmatrix} R = \begin{bmatrix} \lambda_1 & & \\ & \lambda_2 & \\ & & \lambda_3 \end{bmatrix} \tag{13}$$

The eigenvalues, and any combination thereof, are the tensor rotational invariants. [36] makes extensive use of one such rotational invariant, specifically, the standardized source magnitude. It is defined as:

$$\mu_{nss} = \sqrt{-\lambda_2^2 - \lambda_1\lambda_3} \tag{14}$$

In contrast to magnitude of the tensor (Frobenius norm) $|\,B\,|$, ref. [36] notes that it is entirely isotropic in the vicinity of a dipole source. It is perfect for inverse operation (total magnetic field calculation).

As a result, the previous formulation may be used to determine the horizontal invariants of the FTG map the edge of units (Figure 4). They are characterized as:

$$I_{h1} = \sqrt{B_{xz}^2 + B_{yz}^2} \tag{15}$$

$$I_{h2} = \sqrt{B_{xy}^2 + \left(\frac{B_{yy} - B_{xx}}{2}\right)^2} \tag{16}$$

The gradient tensor determinant of 2-D structures, according to [10], is equal to zero. The first system for which the x-axis corresponds with the strike direction will be obtained if the coordinate system is then rotated such that it lines up with the striking direction., the element $\frac{\partial^2 g}{\partial x^2}$ becomes zero. Although this does not really occur in practice, the optimal striking direction can be identified if the matrix is rotated around the z-axis to decrease this component. You may compute this rotation angle [44–46] (Figure 5) using:

$$\theta = \frac{1}{2}\arctan\left(2\frac{B_{xy}\left(B_{xx} + B_{yy}\right) + B_{xz}B_{yz}}{B_{xx}^2 - B_{yy}^2 + B_{xz}^2 - B_{yz}^2}\right) \tag{17}$$

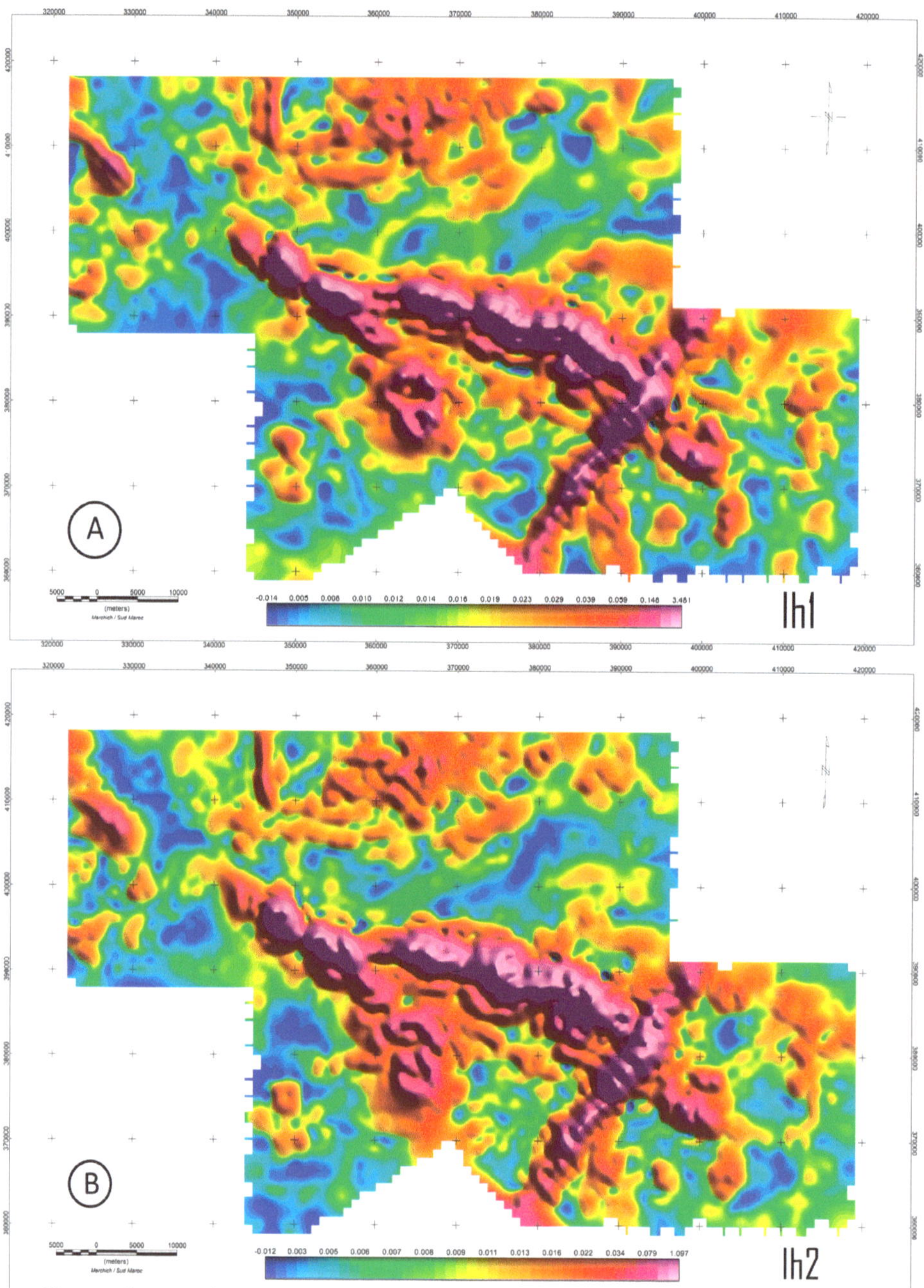

Figure 4. Processed Horizontal invariants from calculated Gradient Tensor of Bou Azzer Inlier. (**A**) Lh1. (**B**) Lh2.

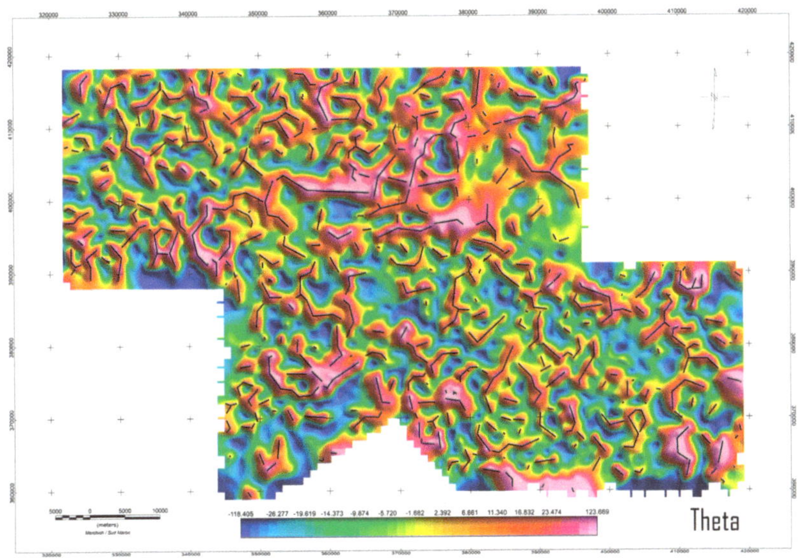

Figure 5. Theta map further strike lineaments (black lines) performed from the calculated Gradient Tensor and Invariants of Bou Azzer Inlier.

4. Results and Discussion

4.1. Bou Azzer Magnetic Field

Given the great geological and structural complexity of the region, the residual magnetic field map shows a variety of anomalies. Thus, after examining the residual Aeromagnetic map of Bou Azzer inlier (Figure 2A) we notice the presence of a large bipolar anomaly-oriented WNW-ESE, which runs along the central part of our study area, with a length exceeding 10 km. It is a linear anomaly with a positive pole more developed than the negative pole. In such a context at this latitude and with an inclination of the magnetic field which is around 45°, it is obvious that this structure has an NNE dip.

A second linear bipolar anomaly is situated in the eastern section of the inlier. It forms an elongated feature stretching in the northeast to southwest direction, covering approximately ten kilometers. It signifies a consistent continuation beyond the main geological formation. An additional set of anomalies was identified within the research area. These anomalies exhibit circular shapes and belong to a distinct group. The initial anomaly is positioned to the south of the central WNW-ESE structure within the geological formation. Its circular appearance suggests it might be a result of a basic magmatic intrusion. The second anomaly is situated to the southeast of the study region. Currently, it is challenging to differentiate among the various anomalies since all these formations exhibit significant consistency on both sides of their originating sources. As a result, the utilization of enhancement filters becomes crucial for accurately interpreting the aeromagnetic data. Hence, to ensure accurate comprehension, it is vital at this phase to employ a range of filters aimed at enhancing the analysis. This might involve attenuating prolonged wavelength anomalies in some cases or amplifying it in others.

Examination of the RTP map (Figure 2B), allowed for the detection of several linear and circular formations with varying amplitudes. In furtherance of perceiving their origin, the contours lines of the positive RTP data values were overlapped on the geological Bou Azzer inlier map (Figure 6). We will address each anomaly separately in the following sections. We will begin by interpreting the positive anomalies and then move on to the negative anomalies and those with lower intensity.

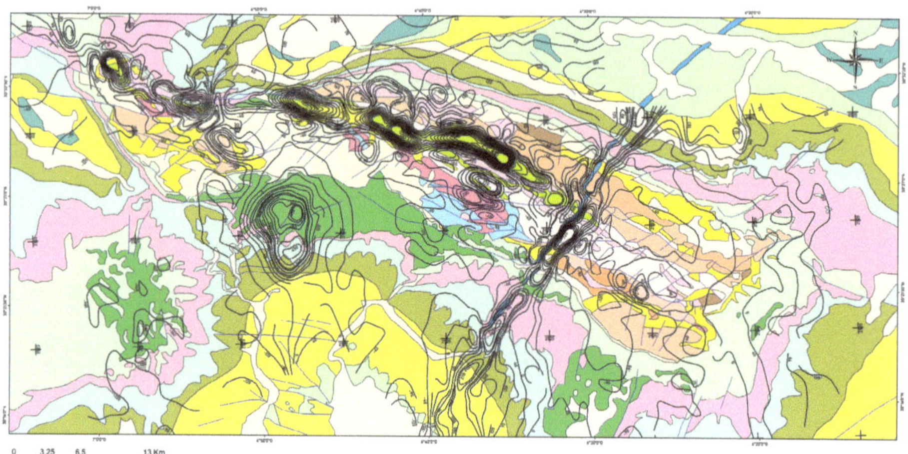

Figure 6. Contours lines of the positive values of the RTP data overlapped on the geological map of the Bou Azzer inlier (same legend as Figure 1C).

BAOA: This anomaly (Figure 2B) corresponds to the ophiolite complex of Bou Azzer which runs along the central part of our inlier along a WNW-ESE axis, showing a value greater than 2240 nT. The elevated magnetic susceptibility is attributed to the rock's nature. As previously mentioned, the ophiolitic complex has undergone near-complete serpentinization, accounting for the heightened magnetic intensity observed within this formation. This axis is evidently shifted by several strike-slip faults.

FZDA: This anomaly (Figure 2B) corresponds to the famous Foum-zguid mafic dyke. It displays a value greater than 2000 nT (Figure 2A,B). This dyke is formed mainly by dolerites which explains its high magnetic susceptibility. The FZ dyke extends over more than 200 km along a linear NE-SW fracture. Its lower Jurassic age (Lias), suggests that it was formed during the opening of Tethys. This large dyke is an integral portion of the Central Atlantic Magmatic Province (CAMP) [47,48].

JBA: This anomaly showing a value of about 2100 nT (Figure 2B) is linked to the basic magmatism of Jbel Boho, then located south of the ophiolitic complex, this circular anomaly has been the subject of several studies. This magmatism is known for its richness in rare earth and copper which makes it one of the important districts in the inlier. The igneous formation includes a central syenitic core surrounded by extensive flows of volcanic lava, characterized as trachytes, andesites and basalts, which extend for tens of kilometers from the complex's center. These volcanic stones are found within the lower portions of a sequence of carbonate rocks that were deposited at the same time [49].

BDA: This anomaly displaying a value of about 200 nT (Figure 2B) located SE of the inlier to the east of the FZ dyke, corresponds to the basic complex formed mainly by basalts which belongs to the Tachdamt-Bleïda platform, this anomaly sticks perfectly with the famous mining district of Bleïda marked by the presence of copper mineralization in the form of lenses.

AD: The negative magnetic anomaly (Ad) displaying a value of about −150 nT (Figure 2B), unquestionably represents the Adoudounian sedimentary cover. The negative values reflect the deep aspect of the said cover, it is a thick sedimentary filling having originated at the level of a deep sedimentary basin. This series, more than 2000 m thick here, rests unconformably on the series of the Ouarzazate Group [16].

GN: The negative Gn anomaly displaying a value of about −400 nT (Figure 2B) located south of the ophiolitic complex is related to granitoids (quartz diorite) of upper cryogenic age. They occurred during the Pan-African deformation [15].

Zn: This anomaly displaying a value of about -50 nT (Figure 2B) located in the northwest fringe of the inlier and which is in the form of an arc with weak magnetic field values corresponds to the northeast limit of the Zenaga inlier (Zn).

4.2. Gradient Tensor Analysis of Bou Azzer Inlier

The approach employed (GTA) still performant in highlighting shallow formations in all directions (Figures 4, 5 and 7), thus its application allows us to highlight a basic complex (Bc) of the Tonian—Cryogenian series of Tachdamt-Bleïda which lies in the center between the ophiolitic complex of Bou Azzer and Jbel Boho. It is mainly composed of basalts, which explains its high magnetic susceptibility. Note also that a fragment of serpentinite known as Takeroumt's (Tk) was also better understood by applying this gradient, the latter is associated with the Cryogenian oceanic subduction system. We also note the continuous interruption of the Foum Zguid dyke all along the NE-SW axis, which corroborates with the geological map, thus the responses obtained at the level of the Horizontal invariants reflect the deep character of said dyke. It is noted that the application of GTA allowed us to attenuate the noise effect.

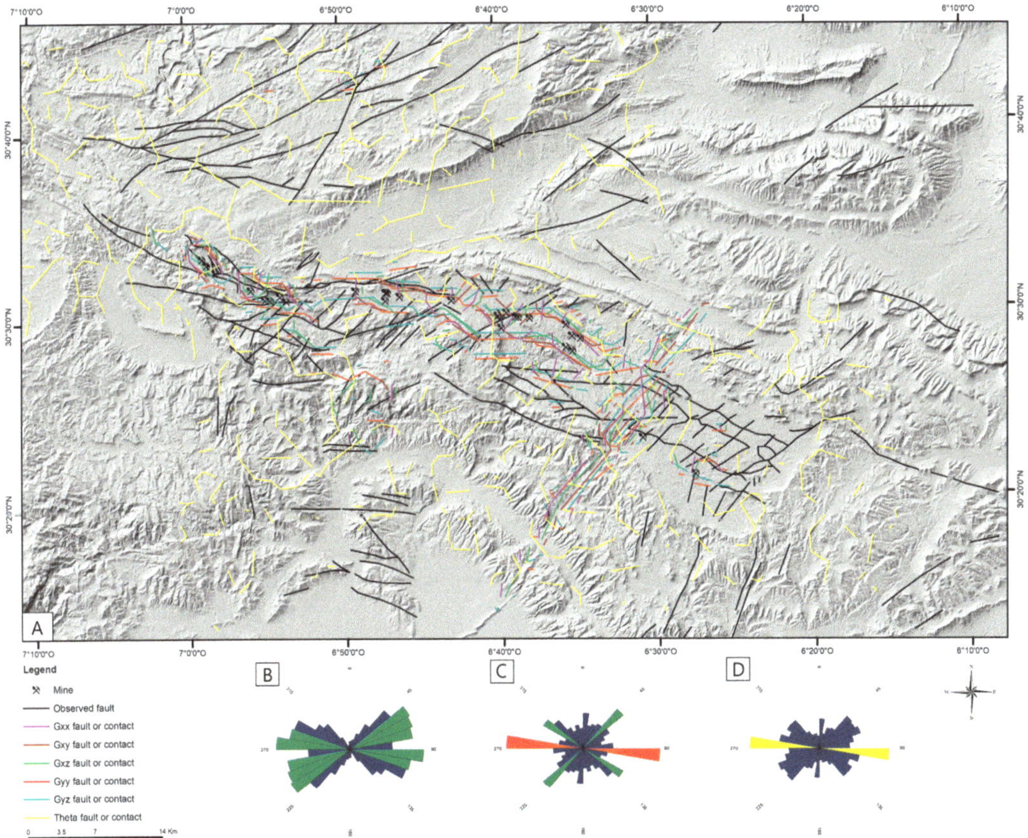

Figure 7. (**A**) Global map of Bou Azzer Lineaments obtained by Gradient tensor Analysis. (**B**) Rose diagram of observed faults. (**C**) Rose diagram of magnetic lineaments obtained by Gradient tensor Analysis and (**D**) Rose diagram showing the synthesis contacts in the study area.

Note that it is difficult at this stage to carry out a precise mapping of the study area given the great contrast that appears at the level of the gradient tensor maps, but one can easily detect a certain number of tectonic events having affected the study area. On the one hand, there is a slight offset at the level of serpentinites at the level of the Ait Ahmane sector. This is due to the presence of a fracture N70° with sinistral lateral rejection, it is acts of Irthem fault (Figures 4 and 7). On the other hand, we have detected at the level of the Foum Zguid dyke a fracture oriented N120 with a sinistral component which corresponds to the AMAA. We note that throughout the ophiolitic complex several fractures appear (Figure 7), they are mainly oriented N40 to N60 with a left lateral strike-slip movement. Note that throughout the northern part of the ophiolitic complex, we notice the presence of a strong negative vertical gradient which can probably be attributed to the Anti-Atlas Major Accident (Figure 3).

Meanwhile, the GTA enables us to separate closely spaced bodies and allows clear visualization of the plunging bodies, which will allow us to determine the plunging direction. The aforementioned GTA will help locate the maxima above the causative sources. Therefore, it will ensure a good visualization of magmatic intrusions, especially those of a basic nature as present in the study area (Figure 7A).

We note that the aforementioned study region is distinguished by the presence of a famous ophiolitic complex, with 70% serpentinization. The obtained results (Figures 4, 5 and 7A) easily allow one to identify the anomalies of high magnetic susceptibility. Through the analysis of the map, we observe that the Bou Azzer complex represents a solitary serpentine structure along the WNW-ESE axis, confirming the fragmented nature of said ophiolitic complex. It is worth noting that the aforementioned complex exhibits a northward plunging direction. Similarly, we observe a lateral continuity towards the west of the Takeroumt's serpentinite fragment.

Also, GTA allows the balancing of anomalies originating from both shallow and deep origins and to accentuate specific features of interest (Figures 5 and 7A), such as the edges of magnetic bodies. GTA can emphasize the edges of the anomalies identified in the Bou Azzer sector. We would like to point out that the cobaltiferous and nickeliferous mineralization is located at the contact between the ophiolitic complex, in other words, between high-amplitude anomalies and the surrounding rocks. Hence, it is important to map these contacts. This method exhibits remarkable traits, as it generates peak amplitudes at the edges of the source and equalizes signals originating from both shallow and deep sources. The application of this method extends to both synthetic and real data. Its effectiveness is assessed by comparing it to other contour detection methods found in previous literature that rely on derivatives.

The black lines (Figures 3 and 5) and the colored lines (Figure 7) demonstrate the undeniable effectiveness of our method in accurately indicating the location of the boundaries of causative bodies, even for anomalies caused by numerous intrusive sources. It is worth noting that the northern contact between serpentinites and basic rocks in the uplifted block shows a straight alignment, as does the southern contact. However, we must then emphasize that the most significant deposits are located on the southern interaction among quartz diorite and serpentinite. The application of GTA in the Bou Azzer uplifted block has yielded effective results. Thus, Figure 4 displays the amplitude maxima of horizontal invariants along the edges of the ophiolitic complex, marked by a tectonic contact with the surrounding rocks. In this case, regarding Figure 5, it becomes easy to identify the geological lineaments with theta map, enabling a better comprehension of the area's tectonic history. The effect of depth has been observed in the extension of the ophiolitic complex in the north-western part of the uplifted block. From a structural standpoint, the major Anti-Atlas fault-oriented WNW-ESE is more clearly visible on the gradient tensor maps (Figure 3), snaking through the northern portion of the aforementioned ophiolitic complex. We emphasize that the contact between the basic magmatism of Jbel Boho and the surrounding rocks is subcircular (Figure 4). Therefore, we recall that the northern contact is marked by the presence of dolomitic beds from the lower Adoudounian, while the southern contact is marked by the presence of dolomites. For the Foum Zguid dyke, we note that the maxima are located on either side of the center of said dyke [28]. Then said maxima show an elongation along the NE-SW direction.

4.3. Structural Analysis

Examination of the first rose diagram (Figure 7B) shows that the observed geological faults extracted from the geological map are globally dispersed among main directions ENE-WSW, E-W and WNW-ESE. In return, the rose diagram relating to the magnetic lineaments shows that the latter are aligned in lightly diverse orientations, NE-SW, NW-SE and fewer ostensibly N-S. The noticeable contrast in these alignments can be associated with the underlying structural composition of the basement resulting from the magnetic interpretation. These structures are entirely hidden beneath Quaternary sediments, and consequently, they are not depicted on the geological map. Moreover, the coverage of the geophysical data exceeds the limits of the Bou Azzer inlier.

The NE-SW direction particularly represents strike-slip faults that cause a notable offset of the ophiolitic body.

The WNW-ESE to E-W (Figure 7C) directions are particularly evident in the inlier central zone, this is the main orientation of the Bou Azzer ophiolitic complex, these directions represent the faults inherited from the Pan-African orogeny and which were reactivated in the long term until the uplift by Variscan deformation [15,16]. It is a period marked by a generalized shortening, at the level of the central and eastern Anti-Atlas, so we are witnessing a reactivation of the Precambrian structures along the Major Anti-Atlas Accident which led to the raising of the inlier along a NW-SE axis [50].

The structural rose diagram (Figure 7D) shows that the lineaments display an E-W, NE-SW, WNW-ESE main trend and N-S less important direction.

The obtained results demonstrate that the studied region is prominent owing to its high structural complexity resulting from polyphase tectonics at the level of the Bou Azzer inlier.

4.4. Tectonic Implications

Regarding the structural analysis of the Bou Azzer inlier, we were able to highlight several fault families representing different tectonic episodes that affected the study area. It is noted that this variety of direction is the result of polyphase tectonics. As mentioned above, the obtained results revealed the following structures:

- The family of faults showing directions N°020 and N°030 particularly concerns the old basement [51] (Figure 7B,C).
- Faults N°020, N°040, N°060, N°070, N°100, N°120, N°130 and N°140 mainly affect Upper Proterozoic formations [52] (Figure 7B,C).
- The Infracambrian and Paleozoic cover formations are affected by families, N°020, N°030, N°060 and N°070 [15] (Figure 7B,C).

Studies conducted by various authors [15,16,51,52] have concluded that:

The main Pan-African B1 phase is marked mainly by the N030° to N050° dextral faults and by the N140 to N160° sinistral faults. This phase is responsible for the exhumation of the ophiolitic complex by the process of obduction. This is a WNW-ESE-oriented compression phase.

The late main Pan-African phase B2 is distinguished by a significant tightening at the level of the inlier, thus we find two large families of faults: the family N°150° to 160° dextral strike-slip, and a family N°70 sinistral strike-slip. The tightening is oriented N020° to N030°.

The post Pan-African period is marked by a distensive phase, giving rise to faults oriented N70°. This phase is marked by the formation of grabens with detrital filling. This phase is followed by a major extensive phase marked by normal synsedimentary faults N20° and N70°.

The Hercynian phase corresponds to a period of reactivation of the N70° faults. This is a compressive event marked by a sinistral movement of these faults. The latter have dextral inverse NW-SE conjugates.

4.5. Mining Implications

In the Bou Azzer inlier, the structural study combining geological and geophysical data (GTA) has proven effective in analyzing the structural and geodynamic evolution of the inlier. Note that the majority of hydrothermal mineral deposit types seem to have originated from extensive hydrothermal systems often concentrated near significant crustal features (faults and crustal discontinuities) [53]. This is the case for the mineralization at the level of the Bou Azzer inlier which displays an intimate relationship with the Ophiolitic complex of Bou Azzer and surrounding rocks. This association suggests that a deeply embedded structure served as a conduit for concentrating mineral-rich fluids through the crust. Nevertheless, these substantial structures may not always be evident at the surface (e.g., the Carlin Trend) [54]. Therefore, the significance of employing gradient tensor analysis lies in its ability to identify these underlying formations that are not visible on the geological map.

The mineralized structures within the Bou Azzer mining district can be categorized into five primary orientations: East-West, West-Northwest to East-Southeast, Northeast-Southwest, Northwest-Southeast, and North-South. These orientations align with the magnetic lineaments that have been identified.

Cobaltiferous and nickeliferous mineralization in the Bou Azzer mining district preferentially developed in old tectonic breccia bodies located between serpentinites and magmatic intrusions (quartz diorites). This configuration testifies to the existence of old normal faults WNW-ESE having participated in the exhumation of deep mantle rocks. The contact between seawater and magmatic bodies leads to hydrothermal reactions at the origin of mineralogical transformations leading to the formation of both cobalt and nickel.

The WNW-ESE and ENE-WSW families represent the major pan-African accidents. They are sinistral in nature and linked to the compressive episode affecting the entire Anti-Atlas chain. The WNW-ESE trending faults represent drains for the circulation of hydrothermal fluids. Furthermore, the NE-SW and E-W trending lineaments constitute the structural framework in which the mineralization is situated.

It should be noted that the vein system at the level of the cobaltiferous deposit of Bou Azzer presents numerous movement criteria which diverge according to the orientation of the vein. Measurements taken in situ and ex situ have shown that the veins tend mainly from N10° to N70° E with a main dip towards W or NW, respectively, and values close to 60–80° (Figure 8). Indeed, the veins oriented N70° E are associated with a left lateral movement [55] (Figure 9A), while those oriented N-S present a right lateral movement [55] (Figure 9B).

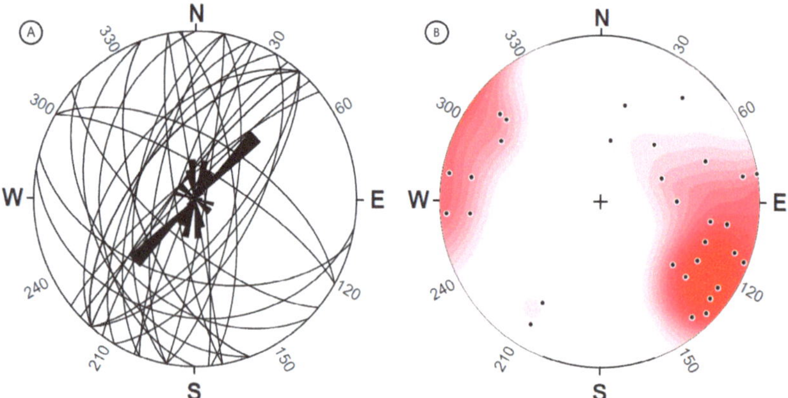

Figure 8. (**A**) Stereonet diagram of Bou Azzer veins measured on the field (Schmidt stereonet, lower hemisphere). (**B**) Stereonet poles contour diagram of Bou Azzer veins measured orientation on the field (Schmidt stereonet, lower hemisphere).

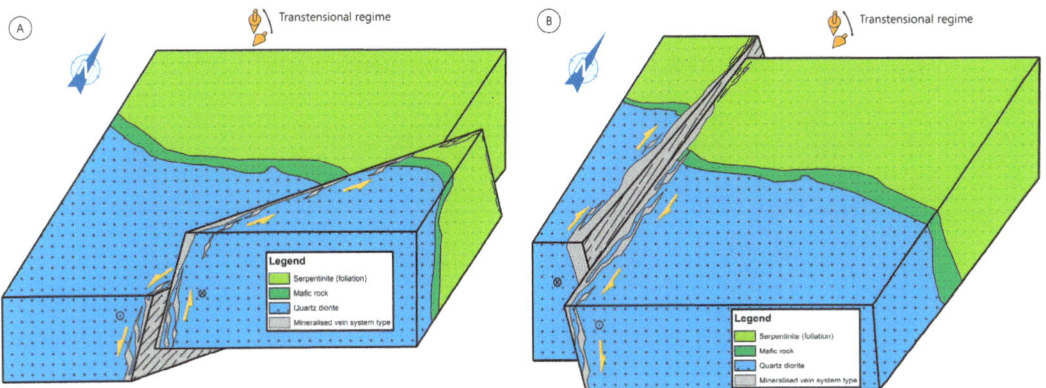

Figure 9. Schematic diagram describing the conjugate character of the vein system at Bou Azzer. (**A**) NE-SW systematic normal sinistral motion. (**B**) N-S normal dextral motion. Modified after [55].

The copper mineralization at the level of the Bleïda mining district located SE of the inlier is associated with brittle, secant faults on the Pan-African structures. These faults are in the N070° direction with left-lateral movement [56] (Figure 10).

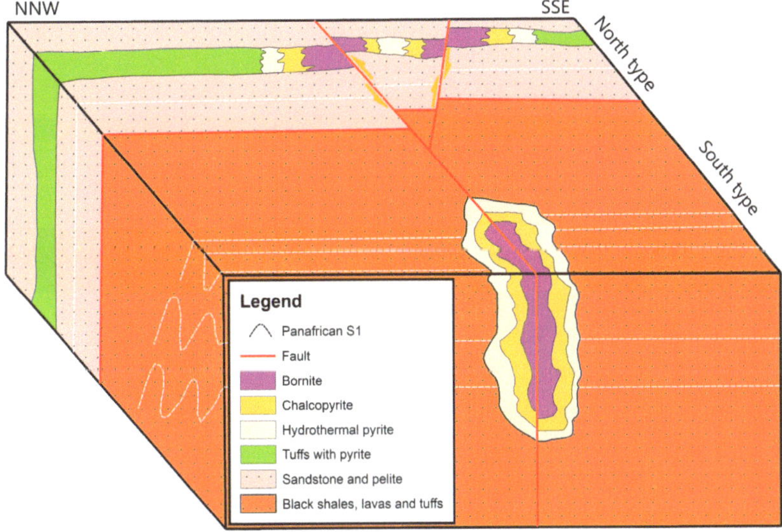

Figure 10. Schematic 3D block of Bleïda-type mineralization modified after [56].

Faults trending N110 at 130° are associated with a replay zone of Pan-African ductile structures [56] (Figure 10). These faults would constitute a feeder zone for copper-rich fluids then circulating in the N070° faults and precipitating copper sulfides along these faults.

5. Conclusions

The GTA conducted in the central Anti-Atlas, particularly in the Bou Azzer inlier, using high resolution aeromagnetic data, proved beneficial and allowed us to address several issues. It identified several favorable areas for mineralization, and helped to solve part of the mystery regarding the polarity of the subduction plane.

The geophysical study with GTA in the Bou Azzer inlier revealed the presence of several positive and negative anomalies. The main positive anomaly, occupying the core region of the investigated area, is directly related to serpentinized oceanic rocks, while the other positive anomalies are associated with basic intrusions. The detected negative anomalies are closely related either to quartz diorites or to the Adoudounian sedimentary cover.

In terms of structural analysis, the investigation conducted using geological and geophysical data was able to confirm the results obtained in previous studies. Several fault families were identified, namely WNW-ESE and ENE-WSW families associated to the Pan-African episode, and the NE-SW family related to the post-orogenic episode, thus confirming the previously mentioned polyphase tectonics.

Regarding the mineralized structures, this study demonstrates that the vein system is undoubtedly controlled by the NE-SW faults, while the massive mineralization follow the WNW-ESE family. The Bou Azzer inlier lives up to its reputation as a premier metallogenic province, as the conducted study highlighted several zones that showed promising indicators, encouraging future detailed prospecting campaigns to confirm the obtained results.

The use of findings from previous studies like [20,24,57] has proven to be effective, as it has highlighted several promising zones characterized by high structural complexity and strong alterability. We have detected a zone closely related to mantle rocks known for their cobalt enrichment. Another zone has also attracted interest, as it is associated with the basic complex of the Tachdamt-Bleïda platform, where a copper mineralization in the form of lenses is present.

Therefore, we recommend tightening the grid spacing in future geophysical surveys to achieve higher resolution, and conducting reconnaissance drilling in the areas deemed favorable for mineralization. Additionally, we suggest applying other geophysical methods such as gravimetry, electromagnetism, and electrical surveys to accurately cross-reference different geophysical signatures.

Author Contributions: Conceptualization, A.S., S.C., S.S. and A.M. (Ahmed Manar); methodology, A.S., S.S. and A.M. (Ahmed Manar); software, A.S., S.S., A.M. (Ahmed Manar) and A.M. (Abdelhalim Miftah); validation, A.S., S.C., S.S., A.M. (Ahmed Manar), Z.B. and A.M. (Abdelhalim Miftah); formal analysis, A.S., S.C. and S.S.; investigation, A.S.; resources, A.S.; data curation, A.S.; writing—original draft preparation, A.S., S.C. and S.S.; writing—review and editing, A.S., S.C., S.S., A.M. (Ahmed Manar), Z.B. and A.M. (Abdelhalim Miftah); visualization, A.S., S.S., A.M. (Ahmed Manar) and A.M. (Abdelhalim Miftah); supervision, S.C., S.S., A.M. (Ahmed Manar) and Z.B.; project administration, A.S.; funding acquisition, A.S. All authors have read and agreed to the published version of the manuscript.

Funding: This research received no external funding.

Institutional Review Board Statement: Not applicable.

Informed Consent Statement: Not applicable.

Data Availability Statement: The data used to support the findings of this study are available from the corresponding author upon request.

Conflicts of Interest: The authors declare no conflict of interest.

References

1. Grauch, V.J.S.; Cordell, L. Limitations of Determining Density or Magnetic Boundaries from the Horizontal Gradient of Gravity or Pseudogravity Data. *Geophysics* **1987**, *52*, 118–121. [CrossRef]
2. Foss, C.A. Mapping Basement Relief with Airborne Gravity Gradiometry. *ASEG Ext. Abstr.* **2001**, *2001*, 1–4. [CrossRef]
3. Oruç, B.; Sertçelik, I.; Kafadar, Ö.; Selim, H.H. Structural Interpretation of the Erzurum Basin, Eastern Turkey, Using Curvature Gravity Gradient Tensor and Gravity Inversion of Basement Relief. *J. Appl. Geophys.* **2013**, *88*, 105–113. [CrossRef]
4. Won, I.J.; Murphy, V.; Hubbard, P.; Oren, A.; Davis, K. Geophysics and Weapons Inspection. To Dig or Not to Dig? That Is (at Least) One Question. *Lead. Edge* **2004**, *23*, 658–662. [CrossRef]
5. de Ridder, S.A.L.; Biondi, B.L. Ambient Seismic Noise Tomography at Ekofisk. *Geophysics* **2015**, *80*, B167–B176. [CrossRef]
6. De Ridder, S.; Dellinger, J. Ambient Seismic Noise Eikonal Tomography for Near-Surface Imaging at Valhall. *Lead. Edge* **2011**, *30*, 506–512. [CrossRef]

7. De Ridder, S.A.L.; Biondi, B.L. Near-Surface Scholte Wave Velocities at Ekofisk from Short Noise Recordings by Seismic Noise Gradiometry. *Geophys. Res. Lett.* **2015**, *42*, 7031–7038. [CrossRef]
8. Murphy, C.A.; Brewster, J. Target Delineation Using Full Tensor Gravity Gradiometry Data. *ASEG Ext. Abstr.* **2007**, *2007*, 1–3. [CrossRef]
9. Murphy, C.A.; Dickinson, J.L. Exploring Exploration Play Models with FTG Gravity Data. In Proceedings of the 11th SAGA Biennial Technical Meeting and Exhibition; European Association of Geoscientists & Engineers, Swaziland, South Africa, 16–18 September 2009; p. cp-241-00020.
10. Beiki, M.; Pedersen, L.B. Eigenvector Analysis of Gravity Gradient Tensor to Locate Geologic Bodies. *Geophysics* **2010**, *75*, I37–I49. [CrossRef]
11. Mataragio, J.; Kieley, J. Application of Full Tensor Gradient Invariants in Detection of Intrusion-Hosted Sulphide Mineralization: Implications for Deposition Mechanisms. *First Break* **2009**, *27*, 95–98. [CrossRef]
12. Murphy, C.; Dickinson, J.; Salem, A. Depth Estimating Full Tensor Gravity Data with the Adaptive Tilt Angle Method. *ASEG Ext. Abstr.* **2012**, *2012*, 1–3. [CrossRef]
13. Bouougri, E.H.; Lahna, A.A.; Tassinari, C.C.G.; Basei, M.A.S.; Youbi, N.; Admou, H.; Saquaque, A.; Boumehdi, M.A.; Maacha, L. Time Constraints on Early Tonian Rifting and Cryogenian Arc Terrane-Continent Convergence along the Northern Margin of the West African Craton: Insights from SHRIMP and LA-ICP-MS Zircon Geochronology in the Pan-African Anti-Atlas Belt (Morocco). *Gondwana Res.* **2020**, *85*, 169–188. [CrossRef]
14. Hefferan, K.; Soulaimani, A.; Samson, S.D.; Admou, H.; Inglis, J.; Saquaque, A.; Latifa, C.; Heywood, N. A Reconsideration of Pan African Orogenic Cycle in the Anti-Atlas Mountains, Morocco. *J. Afr. Earth Sci.* **2014**, *98*, 34–46. [CrossRef]
15. Blein, O.; Baudin, T.; Chèvremont, P.; Soulaimani, A.; Admou, H.; Gasquet, P.; Cocherie, A.; Egal, E.; Youbi, N.; Razin, P.; et al. Geochronological Constraints on the Polycyclic Magmatism in the Bou Azzer-El Graara Inlier (Central Anti-Atlas Morocco). *J. Afr. Earth Sci.* **2014**, *99*, 287–306. [CrossRef]
16. Soulaimani, A.; Ouanaimi, H.; Saddiqi, O.; Baidder, L.; Michard, A. The Anti-Atlas Pan-African Belt (Morocco): Overview and Pending Questions. *Comptes Rendus-Geosci.* **2018**, *350*, 279–288. [CrossRef]
17. El Hadi, H.; Simancas, J.F.; Martínez-Poyatos, D.; Azor, A.; Tahiri, A.; Montero, P.; Fanning, C.M.; Bea, F.; González-Lodeiro, F. Structural and Geochronological Constraints on the Evolution of the Bou Azzer Neoproterozoic Ophiolite (Anti-Atlas, Morocco). *Precambrian Res.* **2010**, *182*, 1–14. [CrossRef]
18. D'Lemos, R.S.; Inglis, J.D.; Samson, S.D. A Newly Discovered Orogenic Event in Morocco: Neoproterozic Ages for Supposed Eburnean Basement of the Bou Azzer Inlier, Anti-Atlas Mountains. *Precambrian Res.* **2006**, *147*, 65–78. [CrossRef]
19. Souiri, M.; Aissa, M.; Góis, J.; Oulgour, R.; Mezougane, H.; El Azmi, M.; Moussaid, A. Application of Multivariate Statistics and Geostatistical Techniques to Identify the Distribution Modes of the Co, Ni, As and Au-Ag Ore in the Bou Azzer-East Deposits (Central Anti-Atlas Morocco). *Econ. Environ. Geol.* **2020**, *53*, 363–381. [CrossRef]
20. Leblanc, M. Proterozoic Oceanic Crust at Bou Azzer. *Nature* **1976**, *261*, 34–35. [CrossRef]
21. Hodel, F.; Triantafyllou, A.; Berger, J.; Macouin, M.; Baele, J.M.; Mattielli, N.; Monnier, C.; Trindade, R.I.F.; Ducea, M.N.; Chatir, A.; et al. The Moroccan Anti-Atlas Ophiolites. Timing and Melting Processes in an Intra-Oceanic Arc-Back-Arc Environment. *Gondwana Res.* **2020**, *86*, 182–202. [CrossRef]
22. Admou, H.; Juteau, T. Decouverte d'un Systeme Hydrothermal Oceanique Fossile Dans l'ophiolite Antecambrienne de Khzama (Massif Du Siroua, Anti-Atlas Marocain). *Comptes Rendus l'Academie Sci. -Ser. IIa Sci. la Terre des Planetes* **1998**, *327*, 335–340. [CrossRef]
23. Samson, S.D.; Inglis, J.D.; D'Lemos, R.S.; Admou, H.; Blichert-Toft, J.; Hefferan, K. Geochronological, Geochemical, and Nd-Hf Isotopic Constraints on the Origin of Neoproterozoic Plagiogranites in the Tasriwine Ophiolite, Anti-Atlas Orogen, Morocco. *Precambrian Res.* **2004**, *135*, 133–147. [CrossRef]
24. Walsh, G.J.; Benziane, F.; Aleinikoff, J.N.; Harrison, R.W.; Yazidi, A.; Burton, W.C.; Quick, J.E.; Saadane, A. Neoproterozoic Tectonic Evolution of the Jebel Saghro and Bou Azzer-El Graara Inliers, Eastern and Central Anti-Atlas, Morocco. *Precambrian Res.* **2012**, *216–219*, 23–62. [CrossRef]
25. Inglis, J.D.; D'Lemos, R.S.; Samson, S.D.; Admou, H. Geochronological Constraints on Late Precambrian Intrusion, Metamorphism, and Tectonism in the Anti-Atlas Mountains. *J. Geol.* **2005**, *113*, 439–450. [CrossRef]
26. Gasquet, D.; Levresse, G.; Cheilletz, A.; Azizi-Samir, M.R.; Mouttaqi, A. Contribution to a Geodynamic Reconstruction of the Anti-Atlas (Morocco) during Pan-African Times with the Emphasis on Inversion Tectonics and Metallogenic Activity at the Precambrian-Cambrian Transition. *Precambrian Res.* **2005**, *140*, 157–182. [CrossRef]
27. Fekkak, A.; Ouanaimi, H.; Michard, A.; Soulaimani, A.; Ettachfini, E.M.; Berrada, I.; El Arabi, H.; Lagnaoui, A.; Saddiqi, O. Thick-Skinned Tectonics in a Late Cretaceous-Neogene Intracontinental Belt (High Atlas Mountains, Morocco): The Flat-Ramp Fault Control on Basement Shortening and Cover Folding. *J. Afr. Earth Sci.* **2018**, *140*, 169–188. [CrossRef]
28. Soulaimani, A.; Chakiri, S.; Soulaimani, S.; Manar, A.; Bejjaji, Z.; Miftah, A.; Zerdeb, M.A.; Zidane, Y.; Boualoul, M.; Muzirafuti, A. Semi-Automatic Image Processing System of Aeromagnetic Data for Structural and Mining Investigations (Case of Bou Azzer Inlier, Central Anti-Atlas, Morocco). *Appl. Sci.* **2022**, *12*, 11270. [CrossRef]

29. Smith, W.H.F.; Wessel, P. Gridding with Continuous Curvature Splines in Tension. *Geophysics* **1990**, *55*, 293–305. [CrossRef]
30. Nelson, J.B.; Marcotte, D.L.; Hardwick, C.D. Comments on "Magnetic Field Gradients and Their Uses in the Study of the Earth's Magnetic Field" by Harrison and Southam. *J. Geomagn. Geoelectr.* **1992**, *44*, 367–370. [CrossRef]
31. Hardwick, C.D. Gradient-Enhanced Total Field Gridding. In Proceedings of the 1999 SEG Annual Meeting, Houston, TX, USA, 31 October–5 November 1999. [CrossRef]
32. Hardwick, C.D. Aeromagnetic Gradiometry in 1995. *Explor. Geophys.* **1996**, *27*, 1–11. [CrossRef]
33. Schneider, M.; Stolz, R.; Linzen, S.; Schiffler, M.; Chwala, A.; Schulz, M.; Dunkel, S.; Meyer, H.G. Inversion of Geo-Magnetic Full-Tensor Gradiometer Data. *J. Appl. Geophys.* **2013**, *92*, 57–67. [CrossRef]
34. Stolz, R.; Schiffler, M.; Becken, M.; Thiede, A.; Schneider, M.; Chubak, G.; Marsden, P.; Bergshjorth, A.B.; Schaefer, M.; Terblanche, O. SQUIDs for Magnetic and Electromagnetic Methods in Mineral Exploration. *Miner. Econ.* **2022**, *35*, 467–494. [CrossRef]
35. Blakely, R.J. *Potential Theory in Gravity and Magnetic Applications*; Cambridge University Press: Cambridge, UK, 1995. [CrossRef]
36. Clark, D.A. New Methods for Interpretation of Magnetic Vector and Gradient Tensor Data I: Eigenvector Analysis and the Normalised Source Strength. *Explor. Geophys.* **2012**, *43*, 267–282. [CrossRef]
37. Hinze, W.J.; Von Frese, R.R.B. Magnetics in Geoexploration. *Proc. Indian Acad. Sci. -Earth Planet. Sci.* **1990**, *99*, 515–547. [CrossRef]
38. Fitzgerald, D.; Argast, D.; Holstein, H. Further Developments with Full Tensor Gradiometry Datasets. *ASEG Ext. Abstr.* **2009**, *2009*, 1. [CrossRef]
39. Argast, D.; FitzGerald, D.; Holstein, H.; Stolz, R.; Chwala, A. Compensation of the Full Magnetic Tensor Gradient Signal. *ASEG Ext. Abstr.* **2010**, *2010*, 1–4. [CrossRef]
40. Kolecki, J.C. *An Introduction to Tensors for Students of Physics and Engineering*; NASA: Washington, DC, USA, 2002.
41. Heath, P.; Heinson, G.; Greenhalgh, S. Some Comments on Potential Field Tensor Data. *Explor. Geophys.* **2003**, *34*, 57–62. [CrossRef]
42. Wilson, G.A.; Cuma, M.; Zhdanov, M.S. Large-Scale 3D Inversion of Airborne Potential Field Data. In Proceedings of the 73rd EAGE Conference and Exhibition Incorporating SPE EUROPEC 2011, Vienna, Austria, 23–27 May 2011. [CrossRef]
43. Holstein, H.; Fitzgerald, D.; Willis, C.P.; Foss, C. Magnetic Gradient Tensor Eigen-Analysis for Dyke Location. In Proceedings of the 73rd EAGE Conference and Exhibition Incorporating SPE EUROPEC 2011, Vienna, Austria, 23–27 May 2011; Volume 5, pp. 3330–3334. [CrossRef]
44. Wijns, C.; Perez, C.; Kowalczyk, P. Theta Map: Edge Detection in Magnetic Data. *Geophysics* **2005**, *70*, L39–L43. [CrossRef]
45. Dilalos, S.; Alexopoulos, J.D.; Vassilakis, E.; Poulos, S.E. Investigation of the Structural Control of a Deltaic Valley with Geophysical Methods. The Case Study of Pineios River Delta (Thessaly, Greece). *J. Appl. Geophys.* **2022**, *202*, 104652. [CrossRef]
46. Eldosouky, A.M.; Pham, L.T.; Mohmed, H.; Pradhan, B. A Comparative Study of THG, AS, TA, Theta, TDX and LTHG Techniques for Improving Source Boundaries Detection of Magnetic Data Using Synthetic Models: A Case Study from G. Um Monqul, North Eastern Desert, Egypt. *J. Afr. Earth Sci.* **2020**, *170*, 103940. [CrossRef]
47. Palencia-Ortas, A.; Ruiz-Martínez, V.C.; Villalaín, J.J.; Osete, M.L.; Vegas, R.; Touil, A.; Hafid, A.; Mcintosh, G.; van Hinsbergen, D.J.J.; Torsvik, T.H. A New 200 Ma Paleomagnetic Pole for Africa, and Paleo-Secular Variation Scatter from Central Atlantic Magmatic Province (CAMP) Intrusives in Morocco (Ighrem and Foum Zguid Dykes). *Geophys. J. Int.* **2011**, *185*, 1220–1234. [CrossRef]
48. Bouiflane, M.; Manar, A.; Medina, F.; Youbi, N.; Rimi, A. Mapping and Characterization from Aeromagnetic Data of the Foum Zguid Dolerite Dyke (Anti-Atlas, Morocco) a Member of the Central Atlantic Magmatic Province (CAMP). *Tectonophysics* **2017**, *708*, 15–27. [CrossRef]
49. Benaouda, R.; Holzheid, A.; Schenk, V.; Badra, L.; Ennaciri, A. Magmatic Evolution of the Jbel Boho Alkaline Complex in the Bou Azzer Inlier (Anti-Atlas/Morocco) and Its Relation to REE Mineralization. *J. Afr. Earth Sci.* **2017**, *129*, 202–223. [CrossRef]
50. Soulaimani, A.; Jaffal, M.; Maacha, L.; Kchikach, A.; Najine, A.; Saidi, A. Magnetic Modelling of the Bou Azzer-El Graara Ophiolite (Central Anti-Atlas, Morocco). Geodynamic Implications of the Panafrican Reconstruction. *Comptes Rendus-Geosci.* **2006**, *338*, 153–160. [CrossRef]
51. Emran, A.; Chorowicz, J. La Tectonique Polyphasée Dans La Boutonnière Précambrienne de Bou Azzer (Anti-Atlas Central, Maroc): Apports de l'imagerie Spatiale Landsat-MSS et de l'analyse Structurale de Terrain / The Panafrican Tectonic Events in the Eroded Anticline of Bou Azzer. *Sci. Géologiques. Bull.* **1992**, *45*, 121–134. [CrossRef]
52. Soulaimani, A.; Michard, A.; Ouanaimi, H.; Baidder, L.; Raddi, Y.; Saddiqi, O.; Rjimati, E.C. Late Ediacaran-Cambrian Structures and Their Reactivation during the Variscan and Alpine Cycles in the Anti-Atlas (Morocco). *J. Afr. Earth Sci.* **2014**, *98*, 94–112. [CrossRef]
53. Jaques, A.L.; Jaireth, S.; Walshe, J.L. Mineral Systems of Australia: An Overview of Resources, Settings and Processes. *Aust. J. Earth Sci.* **2002**, *49*, 623–660. [CrossRef]
54. Crafford, A.E.J.; Grauch, V.J.S. Geologic and Geophysical Evidence for the Influence of Deep Crustal Structures on Paleozoic Tectonics and the Alignment of World-Class Gold Deposits, North-Central Nevada, USA. *Ore Geol. Rev.* **2002**, *21*, 157–184. [CrossRef]
55. Tourneur, E.; Chauvet, A.; Kouzmanov, K.; Tuduri, J.; Paquez, C.; Sizaret, S.; Karfal, A.; Moundi, Y.; El Hassani, A. Co-Ni-Arsenide Mineralisation in the Bou Azzer District (Anti-Atlas, Morocco): Genetic Model and Tectonic Implications. *Ore Geol. Rev.* **2021**, *134*, 104128. [CrossRef]

56. Bourque, H. Le Cuivre de l'Anti-Atlas, Un Problème Complexe: Synthèse Des Occurrences Cuprifères de La Boutonnière de Bou Azzer-El Graara et Nouvelles Données (Anti-Atlas, Maroc). Ph.D. Thesis, Université d'Orléans, Orléans, France, 2016.
57. Jaffal, M.; Soulaimani, A.; Ilmen, S.; Anzar, F.; Kchikach, A.; Manar, A.; Soulaimani, A.; Maacha, L.; Bajddi, A. Insights into the Deep Structure of the Bou Azzer-El Graara Inlier (Central Anti-Atlas, Morocco): Inference from High-Resolution Magnetic Data, and Geodynamic Implications. *Tectonophysics* **2023**, *856*, 229865. [CrossRef]

Article

Spatiotemporal Variation of Land Surface Temperature Retrieved from FY-3D MERSI-II Data in Pakistan

Bilawal Abbasi [1], Zhihao Qin [1,*], Wenhui Du [1], Jinlong Fan [2], Shifeng Li [1] and Chunliang Zhao [1]

1   MOA Key Laboratory of Agricultural Remote Sensing, Institute of Agro-Resources and Regional Planning, Chinese Academy of Agricultural Sciences, Beijing 100081, China

2   National Satellite Meteorological Center, Beijing 100081, China

*   Correspondence: qinzhihao@caas.cn; Tel.: +86-135-2135-0214

Abstract: The concept of land surface temperature (LST) encompasses both surface energy balance and land surface activities. The study of climate change greatly benefits from an understanding of the geographical and temporal fluctuations of LST. In this study, we utilized an improved version of the TFSW algorithm to retrieve the LST from the Medium resolution spectral imager II (MERSI-II) data for the first time in Pakistan. MERSI-II is a payload for the Chinese meteorological satellite Fengyun 3D (FY-3D), and it has the capability for use in various remote sensing applications such as climate change and drought monitoring, with higher spatial and temporal resolutions. Once the LSTs were retrieved, accuracy of the LSTs were investigated. Later, LST datasets were used to detect the spatiotemporal variations of LST in Pakistan. Monthly, seasonal, and annual datasets were utilized to detect increasing and decreasing LST trends in the regions, with Mann–Kendall and Sen's slope estimator tool. In addition, we further revealed the long-term spatiotemporal variations of LST by utilizing Moderate Resolution Imaging Spectrometer (MODIS) LST observations. The cross-validation analysis shows that the retrieved LST of MERSI-II was more consistent with the MODIS MYD11A1 LST product compared to the MYD21A1. The spatial distribution of LSTs demonstrates that the mean LST exhibits a pattern of spatial variability, with high values in the southern areas and low values in the northern areas; there are areas that do not follow this trend, possibly due to reasons of elevation and types of land cover also influencing the LST's spatial distribution. The annual mean LST trend increases in the northern regions and decreases in the southern regions, ranging between −0.013 and 0.019 °C/year. The trend of long-term analysis were also consistent with MERSI-II, excepting region II, with increasing effects. This study will be helpful for various environmental and climate change studies.

Keywords: remote sensing; spatiotemporal; land surface temperature; MODIS; MERSI-II

Citation: Abbasi, B.; Qin, Z.; Du, W.; Fan, J.; Li, S.; Zhao, C. Spatiotemporal Variation of Land Surface Temperature Retrieved from FY-3D MERSI-II Data in Pakistan. *Appl. Sci.* **2022**, *12*, 10458. https://doi.org/10.3390/app122010458

Academic Editors: Giovanni Randazzo, Stefania Lanza and Anselme Muzirafuti

Received: 2 September 2022
Accepted: 12 October 2022
Published: 17 October 2022

1. Introduction

Land surface temperature (LST) is one of the key parameters in the land surface physical process on regional and global scales, and an accurate LST is essential for the study, such as agricultural drought monitoring, climate change [1,2], estimating air temperature and evapotranspiration [3,4], analyzing land cover change, and urbanization [5,6]. Traditionally, ground measurements cannot practically achieve an aim that measures the LST over wide and continuous areas. With the development of remote sensing from space, satellite remote sensing is a powerful and effective method for analyzing the spatiotemporal variation of LST over the entire globe with a sufficiently high spatial–temporal resolution. It can provide multi-temporal, multi-spectral, and real-time data. In this way, remote sensing is become more applicable and highly preferred compared to traditional ground measurements.

The spatiotemporal variation of LST is important for detecting changes in land surface characteristics and understanding climate change. This study utilizes MERSI-II sensor data, which provides a unique opportunity for monitoring the changes in LST at spatial and temporal scales. The MERSI-II sensor is carried by the Chinese FY-3D polar-orbiting meteorological satellite, which has a higher temporal and spatial resolution than the MODIS satellite, and can be utilized as a backup data source for global LST measurements. MERSI-II is equipped with two thermal infrared (TIR) channels (bands 24 and 25), and provides a spatial resolution of up to 250 m, which confirms the basic requirements for LST estimation. Until now, a few studies have been published on LST retrieval from FY-3D MERSI-II data [7–9]. The SW algorithm is categorized as a multi-channel algorithm because of its superior accuracy when compared to in situ data; the algorithm is considered one of the most mature methods for retrieving LST using TIR remote sensing, and it has a wide application range [10,11]. LST products are generally generated for various satellite sensors using various algorithms based on the intended research aims, and the retrieval of LST utilizing TIR data has made significant development [12]. In this study, we introduced the application potential of the recently listed Two-Factor Split Window (TFSW) algorithm modified by Du et al. [7] for the MERSI-II instrument, to analyze the spatiotemporal variation of LST and validation in Pakistan for the first time. The average mean absolute error and R^2 of the TFSW algorithm was 1.97 K and 0.98, respectively, retrieved from MERSI-II data [7].

On a global scale, the mean temperature was 1.2 ± 0.1 °C and 1.39 °C (in Asia); for 2020, this was above the 1850–1900 baseline data. Pakistan is highly vulnerable to climate change [13,14]. The most prominent devastating consequences of climate change are increases of floods, storms, droughts, and heatwaves [13]. The research study conducted by Adnan et al. [15] for Pakistan shows a total change of a temperature warming trend up to 0.2 °C since the beginning of the last century by incorporating a long-term time series (1876–1993) using reconstructed temperature data. During 1901–2007, the overall increase in the area-weighted mean temperature of Pakistan was 0.64 °C, and it has been continuously rising at 0.06 °C per decade [16].

Until now, studies about LST monitoring at the country level are rare, and a few studies have been published regarding LST monitoring at small regional scales such as Mumtaz et al. [17], who examined the spatiotemporal changes in land use land cover (LULC) and investigated its effects on LST using 20 years of data and a CA-Markov model specially for the two urban cities. Arshad et al. [18] reported that the urban land cover of Faisalabad city in Pakistan caused higher regional temperatures according to the result of LULC changes. Similarly, Saleem et al. [19] considered the normalized difference vegetative index (NDVI) as a primary factor of LULC change to identify the role of urban land cover in the context of LST change for the cities of Lahore, Faisalabad, and Multan in Pakistan. From the above literature, it has been observed that most studies have been related to urbanization, and LULC and its contribution to the LST at city levels except for at the country level.

As global LST has been rising, the LST can no longer be believed to be static. As a result, the country requires a focus of LST monitoring to support various aspects of climate change and environmental studies in the future. Hence, the objective of the study is (i) the MERSI-II data that are available, which are important for LST; these (ii) data are cross-validated with MODIS data and then (iii) applied to assess the spatiotemporal analysis of LST during the years from 2018 to 2021 for different regions of Pakistan.

2. The Study Region and the Data

2.1. The Study Region

Pakistan is counted as an agricultural country with differential topography and climatology. Figure 1 shows spatial distribution of landcover types, elevation and five sub-regions of the country. According to the land utilization report generated by the Statistics Bureau of Pakistan [20], an area of 34.15 million hectares (Mha) is cultivable/agricultural land, while the rest of the 23.60 Mha area are not available for cultivation. Two-thirds of the area in the country lies in an arid zone, which provokes slight climatic changes such as an increase of floods, heatwaves, and droughts, etc. [13]. With regard to climate, during the winter season, the average temperature recorded is between 2 and 23 °C throughout Northern Pakistan (Upper Indus Basin), while it is 14–20 °C in Southern Pakistan (Lower Indus Basin) [21,22]. During the summer season, the average maximum temperature can reach up to a range of 42–44 °C in the southern areas, whereas the northern areas can experience an average temperature of between 23 and 49 °C [21]. The total annual precipitation varies from greater than 2000 mm in the north to less than 250 mm in the south [21,23]. Pakistan has two major precipitation systems that produce rainfall. The monsoon system from the Bay of Bengal releases rain in the southeast and east from June, and continue until September [24]. Western disturbances from December to March brought precipitation to northern and western areas of Pakistan [24].

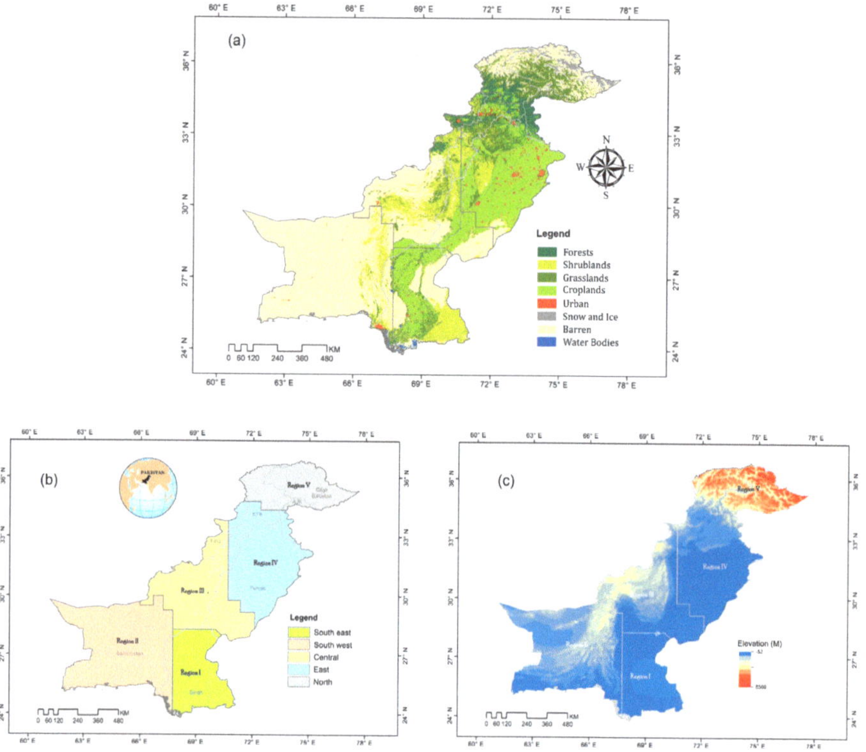

Figure 1. Geographical location of the study region. (**a**) Land cover types in Pakistan during 2019, derived from the MODIS MCD12Q1 Version 6 data product. (**b**) Five homogeneous and contiguous regions were developed, which are similar to the climatic regions developed by [13]. (**c**) Digital Elevation Model based on the ASTER 30 m Digital Elevation Model.

2.2. The FY-3D MERSI-II Data

FY-3D MERSI-II satellite data have very high potential usages for various research fields. Climate change, water vapor estimation, numerical weather forecast, space weather prediction, and ecosystem monitoring are just a few of the many applications for the FY-3D MERSI-II satellite data [25]. The above-mentioned applications would benefit from a method that accurately and quickly retrieves LST from the MERSI-II data (i.e., the TFSW algorithm) [7]. Table 1 and Figure 2 compare the required bands of MERSI-II with the corresponding ones of MODIS for LST estimation. MERSI-II is similar to MODIS in certain ways, particularly in terms of wavelength frequency and band parameters. MERSI-II is a modified version of the first-generation same track type satellite sensors in terms of data quality, band number, resolution, and other factors, in order to increase the detection range for various application areas. According to the TFSW algorithm characteristics, Bands 24 and 25 of MERSI-II are selected for LST estimation. The spectral response function of the MODIS (Band 31/32) and MERSI-II (Band 24/25) TIR bands are shown in Figure 2. From Table 1 and Figure 2, it can be clearly observed that the TIR band range of MERSI-II and MODIS is similar, and that the central wavelength is also similar to some extent. It is also verified that the MODIS data can be used for validation purposes. This study uses FY-3D MERSI-II Level-1B radiance data downloaded from the Fengyun Satellite Data Archiving Portal (http://satellite.nsmc.org.cn) (accessed on 20 April 2022).

Table 1. FY-3D sensor characteristics that correspond to the MODIS TIR Bands are commonly used for retrieving land surface temperature.

Instrument	Country	Launch Time	Sensor	Channel	Wavelength (μm)	Resolution
FY-3D	China	2017	MERSI-II	B24 B25	10.30–11.30 11.50–12.50	1000 m
MODIS	USA	1999	Aqua	B31 B32	10.78–11.28 11.77–12.27	1000 m

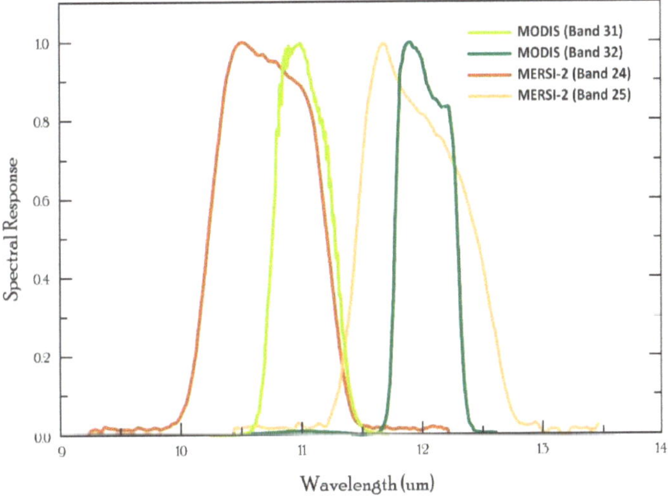

Figure 2. Spectral response functions of the MERSI-II and MODIS TIR channels.

2.3. The MODIS Data

Terra and Aqua are the sensors of the MODIS satellite, which collects data at a high frequency of two observations per day, accessed from portal (https://lpdaac.usgs.gov/products/mod11a1v061/) (accessed on 6 May 2022). Its LST products are among the most mature methods, and have been evaluated in numerous previous studies [26–30]. As we know, the MERSI-II was launched in recent years, and it provide data from 2018, which is not enough to understand the spatiotemporal variations of LST in a broader way at historical scales. Therefore, the MODIS 1 km LST products (MYD11A1 and MYD21A1) were used in this study for two purposes, the first being to cross-validate the LST retrieved from MERSI-II, and second, to evaluate the long-term LST trend for the period between 2005 and 2018. We chose observations from the Terra satellite that provide an estimate of the daily maximum and minimum temperatures. The LST observations were monitored carefully for visual interpretation to ensure that the average LST pixels were within the acceptable quality range for accurate analysis.

3. Materials and Methods

3.1. The TFSW Algorithm for LST Retrieval

The radiance is measured at the top of the atmosphere (TOA) in the TIR bands as follows, using the radiative transfer theory, assuming a cloud-free atmosphere in thermodynamic equilibrium [31]:

$$B_i(T_i) = \varepsilon_i B_i(T_s)\tau_i + R\uparrow atm_i + (1 - \varepsilon_i)R\downarrow atm_i\tau_i \tag{1}$$

where B_i is the Plank function, T_i is the brightness temperature (BT), ε_i is the emissivity, $B_i(T_s)$ is the radiance measured when the surface is a blackbody with a surface temperature of T_s, τ_i is the total atmospheric transmittance (AT), $R\uparrow atm_i$ is the upwelling radiance, and $R\downarrow atm_i$ is the downwelling radiance. The SW technique represents LST as a straightforward linear or nonlinear combination of the TOA brightness temperatures of two adjacent bands based on differential water vapor absorption in two neighboring TIR bands. This method does not require real-time atmospheric profiles [32]. Due to its practicality and efficiency, the SW algorithm is therefore frequently utilized in LST retrieval. The SW algorithm is one of the most widely utilized delegate algorithms that has produced various satellite LST products [33]. Wan [34] developed a more sophisticated generalized split window approach for the MODIS Collection 6 LST product.

Qin et al. [35] developed a TFSW algorithm for LST retrieval from the NOAA-AVHRR Bands 4 and 5, completely based on the mathematical derivation from the thermal radiance transfer equation. The TFSW algorithm uses a series of linear combinations of two apparent temperatures, the corresponding band-averaged emissivity, and transmittance, to estimate the MERSI-II LST as follows:

$$T_s = A_0 + A_1 T_{24} - A_2 T_{25} \tag{2}$$

$$A_0 = a_{24}[D_{25}(1 - C_{24} - D_{24})/(C_{24}D_{25} - C_{25}D_{24})] - a_{25}[D_{24}(1 - C_{25} - D_{25})/(C_{24}D_{25} - C_{25}D_{24})] \tag{3}$$

$$A_1 = 1 + D_{24}/(C_{24}D_{25} - C_{25}D_{24}) + b_{24}[D_{25}(1 - C_{24} - D_{24})/(C_{24}D_{25} - C_{25}D_{24})] \tag{4}$$

$$A_2 = D_{24}/(C_{24}D_{25} - C_{25}D_{24}) + b_{25}[D_{24}(1 - C_{25} - D_{25})/(C_{24}D_{25} - C_{25}D_{24})] \tag{5}$$

$$C_{24} = \varepsilon_{24}\tau_{24} \tag{6}$$

$$D_{24} = [1 - \tau_{24}][1 + (1 - \varepsilon_{24})\tau_{24}] \tag{7}$$

$$C_{25} = \varepsilon_{25}\tau_{25} \tag{8}$$

$$D_{25} = [1 - \tau_{25}][1 + (1 - \varepsilon_{25})\tau_{25}] \tag{9}$$

where T_s is the retrieved FY-3D MERSI-II LST (K), A_0, A_1, and A_2 are the internal parameters, T_i is the brightness temperature (BT) of the FY-3D MERSI-II TIR band i (i = 24 or 25), τ_i is the AT for band i, ε_i is Land Surface Emissivity (LSE) for the band i, and a_i, and b_i are the constants for the band i. The constants for the two FY-3D MERSI-II TIR bands are determined as follows:

$$L_i = B_i(T)/[\partial B_i(T)/\partial T] \approx B_i(T)/\{[B_i(T + \Delta T) - B_i(T)]/\Delta T\} \tag{10}$$

where L_i is the derivative of the Planck function $B_i(T)$ with temperature T for the FY-3D MERSI-II TIR band i, which could be approximated as $[B_i(T + \Delta T) - B_i(T)]/\Delta T$ for the determination of the constants as follows:

$$L_i = a_i + b_i T_i \tag{11}$$

where a_i and b_i are the constants for the band i. It assumes L_i with T from $-50\,^{\circ}C$ to $50\,^{\circ}C$ for the two FY-3D MERSI-II TIR bands, and then the constants are finally determined as follows: $a_{24} = -53.477$, $b_{24} = 0.3951$; $a_{25} = -57.087$, and $b_{25} = 0.4292$. With these constants, it is able to retrieve LST from the FY-3D MERSI-II TIR bands under the conditions where the two required parameters AT and LSE are estimated. This algorithm is just for the cloud-free pixels' LST estimation. Thus, the cloud detection product was used as the cloud mask layer to filter clear-sky pixels.

The LST retrieval algorithm (Equation (2) requires three major parts: (1) radiation calibration for the computation of brightness temperature, (2) water vapor content, and (3) atmospheric transmittance based on water vapor. A more detailed description of these parameters and its calculation methodology required for the improved TFSW algorithm can be found in [7,25].

3.2. Calculation of the LSE for FY-3D MERSI-II

After the emissivity of bare soil εs_i for the FY-3D MERSI-II TIR bands 24 and 25 were calculated from the bare soil emissivity of the Advanced Spaceborne Thermal Emission and Reflection Radiometer (ASTER) TIR bands 13 εs_{13} and 14 εs_{14}, the land surface emissivity ε_i of each FY-3D MERSI-II pixel could be further estimated with the mixed pixel emissivity model, the FY-3D MERSI-II vegetation emissivity εv_i, and the corresponding vegetation cover fraction P_v had been calculated from the FY-3D MERSI-II NDVI data. According to Du et al. [7] the emissivity for the FY-3D MERSI-II data can be calculated as follows:

$$\varepsilon_i = \varepsilon v_i \, P_v + \varepsilon s_i \, (1 - P_v) \tag{12}$$

where ε_i is the LSE of the FY-3D MERSI-II data and εv_i is the vegetation emissivity used to estimate the LSE for FY-3D MERSI-II TIR bands 24 and 25, which has the value as $\varepsilon v_{24} = 0.982$ and $\varepsilon v_{25} = 0.984$, respectively; εs_i is the bare soil emissivity for LSE estimation. The FY-3D MERSI-II multi-year average bare soil emissivity image could be obtained from the ASTER (Global Emissivity Dataset) GED database's mean emissivity to replace the bare soil emissivity constant usually used in the VCM method; this can represent changes in bare soil's emissivity in space, enhancing the precision of emissivity inversion.

In terms of the following, LSE for a pixel can be simply understood as a weighted mix of vegetation and bare soil emissivity [36], as follows:

$$\varepsilon_i = \varepsilon v_i \, P_v + \varepsilon s_i \, (1 - P_v) \tag{13}$$

where ε_i is the emissivity, εs_i and εv_i represent the emissivity of bare soil and vegetation in channel i, and P_v is the proportional of vegetation, which is calculated as follows:

$$P_v = [(NDVI_a - NDVI_{min})/(NDVI_{max} - NDVI_{min})] \tag{14}$$

$NDVI_{min}$ represents the value of bare soil and $NDVI_{max}$ represents the value of vegetation.

The emissivity values of each surface type (i.e., bare soil and vegetated portion) were traditionally determined using spectral library data, which is accurate for the vegetated portion's high and low-contrast emissivity. However, soil emissivity in the TIR bands may vary significantly due to the various mineral components, soil moisture content, and surface roughness. Therefore, it is necessary to estimate each pixel's emissivity of the soil type component. A detailed description can be found in the works of Wang et al. [37,38], to utilize the ASTER GED Product for thermal bands.

3.3. Comparison

The correctness of the TFSW algorithm was tested using a number of statistical techniques, including the Pearson correlation coefficient (R), the root mean square deviation ($RMSD$), and bias. The equations for these statistical metrics are as follows:

$$R = \frac{\sum_{i=1}^{n}(x_i - x)(y_i - y)}{\sqrt{\sum(x_i - x)^2}\,\sqrt{\sum(y_i - y)^2}} \tag{15}$$

where x_i are the estimated data, y_i is the MODIS LST product from the ith time, x is the overall average of the estimated data, y is the average of the MODIS data over the study period, and R is the Pearson correlation. In this scenario, R is between minus and plus 1.

$$RMSD = \sqrt{\frac{\sum_{i=1}^{N}(Estimated_i - MODIS_i)}{N}} \tag{16}$$

$$Bias = \frac{\sum_{i=1}^{N}(Estimated_i - MODIS_i)}{N} \tag{17}$$

where N represents the number of pixels. *RMSD* was performed to see how well the particular variable correlation explained the observed variance. The difference between the actual and estimated variables is referred to as bias.

3.4. Linear Analysis of LST Variation in Pakistan

Monthly, seasonal, and yearly data were processed from the daily data to identify LST fluctuations in Pakistan. The following equation was used to calculate the mean temperatures at various temporal scales:

$$LST_m = \frac{\sum_{i=1}^{N}(T_i)}{N} \tag{18}$$

where LST_m represents the mean of the month, season, or year T_i satellite overpass times, and N is the number of times. We comprehensively analyzed the spatial variation of LST through the annual, seasonal, and monthly variation rates of LST, and focused on the regions with an obvious variation trend from the 2018 to 2021 years' data. In this study, the spring (March to May), summer (June to August), autumn (September to November), and winter (December to February) seasons were utilized to classify Pakistan's seasons into four categories. The linear monthly, seasonally, and yearly changes of temperature were calculated as:

$$Change\ in\ LST = LST_{2018} - LST_{2021} \tag{19}$$

3.5. Trend Analysis

The next step was to compute the spatiotemporal LST distribution after comparing the data's quality. An examination of the trends using statistics was then carried out. Many statistical trend analysis techniques, including linear trend analysis, polynomial trend analysis, harmonic trend analysis, and Mann–Kendall (MK) trend analysis, have already been employed in related investigations [39]. In order to evaluate the statistical significance of the trends and to precisely estimate the amplitude of the changes in LST, the Sen's slope estimator and the non-parametric MK methods were utilized in this study. Numerous research projects have used the non-parametric MK test because it does not rely on any assumptions about the distribution of the data or the linearity of the trend [40]. The alternative hypothesis indicates that there is a monotonic decreasing or increasing trend, whereas the null hypothesis in the calculation of this test asserts that there is no trend in LST data over time. The equations to use the MK test are as follows:

$$S = \sum_{k=1}^{n-1} \sum_{j=k+1}^{n} sgn(X_j - X_k) \tag{20}$$

$$sgn(X_j - X_k) = \begin{cases} 1, & if\ X_j > X_k \\ 0, & if\ X_j = X_k \\ -1, & if\ X_j < X_k \end{cases} \tag{21}$$

$$Var(S_{mk}) = \frac{n(n-1)(2n+5) - \sum_{p=1}^{q} t_p(t_p - 1)(2t_p + 5)}{18} \tag{22}$$

X_j and X_k are two consecutive data values in an n-dimensional data set. The term *"sgn"* represent sign function. For a normal distribution, E(S) = 0 with the variance, and the S statistic behaves roughly the same. Here, q is the total number of affiliated groups in the data set, and t_p is the number of input values contained within the p-th affiliated group. By utilizing above equations, Z_{mk} is calculated as follows:

$$Z_{mk} = \begin{cases} \frac{S-1}{\sqrt{Var(S_{mk})}}, & if\ S > 0 \\ 0, & if\ S = 0 \\ \frac{S+1}{\sqrt{Var(S_{mk})}}, & if\ S < 0 \end{cases} \tag{23}$$

A statistically reliable trend identification tool is Z_{mk}. A Z_{mk} positive sign indicates an upward trend, while a Z_{mk} negative sign indicates a downward trend. The null hypothesis is rejected if the absolute value of $Z_{mk} >$ Z1/2, where Z1/2 is obtained from the normal cumulative distribution tables.

The estimator of Sen [41], which is defined by Equations (24) and (25), was used to calculate the real slope of an existing trend.

$$Q_i = \frac{X_j - X_k}{j - k}\ for\ all\ j > k \tag{24}$$

There will be $N = n\ (n - 1)/2$ slope Q_i estimations if the time series has $n\ X_j$ values. The median of these N values of Q_i is Sen's slope estimator which calculated as follows:

$$Q_{med} = \begin{cases} Q_{[\frac{N+1}{2}]}, & if\ N\ is\ odd \\ \frac{1}{2}\left(Q_{[\frac{N}{2}]} + Q_{[\frac{N+1}{2}]}\right), & if\ N\ is\ even \end{cases} \tag{25}$$

3.6. Procedure of the Study

Figure 3 shows the research procedure and technical methods, including the detailed steps for retrieving LST from the MERSI-II data. The following are the steps that this study takes as its main approach: We start by obtaining Level 1 radiance data from the FY-3D satellite data portal. Second, we estimate the LST and its required parameters of the TFSW algorithm. Third, we conducted a cross-validation analysis on the retrieved LST to determine the algorithm's applicability and usefulness. Finally, the application of the algorithm was applied for the detailed spatial and temporal analyses of LST for Pakistan.

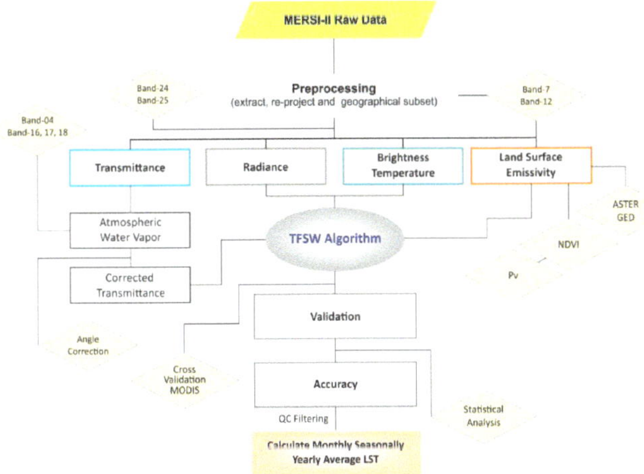

Figure 3. The framework depicting the phases required for LST estimation and spatiotemporal analysis.

4. Results

4.1. Cross-Validation between MERSI-II, and MYD11 and MYD21 MODIS LST Products

The validation of LST based on the measured data is a difficult task. Wan et al. [42] put his extensive efforts toward correctly verifying the MODIS LST products with the measured data. His findings reveal that the MODIS MYD11A1 LST product in the experimental area are accurate to within 1 °C. Hence, for relative error analysis, we used the MYD11A1 and MYD21A1 LST products, and the MERSI-II-estimated LST data. The LST of two different raster datasets (MYD11A1/MYD21A1 and MERSI-II) were extracted by using the shapefiles of different sub-regions over the study area; the results of differences in LST between the MERSI-II, and the MODIS MYD11A1 and MYD21A1 products are shown in Figures 4 and 5.

It was observed that the RMSD between MERSI-II, MYD11A1 and MYD21A1 ranged from 1.71 to 5.31 °C throughout the year, over the sub-regions of Pakistan. The highest RMSDs were recorded in region IV in both the LST products, and they were the lowest in region V, which was an average of annual record. Regarding bias, region II showed the highest negative trend when compared to MYD11A1, and the highest positive trend of bias were observed in region IV in a comparison of both products. The highest error in RMSE was possibly due to various reasons, such as heterogeneous surfaces producing larger errors due to uneven surface features. The LST retrieved from remote sensing data was the instantaneous LST at the moment of sensor imaging. The imaging time of the original images used for the retrieval of MERSI-II LST and MYD11AI and MYD21A1 LST differed, and the sensor calibration accuracy also differed. Thus, the LST results of the same area slightly differed due to retrieval using different remote sensing data sources.

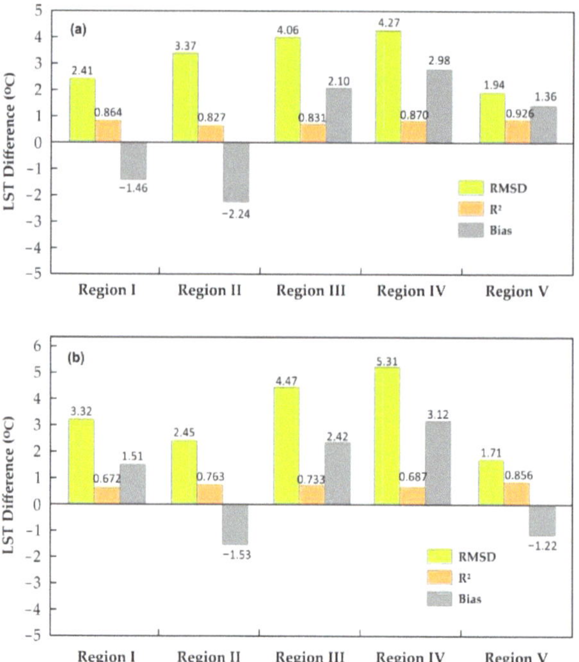

Figure 4. Cross-validation errors between LSTs of MERSI-II, and MODIS (**a**) MYD11A1 and (**b**) MYD21A1 LST products over five sub-regions of Pakistan.

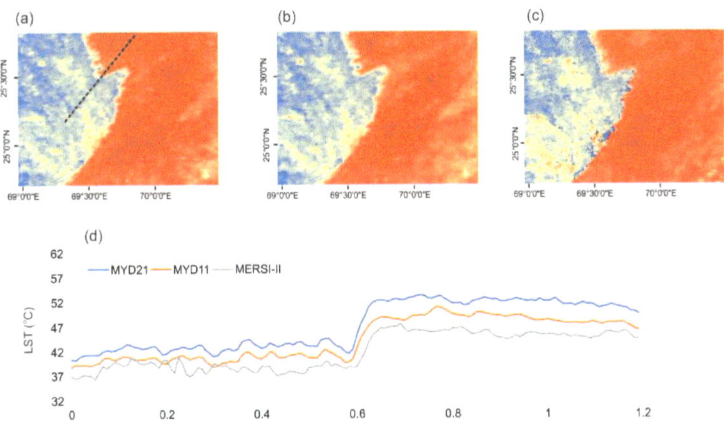

Figure 5. Spatial distribution of (**a**) MYD11A1, (**b**) MYD21A1, and (**c**) MERSI-II LST data. The dotted line represents the profile extraction location. (**d**) Latitudinal profile of three different LST data.

The LST distributions in all datasets were almost same under clear sky conditions (as shown in Figure 5), with only minor deviations being observed in MERSI-II, possibly the stripe noise problem of the sensor. According to the latitudinal profile, it can observed that the LST of MERSI-II was close the MYD11A1 LST product as compared to the MYD21A1, and the overall RMSD errors were also smaller when compared with the MYD11A1 product. However, the image acquisition times of both sensors were similar to some extent, but not identical, and the MODIS LST product itself contained some minor errors when compared to the ground data. In this regard, when we compared MERSI-II's retrieved LST with the MODIS LSTs products, the results are merely a spatial analysis of the LSTs, and relative errors are not perfectly precise.

4.2. Spatial Distribution of the Annual Mean LST

The spatial distribution of the annual mean LSTs of Pakistan is almost the same in 2018 and 2021; small variations were observed, as the time span (i.e., 4 years) is not very long. The annual mean LST of Pakistan ranged between −15 and 55 °C (Figure 6). A bar chart was developed to demonstrate the linear change of the annual mean LST between 2018 and 2021 over the sub-regions of the study area, and differential variations in pixel scale between two raster datasets were performed to validate substantial upward or downward variations. It was observed from the annual mean LST that 2018 was the coldest year (33.63 °C), while 2021 was the warmest (34.42 °C). The temperature difference between the coldest and warmest years was about 0.79 °C.

When the geographical characteristics of the location are considered, it was discovered that the low elevation increased thermal insulation, resulting in a greater LST. In terms of geographical cover, Pakistan's greatest desert, Thar, is located around the Indian border in the southeast. The desert covers an area of 175,000 square kilometers, including significant regions of Pakistan and India. It is Pakistan's largest desert, and Asia's only subtropical desert that absorbs sun radiation. During the analysis of the spatial distribution of annual mean LST throughout the sub-regions, the region II exhibits the highest annual mean LST, varying between 41 and 42 °C. The same scenario was observed in region I. Due to various geographical surface features, region III is slightly different, with a mean LST of 36 °C. Region IV is the transitional zone between the temperate and severe climate zones; the annual mean LST of this region is around 31 °C, but it varies greatly across the landscape. The lowest LST values were discovered in region V, which were decreased by up to −22 °C; however, in the southern parts of this region, the temperature was increased up to 39 °C.

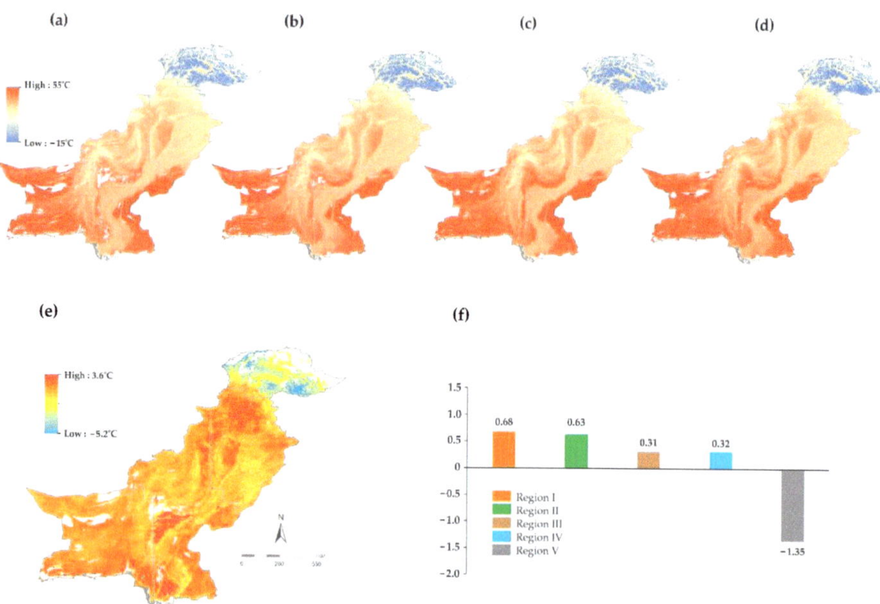

Figure 6. Spatial distribution of annual mean LST in Pakistan. (**a–d**) Mean LSTs of 2018, 2019, 2020, and 2021. (**e**) LST difference between 2018 and 2021 on spatial scales. (**f**) Bar chart for mean LST difference for sub-regions of Pakistan.

We calculated the linear mean LST difference by subtracting the data values from two raster datasets. The warming and cooling of the LST trend were shown by positive and negative values. Figure 6e depicts the results after subtracting the initial and final raster data values at the pixel scale, while Figure 6f depicts the mean annual difference. Northern Pakistan is cooled more than Southern Pakistan, resulting in the highest cooling effect in the north and a moderate cooling effect in the south. However, many locations of region IV have an LST difference of less than 0 °C, which is primarily cooling. The largest warming trend (0.68 °C) was identified in region I, whereas the minimum warming trend (0.32 °C) was recorded in region IV. The explanation for the cooling trend in region V is unknown, possibly due to the missing data of thin clouds. The warming tendencies in regions I and II, on the other hand, are due to changes in their geographical land cover types.

4.3. Trend Analysis of the Annual Mean LST

The MK trend test results (Table 2) of LST for the Pakistan showed positive and negative trends during the time span of 2018 to 2021. In this section, the positive values represent increasing, and the negative values represent decreasing trends in the regions, the *p*-value decides the level of significance, and the slope indicates the magnitude of LST. The highest increasing trend (0.019 °C/year) at a 10% level of significance was observed in region V. The other regions also showed increasing trends, but these were insignificant. The lowest decreasing trend was observed in region I at a 5% level of significance, with a magnitude of −0.013 °C/year, which was lowest among the other regions. However, as evidenced from the minimum and maximum values of Table 2 and Figure 7, there are lot of areas that experience significant increasing and decreasing trends, but these are very small in area. Figure 7 shows the spatial distribution of the MK parameters. The most important significant trends can be observed from Figure 7d, which shows increasing trends in the northern areas and decreasing in the southern areas. The slope of the LST per year throughout the study region ranged between −0.013 and 0.019 °C/Year.

Table 2. MK trend test (Z) statistics, Sen's slope, and *p* value for the LST trends analysis.

Region	MK Z-Score			Slope (°C/Year)	*p*-Value
	Min	Max	Mean		
I	−4.97	3.57	−1.86 **	−0.013	0.05
II	−3.97	3.14	0.23	0.003	0.63
III	−3.28	3.58	0.23	0.003	0.61
IV	−3.39	3.51	0.07	0.001	0.71
V	−2.25	3.96	2.10 *	0.019	0.10

Note: Where "*" and "**" indicate the significance of the test at 5% and 10% levels of significance.

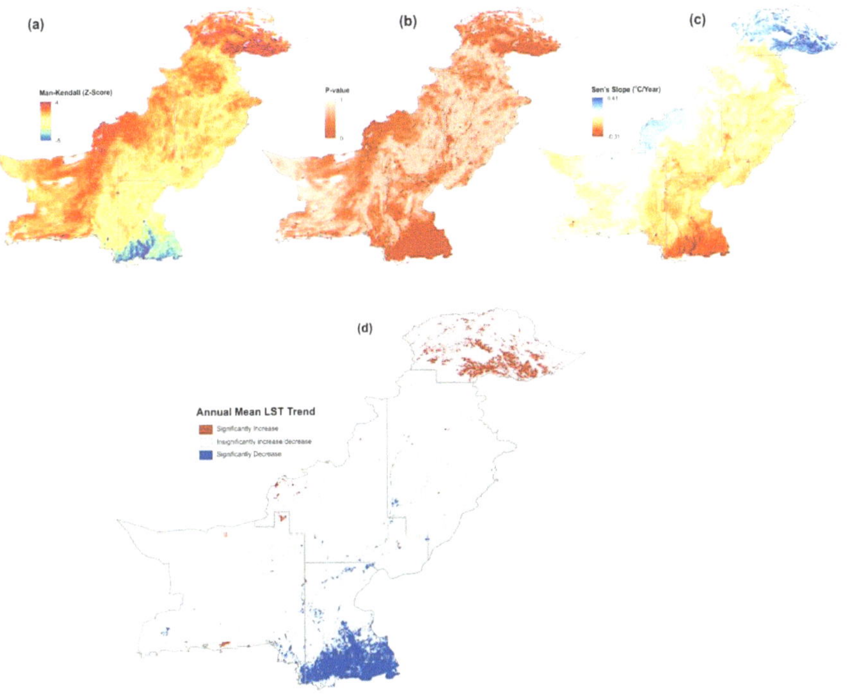

Figure 7. Spatial distribution of (**a**) Mann–Kendall, (**b**) Sen's Slope, (**c**) Level of Significance (*p*-value), (**d**) Statistically significant data at 99% level of confidence.

4.4. Spatial Distribution of Seasonal LST and Trend Analysis

The spatial distribution of the seasonal mean LST rises with decreasing latitude, similar to the yearly mean LST, with low values in the northern and high values in the southern regions of the country. However, there are differences between each seasonal dataset. Because the Earth's axis is tilted relative to the orbital plane, the solar elevation angle fluctuates throughout the year, making winter cold and summer hot. To further reveal the LST changes, seasonal means were processed from the daily averages of the LST data to calculate the trend analysis through the MK test, along with the slope (Table 3). Based on the classification of the four seasons throughout the study region, seasonal variation has an important effect on LST across Pakistan.

Table 3. The mean LST difference along with standard deviation at seasonal scales.

Season	Region	Z-Score	Slope (°C)	p-Value
Spring	I	0.93	0.03	0.40
	II	−0.14	−0.01	0.52
	III	−0.71	−0.03	0.44
	IV	−0.19	−0.01	0.70
	V	−0.95	−0.06	0.30
Summer	I	1.22	0.04	0.32
	II	2.46 **	0.07	0.05
	III	2.31 ***	0.07	0.1
	IV	2.22 **	0.08	0.05
	V	3.00 *	0.20	0.01
Autumn	I	−1.06 **	−0.08	0.05
	II	0.42	0.02	0.61
	III	0.84	0.04	0.43
	IV	0.84	0.03	0.44
	V	1.93 *	0.15	0.10
Winter	I	1.54	0.11	0.18
	II	1.42	0.12	0.20
	III	2.03 **	0.17	0.05
	IV	1.95 ***	0.13	0.10
	V	1.88 ***	0.20	0.10

Note: Where "*", "**" and "***" indicate significance of the test at 1%, 5%, and 10% levels of significance.

The results of the spatial–temporal changes of LST over different seasons are given in Figure 8. The overall mean of the seasonal LST varies from −26 to 60 °C throughout Pakistan. During the overall seasons, the highest LSTs were observed in region II, particularly in the summer season, which reached up to 50.86 °C, which is the highest among the other regions and seasons. This result can be attributable to the fact that the annual rainfall is below average. Furthermore, barren ground, which is more susceptible to LST changes than vegetation-covered land, could be another factor for the rise in LST. Similarly, the lowest LST were observed in region V in all seasons, especially in the winter season, where the mean LSTs of this region were decreased up to −9.7 °C; there are some places that contain higher LSTs, especially in the southern parts of that region. During the autumn and spring seasons, similar trends were observed, followed by winter and summer, especially in regions II and V. The mean LSTs of the spring and autumn season reached up to 42 and 46 °C, respectively, in region II.

The results of the trend analysis and Sen's slope at seasonal scales were presented in Table 3. The highest increasing trend (0.20 °C) at a 1% level of significance ($p < 0.01$) was observed in region V during the summer season; there were also increasing events in other seasons, except for the spring season. Another warming event was observed in region II during the summer season. Contrary, the highest cooling trend was observed in region I during the autumn season (−0.08 °C) which is highest among all seasons and regions. The factors attributed to this cooling trend could be the Khirthar mountainous regions, which cover an area of about 9000 square kilometers, with the highest peak of 2260 m. Due the high altitude, the LST of this region does not significantly increase in 2018 and 2021. When we investigate the low altitude/flat surfaces, the only region IV is a region where the change rate of LST significantly increases significantly in the summer seasons throughout the season, and the highest heating trend has a magnitude of 0.08 °C.

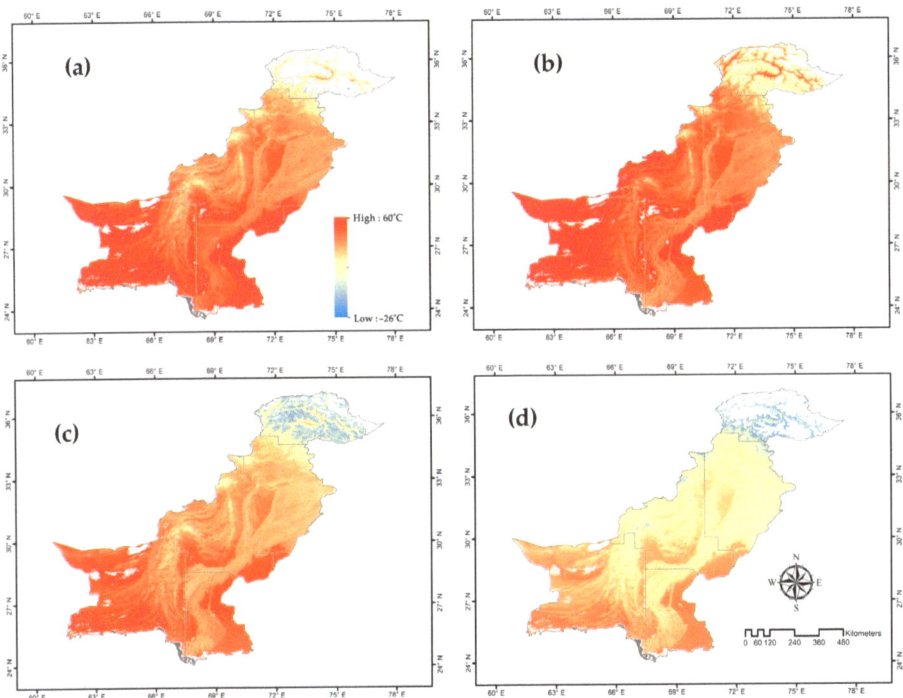

Figure 8. Spatial distribution of seasonal mean LSTs between 2018 and 2021 in Pakistan. (**a**) Spring, (**b**) summer, (**c**) winter, (**d**) autumn.

4.5. Monthly Average Change Analysis

Figures 9 and 10 provide the overall monthly LST spatial distribution and the difference of LST between 2018 and 2021. The spatial variation of LST can easily determined using the monthly datasets of FY-3D MERSI-II data, and the LSTs were gradually increased from January, reached their peak in June to August, and then they declined throughout Pakistan. The highest mean LSTs during 2018 and 2021 were recorded in June (43.6 °C), and the lowest in January (14.3 °C). In terms of sub-regions, the monthly LST variations were followed by annual and seasonal trends. Throughout the monthly analysis, region II contained the highest LST and region V, the lowest; these highest and lowest LST values possibly differed due to the latitudinal change and the geographical land cover types. According to Figure 10, a warming trend were observed in region II throughout the monthly results except for the January (−0.93) and December (−0.83) months, which resulted in that the LSTs in 2021 were higher than 2018. However, region V does not follow decreasing or increasing trends; there are many cooling and warming trends observed because of the surface heterogeneity and snow cover. Regarding region IV, there are some events in the monthly analysis that follow the cooling trends compared to the 2018 and 2021, but they are not a priority for discussion. The monthly change analysis was calculated by subtracting the monthly rasters between 2018 and 2021. The MK trends analysis was never applied over a monthly basis due to the limited availability of temporal data.

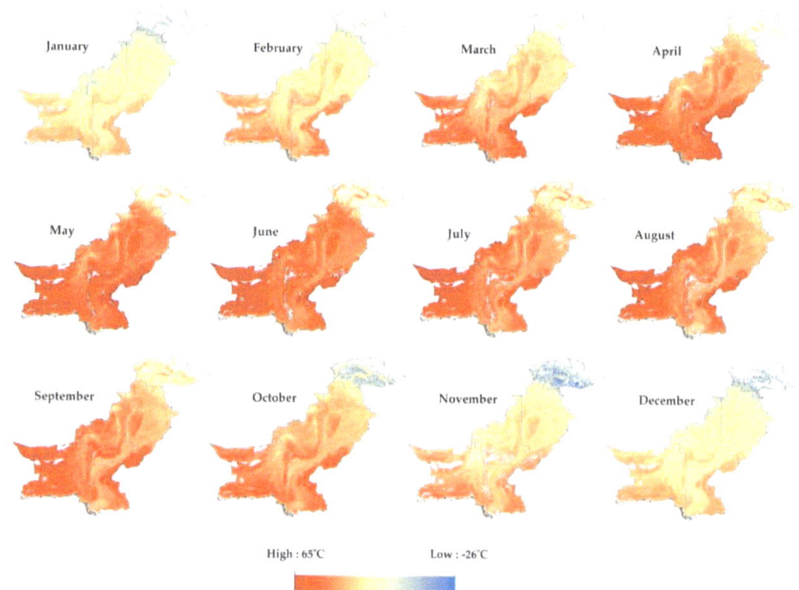

Figure 9. Spatial distribution of mean LSTs of 2018 and 2021 at monthly scales in Pakistan.

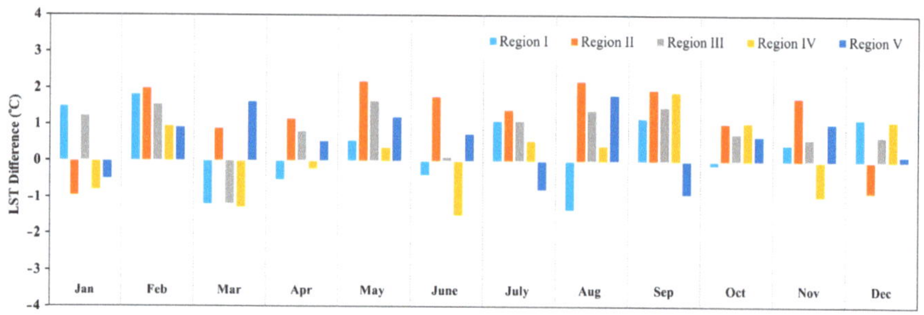

Figure 10. The monthly mean variations of LST over different sub-regions of Pakistan during 2018 to 2021.

4.6. Long-Term Interannual Variations of LST

Long-term interannual variations of LST are crucial for the investigation of the LST warming and cooling trends. From the above analyses, we observed that the FY-3D MERSI-II data and the TFSW algorithm performed well for understanding the LST variations throughout Pakistan in 2018 and 2021; the validations results also supported their usefulness. However, the FY-3D MERSI-II is a recently launched satellite; to understand the spatial–temporal variations of LSTs to a broad extent requires the crucial contribution of historical data (i.e., over 10 years). The MODIS MYD11A1 product is fully mature and is a widely used LST product at global scales. To understand the interannual LST variations in Pakistan, we used MODIS LST data from 2005 to 2018.

The long-term spatial patterns of LST (2005 and 2018) are like the above mentioned annual mean LST trends of MERSI-II. The long-term annual mean LST varies from −29 to 55 °C across Pakistan (Figure 11). The spatial–temporal variation between 14 annual LST raster datasets were performed to demonstrate the change that occurred in Pakistan through MK trend analysis and Sen's slope estimator, and a line chart was introduced to show a mean LST and histogram for an individual year. Figure 11 shows how the mean annual LST

(°C) varied across the country over this time. The coldest year (34.16 °C) was 2012, and the warmest year (35.53 °C) was 2016. The coldest and warmest years were separated by about 1.37 °C. Similar to the MERSI-II annual mean LST, with increasing latitude, the MODIS annual LST likewise decreased; it had low values in the northern areas and high values in the southern areas. The MODIS annual LST also decreased with increasing latitude: it had low values in the northern regions and high values in the southern regions. The LST trend for every year (from 2005 to 2018) followed this increasing and decreasing trend throughout Pakistan.

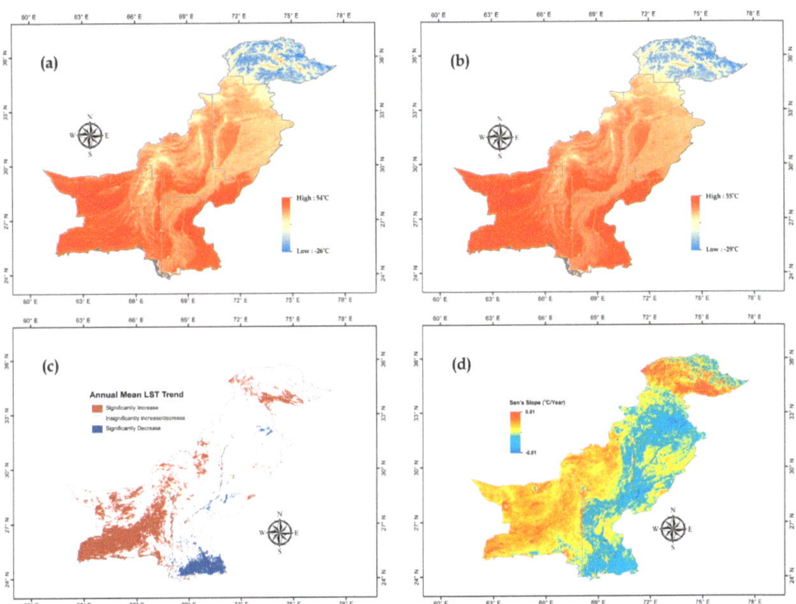

Figure 11. The spatial distribution of the mean LSTs and trend variables. (**a**) Annual LST in 2005, (**b**) annual LST in 2018, (**c**,**d**) significant increase and decrease along with Sen's slope from 2005 to 2018.

The MK trend results show almost similar trends, as we have seen for MERSI-II, but the slopes per year of the annual mean LST are very small, as displayed in Figure 11c,d. The slopes were found to be between −0.01 and 0.01 °C/year throughout Pakistan. In regions II and V, there was a considerable increasing tendency at a 5% level of significance, with a slope of 0.0046 and 0.0047 °C/year. Region I exhibited a considerable decreasing trend of roughly −0.0024 °C/year. The various increasing and decreasing trends in other regions of the study are insignificant. The results of the long-term interannual variations of MODIS are consistent with an annual mean LST of the FY-3D MERSI-II data. The positive slope and the positive correlation of the LSTs in regions I and II are directly related to their geographical land cover types, which have already been discussed in the above sections.

The greatest interannual differences were found in regions I, II, and V. In contrast, there was less interannual variability in region IV. The greatest interannual disparities may be due to changes in the surfaces' vegetation cover. This is because the thermal characteristics of the Earth's surfaces with high vegetation cover differ from those of the Earth's surface with low vegetation cover. The increase in vegetation cover helps to increase the soil moisture content, which helps to prevent soil erosion and desertification. Surface temperatures in the equatorial and temperate regions are often higher in areas with low vegetation cover than in areas with high vegetation cover.

Furthermore, Figure 12 shows the mean LST of individual year and histogram of years between 2005 and 2018, and it was observed that the LST trend was almost same, with small deviations throughout the years. The country's maximum areas of LST were approximately 28 to 50 °C.

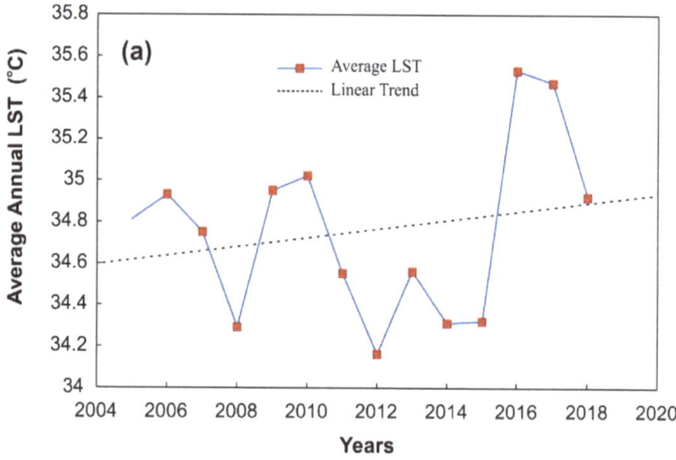

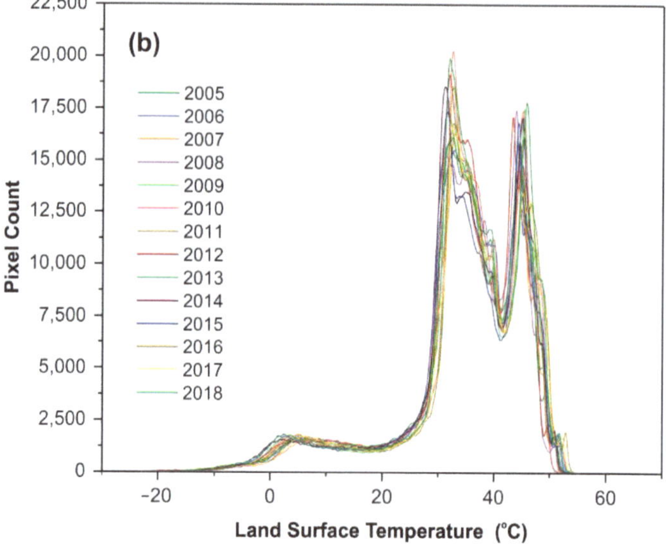

Figure 12. (a,b) Interannual Mean LST, including its frequency histogram for the years of 2005 to 2018 throughout the Pakistan.

5. Discussion

The TFSW algorithm's accuracy and adaptability were evaluated by comparing the LSTs of MERSI-II with MYD11A1 and MYD21A1 LST over various study regions. The LST difference ranged between 1.74 and 5.2 °C over the different regions of Pakistan. The MERSI-II LST performs well with the MODIS MYD11A1 LST product; the error range was between 1.94 and 4.27 °C, which is slightly higher but close to the several published studies [7,12,43,44]. The regions that experience the highest LST difference could possibly be due to the surface types, and the presence of humidity produces larger uncertainties [44]. Regarding the LST distribution, the highest LSTs were observed in the southern areas, and the lowest in the northern areas. The amount of solar radiation energy received by the ground decreases as latitude increases. As a result, the annual mean LST distribution pattern is characterized by high values in the southern regions and low values in the northern parts. There are, however, some locations that do not follow this trend. Because the areas where the Indus River flows, particularly in the country's central lower latitudes, are largely productive, resulting in moderate heat preservation on the ground, it has a lower annual mean LST than its nearby places. The great mountainous region occurred in northern areas, which is Pakistan's greatest altitude region, with an average elevation of above 5000 m. Because the location is at a high elevation, the air near the surface is thin, resulting in poor heat retention on the ground and a lower annual mean LST than in other parts of the country. According to the MK trend analysis, the highest warming impacts (0.019 °C/Year) were noted in the northern areas, with a high level of significance. This scenario was also observed in seasonal trends as well as long-term analysis trends. The reasons for this increasing are possibly the impacts of global warming and climate change [45]. By taking into account meteorological data from two sites, Ali et al. [46] looked into changes in temperature in the Upper Indus Basin. In line with this investigation, they observed that the mean temperature increased by 0.63 °C in Skardu, whereas it recently declined by 0.137 °C in Gilgit. The source of the decreasing trends in southern region I is unknown. There are strong increasing trends through seasonal and monthly analyses in regions I and II. The lack of rain and growing drought have had a substantial impact on the temperature fluctuations of Southern Pakistan. The hottest temperatures have had a significant detrimental influence on Pakistan's primary crops, such as wheat and cotton. Early in the spring season, a minor heat wave (13 days over the usual 2–3 °C temperatures) resulted in a 28 percent decline in wheat grain yield.

When evaluating the results, it is important to take into account the limitations of the current study. The remote sensing data are not enough to obtain precise LST trends. However, the various factors that affect LSTs are still not considered in this study, such as elevation, land cover types, and water vapor; even the NDVI has a considerable influence on LSTs [47]. To ensure the reliability of the results, future studies should employ LST data, including elevation, land cover types, water vapor, etc.

6. Conclusions

During MERSI-II's operation in orbit, China's satellite remote sensing observation capabilities were increasing. This sensor is the first imaging instrument in the world that can acquire thermal infrared data with a spatial resolution of 250 m and 1000 m. The MERSI-II data can be used to precisely recover the land surface temperature using the recently developed TFSW algorithm. The Moderate Spectral Resolution Atmospheric Transmittance Model simulation dataset served as the source for the SW algorithm coefficients (MOD-TRAN). In a comparison of the retrieved LST and the MODIS MYD11A1 and MYD21A1 LST product, it was found that the MERSI-II LST is more reliable and achieves a better accuracy in a comparison with MYD11A1.

In order to study the spatiotemporal changes in LSTs, the FY-3D MERSI-II LST data have never before been employed in Pakistan on this scale. MK trend analysis and Sen's slope estimator were applied to identify significant changes. Due to the limited availability of the MERSI-II data (i.e., 4 years), the MODIS MYD11AI product was utilized to understand the interannual variations of LSTs in Pakistan. Considering both satellites, these conclusions were noticed: (1) The annual mean LST in Pakistan often follows a pattern with high values in the southern parts and low values in the northern regions, which is consistent with the rule that solar radiation intensity decreases with increasing latitude. However, the spatial distribution of LST is also influenced by elevation and the types of land cover. (2) The spatial distribution of the annual mean LST was almost similar during the study period. The trend analysis shows that the northern areas (region v) have increasing LST trends. The LST changed on average by 0.019 °C/year, increasing in the northern parts and decreasing by 0.013 °C/Year in the southern regions. With regard to the sub-regions, a maximum LST were observed in region II, and a greater warming trend were also observed in this region, with a rapidly increasing rate of 0.0046 °C/year when utilizing long-term data, with the *p*-value also being consistent with the slope.

Due to the influence of the stripe noise in the MERSI-II TIR pictures, the validation of the recovered LST utilized may be a little inaccurate in comparison. Even yet, some in-depth and in situ confirmations are still needed. In order to fully utilize the MERSI-II TIR photos, further work should concentrate on stripe noise removal, as well as confirmation with ground-based stations, particularly in the American regions. Additionally, LST variations are crucial elements that affect and that restrain climate change in various ways. Governmental institutions and groups that are concerned with the environment must therefore pay close attention to changes in the LST, in order to define climate change.

Author Contributions: Conceptualization, B.A. and Z.Q.; methodology, B.A.; software, J.F.; validation, B.A., W.D. and S.L.; formal analysis, B.A.; investigation, Z.Q.; resources, C.Z.; data curation, W.D.; writing—original draft preparation, B.A. and Z.Q.; writing—review and editing, W.D., Z.Q. and C.Z.; visualization, B.A.; supervision, Z.Q. and J.F.; project administration, Z.Q.; funding acquisition, Z.Q. All authors have read and agreed to the published version of the manuscript.

Funding: This study was supported by the National Key Research and Development Program of China (Grant No.: 2019YFE0127600), and the National Natural Science Foundation of China (Grant No.: 41771406, 41921001).

Institutional Review Board Statement: Not applicable.

Informed Consent Statement: Not applicable.

References

1. Yao, R.; Wang, L.; Huang, X.; Gong, W.; Xia, X. Greening in Rural Areas Increases the Surface Urban Heat Island Intensity. *Geophys. Res. Lett.* **2019**, *46*, 2204–2212. [CrossRef]
2. Zhou, C.; Wang, K. Land Surface Temperature over Global Deserts: Means, Variability, and Trends. *J. Geophys. Res. Atmos.* **2016**, *121*, 14344–14357. [CrossRef]
3. Bhattarai, N.; Mallick, K.; Stuart, J.; Vishwakarma, B.D.; Niraula, R.; Sen, S.; Jain, M. An Automated Multi-Model Evapotranspiration Mapping Framework Using Remotely Sensed and Reanalysis Data. *Remote Sens. Environ.* **2019**, *229*, 69–92. [CrossRef]
4. Lu, N.; Liang, S.; Huang, G.; Qin, J.; Yao, L.; Wang, D.; Yang, K. Hierarchical Bayesian Space-Time Estimation of Monthly Maximum and Minimum Surface Air Temperature. *Remote Sens. Environ.* **2018**, *211*, 48–58. [CrossRef]
5. Randazzo, G.; Cascio, M.; Fontana, M.; Gregorio, F.; Lanza, S.; Muzirafuti, A. Mapping of Sicilian Pocket Beaches Land Use/Land Cover with Sentinel-2 Imagery: A Case Study of Messina Province. *Land* **2021**, *10*, 678. [CrossRef]
6. Randazzo, G.; Italiano, F.; Micallef, A.; Tomasello, A.; Cassetti, F.P.; Zammit, A.; D'Amico, S.; Saliba, O.; Cascio, M.; Cavallaro, F.; et al. WebGIS Implementation for Dynamic Mapping and Visualization of Coastal Geospatial Data: A Case Study of BESS Project. *Appl. Sci.* **2021**, *11*, 8233. [CrossRef]
7. Du, W.; Qin, Z.; Fan, J.; Zhao, C.; Huang, Q.; Cao, K.; Abbasi, B. Land Surface Temperature Retrieval from Fengyun-3d Medium Resolution Spectral Imager Ii (Fy-3d Mersi-Ii) Data with the Improved Two-Factor Split-Window Algorithm. *Remote Sens.* **2021**, *13*, 72. [CrossRef]
8. Tang, K.; Zhu, H.; Ni, P.; Li, R.; Fan, C. Retrieving Land Surface Temperature from Chinese FY-3D MERSI-2 Data Using an Operational Split Window Algorithm. *IEEE J. Sel. Top. Appl. Earth Obs. Remote Sens.* **2021**, *14*, 6639–6651. [CrossRef]
9. Wang, H.; Mao, K.; Mu, F.; Shi, J.; Yang, J.; Li, Z.; Qin, Z. A Split Window Algorithm for Retrieving Land Surface Temperature from FY-3D MERSI-2 Data. *Remote Sens.* **2019**, *11*, 2083. [CrossRef]
10. Jin, M.; Li, J.; Wang, C.; Shang, R. A Practical Split-Window Algorithm for Retrieving Land Surface Temperature from Landsat-8 Data and a Case Study of an Urban Area in China. *Remote Sens.* **2015**, *7*, 4371–4390. [CrossRef]
11. Ndossi, M.; Avdan, U. Inversion of Land Surface Temperature (LST) Using Terra ASTER Data: A Comparison of Three Algorithms. *Remote Sens.* **2016**, *8*, 993. [CrossRef]
12. Li, Z.-L.; Tang, B.-H.; Wu, H.; Ren, H.; Yan, G.; Wan, Z.; Trigo, I.F.; Sobrino, J.A. Satellite-Derived Land Surface Temperature: Current Status and Perspectives. *Remote Sens. Environ.* **2013**, *131*, 14–37. [CrossRef]
13. Jamro, S.; Dars, G.H.; Ansari, K.; Krakauer, N.Y. Spatio-Temporal Variability of Drought in Pakistan Using Standardized Precipitation Evapotranspiration Index. *Appl. Sci.* **2019**, *9*, 4588. [CrossRef]

14. Haroon, M.A.; Zhang, J.; Yao, F. Drought Monitoring and Performance Evaluation of MODIS-Based Drought Severity Index (DSI) over Pakistan. *Nat. Hazards* **2016**, *84*, 1349–1366. [CrossRef]
15. Adnan, S.; Ullah, K.; Gao, S.; Khosa, A.H.; Wang, Z. Shifting of Agro-Climatic Zones, Their Drought Vulnerability, and Precipitation and Temperature Trends in Pakistan. *Int. J. Climatol.* **2017**, *37*, 529–543. [CrossRef]
16. Afzaal, M.; Haroon, M.A. Interdecadal Oscillations and the Warming Trend in the Area-Weighted Annual Mean Temperature of Pakistan. *Pakistan J. Meteorol.* **2009**, *6*, 13–19.
17. Mumtaz, F.; Tao, Y.; De Leeuw, G.; Zhao, L.; Fan, C.; Elnashar, A.; Bashir, B.; Wang, G.; Li, L.L.; Naeem, S.; et al. Modeling Spatio-Temporal Land Transformation and Its Associated Impacts on Land Surface Temperature (LST). *Remote Sens.* **2020**, *12*, 2987. [CrossRef]
18. Arshad, A.; Zhang, W.; Zaman, M.A.; Dilawar, A.; Sajid, Z. Monitoring the Impacts of Spatio-Temporal Land-Use Changes on the Regional Climate of City Faisalabad, Pakistan. *Ann. GIS* **2018**, *25*, 57–70. [CrossRef]
19. Saleem, M.S.; Ahmad, S.R.; Shafiq-Ur-Rehman; Javed, M.A. Impact Assessment of Urban Development Patterns on Land Surface Temperature by Using Remote Sensing Techniques: A Case Study of Lahore, Faisalabad and Multan District. *Environ. Sci. Pollut. Res.* **2020**, *27*, 39865–39878. [CrossRef] [PubMed]
20. Pakistan Bureau of Statistics. Pakistan Statistical Year Book for Land Utilization. 2011. Available online: http://www.pbs.gov.pk/content/pakistan-statistical-year-book-2011 (accessed on 20 April 2016).
21. Zaman, Q.U.; Rasul, G. Agro-Climatic Classification of Pakistan. *Q. Sci. Vis.* **2004**, *9*, 59–66.
22. Mahessar, A.A.; Qureshi, A.L.; Dars, G.H.; Solangi, M.A. Climate change impacts on vulnerable Guddu and Sukkur Barrages in Indus River, Sindh. *Sindh Univ. Res. J.* **2017**, *49*, 137–142.
23. Krakauer, N.Y.; Lakhankar, T.; Dars, G.H. Precipitation Trends over the Indus Basin. *Climate* **2019**, *7*, 116. [CrossRef]
24. Rehman, S.; Shah, M.A. Rainfall Trends in Different Climate Zones of Pakistan. *Pak. J. Meteorol.* **2012**, *9*, 37–47.
25. Abbasi, B.; Qin, Z.; Du, W.; Fan, J.; Zhao, C.; Hang, Q.; Zhao, S.; Li, S. An Algorithm to Retrieve Total Precipitable Water Vapor in the Atmosphere from FengYun 3D Medium Resolution Spectral Imager 2 (FY-3D MERSI-2) Data. *Remote Sens.* **2020**, *12*, 3469. [CrossRef]
26. Hulley, G.; Veraverbeke, S.; Hook, S. Thermal-Based Techniques for Land Cover Change Detection Using a New Dynamic MODIS Multispectral Emissivity Product (MOD21). *Remote Sens. Environ.* **2014**, *140*, 755–765. [CrossRef]
27. Li, H.; Sun, D.; Yu, Y.; Wang, H.; Liu, Y.; Liu, Q.; Du, Y.; Wang, H.; Cao, B. Evaluation of the VIIRS and MODIS LST Products in an Arid Area of Northwest China. *Remote Sens. Environ.* **2014**, *142*, 111–121. [CrossRef]
28. Duan, S.-B.; Li, Z.-L.; Li, H.; Göttsche, F.-M.; Wu, H.; Zhao, W.; Leng, P.; Zhang, X.; Coll, C. Validation of Collection 6 MODIS Land Surface Temperature Product Using in Situ Measurements. *Remote Sens. Environ.* **2019**, *225*, 16–29. [CrossRef]
29. Yao, R.; Wang, L.; Wang, S.; Wang, L.; Wei, J.; Li, J.; Yu, D. A Detailed Comparison of MYD11 and MYD21 Land Surface Temperature Products in Mainland China. *Int. J. Digit. Earth* **2020**, *13*, 1391–1407. [CrossRef]
30. Li, H.; Li, R.; Yang, Y.; Cao, B.; Bian, Z.; Hu, T.; Du, Y.; Sun, L.; Liu, Q. Temperature-Based and Radiance-Based Validation of the Collection 6 MYD11 and MYD21 Land Surface Temperature Products Over Barren Surfaces in Northwestern China. *IEEE Trans. Geosci. Remote Sens.* **2021**, *59*, 1794–1807. [CrossRef]
31. Li, Z.; Petitcolin, F.; Zhang, R. A Physically Based Algorithm for Land Surface Emissivity Retrieval from Combined Mid-Infrared and Thermal Infrared Data. *Sci. China Ser. E Technol. Sci.* **2000**, *43*, 23–33. [CrossRef]
32. Tang, B.-H. Nonlinear Split-Window Algorithms for Estimating Land and Sea Surface Temperatures From Simulated Chinese Gaofen-5 Satellite Data. *IEEE Trans. Geosci. Remote Sens.* **2018**, *56*, 6280–6289. [CrossRef]
33. Wan, Z.; Dozier, J. A Generalized Split-Window Algorithm for Retrieving Land-Surface Temperature from Space. *IEEE Trans. Geosci. Remote Sens.* **1996**, *34*, 892–905. [CrossRef]
34. Wan, Z. New Refinements and Validation of the Collection-6 {MODIS} Land-Surface Temperature/Emissivity Product. *Remote Sens. Environ.* **2014**, *140*, 36–45. [CrossRef]
35. Qin, Z.; Dall Olmo, G.; Karnieli, A.; Berliner, P. Derivation of Split Window Algorithm and Its Sensitivity Analysis for Retrieving Land Surface Temperature from {NOAA}-Advanced Very High Resolution Radiometer Data. *J. Geophys. Res. Atmos.* **2001**, *106*, 22655–22670. [CrossRef]
36. Sobrino, J.A.; Jimenez-Muoz, J.C.; Soria, G.; Romaguera, M.; Guanter, L.; Moreno, J.; Plaza, A.; Martinez, P. Land Surface Emissivity Retrieval From Different {VNIR} and {TIR} Sensors. *IEEE Trans. Geosci. Remote Sens.* **2008**, *46*, 316–327. [CrossRef]
37. Wang, H.; Yu, Y.; Yu, P.; Liu, Y. Land Surface Emissivity Product for NOAA JPSS and GOES-R Missions: Methodology and Evaluation. *IEEE Trans. Geosci. Remote Sens.* **2020**, *58*, 307–318. [CrossRef]
38. Wang, C.; Duan, S.-B.; Zhang, X.; Wu, H.; Gao, M.-F.; Leng, P. An Alternative Split-Window Algorithm for Retrieving Land Surface Temperature from Visible Infrared Imaging Radiometer Suite Data. *Int. J. Remote Sens.* **2018**, *40*, 1640–1654. [CrossRef]
39. Mann, H.B. Nonparametric Tests Against Trend. *Econometrica* **1945**, *13*, 245. [CrossRef]
40. Sen, P.K. Estimates of the Regression Coefficient Based on Kendall's Tau. *J. Am. Stat. Assoc.* **1968**, *63*, 1379–1389. [CrossRef]
41. Gocic, M.; Trajkovic, S. Analysis of Changes in Meteorological Variables Using Mann-Kendall and Sen's Slope Estimator Statistical Tests in Serbia. *Glob. Planet. Chang.* **2013**, *100*, 172–182. [CrossRef]
42. Wan, Z.; Zhang, Y.; Zhang, Q.; Li, Z.-L. Quality Assessment and Validation of the MODIS Global Land Surface Temperature. *Int. J. Remote Sens.* **2004**, *25*, 261–274. [CrossRef]

43. Hewison, T.J.; Wu, X.; Yu, F.; Tahara, Y.; Hu, X.; Kim, D.; Koenig, M. GSICS Inter-Calibration of Infrared Channels of Geostationary Imagers Using Metop/IASI. *IEEE Trans. Geosci. Remote Sens.* **2013**, *51*, 1160–1170. [CrossRef]

44. Kalma, J.D.; McVicar, T.R.; McCabe, M.F. Estimating Land Surface Evaporation: A Review of Methods Using Remotely Sensed Surface Temperature Data. *Surv. Geophys.* **2008**, *29*, 421–469. [CrossRef]

45. Khan, F.; Ali, S.; Mayer, C.; Ullah, H.; Muhammad, S. Climate Change and Spatio-Temporal Trend Analysis of Climate Extremes in the Homogeneous Climatic Zones of Pakistan during 1962–2019. *PLoS ONE* **2022**, *17*, e271626. [CrossRef] [PubMed]

46. Ali, S.H.B.; Shafqat, M.N.; Eqani, S.A.M.A.S.; Shah, S.T.A. Trends of Climate Change in the Upper Indus Basin Region, Pakistan: Implications for Cryosphere. *Environ. Monit. Assess.* **2019**, *191*, 51. [CrossRef] [PubMed]

47. Rani, S.; Mal, S. Trends in Land Surface Temperature and Its Drivers over the High Mountain Asia. *Egypt. J. Remote Sens. Sp. Sci.* **2022**, *25*, 717–729. [CrossRef]

Article

Hydro Statistical Assessment of TRMM and GPM Precipitation Products against Ground Precipitation over a Mediterranean Mountainous Watershed (in the Moroccan High Atlas)

Myriam Benkirane [1,*], **Nour-Eddine Laftouhi** [1], **Saïd Khabba** [2,3] **and África de la Hera-Portillo** [4]

[1] GeoSciences Laboratory, Geology Department, Faculty of Sciences Semlalia, Cadi Ayyad University (UCA), Marrakech 40000, Morocco

[2] LMFE, Faculty of Sciences Semlalia, University Cadi Ayyad, Marrakech 40000, Morocco

[3] Center for Remote Sensing Applications (CRSA), Mohammed VI Polytechnic University (UM6P), Benguerir 43150, Morocco

[4] Centro Nacional Instituto Geológico y Minero de España (CSIC), Ríos Rosas 23, 28003 Madrid, Spain

[*] Correspondence: myriam.benkirane@edu.uca.ac.ma or myriam14.benkirane@gmail.com; Tel.: +212-662-088741

Abstract: The tropical Rainfall Measuring Mission TRMM 3B42 V7 product and its successor, the Global Precipitation Measurement Integrated Multi-satellitE Retrievals for GPM IMERG high-resolution product GPM IMERG V5, have been validated against rain gauges precipitation in an arid mountainous basin where ground-based observations of precipitation are sparse, or spatially undistributed. This paper aims to evaluate hydro-statically the performances of the TRMM 3B42 V7 and GPM IMERG V05 satellite precipitations products SPPs, at multiple temporal scales, from 2014 to 2017. SPPs are compared with the gauge station and show good results for both statistical and contingency metrics with notable values R > 0.94. Moreover, the rainfall-runoff events implemented on the hydrological model were performed at 3-hourly time steps and showed satisfactory results based on the obtained Nash–Sutcliffe criteria ranging from 94.50% to 57.50%, and from 89.3% to 51.2%, respectively. The TRMM product tends to underestimate and not capture extreme precipitation events. In contrast, the GPM product can identify the variability of precipitation at small time steps, although a slight underestimation in the detection of extreme events can be corrected during the validation steps. The proposed method is an interesting approach for solving the problem of insufficient observed data in the Mediterranean regions.

Keywords: TRMM 3B42 V7; GPM IMERG V5; rain gauge; Mediterranean climate; hydrological modeling

Citation: Benkirane, M.; Laftouhi, N.-E.; Khabba, S.; Hera-Portillo, Á.d.l. Hydro Statistical Assessment of TRMM and GPM Precipitation Products against Ground Precipitation over a Mediterranean Mountainous Watershed (in the Moroccan High Atlas). *Appl. Sci.* 2022, 12, 8309. https://doi.org/10.3390/app12168309

Academic Editors: Giovanni Randazzo, Stefania Lanza and Anselme Muzirafuti

Received: 19 July 2022
Accepted: 17 August 2022
Published: 19 August 2022

Publisher's Note: MDPI stays neutral with regard to jurisdictional claims in published maps and institutional affiliations.

1. Introduction

Precipitation is a major force in global climate change and plays a vital role in hydrological and meteorological applications [1]. As a significant phenomenon in nature, precipitation has complex characteristics of spatiotemporal variations. It is one of the critical components of the global exchange of the surface material, the hydrological cycle, and disaster prevention [2,3].

The variability of precipitation in mountainous areas directly affects local agriculture and the ecological environment [4,5]. Moreover, the heavy precipitation events that occurred in mountainous areas frequently generate flash floods [6]. Therefore, the acquisition of reliable precipitation information in mountainous areas is of great significance to social and economic development and related scientific research [7]. Rain gauge observation could provide a moderately accurate method for a point-based precipitation measurement. However, rain gauges in mountainous regions are often scarce, irregular, and sometimes unavailable [4,8]. Thus, in the applications that need high spatiotemporal resolution precipitation data, such as flood disaster forecasts, gauge data are regularly insufficient [9,10].

Contrary to rain gauge precipitations, satellite remote sensing has the advantages of thoroughly scanning the entire study region and convenient access to the data [11], providing an alternate way to monitor precipitation at regional and global scales [12,13].

In recent decades, a series of high spatiotemporal resolutions satellite precipitation products (SPPs), have been produced with the development of various space-borne and related satellite-based precipitation retrieval algorithms, such as Artificial Neural Networks (PERSIANN) [14], National Oceanic and the Atmospheric Administration/Climate Prediction Center (NOAA/CPC) morphing technique (CMORPH) [15,16], as well as the Climate Hazards Group InfraRed Precipitation with Station data (CHIRPS) [17], the Tropical Rainfall Measuring Mission (TRMM) Multi-satellite Precipitation Analysis (TMPA) [18], and the Integrated Multi-satellitE Retrievals for (GPM) mission (IMERG) [19].

Compared to these satellite precipitation products, the TRMM 3B42V7 precipitation product performance is higher than other products, especially in estimating extreme precipitation events in several areas worldwide [20,21]. TRMM was launched in November 1997 by the National Aeronautics and Space Administration (NASA) with the Japanese Aerospace Exploration Agency (JAXA) collaboration. The TRMM Version-7 offers quasi-global coverage (50° N–50° S) precipitation estimates at a high spatial resolution of (0.25° × 0.25°) and temporal resolution of 3 h [18].

Given the excellent successes of the TRMM, the GPM Core Observatory satellite was set in motion by NASA and JXAX as a successor of TRMM in February 2014. Compared with TRMM, the potential of GPM to detect liquid and solid precipitation is improved by carrying a space-borne dual-frequency precipitation radar [22]. Additionally, the GPM Core Observatory carrying a conical scanning multichannel microwave imager offers a wider measurement range [19]. The lately released IMERG further expands quasi-global coverage from (50° N–50° S) to (60° N–60° S) and provides precipitation estimates with a more satisfactory spatial resolution of (0.1° × 0.1°) and temporal resolution of 30 min [23].

Since the deliverance of IMERG products, many studies have been conducted to evaluate and compare the performance of TMPA and IMERG products regarding rain gauges observations in many regions, such as the USA [24], Brazil [25], Japan [26], China [27], South Korea [28], Malaysia [29], Pakistan [30], South America [31], Cyprus [32], Egypt [33], and Morocco [34,35]. However, most of these studies indicate that (IMERGV5) had greater performance in the characterization of precipitation variability and precipitation detection aptitude, with only slight improvements compared to TMPA products.

This study evaluated statistically and hydrologically the precipitation estimates of the (3B42 V7) and (IMERG V05) satellites regarding ground-based precipitation monitoring over the Zat Mountain semi-arid watershed located in the Moroccan High Atlas. The purpose of this study is to solve a major problem due to the unavailability of precipitation measurement stations, which leads to a considerable lack of data, and therefore difficulties to work on scientific aspects such as flood forecasting and water management. The aims of this paper are (1) to evaluate and statistically compare the performance of the precipitation products (3B42 V7) and (IMERG V05) at several temporal scales in the Zat basin, and (2) to assess the precipitation detection capability of the satellite sensors (3B42 V7) and (IMERG V05), and (3) to be able to evaluate the ability of the SPPSs to reproduce rainfall events and to demonstrate their ability to provide meaningful information in hydrological modeling and flood forecasting.

2. Materials and Methods

2.1. Study Area

The Tensift basin is considered one of the main basins of Morocco, covering an area of 20,450 km^2 around Marrakech city, from the High Atlas Mountains to the Atlantic coast. This watershed is characterized by a semi-arid climate expressed by low rainfall, and high evaporation. Most of the Tensift flow comes from its five main tributaries, which have their source on the northern slopes of the High Atlas, which includes the Zat basin.

Zat watershed is a sub-basin of the Tensift catchment, and is also an Atlas tributary located on the left bank of the Tensift river and situated in the Moroccan High Atlas Mountains (Mount Toubkal, the highest mountain in North Africa) in the South EST of Marrakech city. Geographically, the sub-basin is located between latitude 31°30′ and 31°45′ North and longitude 7°30′ and 7°45′ West. It is drained by the Zat River, which measures 89 km, often has very steep slopes with an average of 19%, and covers a total area of about 520 km^2 (Figure 1). The hypsometry of the catchment varies from 3777 m (above sea level) upstream to the Taferiat station where the outlet is at an altitude of 756 m [36].

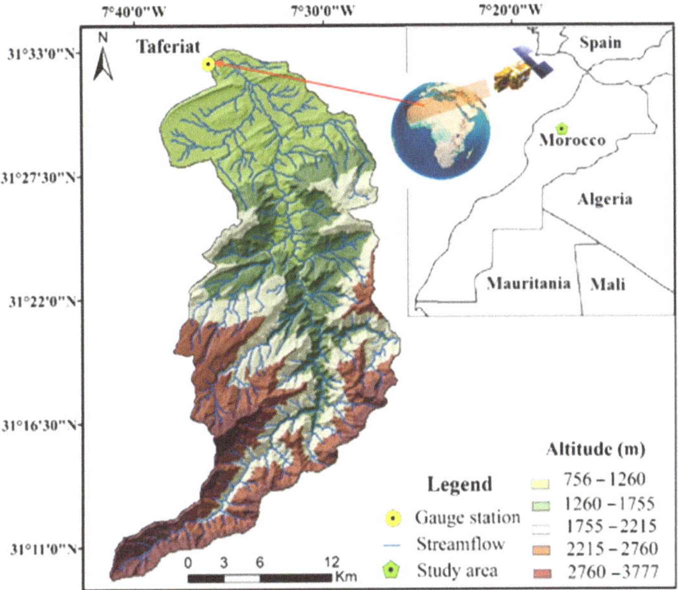

Figure 1. The geographical location of the Zat basin and rain gauge station used in the study.

This sub-basin is characterized by a Mediterranean climate strongly influenced by altitude. The Taferiat hydrometric station controls the discharge of the Zat basin, and also serves as a rain gauge. It receives an annual rainfall average of 523 mm/year; precipitation is mainly concentrated during the rainy period from October to April and a hot and dry period from May to September. Therefore, this study region is subject to frequent flash floods and droughts [36].

2.2. Dem

The terrain pre-processing started with the reconditioning of the Digital Elevation Model (DEM). The DEM tiles of resolution is approximately 30 m. It was downloaded from the United States Geological Survey (USGS). This DEM was clipped along the border of the basin using the polygon shapefile of the county downloaded from ESRI (Figure 1).

2.3. Rain Gauge Data

Rain gauge measurements are 10-min time scales precipitation data, collected from the only meteorological station of this basin shown in (Figure 1) located at the outlet of Zat basin, covering a period from 2014 to 2017. Data sets were provided by the Tensift Hydraulic Basin Agency. These data were used as a benchmark for evaluating TRMM 3B42 V7 and GPM IMERG V05 SPPs. All that was provided by this station were subject to strict quality control such as climate limit value inspection, and station extreme value inspection [37]. In addition, the daily, monthly, and yearly precipitation values were accumulated from 10-min observations.

2.4. Satellite Precipitation Data

This study evaluated two different satellite precipitation products SPPs, the TRMM 3B42 V7, and the GPM IMERG V05 compared to the gauge station at different time scales, from 1 September 2014 to 31 August 2017, in order to assess their reliability, and define their capabilities.

The period was chosen according to the available precipitation and discharge data. The Taferiat station was installed in 2012, and the available flow data extends to 2017 only. This limited the temporal interval of the study. Moreover, the GPM satellite was launched in 2014, which explains the choice of the beginning year of the data series.

2.4.1. TRMM 3B42 V7

The TRMM 3B42 V7 precipitation products were generated by using the TRMM 3B42 Version 7 algorithm [18]. It was designed to combine various microwaves MW, and infrared IR satellite-based precipitation estimates with gauge adjustments observations to provide 3-hourly quasi-global quantitative precipitation estimates. The 3B42 V7 product was derived by bias adjusting the near-real-time product with the GPCC monthly gauge-analysis precipitation data set, and it has a two-month latency [1]. The product can produce rational precipitation estimates in a 0.25° spatial resolution with a quasi-global coverage (50° S–50° N). In this study, the TRMM 3B42 V7 daily precipitation product was acquired from the Precipitation Measurement Mission (PMM) website (https://giovanni.gsfc.nasa.gov/giovanni/ (accessed on 16 May 2022)).

2.4.2. GPM IMERG V5

The GPM project is the result of a collaboration between (NASA) and (JAXA). The GPM satellite carries two primary sensors: the multi-channel GPM Microwave Imager (GMI), and the Dual-frequency Precipitation Radar (DPR). This product is expected to provide the next-generation global observations of rain and snow and improve weather and precipitation forecasts through the assimilation of instantaneous precipitation information [26]. The IMERG V05 uses the Goddard Profiling Algorithm to retrieve precipitation estimates from the GPM constellation using various precipitation-relevant satellite PMW sensors. Thereafter, the precipitation estimates are gridded and inter-calibrated into the GPM combined instrument product, further interpolated, and re-calibrated by the CPC Morphing-Kalman Filter Lagrangian time interpolation and the PERSIAN-Cloud Classification System recalibration schemes. IMERG is the Level 3 precipitation estimation algorithm of GPM, which provides three different daily IMERG products, which include IMERG Day 1 Early Run (near-real-time with a latency of 6 h), IMERG Day 1 Late Run (reprocessed near-real-time with a latency of 18 h) and IMERG Day 1 Final Run (gauged-adjusted with a latency of four months) products [38]. In this study, we evaluated the latest released GPM IMERG Version 5 (IMERG V5), the dataset produced at NASA Goddard Earth Sciences (GES). The IMERG precipitation products have a relatively finer spatial 0.1° spatial resolution with spatial coverage from 60° S to 60° N and temporal (half-hourly) resolution. The daily precipitation data were accumulated to obtain monthly and annual precipitation. The GPM (IMERG V05 final run) precipitation data were downloaded from the PMM website (https://giovanni.gsfc.nasa.gov/giovanni/ accessed on 20 May 2022)).

2.5. Statistical Evaluation of Satellite Precipitation Products

Different methods were used to compare the 3B42 V7 and IMERG V05 products with the gauge precipitation data from the Taferiat station, depending on temporal evolution by considering (sub-hourly, daily, monthly, and yearly) time steps. However, both studied satellite products have different spatial and temporal resolutions of (0.25°/3 h 3B42 V7 and (0.1°/30 min for IMERG V05), respectively, while the rain gauge station is located at the basin outlet and provides 10-min precipitation measurements of precipitation and runoff. For a reliable comparison, the used method is to plot the point precipitation data from the gauges to the same grid scale as the SPPs, either by spatial interpolation or simply by calculating the average. In addition, ref. [39] pointed out that the interpolation can lead to uncertainties due to systematic errors of the gauge density. Therefore, in our case, a direct comparison was used, extracting the numerical data from the SPPs and measuring the precipitation, adjusting them simultaneously to compare them to each other.

Furthermore, to evaluate the ability of the SPPs to estimate extreme rainfall events, it was decided to use them as input data in the HEC-HMS hydrological model. Indeed, the rainfall measurement station is not precise and poorly distributed spatially, especially in the mountainous regions of the High Atlas, which is a real issue for research work on hydrological modeling and flood forecasting. Consequently, it is important to evaluate the rainfall estimated by the satellites to demonstrate their ability to provide significant information and to approve their use as an alternative source of rainfall measurement data, especially during extreme precipitation events.

2.5.1. Continuous Statistical Indices

The assessment and comparison of the SPPs was conducted on the basis of the overall assessment (continuous statistical measures) and the precipitation detection capability (categorical statistical measures).

In addition, the capability of the SPPs to measure different precipitation intensity classes and flood periods was evaluated as well (Table 1). Four statistical measures were selected: correlation coefficient (CC), root mean square error (RMSE), and bias (bias), which were calculated to statistically evaluate the two PPS products.

Table 1. Statistical metrics for evaluating IMERG V5 and 3B42 V7 products.

Statistical Index	Units	Reference Values	Equation	Reference
Correlation Coefficient (CC)	Ratio	1	$CC = \dfrac{\sum_{i=1}^{N}\left(Pi-\bar{P}\right)\left(Si-\bar{S}\right)}{\sqrt{\sum_{i=1}^{N}\left(Pi-\bar{P}\right)^{2}\sum_{i=1}^{n}\left(Si-\bar{S}\right)^{2}}}$	
Root Mean Square Error (RMSE)	mm	0	$RMSE = \sqrt{\dfrac{\sum_{i=1}^{N}\left(Pi-Si\right)^{2}}{N}}$	[40]
Bias	mm	0	$Bias = \dfrac{\sum_{i=1}^{N}\left(Pi-Si\right)}{\sum_{i=1}^{N} N}$	
Probability of Detection (POD)	Ratio	1	$POD = \dfrac{a}{a+c}$	
False Alarm Ratio (FAR)	Ratio	0	$FAR = \dfrac{b}{a+b}$	
Critical Success Index (CSI)	Ratio	1	$CSI = \dfrac{a}{a+b+c}$	[41]
Frequency Bias Index (FBI)	Ratio	1	$FBI = \dfrac{a+b}{a+c}$	

Where N represents the number of samples; Si and $\bar{S}$ are gauge observations and their average; Pi and $\bar{P}$ represent satellite estimates and their average, respectively.

Additionally, a denotes the number of rainfall events that were observed and detected; c, is the number of rainfall events that failed to be detected by the satellite; b, denotes the number of rainfall events detected by the satellite that did not occur; the threshold for identifying a precipitation event is 0.2 mm/day.

2.5.2. Categorical Statistical Indices

To evaluate the precipitation detection capability of IMERG V05 and 3B42 V7 products, four categorical statistical indices were calculated to assess the ability of PPSs. The most common measures, counting Probability of Detection (POD), False Alarm Ratio (FAR), Critical Success Index (CSI), and Frequency Bias Index (FBI) are used in this study. The values of all categorical statistical measures are between 0 and 1.

The POD indicated the ratio of the number of precipitation events correctly detected by satellites among all real precipitation events. The FAR is the ratio of false alarming precipitation events to the total number of detected precipitation events. The FBI represents the fraction of falsely detected precipitation events (false alarm) compared to the total number of detected precipitation events, and indicates whether the dataset tends to overestimate (FBI > 1) or underestimate (FBI < 1) precipitation events. The CSI reported the number of correct predictions of a rain event divided by the total number of successes, false alarms, and failures.

2.6. Hydrological Model

The Hydrologic Engineering Centre's Hydrologic Modeling System (HEC-HMS) is designed to simulate the rainfall-runoff processes of dendritic watershed systems. It is a deterministic, semi-distributed, event-based/continuous, mathematically-based (conceptual) model. It is able to model in a wide variety of geographical regions and different climatic contexts, such as arid and semi-arid mountainous climates, several studies having been carried out in the Tensift region. The HEC-HMS model was chosen mainly because it was previously validated in the same study area, and because it provided good and realistic results [36,42,43] in order to assess the applicability of the model in such an environment, using different approaches (event-based and continuous modeling).

The HEC-HMS model consists of four components: the basin model, the meteorological model, the control specifications, and the input data. The basin model provides information about the physical properties of the model, such as basin areas and stream reach connectivity. Similarly, the meteorological model includes information related to precipitation data. The control specification section contains information related to the timing of the model, the timing of a storm event, and the type of time interval (second, minute, hour, or day) that is required to be used in the model. Finally, the input data component contains the parameters or boundary conditions for the basin and weather models. The main input data used for this study are satellite and in situ precipitation and observed flow, as well as the different basin characteristics (number of curves, soil, LULC) resulting from the HEC-GeoHMS process. The method used in this paper includes the SCS-CN (Soil Conservation Service) curve number, the Clark unit hydrograph, and the base flow recession, which are necessary to determine the hydrological loss rate, runoff transformation, and base flow. The model was used in a lumped way (as we only have one measuring station at the downstream of the basin, which does not allow to do semi-distributed modeling). This approach aims to calibrate three rainfall events for each product, and then three more for observed rainfall. A total of 15 events were calibrated, 12 of which were satellite products, and 3 of which were observed data. The validation used a 3H time step of precipitation by implementing the model with different sources of precipitation data, such as observed and satellite precipitation with observed runoff, to evaluate the ability of SPPs to reproduce rainfall events from 2014 to 2017.

3. Results

3.1. Statistical Evaluation

The SPPs were statically compared against the ground observations to evaluate their accuracy and reliability. Table 2 lists the evaluation results of statistical metrics (CC, RMSE, and Bias) thought the entire study period over the Zat basin.

Table 2. Statistical metrics results of 3B42 V7 and IMERG V5 precipitation estimates at multiple time scales from 2012 to 2017.

	TRMM				GPM			
	3 h	**Daily**	**Monthly**	**Yearly**	**3 h**	**Daily**	**Monthly**	**Yearly**
CC	0.12	0.38	0.79	0.94	0.4	0.59	0.81	0.86
RMSE	1.41	0.9	2.15	16.75	1.35	3.26	1.69	21.1
Bias	0.22	0.22	0.85	0.21	0.25	1.52	1.37	1.49

The rainfall time series of the two selected satellite products and the rain gauge at different timescales in the Zat basin are illustrated in Table 2. In general, the 3B42 V7 and IMERG V5 products present similar chronological precipitation patterns to those of the gauge. However, it can be seen that the product 3B42 V7 underestimates the 3 h and the daily precipitation, while the product IMERG V05 showed good performance on the 3 h and the daily timescale (Table 2). Regarding the monthly and annual time scale precipitation series, the product 3B42 V7 and IMERG V05 clearly showed good results.

3.2. Contingency Statistics

The categorical statistical metrics of 3B42 V7 and IMERG V5 at different time scales are shown in Table 3.

Table 3. Contingency statistical metrics results of 3B42 V7 and IMERG V5 precipitation estimates at multiple time scales from 2014 to 2017.

	TRMM				GPM			
	3 h	Daily	Monthly	Yearly	3 h	Daily	Monthly	Yearly
POD	0.13	0.39	0.89	1	0.58	0.82	1	1
FAR	0.79	0.66	0.08	0	0.83	0.81	0.07	0
CSI	0.08	0.22	0.82	1	0.14	0.18	0.92	1
FBI	0.65	1.17	0.97	1	3.63	4.37	1.07	1

The precision of 3B42 V7 and IMERG V5 at 3 h, daily, monthly and annual scales were compared and analyzed. IMERG V05 demonstrated better performance than 3B42 V7 in detecting small time scale precipitation events, with high values of POD and CSI (0.13, 0.39 vs. 0.58, 0.82) and (0.08, 0.22 vs. 0.14, 0.18), respectively (Table 3), as well as reasonably high values of FAR and FBI (0.79, 0.66 vs. 0.83, 0.81) and (0.65, 1.17 vs. 3.63, 4.37), respectively.

The performance of the categorical statistical measures at the monthly and annual levels are shown in (Table 3). 3B42 V7 and IMERG V05 produced good results for rainfall estimation, with POD values and CSI values (0.99, 1 vs. 1, 1) and (0.07, 0 vs. 1.07, 1), respectively. Similarly, for the FAR and FBI, the results obtained were close to the perfect values, (0.08, 0 vs. 0.08, 0) and (1, 1 vs. 1.09, 1), respectively.

In general, IMERG V05 is better at detecting precipitation events, in particular at capturing precipitation traces and solid precipitation at a 3-hourly and daily scale, while 3B42 V7 can estimate precipitation on a large time scale.

3.3. Hydrological Evaluation of Discharge Simulation Using Two SPPs

The HEC-HMS model was used to calibrate and validate the 3 h of rainfall events from (1 September 2014) to (31 August 2017), at the level of the Zat basin, using the rainfall and runoff data from the Taferiat gauge station and satellite precipitation products. The four episodes that we chose to present are the most representative and complete of the data series.

The hydrological calibration and validation were carried out according to two different scenarios.

The obtained calibration and validation results are very satisfactory; the Nash–Sutcliffe coefficients obtained for calibration and validation are on average 88.20% and 57.50%, respectively (Tables 4 and 5). The episode calibrations were performed by manual adjustment of the parameters in a way that does not lead to the deviation of the parameters from their real physical meaning, and which allows for a better understanding of each calibration parameter. This method requires a lot of time and effort to understand the behavior of each parameter and approximate it to the natural condition of the event occurrence. On the other hand, the objective function optimization method is simple and practical for function-based investigations but may ignore the real physical meaning of the parameters. Therefore, this paper combines the two methods to adjust the model parameters.

Table 4. Summary range of calibrated parameter values.

	Model Parameters	Calibration Ranges
Loss parameters	Initial Abstraction (mm)	-
	Curve Number (CN)	46–83
	Impervious (%)	0–10
Transform parameters	Time of concentration (HR)	0.1–5.5
	Storage Coefficient (HR)	2.6–25.6
Baseflow parameters	Initial discharge (m^3/s)	0.3–2.8
	Recession constant	0.6–0.8
	Ratio	0.3–0.5

Table 5. Episodes calibration settings.

	Id	Events	Curve Number	Time of Concentration	Recession Constant	Nash–Sutcliffe	RMSE
Calibration	Gauge precipitation	20 November 2014	51	0.1	0.6	0.88	0.4
		21 March 2016	60	0.6	0.55	0.88	0.3
		3 May 2016	63	2	0.29	0.83	0.4
		16 December 2016	60	9	0.3	0.58	0.7
	3B42V7	20 November 2014	54	0.6	0.6	0.79	0.5
		21 March 2016	67	0.1	0.6	0.67	0.6
		3 May 2016	65	10	0.3	0.77	0.5
		16 December 2016	61	4	0.6	0.64	0.6
	IMERGV5	20 November 2014	52	0.1	0.6	0.84	0.5
		21 March 2016	50	0.9	0.6	0.84	0.4
		3 May 2016	44	3.1	0.3	0.79	0.5
		16 December 2016	62	6.15	0.38	0.62	0.6
	Mean					**0.76**	

In this paper, eleven parameters were calibrated by maintaining the maximum and minimum intervals of calibration parameters based on the literature. The intervals of the calibrated parameter used are illustrated in Table 4.

Scenario 1 calibration: simulation and calibration using precipitation from both satellite products with observed fluxes, by adjusting the model parameters values until the model results acceptably match the observed data.

Scenario 2 validation: due to the limited sample size, the model was validated using the leave-one-out resampling approach; for the n flood events, each event i is successively removed, in order to find the relationship between the root-soil moisture measured by the time domain reflectometry "TDR" tool and the two models' most sensitive calibration parameters (Curve Number "CN", and time of concentration "TC"). A new CN was then re-estimated (Calculate CN) using the remaining n-1 episode. The CN calibration and the Tc parameters for an event i are set to the median of the calibrated parameters for the n-1 episodes. The calculated CN values obtained by this procedure are then used to model flood event i, and the simulated discharge is compared to the observed discharge. The validation results for the 15 events are presented in Table 6, indicating better model performance when using the SCS-CN model and taking into account soil moisture, with Nash coefficients between 0.51 and 0.82, using the leave-one-out procedure [43].

Table 6. Episodes validation results.

	Id	Events	Calculated CN	Nash-Sutcliffe	RMSE
	Gauge precipitation	20 November 2014	60.39	0.58	0.6
		21 March 2016	56.98	0.64	0.6
		3 May 2016	59.03	0.83	0.4
		16 December 2016	55.83	0.51	0.7
Validation	3B42V7	20 November 2014	59.2	0.52	0.7
		21 March 2016	65.9	0.56	0.7
		3 May 2016	63.05	0.61	0.6
		16 December 2016	60.18	0.63	0.6
	IMERGV5	20 November 2014	39.5	0.71	0.6
		21 March 2016	50.5	0.74	0.5
		3 May 2016	47	0.73	0.5
		16 December 2016	49	0.57	0.7
	Mean			**0.64**	

3.3.1. Event of 20 November 2014

This event represents a torrential flood; since the flood was generated by extreme precipitation spread over more than 15 days, it is the most intense event in the data set. The maximum flow reached was $(123, 75 \ m^3/s)$. However, the soils were saturated, resulting in high permeability and an increase in the runoff coefficient of the watershed.

The results of the calibration of the 3B42V7 and IMERG V05 rainfall data with the observed flow illustrated in the hydrographs of the Figure 2, show that the simulated flow curves were globally well reproduced for both products at the flood rise and the recession part, although the peaks flow were not reached by both products.

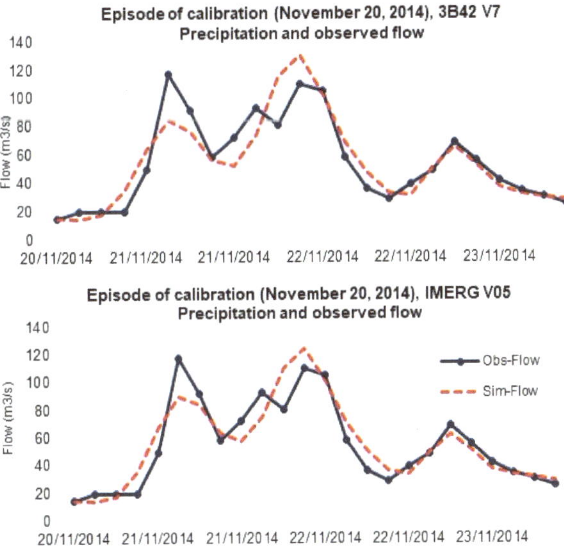

Figure 2. Calibration of the episode 20 November 2014, using 3B42V7 and IMERG V05 rainfall products and observed runoff data as input.

Based on the results the 3B42V7 calibration is satisfactory, as this product has a good capacity to record the high precipitation intensity during rainy episodes.

Furthermore, the IMERG V05 product does not have the ability to capture heavy precipitation; this is well demonstrated in the calibration results. The evaluation criteria are suitable RMSE = 0.5 for both products and a Nash of 79.20% and 70.10% for 3B42V7 and IMERG V05, respectively.

The validation hydrographs results in (Figure 3) were well reproduced for both products. The rise and the recession curves were well reproduced for 3B42V7, noting a slight underestimation of precipitation, but in general, this product well estimates heavy precipitation events. The IMERG V05 was not able to reproduce the validation hydrograph of this event. The rising curve was underestimated in the first pick as it represents the heavier precipitation during this event, and the other two picks were underestimated, the peak flow was not reached. This is because of the inability of this product to properly estimate the heavy precipitation.

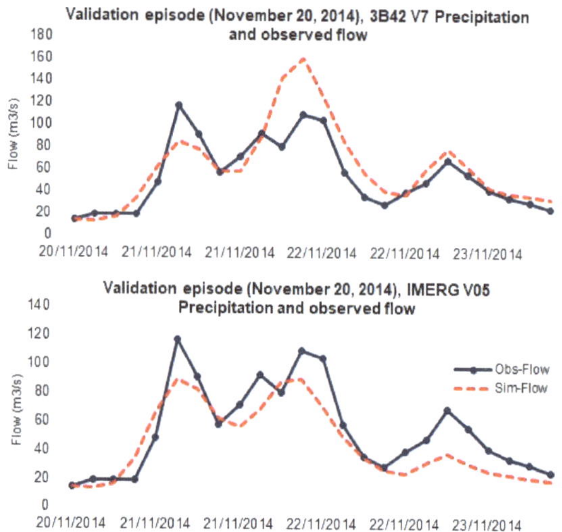

Figure 3. Validation of the episode of 20 November 2014, of 3B42V7 and IMERG V05 rainfall products and observed runoff data as input using the means calibration parameters of the gauge precipitation calibration.

The evaluation criteria are acceptable with an RMSE of 0.7 for both products, and a Nash of 52.20% and 66.50%, respectively.

3.3.2. Event of 21 March 2016

This event represents the typical characteristics of a freshet caused by the melting of snowfall upstream of the Zat watershed. With the progressive increase of temperatures, the snow cover at the summit of the Atlas Mountains starts melting and feeding the streams of the mountainous basins including the study basin. This usually causes significant flooding during the occurrence of moderate rainfall episodes.

The hydrograph of Figure 4 is well calibrated, the simulated flow curves were differently reproduced for both products. Concerning 3B42V7 SPPs, the rise and the recession curves were well reproduced, but the peak flow was underestimated due to the fact that this satellite product is not able to reproduce the low precipitation. However, IMERG V05 hydrograph is well reproduced at the rise, the recession, and the peak flow.

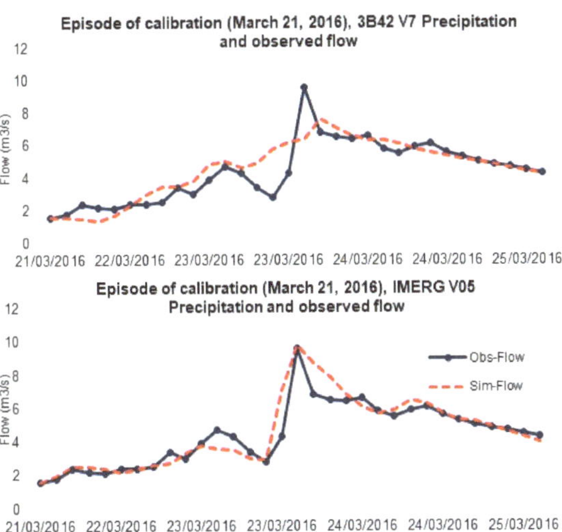

Figure 4. Calibration of the episode 21 March 2016, using 3B42V7 and IMERG V05 rainfall products and observed runoff data as input.

The evaluation criteria are good with an RMSE of 0.3 and 0.3 and a Nash of 66.5% and 83.7% for 3B42V7 and IMERGV5 products.

On the other hand, the hydrographs of validation are also differently reproduced for the two SPPs (Figure 5).

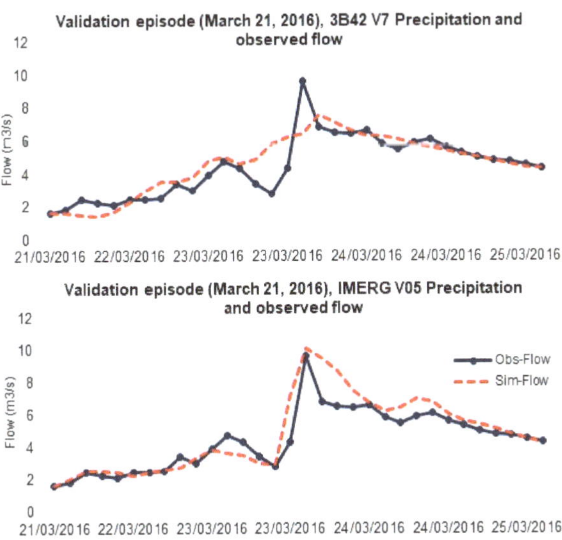

Figure 5. Validation of the episode of 21 March 2016, of 3B42V7 and IMERG V05 rainfall products and observed runoff data as input using the means calibration parameters of the gauge precipitation calibration.

The 3B42V7 underestimates the precipitation; as noticed this episode is generated by the effect of snowmelt, with the occurrence of light precipitation, and for this reason, the following product estimates poorly the slight precipitation as previously indicated, and consequently underestimates the value of precipitation during the event.

However, IMERGV5 has shown good performance in detecting small precipitation events; on the validation hydrograph the simulated curve is well reproduced at all levels. This is due to the fact that this product is capable of estimating small precipitation events with short time steps. This resolution has been well confirmed once more at this event.

The evaluation criteria are good with an RMSE of 0.7 and 0.6 for the product 3B42 V7 and IMERG V05, and a Nash of 56.20% and 74.20. 9%, respectively.

3.3.3. Event of 3 May 2016

The hydrological model results showed a reasonable fit between the shape of the simulated and observed hydrographs. Figure 6 shows a chronological comparison of simulated and observed streamflow at the watershed outlet for a calibration period of 3–5 May 2016 (we limit ourselves to modeling short-duration floods for which the evapotranspiration process is negligible). Although the measured peak flow values did not exactly match the simulated peak flow values for both products, there was a slight improvement for IMERG V05. The model tended to underestimate the streamflow, since the river flow was already high due to snowmelt, and this was added to the runoff caused by the flood, which the model did not take into account.

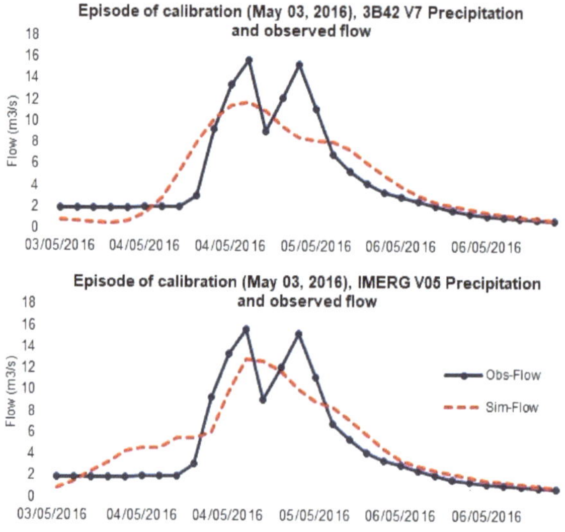

Figure 6. Calibration of the episode 3 May 2016, using 3B42V7 and IMERG V05 rainfall products and observed runoff data as input.

On the other hand, the volumes were well respected and the rise and fall curves were generally well reproduced with a slight improvement on the IMERG V05 side. The results of the evaluation criteria for both products 3B42V7 and IMERG V05 are satisfactory with NSE ranging from 76.80% to 79.30%, respectively.

After following the validation procedures previously mentioned in Section 3.3.4, the comparison of the observed and simulated hydrographs showed that the model underestimates the point flows, due to the non-conclusion of the snowmelt process. However, Figure 7 shows a good trend in the reproduction of the observed and simulated discharge curves, with an NSE of 61.30% for 3B42V7, and 72.90% for IMERG V05, which are good results.

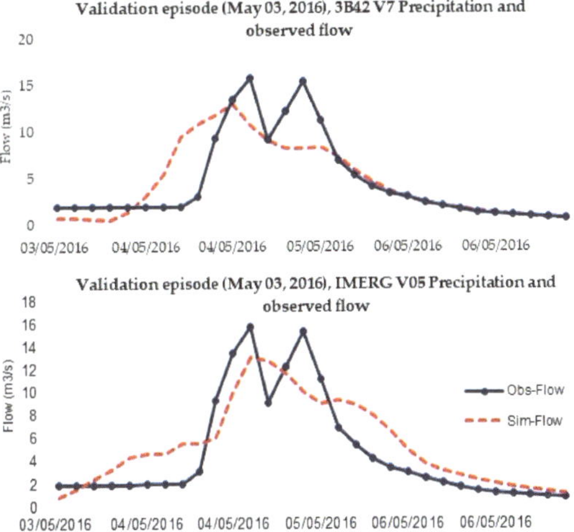

Figure 7. Validation of the episode of 3 May 2016, of 3B42V7 and IMERG V05 rainfall products and observed runoff data as input using the means calibration parameters of the gauge precipitation calibration.

3.3.4. Event of 16 December 2016

The event represents a winter rain-storm characterized by liquid precipitation downstream and snow upstream of the watershed. This type of rain-storm is very frequent during the winter, especially in the high mountains of the Atlas.

The hydrographs of calibration in Figure 8 represent a simulated flow curve quite illustrative; the rising curve and the recession were well reproduced for both products 3B42V7 and IMERG V05, contrary to the peak flow which has not been reached for the 3B42V7. However, the evaluation criteria are acceptable and represent an NSE of 63.50% and 61.90%, respectively.

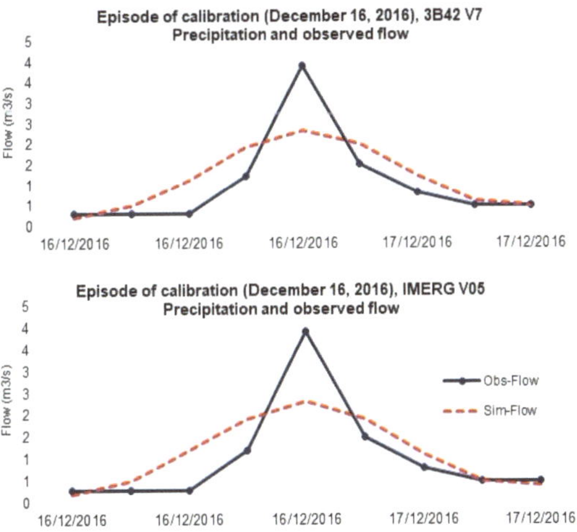

Figure 8. Calibration of the episode 16 December 2016, using 3B42V7 and IMERG V05 rainfall products and observed runoff data as input.

On the other hand, Figure 9 illustrates the validation graphs of each product, although the curves of the simulated flows are not well reproduced for both products, which underestimated winter precipitations since they are a mixture of rainfall and snowfall. The evaluation criteria are acceptable with an NSE of 63.20% for 3B42V7, and 57.20% for IMERG V05.

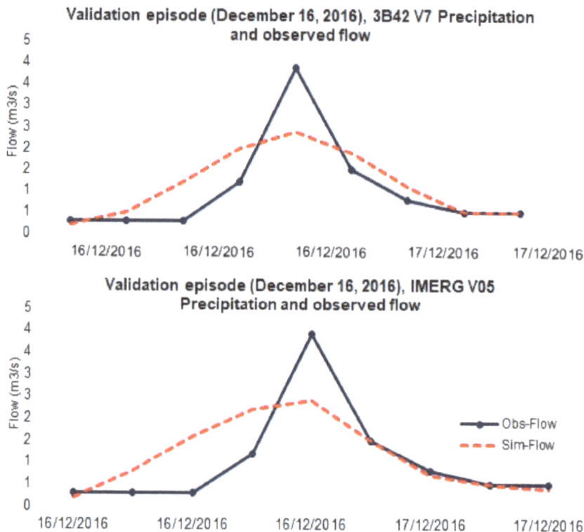

Figure 9. Validation of the episode of 16 December 2016, of 3B42V7 and IMERG V05 rainfall products and observed runoff data as input using the means calibration parameters of the gauge precipitation calibration.

In this paper, an efficient method has been developed for the first time in a country with a Mediterranean climate, on the mountainous watersheds of the Moroccan High Atlas with low density and irregularity of precipitation and flow measurement stations. This is a good method to apply to solve the problem of deficiency of observed data in these regions.

4. Conclusions

SPPs are important precipitation data alternatives, particularly in high mountain watersheds, where measurement gauge stations are poorly distributed or absent. These products will mainly help in the simulation of river flows, flood forecasting, and water resources management in arid to semi-arid regions. This study conducted a complete performance evaluation of two satellite products—the TRMM (3B42 V7) and the GPM—(IMERG V05) using observations (sub-hourly, daily, monthly, and yearly) collected from the only gauge station of the Zat basin, named Taferiat station, and located at the downstream of the watershed. The watershed is characterized by a Mediterranean climate and mountainous topography, and the study was analyzed over 3 years, from 1 September 2014 to 31 August 2017. To evaluate the accuracy of 3B42 V7 and IMERG V05 satellite precipitation products, several quantitative, categorical, and statistical measurements were used: (R, RMSE, Bias) were used to quantitatively analyzed the accuracy of satellite precipitation products, and (POD, CSI, FAR, and FBI) were used to evaluate the precipitation detection capability of satellite precipitation products, and to simulate satisfactorily the flooding events in hydrological model.

The conclusions resulting from this study are summarized as follows:

(1)   3B42 V7 and IMERG V05 products performed well in estimating sub-hourly, daily, monthly, and annual precipitation compared to observed data from the Taferiat station. 3B42V7 underestimated low precipitation events but well estimated heavy precipitation with small time step. However, the monthly and annual precipitation were well captured. While IMERGV05 overestimates heavy precipitation episodes and has a good ability to detect low precipitation in small time step, the monthly and the yearly precipitation are well estimated by this product.

(2) Compared to the ground applications, 3B42V7 and IMERG V5 showed acceptable correlation results at the sub-hourly and daily time scales. However, IMERG V05 performed slightly better than 3B42 V7 for the detection of sub-hourly and daily precipitation at the measuring station. The categorical statistical measures values showed high values of POD and CSI, as well as reasonably high values of FAR and FBI, noting that the results of the categorical measures are good. In general, IMERG V05 is better at detecting precipitation events, in particular at capturing precipitation traces and solid precipitation at a 3-hourly and daily scale, while 3B42 V7 can estimate precipitation on a large time scale.

(3) The hydrological calibration and validation were performed according to two different scenarios; scenario 1 aims to simulate and calibrate events using rainfall from both satellite products with observed flows, while scenario 2 of validation uses the leave-one-out resampling approach; for the n flood events, in order to find the relationship between the root-soil moisture measured and the most sensitive model parameters (CN calibration, and time of concentration "TC"). The obtained results are satisfactory for all calibration and validation parts, the NSE coefficients ranging between 74.75% and 63.31%, respectively. The main point to remember is that the 3B42V7 product does not have a good ability to capture small rainfall events in a short time step, in fact, it underestimates the rainfall. On the other hand, the IMERG V05 product has an excellent capacity to record small rainfall events, which is well demonstrated in the validation graphs. Therefore, it is recommended to use this product for flood modeling and forecasting. The proposed method is an interesting approach to apply for solving the problem of insufficient observed data in the Mediterranean regions. The present manuscript provides a valuable reference for precipitation monitoring and forecasting in mountainous regions characterized by a Mediterranean climate, as well as in basins with few or poorly distributed rainfall stations.

Therefore, the results of this study are of great importance for analyzing the prospect's application of SPPs at different time scales. This paper is one of the first papers developing a comparative approach of satellite rainfall products to observe gauge data in nnthe Moroccan High Atlas; they could indeed serve researchers as a reference work both in Morocco and neighbouring countries with similar climates and areas with irregular or sparse rain gauge networks.

Author Contributions: Conceptualization, M.B. and S.K.; methodology, M.B.; software, M.B.; validation, M.B., N.-E.L. and S.K.; resources, M.B.; data curation, M.B. and S.K.; writing—original draft preparation, M.B.; writing—review and editing, M.B., S.K. and Á.d.l.H.-P.; visualization, N.-E.L.; supervision, N.-E.L. and S.K; project administration, N.-E.L.; funding acquisition, M.B. All authors have read and agreed to the published version of the manuscript.

Funding: This research was supported by the projects: ERANETMED3-062 CHAAMS Global Change: Assessment and Adaptation to Mediterranean Region Water Scarcity and PRIMA-S2-ALTOS-2018 Managing water resources within Mediterranean agrosystems by accounting for spatial structures and connectivities.

Acknowledgments: The authors would like to thank Adam Milewski (Director of Water Resources & Remote Sensing Laboratory (WRRS), Department of Geology, University of Georgia, United States), who thoughtfully revised this manuscript.

Conflicts of Interest: The authors declare no conflict of interest.

References

1. Yuan, F.; Zhang, L.; Win, K.W.W.; Ren, L.; Zhao, C.; Zhu, Y.; Jiang, S.; Liu, Y. Assessment of GPM and TRMM Multi-Satellite Precipitation Products in Streamflow Simulations in a Data-Sparse Mountainous Watershed in Myanmar. *Remote Sens.* **2017**, *9*, 302. [CrossRef]

2. Bollasina, M.A.; Ming, Y.; Ramaswamy, V. Anthropogenic Aerosols and the Weakening of the South Asian Summer Monsoon. *Science* **2011**, *334*, 502–505. [CrossRef] [PubMed]

3. Zhu, G.; He, Y.; Pu, T.; Wang, X.; Jia, W.; Li, Z.; Xin, H. Spatial distribution and temporal trends in potential evapotranspiration over Hengduan Mountains region from 1960 to 2009. *J. Geogr. Sci.* **2012**, *22*, 71–85. [CrossRef]

4. Xia, T.; Wang, Z.-J.; Zheng, H. Topography and Data Mining Based Methods for Improving Satellite Precipitation in Mountainous Areas of China. *Atmosphere* **2015**, *6*, 983–1005. [CrossRef]

5. Jiang, S.; Liu, S.; Ren, L.; Yong, B.; Zhang, L.; Wang, M.; Lu, Y.; He, Y. Hydrologic Evaluation of Six High Resolution Satellite Precipitation Products in Capturing Extreme Precipitation and Streamflow over a Medium-Sized Basin in China. *Water* **2017**, *10*, 25. [CrossRef]

6. Borga, M.; Anagnostou, E.; Blöschl, G.; Creutin, J.-D. Flash floods: Observations and analysis of hydro-meteorological controls. *J. Hydrol.* **2010**, *394*, 1–3. [CrossRef]

7. Germann, U.; Galli, G.; Boscacci, M.; Bolliger, M. Radar precipitation measurement in a mountainous region. *Q. J. R. Meteorol. Soc.* **2006**, *132*, 1669–1692. [CrossRef]

8. Hrachowitz, M.; Weiler, M. Uncertainty of Precipitation Estimates Caused by Sparse Gauging Networks in a Small, Mountainous Watershed. *J. Hydrol. Eng.* **2011**, *16*, 460–471. [CrossRef]

9. Mei, Y.; Anagnostou, E.N.; Nikolopoulos, E.I.; Borga, M. Error Analysis of Satellite Precipitation Products in Mountainous Basins. *J. Hydrometeorol.* **2014**, *15*, 1778–1793. [CrossRef]

10. Yi, L.; Zhang, W.; Wang, K. Evaluation of Heavy Precipitation Simulated by the WRF Model Using 4D-Var Data Assimilation with TRMM 3B42 and GPM IMERG over the Huaihe River Basin, China. *Remote Sens.* **2018**, *10*, 646. [CrossRef]

11. Muzirafuti, A.; Cascio, M.; Lanza, S.; Randazzo, G. UAV Photogrammetry-based Mapping of the Pocket Beaches of Isola Bella Bay, Taormina (Eastern Sicily). In Proceedings of the 2021 International Workshop on Metrology for the Sea; Learning to Measure Sea Health Parameters (MetroSea), Reggio Calabria, Italy, 4–6 October 2021; pp. 418–422. [CrossRef]

12. Chen, C.; Chen, Q.; Duan, Z.; Zhang, J.; Mo, K.; Li, Z.; Tang, G. Multiscale Comparative Evaluation of the GPM IMERG v5 and TRMM 3B42 v7 Precipitation Products from 2015 to 2017 over a Climate Transition Area of China. *Remote Sens.* **2018**, *10*, 944. [CrossRef]

13. Muzirafuti, A.; Boualoul, M.; Barreca, G.; Allaoui, A.; Bouikbane, H.; Lanza, S.; Crupi, A.; Randazzo, G. Fusion of Remote Sensing and Applied Geophysics for Sinkholes Identification in Tabular Middle Atlas of Morocco (the Causse of El Hajeb): Impact on the Protection of Water Resource. *Resources* **2020**, *9*, 51. [CrossRef]

14. Sorooshian, S.; Hsu, K.-L.; Gao, X.; Gupta, H.V.; Imam, B.; Braithwaite, D. Evaluation of PERSIANN System Satellite–Based Estimates of Tropical Rainfall. *Bull. Am. Meteorol. Soc.* **2000**, *81*, 2035–2046. [CrossRef]

15. Joyce, R.J.; Janowiak, J.E.; Arkin, P.A.; Xie, P. CMORPH: A Method that Produces Global Precipitation Estimates from Passive Microwave and Infrared Data at High Spatial and Temporal Resolution. *J. Hydrometeorol.* **2004**, *5*, 487–503. [CrossRef]

16. Guo, J.; Zhai, P.; Wu, L.; Cribb, M.; Li, Z.; Ma, Z.; Wang, F.; Chu, D.; Wang, P.; Zhang, J. Diurnal variation and the influential factors of precipitation from surface and satellite measurements in Tibet. *Int. J. Clim.* **2014**, *34*, 2940–2956. [CrossRef]

17. Funk, C.; Peterson, P.; Landsfeld, M.; Pedreros, D.; Verdin, J.; Shukla, S.; Husak, G.; Rowland, J.; Harrison, L.; Hoell, A.; et al. The climate hazards infrared precipitation with stations—A new environmental record for monitoring extremes. *Sci. Data* **2015**, *2*, 150066. [CrossRef]

18. Huffman, G.J.; Bolvin, D.T.; Nelkin, E.J.; Wolff, D.B.; Adler, R.F.; Gu, G.; Hong, Y.; Bowman, K.P.; Stocker, E.F. The TRMM Multisatellite Precipitation Analysis (TMPA): Quasi-Global, Multiyear, Combined-Sensor Precipitation Estimates at Fine Scales. *J. Hydrometeorol.* **2007**, *8*, 38–55. [CrossRef]

19. Hou, A.Y.; Kakar, R.K.; Neeck, S.; Azarbarzin, A.A.; Kummerow, C.D.; Kojima, M.; Oki, R.; Nakamura, K.; Iguchi, T. The global precipitation measurement mission. *Bull. Am. Meteorol. Soc.* **2014**, *95*, 701–722. [CrossRef]

20. Tong, K.; Su, F.; Yang, D.; Hao, Z. Evaluation of satellite precipitation retrievals and their potential utilities in hydrologic modeling over the Tibetan Plateau. *J. Hydrol.* **2014**, *519*, 423–437. [CrossRef]

21. Ringard, J.; Becker, M.; Seyler, F.; Linguet, L. Temporal and Spatial Assessment of Four Satellite Rainfall Estimates over French Guiana and North Brazil. *Remote Sens.* **2015**, *7*, 16441–16459. [CrossRef]

22. Chandrasekar, V.; Le, M. Evaluation of profile classification module of GPM-DPR algorithm after launch. In Proceedings of the 2015 IEEE International Geoscience and Remote Sensing Symposium (IGARSS), Milan, Italy, 26–31 July 2015; pp. 5174–5177. [CrossRef]

23. Liu, Z.; Ostrenga, D.; Vollmer, B.; Deshong, B.; Macritchie, K.; Greene, M.; Kempler, S. Global Precipitation Measurement Mission Products and Services at the NASA GES DISC. *Bull. Am. Meteorol. Soc.* **2017**, *98*, 437–444. [CrossRef] [PubMed]

24. Gebregiorgis, A.S.; Kirstetter, P.; Hong, Y.E.; Gourley, J.J.; Huffman, G.J.; Petersen, W.A.; Xue, X.; Schwaller, M.R. To What Extent is the Day 1 GPM IMERG Satellite Precipitation Estimate Improved as Compared to TRMM TMPA-RT? *J. Geophys. Res. Atmos.* **2018**, *123*, 1694–1707. [CrossRef]

25. Rozante, J.R.; Vila, D.A.; Chiquetto, J.B.; Fernandes, A.D.A.; Alvim, D.S. Evaluation of TRMM/GPM Blended Daily Products over Brazil. *Remote Sens.* **2018**, *10*, 882. [CrossRef]

26. Kim, K.; Park, J.; Baik, J.; Choi, M. Evaluation of topographical and seasonal feature using GPM IMERG and TRMM 3B42 over Far-East Asia. *Atmos. Res.* **2017**, *187*, 95–105. [CrossRef]

27. Zhong, R.; Chen, X.; Lai, C.; Wang, Z.; Lian, Y.; Yu, H.; Wu, X. Drought monitoring utility of satellite-based precipitation products across mainland China. *J. Hydrol.* **2019**, *568*, 343–359. [CrossRef]

28. Wu, L.; Xu, Y.; Wang, S. Comparison of TMPA-3B42RT Legacy Product and the Equivalent IMERG Products over Mainland China. *Remote Sens.* **2018**, *10*, 1778. [CrossRef]

29. Tan, M.L.; Santo, H. Comparison of GPM IMERG, TMPA 3B42 and PERSIANN-CDR satellite precipitation products over Malaysia. *Atmos. Res.* **2018**, *202*, 63–76. [CrossRef]

30. Hussain, Y.; Satgé, F.; Hussain, M.B.; Martinez-Carvajal, H.; Bonnet, M.-P.; Cárdenas-Soto, M.; Roig, H.L.; Akhter, G. Performance of CMORPH, TMPA, and PERSIANN rainfall datasets over plain, mountainous, and glacial regions of Pakistan. *Theor. Appl. Climatol.* **2018**, *131*, 1119–1132. [CrossRef]

31. Palomino-Ángel, S.; Anaya, J.A.; Botero, B.A. Evaluation of 3B42V7 and IMERG daily-precipitation products for a very high-precipitation region in northwestern South America. *Atmos. Res.* **2019**, *217*, 37–48. [CrossRef]

32. Retalis, A.; Katsanos, D.; Tymvios, F.; Michaelides, S. Validation of the First Years of GPM Operation over Cyprus. *Remote Sens.* **2018**, *10*, 1520. [CrossRef]

33. Saber, M.; Hamaguchi, T.; Kojiri, T.; Tanaka, K.; Sumi, T. A physically based distributed hydrological model of wadi system to simulate flash floods in arid regions. *Arab. J. Geosci.* **2015**, *8*, 143–160. [CrossRef]

34. Milewski, A.; Elkadiri, R.; Durham, M. Assessment and Comparison of TMPA Satellite Precipitation Products in Varying Climatic and Topographic Regimes in Morocco. *Remote Sens.* **2015**, *7*, 5697–5717. [CrossRef]

35. Milewski, A.; Seyoum, W.M.; Elkadiri, R.; Durham, M. Multi-Scale Hydrologic Sensitivity to Climatic and Anthropogenic Changes in Northern Morocco. *Geosciences* **2019**, *10*, 13. [CrossRef]

36. Benkirane, M.; Laftouhi, N.-E.; El Mansouri, B.; Salik, I.; Snineh, M.; El Ghazali, F.E.; Kamal, S.; Zamrane, Z. An approach for flood assessment by numerical modeling of extreme hydrological events in the Zat watershed (High Atlas, Morocco). *Urban Water J.* **2020**, *17*, 381–389. [CrossRef]

37. Chen, H.; Chandrasekar, V.; Tan, H.; Cifelli, R. Rainfall Estimation from Ground Radar and TRMM Precipitation Radar Using Hybrid Deep Neural Networks. *Geophys. Res. Lett.* **2019**, *46*, 10669–10678. [CrossRef]

38. Zhang, Z.; Tian, J.; Huang, Y.; Chen, X.; Chen, S.; Duan, Z. Hydrologic Evaluation of TRMM and GPM IMERG Satellite-Based Precipitation in a Humid Basin of China. *Remote Sens.* **2019**, *11*, 431. [CrossRef]

39. Duan, Z.; Liu, J.; Tuo, Y.; Chiogna, G.; Disse, M. Evaluation of eight high spatial resolution gridded precipitation products in Adige Basin (Italy) at multiple temporal and spatial scales. *Sci. Total Environ.* **2016**, *573*, 1536–1553. [CrossRef]

40. Wu, Y.; Zhang, Z.; Huang, Y.; Jin, Q.; Chen, X.; Chang, J. Evaluation of the GPM IMERG v5 and TRMM 3B42 v7 Precipitation Products in the Yangtze River Basin, China. *Water* **2019**, *11*, 1459. [CrossRef]

41. Li, X.; Zhang, Q.; Xu, C.-Y. Assessing the performance of satellite-based precipitation products and its dependence on topography over Poyang Lake basin. *Theor. Appl. Climatol.* **2013**, *115*, 713–729. [CrossRef]

42. El Khalki, E.M.; Tramblay, Y.; Saidi, M.E.M.; Bouvier, C.; Hanich, L.; Benrhanem, M.; Alaouri, M. Comparison of modeling approaches for flood forecasting in the High Atlas Mountains of Morocco. *Arab. J. Geosci.* **2018**, *11*, 410. [CrossRef]

43. Tramblay, Y.; Bouaicha, R.; Brocca, L.; Dorigo, W.; Bouvier, C.; Camici, S.; Servat, E. Estimation of antecedent wetness conditions for flood modelling in northern Morocco. *Hydrol. Earth Syst. Sci.* **2012**, *16*, 4375–4386. [CrossRef]

Article

An Algorithm to Generate a Weighted Network Voronoi Diagram Based on Improved PCNN

Xiaomin Lu [1,2] and Haowen Yan [1,2,*]

[1] Faculty of Geomatics, Lanzhou Jiaotong University, Lanzhou 730070, China; xiaominlu08@mail.lzjtu.cn
[2] Gansu Provincial Engineering Laboratory for National Geographic State Monitoring, Lanzhou 730070, China
* Correspondence: yanhw@mail.lzjtu.cn; Tel.: +86-136-0931-0452

Abstract: The network Voronoi diagram has been extensively applied in many fields, such as influence area construction, location selection and urban planning, owing to its high accuracy and validity in space division. Taking advantage of parallel processing and auto-wave division of the pulse coupled neural network (PCNN), an algorithm for generating a weighted network Voronoi diagram is proposed in this paper. First, in order to better accommodate the scenes of urban facility points and road networks, the PCNN is improved. Second, the speed of the auto-wave in the improved PCNN is calculated by the weights of the facility points and the attributes of the related road network. Third, the nodes in the road network are considered as neurons, the facility points are projected onto the nearest road segments and the projected points are treated as initial neurons. The initial neurons generate auto-waves simultaneously, and the auto-waves transmit along the shortest path from neurons to other neurons with the calculated speed until all the neurons are fired. During this procedure, the road network and the corresponding space are assigned to the initial neurons and the weighted network Voronoi diagram is constructed. The experiments on the specific region with the real POIs present the feasibility, applicability and efficiency of the algorithm.

Keywords: map generalization; weighted network Voronoi diagram; point cluster simplification; network Voronoi polygon

Citation: Lu, X.; Yan, H. An Algorithm to Generate a Weighted Network Voronoi Diagram Based on Improved PCNN. *Appl. Sci.* **2022**, *12*, 6011. https://doi.org/10.3390/app12126011

Academic Editors: Giovanni Randazzo, Anselme Muzirafuti and Stefania Lanza

Received: 24 April 2022
Accepted: 10 June 2022
Published: 13 June 2022

1. Introduction

As an excellent tool for spatial analysis and spatial optimization, the Voronoi diagram has been applied in various fields [1–3]. Originally, it was defined through a set of n points $P = \{P_1, \ldots, P_n\}$ (termed generator points or generators) on the plane, where the Euclidean distance between an arbitrary point p_i and a generator point p is $d(p, p_i)$ [4–6]. In these terms, the ordinary planar Voronoi diagram is defined as a set of polygons, $\mathrm{Vor} = \{\mathrm{Vor}_1, \ldots, \mathrm{Vor}_n\}$, where the polygon Vor_i is given by

$$Vor_i = \left\{ p \mid d(p, p_i) \leq d(p, p_j), j \neq i, j = 1, \ldots, n \right\} \tag{1}$$

An example of a Voronoi diagram is illustrated in Figure 1. It can be seen that the distance between points is computed by Euclidean distance in the planar Voronoi diagram.

However, in the real-world cases, the accessibility between points that stand for geographic objects are generally constrained by road networks [7,8]. The Euclidean-distance-based Voronoi diagram is less proper as a subdivision of network space. In the literature, although there are many phenomena that occur on a network or alongside a network, they are often analyzed based on the planar Voronoi diagram. Therefore, it is significant to propose methods to build applicable Voronoi diagrams in the network space.

The network Voronoi diagram replaces the plane with network space (such as a road network) compared with the planar Voronoi diagram. Correspondingly, distances defined on a network (such as the shortest path distance between two points) [6,9] are applied instead of the Euclidean distance, which can be defined as:

$$\text{Vor} = \{\text{Vor}_1, \text{Vor}_2, \dots, \text{Vor}_n\} \tag{2}$$

where $\text{Vor}_i = \{P \mid d(p, p_i) \leq d(p, p_j)\}$; $d(p, p_i)$ is the distance between an arbitrary point p_i and a generator point p in the network [10–12]. It is a better choice for spatial division and analysis [6], and it has been widely utilized. Harn et al. (2021) found an optimal partition that minimizes the overall delivery time of all trucks based on the weighted road network Voronoi diagram [13]. Gotoh et al. (2017) proposed and evaluated a searching scheme for a biochromatic reverse k-nearest neighbor with a network Voronoi diagram [14]. Xie et al. (2011) studied the maximal covering spatial optimization based on network Voronoi diagram [15]. Tu et al. (2014) proposed the method for logistics routing optimization based on a network Voronoi diagram [16]. Miller (1994) presented a method for market area delimitation [17].

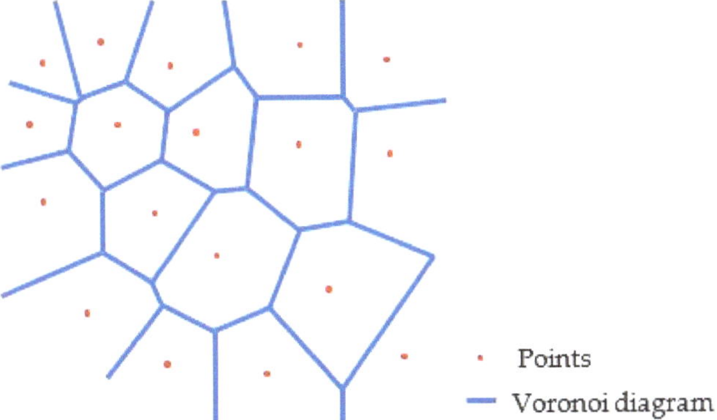

Figure 1. An example of the planar Voronoi diagram.

In this study, we propose an algorithm for dynamically constructing network Voronoi diagrams. The approach attempts to take advantage of the properties of concurrency and auto-wave distribution of the PCNN. The basic idea is that auto-waves are generated from initial neurons and transmit along the shortest path between neurons at a calculated speed until all the neurons are fired and the transmitting of auto-wave stops.

The organization of this paper is as follows. Section 2 introduces the improved PCNN. In Section 3, the procedures of generating the weighted network Voronoi diagram on the basis of the improved PCNN are described. Experiments are illustrated and discussed in Section 4. Finally, conclusions and future works are presented in Section 5.

2. Related Work

Currently, several achievements have been made in the construction of network Voronoi diagrams. Ai et al. (2015) presented a constrained network Voronoi diagram using stream flowing ideas [11]. The method by Tan et al. (2012) calculates the network distance using the shortest-path tree technique [18]. Okabe et al. (2008) formulated six types of generalized network Voronoi diagram [6]. In existing studies, the typical method is the algorithm based on the shortest path distance, in which the "extended shortest path trees" are built and the weight shortest path distance are considered to generate the network Voronoi diagram. Inspired by the extension operators in mathematical morphology used in computing ordinary Voronoi diagrams [19–21], this paper proposes an algorithm for constructing the weighted network Voronoi diagram by integrating the ideas of the shortest path distance and extension operators.

The standard pulse coupled neural network (PCNN) is a type of neural network with a biology background, which is developed for image processing and pattern recognition. It has been used in research fields such as image processing. Deng et al. (2022) obtained smooth and unbroken single pixels based on it [22]. Yang et al. constructed a novel model of sparse representation for image denoising based on improved PCNN [23]. Basar et al. (2022) proposed a novel and robust defocus-blur segmentation scheme consisting of a local ternary pattern (LTP) measured alongside the pulse coupled neural network (PCNN) technique [24]. A new fusion algorithm of infrared and color visible images based on multi-scale transformation and adaptive PCNN was proposed by Shen et al. (2021) [25]. We found that its properties of concurrency and auto-wave distribution are helpful to establishing the network Voronoi diagram [14], especially solving the pairs shortest path problem. The auto-wave distribution can be applied into the situation of the generator growing in all directions simultaneously, which is similar to the previously mentioned extension operators. On the other hand, its characteristics of concurrency can improve the efficiency of the algorithm. However, before the model is applied to the process of constructing the network Voronoi diagram, it should be adjusted in several aspects, because the iteration process is complicated and it is difficult to control the auto-wave transmission in the standard PCNN.

3. Improved PCNN

3.1. Structure of the PCNN

The standard PCNN, also known as the third artificial neural network, is a laterally connected feedback network of pulse coupled neurons without requiring any training [26,27]. The standard neuron model is given by Figure 2 and the following equations [28–30]:

$$F_{ij}[n] = e^{-\alpha_F} F_{ij}[n-1] + V_F \sum_{kl} M_{ijkl} Y_{kl}[n-1] + I_{ij} \tag{3}$$

$$L_{ij}[n] = e^{-\alpha_L} L_{ij}[n-1] + V_L \sum_{kl} W_{ijkl} Y_{kl}[n-1] \tag{4}$$

$$U_{ij}[n] = F_{ij}[n]\left(1 + \beta L_{ij}[n]\right) \tag{5}$$

$$Y_{ij}[n] = \begin{cases} 1 & U_{ij}[n] > E_{ij}[n] \\ 0 & otherwise \end{cases} \tag{6}$$

$$E_{ij}[n+1] = e^{-\alpha_E} E_{ij}[n] + V_E Y_{ij}[n] \tag{7}$$

where the index (i, j) and index (k, l) refer to the current neuron and its neighbors. I denotes the external stimulus. F is feeding input, and L is linking input in iteration n. M and W represent linking synapse weights. Internal activity U is generated by the modulation of F and L through linking strength β. The neuron will be stimulated when the internal activity U is greater than the dynamic threshold E. In addition, V_F, V_L and V_E are normalizing constants; and the parameters α_F, α_L and α_E are the time constants [31].

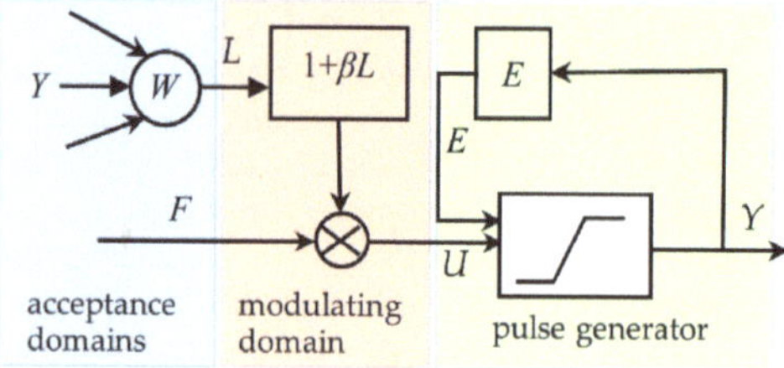

Figure 2. Structure of the PCNN model.

Drawn by its characteristics of auto-wave and concurrency, we planned to incorporate the PCNN into the construction process, but in the standard PCNN model, the iteration process is complex and the transmission method of auto-wave is hard to control. In order to make the PCNN more suitable for the construction of the network Voronoi diagram, an improved PCNN is proposed.

3.2. Structure of the Improved PCNN

The principle of the improved PCNN can be summarized in the following three aspects: (1) all points in the point cluster and all nodes in the road network are considered as the neurons, in which the points in the point cluster are treated as initial neurons; (2) the initial neurons will generate an auto-wave simultaneously, and the auto-wave transmits along the road segments from a neuron to another at a certain speed; (3) the transmission speed of the auto-wave on a certain road segment is determined by the importance of its initial neuron and the attributes of the passed road segments.

The structure of the improved model is shown in Figure 3, and it can be represented with the following equations.

$$E_{p_m}[n+1] = \begin{cases} V_E & \text{if } Y_p[n] = 1 \\ \min(S[n] + W_{qp}[n], E_p[n]) & \text{if } Y_p[n] = 0, Y_q[n] = 1 \text{ and } p \in R_q \\ E_p[n] & \text{otherwise} \end{cases} \tag{8}$$

$$S[n+1] = S[n] + \Delta S[n] \tag{9}$$

$$Y_p[n+1] = \begin{cases} 1 & \text{if } S[n+1] \ge E_p[n+1] \\ 0 & \text{otherwise} \end{cases} \tag{10}$$

$$W_{qp}[n+1] = \begin{cases} V_E & \text{if } Y_p[n+1] = 1 \text{ and } p \in R_q \\ W_{qp}[n] & \text{otherwise} \end{cases} \tag{11}$$

where n is the iteration number and E_p stands for the dynamic threshold of neuron p. V_E represents a constant with a large value. ΔS is the speed of the current auto-wave. S is the distance between the initial neuron and the current iteration. Y_P is determined by the comparison between the current wavelength S and the threshold E_p. W_{pq} represents the connection weight between neuron p and neuron q. R_q is the neighbor set of the neuron q.

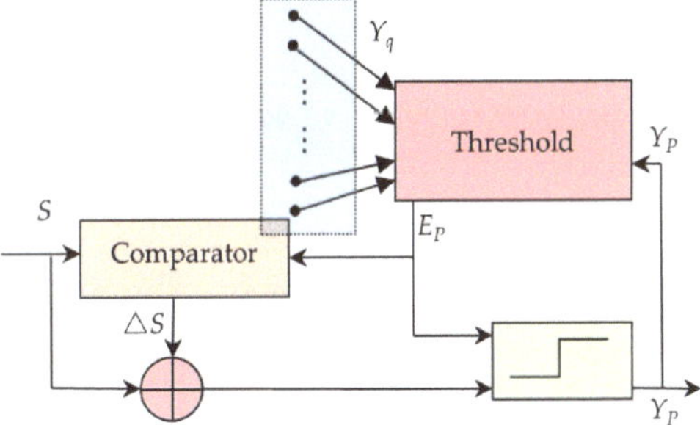

Figure 3. Structure of the improved PCNN.

4. Network Voronoi Diagram Construction Based on the Improved PCNN

The construction of the planar Voronoi diagram can be described as follows: each entity within the Voronoi diagram expands outward to gain growth area and finally encounters the expansion from neighboring entities [16,17]. Similarly, in the process of generating the network Voronoi diagram, each initial neuron generates an auto-wave simultaneously, and the auto-waves compete the growth area by expanding along the shortest path between neurons. The competition will not stop until all the neurons are fired and the transmitting auto-wave stops. As a result, the routes and the area taken by generators delineate their impacting range in network space, which can be considered as network Voronoi diagram of the certain point set.

The concrete steps of the construction can be concluded as: Firstly, the initialization is down so that all facility points can be projected onto the related road segments. Secondly, the projected points act as initial neurons in the improved PCNN, and generate auto-wave simultaneously. The speed of auto-wave is calculated and then the auto-wave transmits along the shortest path between neurons. Finally, the road segments conquered by a specific auto-wave and their related space are identified as its initial neuron's impacting region. Until all the nodes are fired, the network Voronoi diagram of corresponding facility points is successfully generated. The flowchart of the algorithm is shown in Figure 4, and the following describes the procedure in detail.

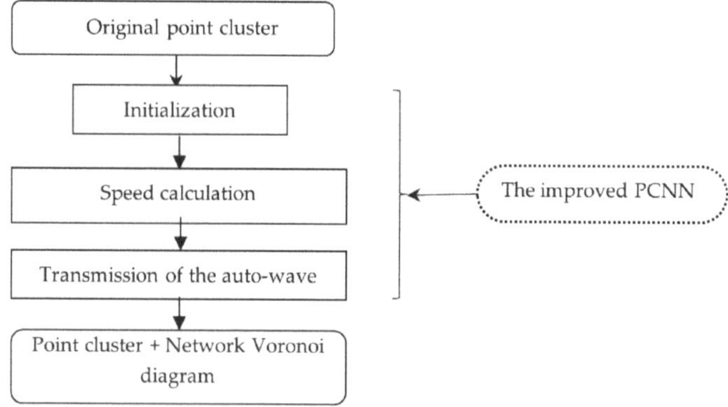

Figure 4. Flowchart of the algorithm.

As shown in Figure 4, the new algorithm includes three procedures:

– Initialization;
– Speed calculation;
– Transmission of the auto-wave.

4.1. Initialization

On the map, the points do not exactly lie on the road segments because the roads are usually represented as lines. In order to match the points with the road segments, the initialization of the improved PCNN model includes points' projection and the establishment of the connection relationship matrix between neurons.

4.1.1. Points' Projection

The first step of the algorithm is projecting all the points in the point cluster onto the road segments, aiming at locating the positions of initial neurons (points in the point cluster) on the road network in the improved PCNN model. As shown in Figure 5, perpendiculars from points P_1–P_5 to the nearest road segments are able to be found. The intersection (marked as P_1'–P_5') between the perpendiculars and the road segments can be regarded as the projection of the points.

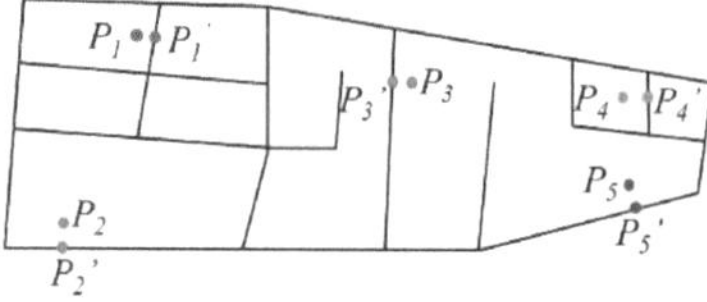

Figure 5. Projection of the points.

4.1.2. Connection Relationship Matrix between Neurons

The key of the algorithm is that the auto-waves emitted from different points broadcast along the shortest path between neurons. In order to find the shortest path between neurons, the connection relationship matrix, which stores the distances between neurons, should be constructed at first, as shown in Formula (12).

$$W = \begin{bmatrix} w_{11} & w_{12} & \cdots & w_{1n} \\ w_{21} & w_{22} & \cdots & w_{2n} \\ \cdots & & & \cdots \\ w_{n1} & w_{n2} & \cdots & w_{nn} \end{bmatrix} \tag{12}$$

where n is the total number of the neurons. The weight between neuron i and j is denoted as w_{ij}, which assigns the length of the road if there is connection between neuron i and j, or infinity if there is no connection between neuron i and j.

4.2. Transmission of the Auto-Wave

After initialization, the auto-wave will be generated from the initial neurons and then transmit along the shortest distance from the current neuron to the next neuron. In this procedure, the transmission speed of the auto-wave is computed based on the importance of initial neurons and attributes of road segments. The speed needs to be calculated at first, and then transmission will be done.

4.2.1. Speed Calculation

In real traffic networks, some roads are one-way and some roads are two-way, which play different roles in the transportation system. Similarly, the facilities themselves also have different importance with respect to different functions, which can affect the impacting regions of facilities. Therefore, these factors should be considered while computing transmission speed. For instance, the auto-wave will transmit quicker to compete for a larger growth area if it is generated from an important point and expand along an important road. These impact factors are illustrated specifically as follows:

- Direction of the roads

The auto-wave can only expand along the road if its expanding direction is the same as the direction of the road. Otherwise, the transmission of the auto-wave should be terminated. For example, if the road merely permits moving from west to east, the auto-wave expanding from east to west must stop on this road. This can be expressed by Formula (13):

$$D_R = F(N_1, N_2) \tag{13}$$

where D_R is the direction factor of the current road segment. $F(N_1, N_2)$ is a direction function which equals to 1 if the auto-wave is the same as the direction of the road and equals to 0 otherwise.

- Grades of the roads

As we all know, the traffic capacities of roads are different. Generally, a road of a higher grade has a higher traffic capacity, and the auto-wave can transmit quicker along these kinds of roads correspondingly. Formula (14) is defined to represent the differences:

$$G_R = W(R) \tag{14}$$

where G_R is grade factor of the current road segment, and function $W(R)$ denotes the grade function of the road segment.

- Grades of the points

The higher the grade of a point is, the larger its impact range, which can be represented as Formula (15):

$$G_P = W(N) \tag{15}$$

where G_P is the grade factor of the current point, and function $W(N)$ denotes the importance of the initial neuron.

By considering all above factors, the transmitting speed can be defined as Formula (16), where k is the speed of the auto-wave which is generated from initial neurons, weight 1, and transmits along the road segment of weight 1.

$$L_{RN} = k \times D_R \times G_R \times G_P = k \times F(N_1, N_2) \times W(R) \times W(N) \tag{16}$$

As shown in Figure 6, if the unit speed of the auto-wave (generated by neurons of weight 1 and expanding on the road segment of weight 1) is set as 10 m per iteration, the speed of the auto-wave generated from point P is 20 m per iteration (V_{p1}) when expanding along road segment PN_2, and 0 m per iteration (V_{p2}) when expanding along road segment PN_1. V_{Q1} (the speed of the auto-wave generated from point Q and expanding along road segment QN_6) is 10 m per iteration, and VQ_2, which represents the speed on road segment N_6N_5, is 20 m per iteration.

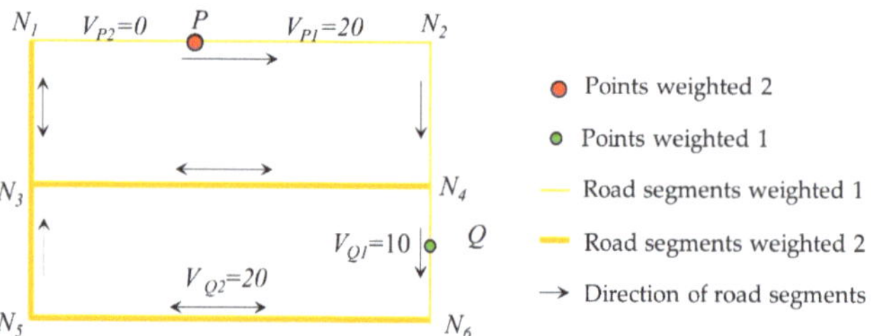

Figure 6. The speed of the auto-wave under constraints.

4.2.2. Transmission of the Auto-Wave

In the improved PCNN, the auto-wave which is generated simultaneously by initial neurons will expand along the shortest path between neurons with the above-calculated speed, until all the neurons are fired and the transmitting auto-wave stops. The concrete steps are described as follows:

(1) Initialization of networks' parameters

The initial iteration number is set as $n = 0$. The projected points in point cluster and all the nodes in road network are regarded as neurons, and the projected points are chosen as initial neurons (b_m) whose initial state is set to be fired ($Y_{bm}[0] = 1$). The initial neurons are regarded as the mother neurons of other neurons ($par(p) = b_m$). V_E is a constant satisfying condition: $V_E > (n-1)\,W_{max}$, in which N is the total number of the neurons, and W_{max} represents the maximum value of the connection weight between neurons. The initial dynamic threshold of iteration is set as $E_P[0] = V_E$, $Y_P[0] = 0$, $S[0] = 0$, $\triangle S[0] = 0$.

(2) Operation of the improved PCNN

Step1: Dynamic threshold of iteration ($E_P[n]$) is calculated by the Formula (8).

Step2: The transmission speed of the auto-wave (L_{Bn}) is updated according to Formula (16), and the current wave velocity is set as $\triangle S[n] = L_{Bn}$.

Step3: The current wavelength is calculated by Formula (9) as: $S[n] = S[n-1] + \triangle S[n]$.

Step4: The current dynamic threshold ($E_P[n]$) is compared with the current wavelength ($S[n]$). If $S[n] \geq E_P[n]$, then $Y_P[n] = 1$, which means neuron P is fired successfully in the nth iteration. Otherwise, $Y_P[n] = 0$ and the current neuron P is not fired yet.

Step5: The connection weight $W_{qp}[n+1]$ is updated by Formula (11).

Step6: Turn to step 1 to repeat the above steps, until all the neurons are fired and the transmitting auto-wave stops.

The initial state of the points and the road network is shown as Figure 7a. If the weights of the point and the road segments are ignored, which means all the points have the same importance and all the road segments are two-way and have the same importance, the end state of auto-wave transmission is shown as Figure 7b. When the weights of the points and the road networks are all considered and the auto-wave transmits with different speed on different road segments, the end state is shown in Figure 7c. In Figure 7c, the auto-wave generated from the projection points expands along road segments at the speed calculated by Formula (9). It will transmit along the shortest path between neurons until all the neurons are fired. In Figure 7b,c, the black dotted lines represent the road segments which were not passed by the auto-wave. The appearance of this type of road segment is because the auto-wave travels along the shortest path between neurons in the improved PCNN. As soon as the neurons ahead are fired (i.e., the shortest path toward the current neurons is found already), the current auto-wave that travels toward them will stop. For example, in Figure 7c, the auto-wave generated from point P_3' (marked in purple) stops at point r_2, because at that very moment the neuron ahead of the auto-wave (r_2) is fired by another auto-wave (marked in red). In the end, $r_1 r_2$ is the road segment that is never reached by any auto-wave. This kind of road segment is assigned to two initial neurons if the two nodes of the road segment are fired by auto-waves generated from two initial neurons ($r_1 r_2$ for example). Otherwise, the road segments are assigned to the only initial neurons ($r_3 r_4$ for example). According to this method, these untouched road segments are assigned to the corresponding neurons, and the result is as shown in Figure 8. It can be seen that the road segments (marked in different color) and the corresponding space are assigned to different points.

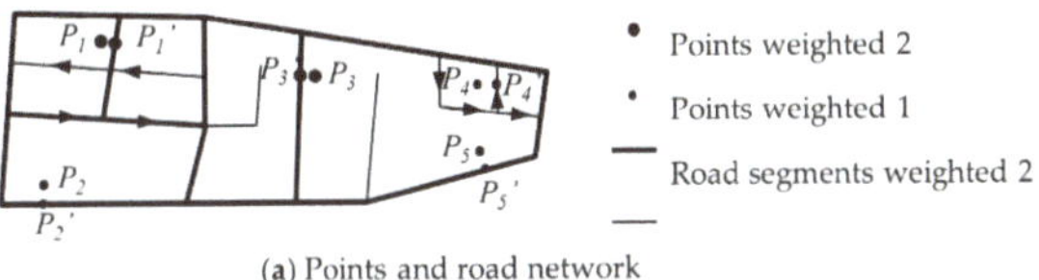

(a) Points and road network

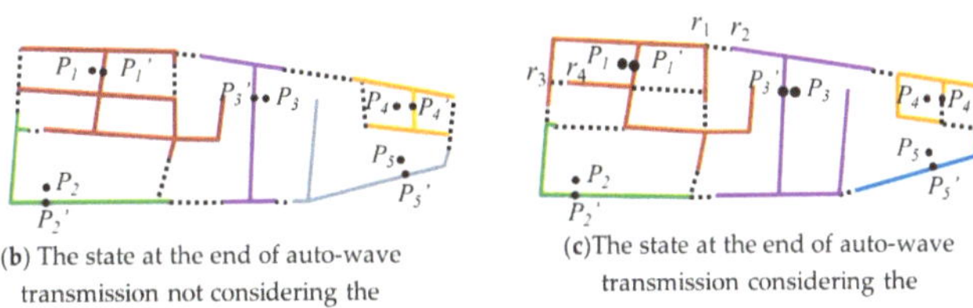

(b) The state at the end of auto-wave transmission not considering the

(c)The state at the end of auto-wave transmission considering the

Figure 7. Transmission of auto-waves.

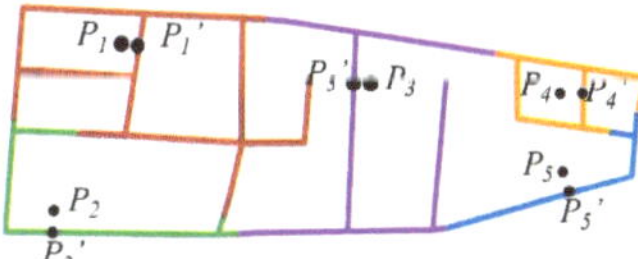

Figure 8. The final state of auto-wave transmission.

4.3. The Construction of the Weighted Network Voronoi Diagram

As described above, the construction of the weighted network Voronoi diagram can be summarized as the following steps:

(1) Preprocess the network data: ① All the nodes are found and regarded as the neurons in the improved PCNN, and the road network is divided by the nodes into road segments. ② The point cluster is projected onto the road segments, and the projection points are set as initial neurons. ③ The connection relationship matrix between neurons is computed.

(2) Perform transmission based on the improved PCNN: The auto-waves generated from the initial neurons expand along the corresponding road segments with the speed calculated by the Formula (9); they look for and travel along the shortest paths between neurons simultaneously. The auto-wave will not stop until all the neurons ahead have been fired.

(3) Repeat (2) until all the neurons are fired and the transmitting auto-waves stop.

(4) Road segments that are not used by any auto-wave are assigned to the corresponding initial neurons. So far, the road network and the corresponding space are assigned to the initial neurons and the weighted network Voronoi diagram has been constructed.

As shown in Figure 8, the network space is distributed to different points according to the weights of the points and the attribution of the road segments by the shortest path principle.

5. Experiment Studies and Discussion

5.1. Experiments

To illustrate the soundness of the proposed algorithm, experiments were conducted on the data of a district in a big city, which consists of 2820 road segments (50 national highways, 44 provincial ways, 892 urban first-grade ways, 1834 urban second-grade ways) and 148 educational institutions in three grades. The initial state is shown in Figure 9a. The state at the beginning of the auto-wave transmission is shown as Figure 9b, in which the auto-waves generated from different initial neurons are marked in different colors. Figure 9c shows the final state of auto-wave transmission, where the road segments that are not traveled along by the auto-waves are assigned to the corresponding initial neurons. Finally, the weighted network Voronoi diagram is generated as shown in Figure 9d.

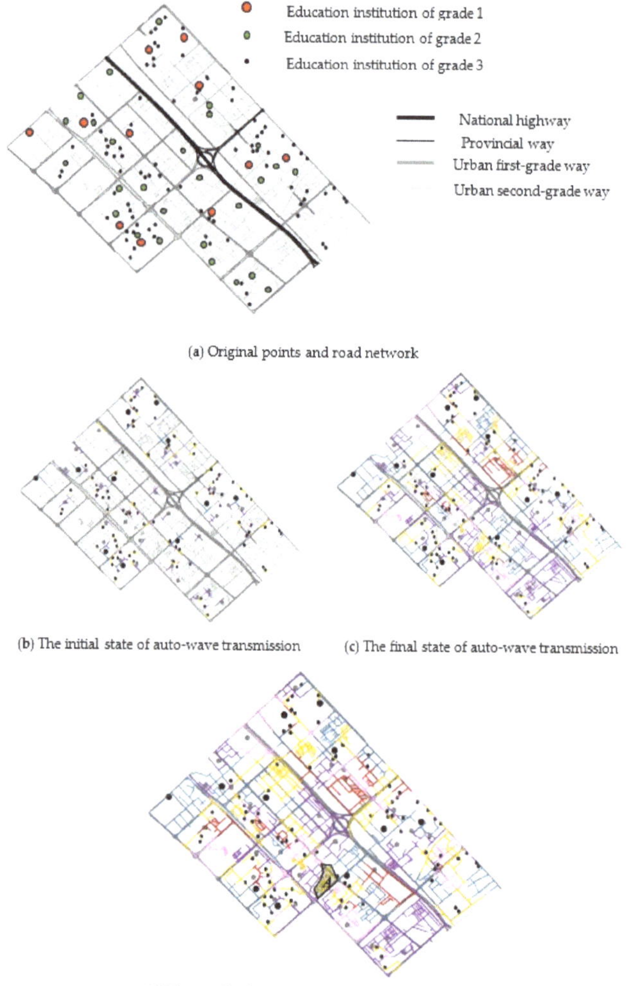

(a) Original points and road network

(b) The initial state of auto-wave transmission

(c) The final state of auto-wave transmission

(d) The weighted network Voronoi diagram of the point cluster

Figure 9. The generation of the weighted network Voronoi diagram of points in the study area.

For comparison, experiments using ordinary weighted Voronoi diagram were performed on the same point cluster (Figure 10).

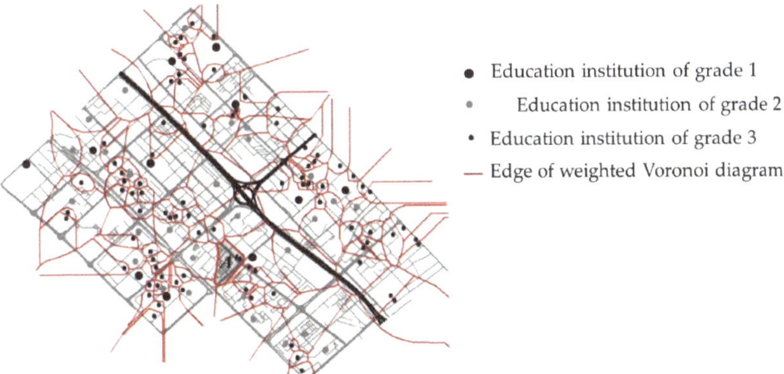

Figure 10. Weighted Voronoi diagram of points in the study area.

5.2. Discussion

It can be seen in Figures 9d and 10 that: (1) the ordinary weighted Voronoi diagram cut the network planar smoothly, and the weighted network Voronoi diagram cut the network planar unevenly, for the road network is usually unevenly distributed and there is a slight chance to connect original facility points by a straight line. (2) The ordinary weighted Voronoi diagram ignores the influence of the road network where the point cluster is located; thus, the final results would be the same even if the point cluster were put into different road networks. However, the weighted network Voronoi diagram is constructed while fully considering the attributes of road networks, and the importance of points, which ends up providing results more suitable to real life. (3) The weighted network Voronoi diagram is close to the ordinary weighted Voronoi diagram when the road network around the point is dense enough. For instance, the shaded areas (A in Figure 9d and A' in Figure 10), which represent the impacting regions of the same points calculated by the two methods, are similar in both shape and size, due to the large density of the road network related to the point. If the following three assumptions stand—(1) the road network density is large enough and any two facility points can be connected by roads similar to straight lines, (2) all the roads in road network are of the same weight and (3) all the roads in road network are two-way—it can be inferred that the weighted network Voronoi diagram will be nearly the same as the ordinary weighted Voronoi diagram. However, in fact, the road networks are usually unevenly distributed, and different roads in road network always have different attributes, such as grades and directions. Therefore, it is doubtless that the weighted network Voronoi diagram is a better choice for the representation of the impact scope of a point cluster rather than the ordinary weighted Voronoi diagram.

A comparison experiment based on the idea of the shortest path was performed on the same data. The weighted network Voronoi diagram was constructed using the tool SANET (http://sanet.csis.u-tokyo.ac.jp/), as shown in Figure 11b. It can be deduced that the method based on the idea of the shortest path ignores some of influence factors, such as the attributes of the related road networks. For instance, the road segments next to the points P_1 and P_2 are of higher grade in Figure 11a, so the auto-waves transmit faster along them, resulting in larger impacting regions of P_1 and P_2 than in Figure 11b. In addition, as shown in Table 1, experiments with different unit speed were performed, and performance parameters of the proposed algorithm for constructing weighted network Voronoi diagrams are described in Table 1. It can be seen that the iteration number is directly related to the transmission speed of the auto-wave, which can be adjusted according to the practical situation. It can be seen that the time complexity of the algorithm mainly depends on the transmission speed of the auto-wave in the improved PCNN, which is able to be adjusted according to the attributes of the point cluster and the related road network to meet the requirements of accuracy and efficiency. For example, the unit speed of the auto-wave can be set a larger value if the average distance between neurons is small, or a smaller value if the higher accuracy is demanded.

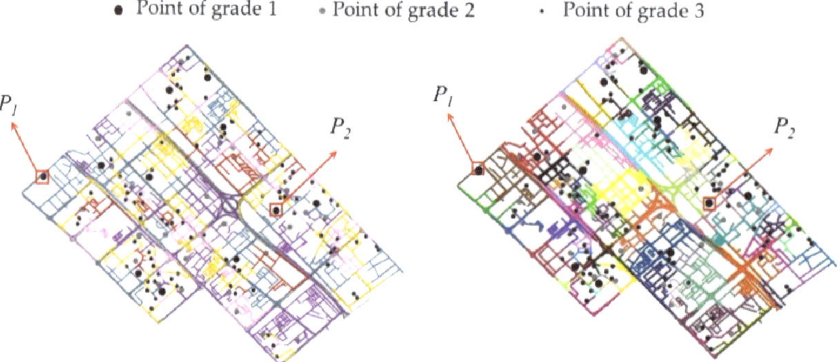

(**a**) Weighted network Voronoi diagram using the model of improved PCNN

(**b**) Weighted network Voronoi diagram based on the idea of the shortest path

Figure 11. A comparison between the weighted network Voronoi diagram using the model of improved PCNN and that based on the idea of the shortest path.

Table 1. Performance parameters of the algorithm for constructing weighted network Voronoi diagrams.

Experiment	Unit Speed of the Auto-Wave (m)	Number of Iterations	Construction Time(s)
Figure 9	1	1323	9.3
	5	321	2.1

6. Conclusions

This paper proposed an algorithm with which to generate weighted network Voronoi diagrams based on improved PCNN. Experiments showed that: (1) owing to the idea of parallel processing and shortest path transmission of the improved PCNN, the construction process is intuitive and conforms to the basic ideas of the original Voronoi diagram but with a different distance metric. (2) The influences of the point cluster and the related road network are taken into account as weights in the computation, and the constructed weighted network Voronoi diagram can better define the boundaries of the service region of the corresponding point.

The improved PCNN and the algorithm can also be extended to linear facilities and polygon facilities; and it will be our future work to introduce more semantic information into the construction, based on which, the weighted network Voronoi diagram for linear and polygon facilities will be constructed.

Author Contributions: X.L. and H.Y. proposed the methodology. X.L. performed the experiments and wrote the draft of the manuscript. H.Y. guided the research and revised the manuscript. All the authors contributed to the development of the proposed generalization algorithm and this manuscript. All authors have read and agreed to the published version of the manuscript.

Funding: This research was funded by the National Nature Science Foundation of China, grant numbers 42161066 and 41801395, and the Nature Science Foundation of Gansu province, grant number 20JR10RA246.

Institutional Review Board Statement: Not applicable.

Informed Consent Statement: Informed consent was obtained from all subjects involved in the study.

Data Availability Statement: Not applicable.

Acknowledgments: The authors are grateful to anonymous reviewers, whose comments and suggestions have helped us to improve the context and presentation of the article.

Conflicts of Interest: The authors declare no conflict of interest.

References

1. Gold, C.M. Review: Spatial Tessellations-Concepts and Applications of Voronoi Diagrams. *Int. J. Geogr. Inf. Sci.* **1994**, *8*, 237–238.
2. Dumic, E.; Bjelopera, A.; Nüchter, A. Dynamic Point Cloud Compression Based on Projections, Surface Reconstruction and Video Compression. *Sensors* **2022**, *22*, 197. [CrossRef] [PubMed]
3. Huang, S.-K.; Wang, W.-J.; Sun, C.-H. A Path Planning Strategy for Multi-Robot Moving with Path-Priority Order Based on a Generalized Voronoi Diagram. *Appl. Sci.* **2021**, *11*, 9650. [CrossRef]
4. Polianskii, V.; Pokorny, F.T. Voronoi Graph Traversal in High Dimensions with Applications to Topological Data Analysis and Piecewise Linear Interpolation. In Proceedings of the KDD'20: The 26th ACM SIGKDD Conference on Knowledge Discovery and Data Mining, Virtual Event, 6–10 July 2020.
5. Zhao, X.; Chen, J.; Wang, J. O-QTM based Algorithm for the Generating of Voronoi Diagram for Spherical Objects. *Acta Geod. Cartogr. Sin.* **2002**, *31*, 158–163.
6. Okabe, A.; Satoh, T.; Furuta, T.; Suzuki, A.; Okano, K. Generalized Network Voronoi Diagrams: Concepts, Computational Methods and Applications. *Int. J. Geogr. Inf. Sci.* **2008**, *22*, 965–994. [CrossRef]
7. Randazzo, G.; Cascio, M.; Fontana, M.; Gregorio, F.; Lanza, S.; Muzirafuti, A. Mapping of Sicilian Pocket Beaches Land Use/Land Cover with Sentinel-2 Imagery: A Case Study of Messina Province. *Land* **2021**, *10*, 678. [CrossRef]
8. Lage, M.d.O.; Machado, C.A.S.; Monteiro, C.M.; Davis, C.A., Jr.; Yamamura, C.L.K.; Berssaneti, F.T.; Quintanilha, J.A. Using Hierarchical Facility Location, Single Facility Approach, and GIS in Carsharing Services. *Sustainability* **2021**, *13*, 12704. [CrossRef]
9. Kiani, F.; Seyyedabbasi, A.; Nematzadeh, S.; Candan, F.; Çevik, T.; Anka, F.A.; Randazzo, G.; Lanza, S.; Muzirafuti, A. Adaptive Metaheuristic-Based Methods for Autonomous Robot Path Planning: Sustainable Agricultural Applications. *Appl. Sci.* **2022**, *12*, 943. [CrossRef]
10. Ai, T.; Yu, W. Algorithm for Constructing Network Voronoi Diagram Based on Flow Extension Ideas. *Acta Geod. Cartogr. Sin.* **2013**, *42*, 760–766.
11. Ai, T.; Yu, W.; He, Y. Generation of constrained network Voronoi diagram using linear tessellation and expansion method. *Comput. Environ. Urban Syst.* **2015**, *51*, 83–96. [CrossRef]
12. Liu, F.; Andrienko, G.; Andrienko, N.; Chen, S.; Janssens, D.; Wets, G.; Theodoridis, Y. Citywide Traffic Analysis Based on the Combination of Visual and Analytic Approaches. *J. Geovisualization Spat. Anal.* **2020**, *4*, 15. [CrossRef]
13. Harn, P.; Zhang, J.; Shen, T.; Wang, W.; Jiang, X.; Ku, W.-S.; Sun, M.-T.; Chiang, Y.-Y. Multiple ground/aerial parcel delivery problem: A Weighted Road Network Voronoi Diagram based approach. *Distrib. Parallel Databases* **2021**, 1–21. [CrossRef]
14. Gotoh, Y.; Okubo, C. A proposition of querying scheme with network Voronoi diagram in bichromatic reverse k-nearest neighbor. *Int. J. Pervasive Comput. Commun.* **2017**, *13*, 62–75. [CrossRef]
15. Xie, S.; Feng, X.; Du, J. Maximal Covering Spatial Optimization Based on Network Voronoi Diagrams Heuristic and Swarm Intelligence. *Acta Geod. Cartogr. Sin.* **2011**, *40*, 778–784.
16. Tu, W.; Li, Q.; Fang, Z. Large Scale Multi-depot Logistics Routing Optimization Based on Network Voronoi Diagram. *Acta Geod. Cartogr. Sin.* **2014**, *43*, 1075–1082.
17. Miller, H.J. Market Area Delimitation within Networks Using Geographic Information Systems. *Geogr. Syst.* **1994**, *1*, 157–173.
18. Tan, Y.; Zhao, Y.; Wang, Y. Power network Voronoi diagram and dynamic construction. *J. Netw.* **2012**, *7*, 675–682. [CrossRef]
19. Grumbly, S.M.; Frazier, T.G.; Peterson, A.G. Examining the Impact of Risk Perception on the Accuracy of Anisotropic, Least-Cost Path Distance Approaches for Estimating the Evacuation Potential for Near-Field Tsunamis. *J. Geovisualization Spat. Anal.* **2019**, *3*, 3. [CrossRef]
20. Chen, J. A Raster-based Method for Computing Voronoi Diagrams of Spatial Objects Using Dynamic Distance Transformation. *Int. J. Geogr. Inf. Sci.* **1999**, *13*, 209–225. [CrossRef]
21. Zhao, R.; Li, Z.; Chen, J.; Gold, C.M.; Zhang, Y. A Hierarchical Raster Method for Computing Voronoi Diagrams Based on Quadtrees. In *Computational Science—ICCS 2002*; Lecture Notes in Computer Science Computational Science-ICCS; Springer: Amsterdam, The Netherland, 2002; pp. 85–92.
22. Deng, X.; Yang, Y.; Zhang, H.; Ma, Y. PCNN double step firing mode for image edge detection. *Multimed. Tools Appl.* **2022**, 1–15. [CrossRef]
23. Yang, G.; Lu, Z.; Yang, J.; Wang, Y. An adaptive contourlet HMM–PCNN model of sparse representation for image denoising. *IEEE Access* **2019**, *7*, 88243–88253. [CrossRef]
24. Basar, S.; Waheed, A.; Ali, M.; Zahid, S.; Zareei, M.; Biswal, R.R. An efficient Defocus Blur Segmentation Scheme Based on Hybrid LTP and PCNN. *Sensors* **2022**, *22*, 2724. [CrossRef] [PubMed]
25. Shen, Y.; Yuan, Y.; Peng, J. Research on Near Infrared and Color Visible Fusion Based on PCNN in Transform Domain. *Spectrosc. Spec. Anal.* **2021**, *41*, 2023.
26. Sun, Y.; Yang, H. Path Planning Based on Pulse Coupled Neural Networks with Directed Constraint. *Comput. Sci.* **2019**, *46*, 28–32.

27. Eckhorn, R.; Reitboeck, H.J.; Arndt, M.; Dicke, P.W. A neural network for feature linking via synchronous activity: Result from cat visual cortex and from simulation. In *Models of Brain Function*; Cambridge University Press: Cambridge, UK, 1989; pp. 255–272.
28. Gao, C.; Zhou, D.; Guo, Y. An Iterative Thresholding Segmentation Model Using a Modified Pulse Coupled Neural Network. *Neural Process. Lett.* **2014**, *39*, 81–95. [CrossRef]
29. Zhao, T.; Wang, J.; Zhang, J. Real-time multimodal transport path planning based on a pulse neural network model. *Int. J. Simul. Process Model.* **2017**, *12*, 356. [CrossRef]
30. Johnson, J.L. Pulse-coupled neural networks. In *Neural Networks and Pattern Recognition*; Academicpp: San Diego, CA, USA, 1998; pp. 1–56.
31. Li, B.; Zhao, K.; Sandoval, E.B. A UWB-Based Indoor Positioning System Employing Neural Networks. *J. Geovisualization Spat. Anal.* **2020**, *4*, 18. [CrossRef]

MDPI AG

Grosspeteranlage 5

4052 Basel

Switzerland

Tel.: +41 61 683 77 34

Applied Sciences Editorial Office

E-mail: applsci@mdpi.com

www.mdpi.com/journal/applsci

9 7 8 3 7 2 5 8 3 2 7 6 7